GENOMIC CONTROL PROCESS
Development and Evolution

science &
technology books

Genomic Control Process: Development and Evolution
Isabelle S. Peter and Eric H. Davidson

Available Resources:

- All figures from the book available as both Power Point slides and .jpeg files

ACADEMIC
PRESS

GENOMIC CONTROL PROCESS
Development and Evolution

Isabelle S. Peter
Eric H. Davidson
California Institute of Technology, Pasadena, CA

Amsterdam • Boston • Heidelberg • London
New York • Oxford • Paris • San Diego
San Francisco • Singapore • Sydney • Tokyo

Academic Press is an imprint of Elsevier

Academic Press is an imprint of Elsevier
32 Jamestown Road, London NW1 7BY, UK
525 B Street, Suite 1800, San Diego, CA 92101-4495, USA
225 Wyman Street, Waltham, MA 02451, USA
The Boulevard, Langford Lane, Kidlington, Oxford OX5 1GB, UK

ISBN : 978-0-12-404729-7

British Library Cataloguing-in-Publication Data
A catalogue record for this book is available from the British Library

Library of Congress Cataloging-in-Publication Data
A catalog record for this book is available from the Library of Congress

For information on all Academic Press publications visit our
website at http://store.elsevier.com/

Typeset by TNQ Books and Journals
www.tnq.co.in

About the Authors

Isabelle S. Peter is Assistant Research Professor and Eric H. Davidson is Norman Chandler Professor of Cell Biology in the Division of Biology and Biological Engineering at the California Institute of Technology, Pasadena, California. Over the last 7 years they have coauthored a series of works on experimental, conceptual, and computational analyses of developmental gene regulatory networks, including their evolutionary significance. The discussions and conceptual explorations occasioned by this collaboration produced the new synthetic views encompassed in this book, building on decades of earlier work summarized in the 2001 and 2006 Academic Press books by Eric H. Davidson.

Contents

Preface

In our time, the sheer volume of published experimental measurements, their scope and technical sophistication, compounded with a proliferation of diverse approaches, objectives, and model systems, has made it particularly difficult to see the conceptual forest for the trees. Yet all of the elegant and sophisticated though disparate and unconnected data sets with which we are confronted represent biological output of the same fundamental operating principles. Each experimental system provides a different window which offers a pathway to these principles. In this book we provide a conceptual framework that we hope will make accessible the principles by which the genomic control system operates developmental and evolutionary process. This framework grows from the realization that the most fundamental causal principles in biology, which distinguish biology from all other sciences, emerge from the existence and function of genomic information. From the genomic sequence are to be recovered the determinants of body plan development in animals. Of course, the processes of biology are subject to the same laws of physics and chemistry as are those of the inanimate world, but it is the genome that mandates biological organization. This is not a metaphor, it is a description of mechanisms that we can now begin to perceive as an unbroken chain of causal connections, leading from the A's, C's, G's and T's of the genomic DNA to the developmental formulation of the elements of the organism.

The new field that is coalescing around the concepts of genomic information processing partakes of principles and evidence from systems biology, developmental molecular biology, various aspects of body plan evolution and phylogenetics, as well as biological engineering and computational modeling. We found it useful to select and incorporate insights from all of these fields, where these illuminate the genomic control of development, without operating wholly within the paradigms of any one of them. In this book we focus on the main characteristics of the genomic control system, which include its hierarchy, its logic processing functions and its structural organization in the form of gene regulatory networks. Such networks encompass at a system level the recognition interactions between transcription factors and DNA sequence that lie at the heart of the whole regulatory process. The general operational properties of genomic regulatory systems are shared across the Bilateria, while diversity in animal forms directly reflects diversity in genomic developmental programs. Focus on the genomic programs controlling development provides a single conceptual lens through which the most disparate phenomena of development and evolution can be viewed, causally understood and interpreted.

A brief précis of the trajectory of our treatment follows:

Chapter 1 is about the molecular biology of sequence-dependent regulation of gene expression in animal development. Here we consider three major levels at which gene expression is controlled, with respect to their roles in the developmental process. The first is transcriptional regulation by sequence-specific transcription factors and their interaction with *cis*-regulatory modules. Secondly, miRNAs modulate transcript prevalence, and a third level operates by means of chromatin modifications installed upon transcription factor-DNA interactions. While the expression of many genes is affected successively by all these mechanisms, control of developmental complexity is executed by transcription factors and their sequence-specific interactions with the regulatory genome. We further conclude that the primary key to the informational process of development lies in the control system regulating the expression of genes encoding transcription factors.

Chapter 2 focusses on the transcriptional control apparatus that operates animal development. Gene regulatory network (GRN) theory defines the principal structural and functional properties of genomic control programs in animals. Here we provide an introductory overview, specifying the components of GRNs, and focusing on higher level design features such as hierarchy, modular organization, and the unidirectionality of these encoded regulatory systems. Two major aspects of GRN output are their generation of regulatory states that in turn determine all downstream genetic functions, and the Boolean nature of spatial gene expression. The genomic regulatory transactions linked together in GRNs are executed by *cis*-regulatory modules, and their combinatorial information processing functions deeply affect GRN organization. This Chapter further includes a first principles quantitative treatment of network dynamics, which rationalizes the measurable kinetics of accumulation of transcriptional products and permits computational assessment of the outputs of regulatory gene cascades. Current GRN theory devolves from multiple earlier roots which we very briefly trace.

Chapter 3 focusses on the means by which the transcriptional regulatory system is deployed in the pre-gastrular development of many kinds of bilaterian animal. The fundamental task in early embryogenesis is to install in the descendants of a single cell initial patterns of spatial regulatory gene expression in respect to the axes of the future body plan. This poses control challenges unlike any others encountered in bilaterian life. We introduce in this Chapter general principles of development couched in terms of regulatory logic, such as the regulatory properties of the egg, inductive signaling, differentiation and cellular morphogenesis. Beyond these, bilaterian evolution has given rise to several distinct developmental strategies for pre-gastrular embryogenesis, specifically those of regularly cleaving small embryos with early onset of transcription; of large embryos undergoing rapid cell division with delayed onset of transcription and cell motility; of embryos utilizing a transcriptionally active syncytium. Here, for each of these strategies, we focus on particular regulatory design features of GRNs encoding embryonic specification, response to maternal spatial inputs, territorial pattern formation, and global control of zygotic gene expression.

Chapter 4 is about animal body part development. Bilaterian body parts are formed during development according to a common scheme. The basic principles, irrespective of body part or organism, emerge from studies that reveal the genomically encoded mechanisms underlying body part formation. Here we consider partial gene regulatory networks that have been solved for a diverse variety of body parts including brains, hearts, limbs, and guts, in *Drosophila* and/or vertebrates. In all of these cases, the process of body part formation begins with the allocation of a progenitor field, defined by installation of an initial regulatory state, and its placement in respect to the axes of the body plan. There follows the progressively finer subdivision of this field into appropriately positioned regulatory state domains that generate the subparts and ultimately the cell types of the body part. Circuitry commonly encountered in these GRNs controls signaling interactions, exclusion functions and boundary formation, and often involves feedback and feed forward regulation.

Chapter 5 considers the differentiation of cell types as the final readout of the preceding spatial specification processes in development of the body plan. Cell types represent the installation of biological and molecular effector functions within the matrix of spatial regulatory states. Here the focus is on the specific genomically encoded wiring utilized for deployment of cohorts of effector genes in terminal differentiation. Two objectives require to be fulfilled: one is the regional activation and permanent maintenance of expression of a small set of driver regulatory genes; and the other is control of large sets of effector genes by these drivers. The first is accomplished by positive feedback circuitry among the driver genes, usually accompanied by exclusion of drivers of alternative cell fates, and the second depends on effector gene *cis*-regulatory architecture. Our examples include *Drosophila* photoreceptors, and differentiation of erythrocytes, somitic muscle, neural crest and placodes in vertebrates. Diverse body parts often deploy similar cell types and this is mediated by modularity of genomic regulatory systems.

In Chapter 6 we analyze diverse approaches to construction of quantitative and logic models of gene regulatory networks. The structural and functional properties of GRNs can be accessed only by use of models. In this Chapter we consider diverse forms of GRN model, focusing on insights into the biology of developmental GRNs that accrue from the generation of abstract mathematical, logical or topological models. Topological models provide genome-centered maps of regulatory linkages, and thus graphically represent the architecture of causal interaction networks. We address the significance of network topology in an extensive comparative structure/function analysis of GRN subcircuit architecture. Insights otherwise unattainable emerge from dynamical mathematical treatment of GRN function. We review informative applications of ODE analysis to developmental GRNs operating in *Drosophila* embryos and in mammalian hematopoietic cells, and consider the regulatory interpretation of morphogen gradients. We turn then to Boolean models which are focused on GRN logic transactions. Several asynchronous models are considered and we end with a summary of the new insights deriving from a comprehensive Boolean logic model of the sea urchin embryo GRN.

Chapter 7 provides interpretations of major macro-evolutionary processes in terms of GRN structure and function. Evolution of the animal body plan is the outcome of change in the encoded genomic regulatory program for development. Major features of Phanerozoic animal evolution relate directly to developmental GRN hierarchy. We first consider rapid and continuously adaptive changes occurring at the species level, in terms of regulation of effector gene expression at the periphery of developmental GRNs. We then discuss developmental processes responsible for the generation of definitive characters of the body plan shared amongst all members of given phyla or classes, which occur at upper levels of GRN hierarchy. Regulatory mechanisms also account for the evolutionary stasis of such developmental characters. Distinct evolutionary mechanisms operate at different levels of GRN hierarchy, ranging from *cis*-regulatory adaptation at individual genes to co-optive redeployment of whole regulatory circuits. Conservation of regulatory circuitry within GRN hierarchy leads to an explanation for the nested organization of shared character sets underlying animal phylogeny.

We have maintained the discussion in this book at two different levels. Throughout, we have stressed conceptual themes while, at the same time, tying these concepts to the detailed experimental results that underpin them (these accounts might well exceed the concerns of some readers and could be skipped over without losing the thread of the discussion). Our hope has been to make this book accessible to a wide range of readers. Those with expert knowledge for whom the details may harbor particular fascination will find the specific information in the figure captions more important. For those entering this area from without, whose preoccupation may be to obtain a sense of the overall conceptual structure of genomic control process, we have included a brief introduction into each of the developmental contexts considered. Although having deep historical roots, much of our conceptual framework is only of very recent vintage, as its formulation has depended on the ever increasing accumulation of high quality experimental data such as we have included here.

It is our great pleasure to acknowledge the help we received from some of the world's most renowned scientists in their respective fields, who were kind enough to devote their time to reviewing various of our chapters, thus providing us with invaluable comments, corrections, suggestions and insights. Our reviewers were as follows: for Chapters 1 and 2, Gary Felsenfeld of the National Institutes of Health, and Ellen Rothenberg of Caltech; for Chapter 3, Ellen Rothenberg; for Chapter 4, Marianne Bronner of Caltech, and Roger Patient of the University of Oxford; for Chapter 5, Marianne Bronner and Oliver Hobert of Columbia University; for Chapter 6, Bertie Göttgens of the University of Cambridge, Ellen Rothenberg, Eric Siggia of Rockefeller University, and Stas Shvartsman of Princeton University; for Chapter 7, Ellen Rothenberg and Doug Erwin of the Smithsonian Institution. We are indebted to Ellen Rothenberg, not only for reviewing well over half of our book and providing continuing scientific feedback, criticism and discussion, but also for her ongoing encouragement and enthusiastic support for this project from the time of its inception to its final stages. We are particularly grateful to Yi Krooss for her skillful and professional graphic design, which can be found in every figure of this book. We appreciate the hundreds of permissions we received, allowing us to reproduce the results constituting these figures and our deepest appreciation goes to all the scientists whose intelligence, hard work and discoveries have led to the state of knowledge that this book concerns. We would like to thank Deanna Thomas in our laboratory, who extracted endless references for our use. At Elsevier, it has been a privilege to work with our editor Christine Minihane, who has been endlessly supportive, with Halima Williams, who cheerfully confronted all of our requirements and demands regarding the practical organization of the project, and finally with Julia Haynes, our Project Manager.

In the long view, this book grows out of many years of work on gene regulatory networks at Caltech. In this sense, all of the many participants of the network project over the last 15 years in this lab and elsewhere have contributed to the foundation on which this work stands. For decades now, Eric Davidson has worked intimately with Jane Rigg, whose perspicacity and insights are as always a pleasure to acknowledge. Our experimental and conceptual research on gene regulatory networks has over the years been primarily funded by the National Institutes of Child Health and Development, and in this context we are particularly grateful for the longstanding support of our Program Director, James Coulombe.

On our cover we have reproduced a painting made about a half century ago by Morris Davidson (Eric Davidson's father, 1898-1979), who was a well-known American artist. In the abstract relations of its patterns, this painting for us evokes the beautiful organization of the natural world.

Isabelle S. Peter and Eric H. Davidson
December 2014
Caltech, Pasadena

Dedications

I would like to dedicate this work to my son Milad, whose existence gave me the inspiration and the courage to follow my dreams further than what I would ever have considered possible.

Isabelle S. Peter

I would like to dedicate this book to those scientists whose work has most deeply shaped my intellectual world, none of whom are still with us. They are Theodor Boveri, Alfred Mirsky, and my partner for those many years, Roy J. Britten.

Eric H. Davidson

Chapter 1

The Genome in Development

1. Views of Development

Considered from an informational perspective, the most interesting and defining feature of animal development is that it generates a continuous increase in complexity of spatial organization. Our early predecessors, from Aristotle to Caspar Friedrich Wolff, perceived this externally, remarking explicitly on the developmental increase in observed morphological complexity, the progressive appearance of more and more complicated body parts. This unique, and for centuries fundamentally mysterious, aspect of development is what struck every thoughtful observer. It is illustrated in Malpighi's beautiful seventeenth-century drawings of chick embryogenesis reproduced in Fig. 1.1 (Malpighi, 1673, 1686), where the growing multiplicity of anatomical features as development progresses is carefully indicated.

Developmental increase in spatial complexity can today easily be perceived and measured in any degree of detail, at any underlying level: as developmental control of spatial gene expression, as progressive definition

Genomic Control Process
http://dx.doi.org/10.1016/B978-0-12-404729-7.00001-0

Figure 1.1

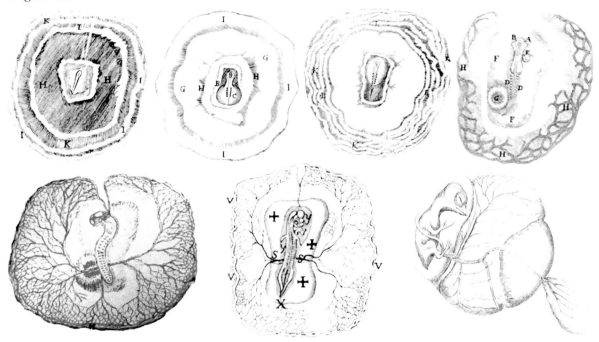

Figure 1.1 Malpighi's early demonstration of increasing organismal complexity in development of the chick embryo. Marcello Malpighi's drawings of a developing chick embryo, according to his own microscopic observations, are reproduced from his 1686 *Opera Omnia*, except for the dorsal view in color which was published by him in 1673 in *Dissertatione Epistologica de Formatione Pulli in Ovo*.

of regulatory gene expression patterns, and as diversification of cellular function in space. These phenomena all depend directly on the spatially mandated regulation of genes, ultimately thousands of them in any given developmental process. Developmental complexity is the direct output of the spatially specific expression of particular gene sets and it is at this level that we can address causality in development. Thus we come to our starting point. Increasing spatial complexity of gene expression directly requires continuing regulatory informational input, and conversely, the complexity of the developing organism is at root defined by these informational requirements.

To view development as the output of informational transactions provides a unifying principle. Its application permits any aspect of the marvelous and diverse phenomena of development to be considered through the same lens. It took the whole of the twentieth century to arrive at an unequivocal appreciation of the fact that the information encoding development is located in the genomic sequences carried in the chromosomes of every cell of the organism (we briefly recap the major milestones along the intellectual way below). Now we have at last attained a framework view of what the genomic control system is, and how it operates.

Many logical spokes of the phenomenological wheel lead inexorably to the genomic regulatory code at its hub. For example, one might start by trying to explain how it could work that frogs beget frogs and dogs beget dogs, and never does one sort of animal produce an embryo that develops into another. There is no escape from the conclusions, first, that each sort of animal possesses a heritable, hence genomic, set of instructions for building the body plan of its own sort; second, that the genomic program suffices to direct the developmental construction of that body plan; it cannot depend on the environment or on magic hormones, or anything else not originating internally in the inherited genome. Or one might consider how it can be that every developing embryo of a given species has exactly the same major parts, in the same

Figure 1.2

Figure 1.2 Identical operation of the developmental program: a field of 72h sea urchin larvae. The larvae are from a single batch of fertilized eggs of *Strongylocentrotus purpuratus*, visualized at 72h. Each contains about 1800 cells, organized canonically individual to individual into differentiated tissues; embryogenesis is now complete and the larvae are about to commence feeding.

spatial relations to one another, executing the same functionalities. If we look for instance at a random field of nearly identical sea urchin larvae of the same age, as shown in Fig. 1.2, it is impossible to escape the implication that the same instructional program has operated to produce each individual. Therefore, again, such programs must exist; they must be identically replicated, hence genomic; and they must suffice to control the nature of developmental events independently and similarly in each organism. But though perhaps more subtle, the most revealing and powerful insight is implied by that unique phenomenological feature of development earliest noticed, the continuing increase in organismal complexity, however that be assessed. This tells us something profound about the nature and use of the developmental regulatory program. It means that it is capable of applying the state it has produced at each stage of development, in every cell, as inputs to formulate new states for the next stage, and the same again for the next stage after that. Since complexity increases, more cells are made to do more diverse things at each stage. But since they always do the right thing for that species, the states at Stage X not only have to be "read" in a given way, but in addition the right, specific, responses have to be installed at Stage X + 1, instructions for which also have to be written in the program. Thus the preexistence of a genomic regulatory program is directly implied by the increase in output complexity during development. This program includes the information to read the regulatory situation at each point in time everywhere in the organism throughout development, and to utilize these newly assessed data to generate the specific new situations needed for developmental progress in exactly the correct manner. It must operate from egg to adult form, integrating over all states and all spatial elements. It is thus no wonder that the regulatory genome used for development includes such an enormous amount of information, as we summarize below.

Development produces the genomically encoded body plans of animals, and evolution is the process by which body plans arise de novo, and heritably alter in deep time. Since causality in development must at root be considered in terms of information processing, and how spatial control of gene expression is mandated by genomic regulatory information, so also the evolution of the body plan. At the informational level of the genomically encoded program for development, evolutionary process is equivalent to the time derivative of developmental process: evolution is change in this same program over deep time (Peter and Davidson, 2011). That is, new morphological forms, new appendages, new central nervous system structures, and newly positioned skeletomuscular systems, arise in evolution by mutational alteration of preexistent developmental programs. Not only is evolutionary process in this specific mechanistic sense derivative of the developmental process, but, as we examine in the final chapter of this book, the structure/function rules that apply to the genomic developmental program also determine the effects of changes that cause evolutionary alterations in outcome. Thus from an informational point of view, no longer can there be any but artificial or arbitrary distinctions separating causality in evolution and causality in development. These areas of bioscience indeed have entirely different historical origins and evidential bases, such as paleontological data on the one hand and cell biological functions on the other. But they are directly convergent at the genomic regulatory information system that controls development.

How the developmental regulatory system works, at the individual gene level and the circuit level, are subjects we are concerned with throughout this book. Development cannot be explained in terms of the regulation of individual genes considered one by one, no matter how complex their regulatory apparatus, nor how important their particular roles. The major informational transactions underlying progressive developmental process control the orchestration of gene expression. This function is executed by gene regulatory circuits encoded in multiple genomic regulatory sequences. Operating in a modular manner according to their architecture, circuits powerfully enhance the regulatory potentiality of the control system in a large number of ways: for example, they provide the opportunity of coordinately controlling multiple genes, positively or negatively in diverse cells; they mediate system responses to outside signals; they are capable of producing diverse alternative outputs under different input circumstances; certain circuits respond sharply in different ways to graded input levels; others generate diverse dynamic behaviors.

Developmentally controlled expression of confined gene sets determines the identity, fate, and function of the fields of cells from which body parts develop, and of their progressively restricted subdomains. Spatially organized gene expression ultimately defines all individual cell types constituting the resulting body part or organ. The essential control feature in this process is thus the regulation of gene expression in time and space in order to accomplish the diversification of cell fates in development. Thus developmental complexity is a function of spatially controlled gene expression. Since all gene expression is initially regulated by transcription factors and their sequence-specific interaction with regulatory DNA, we can consider the control of expression of those genes that encode transcription factors to be the key and fundamental regulatory mechanism causal to the development of body plans. However, as we now know, gene expression is affected at many successive levels downstream of the primary transcriptional control mechanisms, including the regulation of mRNA levels by microRNAs (miRNAs), and the roles of chromatin modifications and long noncoding RNAs (lncRNAs) in gene expression. What is their contribution to the ultimate regulated gene output? Which of these will have to be considered to understand control of spatial gene expression in development? Before we take up larger systems of developmental gene regulatory interactions in the following chapters, we turn to the particular roles of each of these regulatory mechanisms in the genomic control of spatial gene expression.

2. Levels of Control of Gene Expression: Transcriptional Regulation

Though mechanistic aspects of the transcriptional regulatory system in animal cells are dealt with in some detail in Chapter 2, it is useful here at the outset to consider the outlines of the regulatory control system, its components, and functions. Conveniently, transcriptional control is considered in terms of *cis* and *trans* components and their interactions: *cis* referring to regulatory sequence features located on the same DNA molecule as a given gene (*cis*, from the Latin "this side of"), and *trans* referring to regulatory molecules that diffuse to the DNA from elsewhere (*trans*, from the Latin "on the opposite side"). We begin with the *trans*-acting regulatory molecules that form the basis of the animal cell gene control system, transcription factors.

2.1   Basic facts about transcription factors

Transcription factors are encoded by a unique class of genes amounting in the animal genome to only a few percent of the total number of protein-coding genes. In this book, transcription factors are defined strictly as proteins that recognize and bind to specific short DNA sequences, such that their interaction with these sequences, causally affects expression of genes. The recognition of DNA sequence by transcription factors occurs by chemical interactions of the amino acid side chains of the transcription factor protein with base pair residues of DNA functioning as regulatory sequence. The transcription factors thus "read" the genomic sequence, and the most fundamental fact about this mechanism is that it is the sequence recognition function on which informational aspects of all regulatory transactions controlling gene expression depend. Only a relatively small, finite number of protein structural motifs have evolved which have the capacity to recognize and bind to specific DNA sequence. With few exceptions, all animals utilize the same families of transcription factors, representatives of which were evidently present in their common evolutionary ancestor. Transcription factors consist of DNA-binding domains by which these families are defined and effector domains that mediate interactions with other proteins necessary for transcription, and with other transcription factors. The DNA-binding domains are typically highly conserved per family across the Animal Kingdom, while the effector domains evolve more rapidly. Transcription factors execute many functions as we shall see in the course of this book, including gene activation, gene repression, and signal response. They function as diffusible regulatory molecules, that is, they are transcribed in the nucleus, translated in the cytoplasm, and find their target sites in the genomic DNA on reentry into the nucleus, mediated by nuclear localization sites included in all transcription factor protein sequences. Transcription factor presence in a particular cell in development depends on regulation of the genes encoding them. This is consequently the key primary locus of control in development. Transcription factor occupancy of their target sites depends on two parameters, as we examine in detail in Chapter 2: the intrinsic tightness of binding between the transcription factor protein and the DNA target site, and the concentration at which the transcription factor is present in the nucleus. Quantitative and biochemical studies have shown that an important and general aspect of transcription factor–DNA interaction is that all transcription factors include basic domains which cause them to be concentrated nonspecifically in the vicinity of the DNA, facilitating the diffusion-limited discovery of their target sites. The most important general principle of transcription factor regulatory function in animal cells is that they never work alone, but always together with other specifically bound transcription factors and cofactors, since regulation of transcription requires a multiplicity of biochemical effector functions.

2.2   *Cis*-regulatory modules

Cis-regulatory modules (CRMs) are DNA sequence elements that have transcriptional regulatory activity. Some CRMs are promoter sequences which bind the basal transcription machinery, consisting of RNA polymerase II and many cofactors, and determine where transcription of each gene is initiated. Other CRMs function as insulators, isolating regulatory domains of the genome from one another, as we discuss in Chapter 2, but the major class considered here consists of CRMs also known as enhancers. These *cis*-regulatory control elements function to determine when and where transcription occurs during development. Enhancers may be located at some distance from the transcriptional start site and they combinatorially bind multiple transcription factors. Generally, enhancers are a few hundred base pairs in length and they contain multiple short DNA sequence motifs, which directly interact with specific amino acid sequences of the cognate transcription factors. A typical CRM contains multiple binding sites for each of several different transcription factors. The result is that transcription factors bind to DNA densely within a short distance, and this is essential for CRM activity. From an informational point of view, it is crucial to realize that the density of specific transcription factor target sites within CRMs uniquely distinguishes these clustered regulatory sequences from the single sites that occur randomly throughout the genome ("orphan" sites). That is, short transcription factor target sites (typically 6–8 bp, with some redundancy allowed at specific positions) are expected to, and do occur every few thousand base pairs in the genome on a random basis, but regulatory specificity is achieved by the tight clustering of transcription factor binding sites, a configuration highly improbable in random DNA sequence. CRM activity depends on the availability of the relevant transcription factors within the nucleus, on their binding to the CRM, and on the recruitment of additional cofactors by the bound transcription factors. Direct interactions between the regulatory factors bound to the CRMs and the basal transcription apparatus may result in the formation of a DNA loop between a distantly located enhancer and the promoter (see Chapter 2). Depending on the nature of the transcription factors and their associated cofactors, this interaction causes either activation or repression of gene expression. In addition, most genes are regulated not just by one but by several CRMs, and these may be located upstream, downstream, or within the introns of the gene at distances ranging from a few kilobases to more than a megabase from the promoter. Generally, the various CRMs controlling a given gene are functionally distinct, though sometimes their functions may partially overlap. Thus, for genes expressed in multiple cellular domains or at different times during embryonic development, each CRM usually mediates only a given part of the overall expression pattern, so that the overall expression pattern of a given gene is the combined outcome of the regulatory activity of all of its CRMs.

Combinatorial control of CRM function is fundamental to the complexity of gene expression patterns during animal development. For example, the genome of the sea urchin *Strongylocentrotus purpuratus* encodes approximately 300 non-zinc finger transcription factors plus approximately 380 zinc finger proteins, most of which are putative transcription factors. This sums to a total of over 600 possible regulatory genes (Howard-Ashby et al., 2006c). It is interesting to reflect on the enormous regulatory capacity that this encompasses. For instance, assuming that activity of CRMs depends on binding of three different required transcription factors (which as we see below is an underestimate), and that all three factors need to be present for CRM activity, there are >200 million (600^3) different possible regulatory input designs, or about 30 million if we consider the spatial order of binding sites for these transcription factors irrelevant. The theoretical complexity of the transcriptional control system by far exceeds the requirements for programming the distinct spatial and temporal expression patterns observed during animal development. This allows for a certain redundancy in the use of regulatory information. First, not all transcription factors are expressed in specific patterns, rather some are expressed ubiquitously in all cells at all times. Even though such factors may contribute to the regulation of gene expression levels, they cannot themselves be responsible for the specific spatial patterns of gene expression, unless their activity is modified specifically by interaction with a spatially confined cofactor or by spatially confined covalent modification in response to signaling. Furthermore, even in terminally differentiated cell types, there are many more than three transcription factors specifically expressed. These transcription factors are used as drivers for specific batteries of effector genes, and for the control of other specific gene functions in these cells. Yet though each specifically expressed gene requires only a subset of the available regulators, when added together these subsets amount to the total of expressed regulatory genes in that cell type. There are many cell types in an organism, each specifically expressing many genes. As we now know, a majority of the regulatory information utilized in the life of an animal is consumed in the developmental processes that lead to the specification of terminally differentiated cell types in the right places.

2.3 Transcriptional control of spatial gene expression in development

Spatial gene expression is predominantly controlled at the transcriptional level. This becomes evident in systematic analyses of the spatial distribution of transcripts. For several developmental systems, including *Drosophila*, zebrafish and mouse, temporal time courses of spatial gene expression data have been collected for the entire genome, and the images together with detailed annotations of spatial expression patterns have been incorporated in publicly accessible databases (Richardson et al., 2010; Armit et al., 2012; Howe et al., 2013). What these data all demonstrate very impressively is the extent of spatially restricted gene expression. For instance, during early sea urchin development, at least 80% of over 120 expressed transcription factor genes are transcribed in specific domains (Arnone et al., 2006; Howard-Ashby et al., 2006a,b; Howard-Ashby et al., 2006c; Materna et al., 2006; Rizzo et al., 2006; Tu et al., 2006). Thus at each stage, a majority of regulatory genes is transcribed in a way that is dependent on spatial determinants not present in all cells of the organism. The spatial expression patterns have been used to generate embryonic maps which display spatial domains consisting of cells expressing similar sets of genes (for example, Frise et al., 2010). Such embryonic domains are defined by unique combinations of expressed transcription factors.

The large number of CRMs that execute accurate spatial control functions even when isolated from their genomic context underpins the view that spatial gene expression is predominantly regulated transcriptionally. Assays in which individual DNA fragments with putative regulatory function are used to drive reporter gene expression have demonstrated repeatedly that specific DNA sequences encode spatial gene expression patterns. With hardly any exceptions, the mechanism of spatial control has been shown by this means to be transcriptional, and for genes with spatially restricted expression, one or several CRMs can always be found that execute spatial expression functions of the endogenous genes. Concomitantly, most CRMs analyzed without bias on a genome-wide scale have spatial functions. Several recent approaches to the systematic identification of CRMs in the genome have appeared, though their success rates vary (Hardison and Taylor, 2012). Using conservation in noncoding DNA sequence to identify CRMs is an approach that is not biased for particular spatial activity, because it does not rely on the presence of specific transcription factor binding motifs or on chromatin marks in specific cell types. In a recent analysis, >400 highly conserved noncoding sequences in the mouse genome were tested in reporter assays in transgenic mice, demonstrating that about half of these sequences showed regulatory functions during the period of development that was examined, a majority of them in restricted embryonic domains (Visel et al., 2008). Spatial control of gene transcription is therefore not an exception but occurs with high frequency, especially during development. Thus in an extensive experimental assessment of >3500 active enhancers in *Drosophila*, the vast majority are expressed in specific temporal and spatial domains (only 0.8% were expressed ubiquitously during embryogenesis), essentially defining the organism in terms of encoded regulatory information (Kvon et al., 2014). Transcriptional control provides the direct link between genomic sequence and spatial organization in development.

2.4 The fundamental functional experiment in developmental transcription control

In reflecting on the history of this field over the last 30 years, it is quite remarkable how a single experimental paradigm has produced almost all of our direct DNA level knowledge of how spatial regulation of developmental gene expression works. Of course there are a dozen technological variations at every step depending on when the experiment was done over these years, and on the identity of the biological system, but its logical essence remains always the same. A putative regulatory DNA sequence is excerpted from the genome (or made synthetically), and by laboratory molecular biology methods is associated with a "reporter" gene in an appropriate vector, as in Fig. 1.3(A1). The vector should have no significant activity on its own, and no capability of spatial control; these are the functional properties of the regulatory DNA sequence if it is complete. Instead of isolating a fragment of regulatory DNA, a variant

Figure 1.3

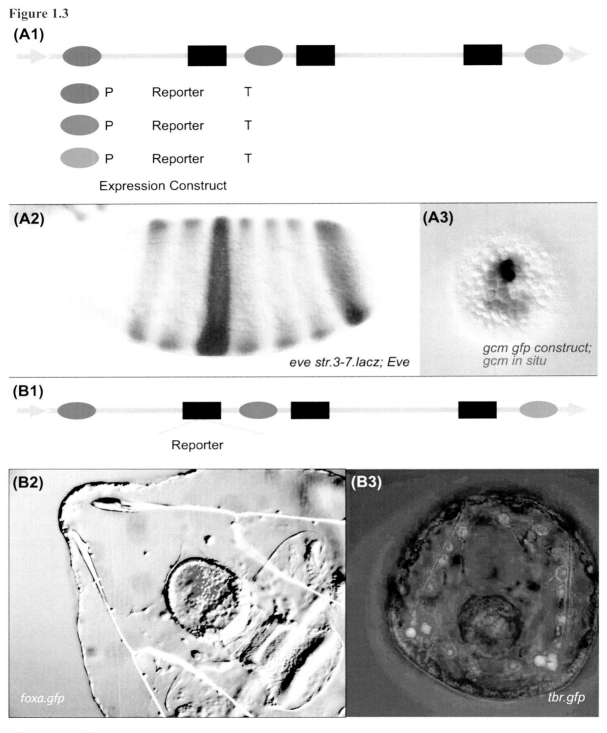

Figure 1.3 The fundamental experimental paradigm demonstrating genomic sequence function in developmental gene regulation. (A), Reporter constructs and experimental validation by introduction into eggs. (A1), Diagram of assembly and structure of expression vectors to be introduced into eggs of a developing animal in order to test spatial function of given *cis*-regulatory modules (CRM). A

typical gene is shown at the top, exons as black rectangles, with three CRM indicated as oval forms. These regulatory DNA fragments can be alternatively (or in combination) excerpted from the gene and associated with a vector containing a basal promoter (P), a reporter gene (Reporter), and a trailer sequence (T) including p(A) addition site. (A2, A3), Examples of spatial performance of such expression constructs which overlaps with expression of endogenous gene, demonstrating accuracy of spatial control mediated by the regulatory DNA fragment. (A2), A famous example from the *evenskipped* gene of *Drosophila*, in which the endogenous 7 stripes of expression of the gene are shown in brown, and the activity of a lacZ reporter in an expression construct present in all genomes of a developing egg is shown in blue *(from Small et al. (1996))*. The expression construct, assembled as in (A1), is driven by a *cis*-regulatory module controlling expression of stripes 3 + 7. The experiment demonstrates that the *cis*-regulatory module accounts perfectly for a discrete part of the spatial expression of the gene overall; other CRM account for the remaining stripes. (A3), A second example from the sea urchin, in which a *cis*-regulatory module from the aboral mesodermal regulatory gene *gcm* has been built into an expression construct and introduced into the egg. Endogenous *gcm* expression is shown in red, and the reporter is activated exclusively in an overlapping clone of these same cells, identified by a blue in situ hybridization stain *(unpublished experiment of A. Ransick, authors' lab)*. (B1), Another type of expression construct built by insertion of a sequence encoding a reporter protein, usually a fluorescent protein, into the gene in its natural sequence environment. This ensures the presence of the endogenous basal promoter and the whole regulatory system. The insertion is done in vitro by site-specific recombination using BAC vectors, and as in (A1) the construct is then introduced into the egg. (B2), (B3), Examples: (B2), Expression of GFP reporter inserted in the endodermal *foxa* gene of a BAC construct and visualized in a 2-week-old feeding larva. Expression is confined to large clones of cells of the gut *(unpublished experiment of M. Cui, authors' lab)*. (B3), Accurate expression of GFP reporter inserted in the skeletogenic sea urchin *tbrain* gene in a BAC construct, in all syncytial skeletogenic cells of a postgastrular embryo *(from Damle and Davidson (2012))*.

is to insert the reporter into a gene within its genomic context which contains most or all of its endogenous regulatory apparatus (Fig. 1.3(B1)). Then, by one means or another according to the gene transfer technique employed for that species, the expression constructs are inserted into eggs before development begins. Thereupon they enter into the host genome, replicate together with it, and are exposed to the same regulatory inputs in the different regions of the embryo as are the endogenous genomes. When and where those inputs to which the test regulatory DNA sequence responds are present, the reporter gene is transcribed. The usefulness of the experiment devolves in part from the character of the reporter. It must have three properties: if it is to be detected as a protein (for example green fluorescent protein) then the protein must be different from any encoded in the host genome, and if as an RNA then the sequence must be distinct from any in the host genome; it should be spatially and or quantitatively detectable at cellular resolution; and the reporter sequences must be completely functionless and nontoxic to the host embryo.

This experimental paradigm is pivotal for the whole field of developmental gene regulation for many fundamental reasons. Such experiments have provided innumerable proofs that given fragments of DNA by themselves suffice to generate specific, developmentally accurate spatial and temporal patterns of reporter gene expression exactly reproducing discrete components of the spatial expression of the gene from which the regulatory DNA had been derived, as illustrated in Fig. 1.3(A2,3) and (B2,3) (for many additional diverse examples, see Davidson, 2006). Therefore, the regulatory DNA sequence included in the construct contains all necessary information for the resulting developmental expression pattern. No additional primary informational transactions can be required for spatial and temporal control of gene expression than those occurring on the exogenous DNA sequence in the expression vector. This experimental paradigm provides the means to functionally define the CRMs responsible for control of given aspects of a gene's function. Furthermore, this type of experiment affords the means to determine the necessity of given modules for given phases of expression, particularly using the method of Fig. 1.3(B) where the consequences of deleting given individual CRMs can be directly determined. As discussed in Chapter 2, many additional problems having to do with the spacing and the functions of multiple modules can be addressed similarly in the context of the natural genomic sequence organization. The significance of any given transcription factor target site, or suspected site, or any particular base pair in a site, or any other intramodular sequence feature can be assessed experimentally, by mutationally altering the regulatory DNA sequence and determining the functional outcome when it is introduced into the egg instead of the normal complete sequence. In the same way, the identity of predicted inputs into a CRM can be tested, either by deleting its known target sites or by synthetic introduction of these sites into expression constructs. Finally, the conservation of regulatory functions can be directly tested by insertion of given control sequences into different species.

From the fundamental experiment come fundamental simplifying principles, all demonstrated by function in the context of development: (1) Information necessary and sufficient to specify any spatial pattern of gene expression in development is resident in genomic DNA sequence. (2) There are no prior events particular to given cell types which alter DNA sequence in order to endow it with spatial control capability, since the regulatory DNA within the expression construct is either amplified in bacteria or produced synthetically. Therefore all that is required, is the primary DNA sequence. (3) Individual regulatory gene expression functions are confined within small discrete pieces of genomic DNA so that "regulatory genome" is not a metaphor, but a physical reality. Therefore, just as required by the logic of development and evolution with which we began this chapter, the naked, inherited genomic sequence does indeed contain the information for specific control of gene expression in development.

3.   Levels of Control of Gene Expression: Noncoding RNAs

Following the primary transcriptional event, which determines the presence or absence of a gene transcript, there are many levels at which subsequent processes can affect gene expression quantitatively and qualitatively. Alternative splicing can produce qualitatively diverse transcripts with varying untranslated or protein-coding sequences, for instance allowing the production of transcription factors with the same DNA-binding domain but different effector domains. It has become clear from transcriptome data that the production of alternative mature mRNAs from the same gene is not exceptional, but occurs in a considerable fraction of genes. Where different isoforms are expressed in different spatial domains of an organism, this is usually due to the utilization of alternative 5' exons, which are serviced by distinct CRMs. Thus, differential isoform distribution is another outcome of transcriptional regulation.

mRNA levels are affected by several mechanisms, including polyadenylation and 5' capping. All mechanisms which contribute to the generation of a mature mRNA and its transport to the cytoplasm are potentially subject to regulation and may affect the availability of the mRNA. A major determinant of mRNA level is the rate of its turnover, which depends on its intrinsic features such as secondary structure and base composition. While default mRNA stability is likely to be similar in most cells of a given organism, local regulation of mRNA turnover may depend sharply on sequence-specific interaction with small noncoding RNAs (ncRNAs), which may or may not be present.

Several mechanisms for control of mRNA levels involve ncRNAs. These ncRNAs are often divided into classes of short (<200 nt; e.g., miRNAs and piwiRNAs (piRNAs)) and long noncoding RNAs (>200 nt; lncRNAs). This distinction is useful as it discriminates the different potential functions of these ncRNAs. While both classes of ncRNAs may affect gene expression, they do so by quite different mechanisms. miRNAs regulate the levels of mRNA and protein by direct sequence-specific interaction with the target mRNA. The role of lncRNAs in the control of gene expression is far less understood, but several cases have been described in which lncRNAs affect gene expression by assisting the introduction of chromatin modifications, as we discuss below in respect to epigenetic functions in gene expression. New lncRNA functions continue to be discovered. The functions of miRNAs have been analyzed systematically, revealing common mechanisms in miRNA-mediated effects on mRNA level. In our quest to include components that directly contribute to the transformation of DNA sequence information into developmental process, we focus in the following mainly on the role of miRNAs in the control of spatial gene expression.

3.1   Mechanisms for miRNA-mediated control of gene expression levels

The structure and function of miRNAs devolve from their unique mode of synthesis. miRNA synthesis is mediated by RNA polymerase II, producing a primary transcript of particular secondary structure. Two complementary RNA sequences within the primary transcript, one of which includes the short miRNA sequence, hybridize to form a double stranded RNA stem structure, with the noncomplementary sequences in between them forming a single-stranded RNA loop. Several processing steps are required for the processing of the mature miRNA. The first of these occurs in the nucleus, mediated by the Drosha/DBCR8 complex, which uncovers the 5′ end of the future miRNA (see Fig. 1.4(A)). After export to the cytoplasm, further processing by the Dicer/TRBP complex gives rise to the mature miRNA, which is 21–23 nucleotides long. miRNAs and Argonaute (Ago) proteins form the RNA-induced silencing complex, which is responsible for mediating posttranscriptional gene regulation (Winter et al., 2009).

There are several mechanisms by which miRNAs regulate gene expression. The recruitment of Ago to specific mRNA targets, which is mediated by the miRNA, may induce cleavage of the mRNA molecule and thereby lower the level of mRNA and protein product. This mechanism depends on extensive complementarity between miRNA and target mRNA sequences, but is not frequently used in animal cells. Another mechanism of miRNA function involves repression of translation, which could lead to lower levels of protein product without directly affecting the levels of the corresponding mRNA. Secondarily however, this will also in the end decrease mRNA transcript levels, since nontranslated mRNA is less stable. A third mechanism also results in an increase in the turnover rate of the target mRNA. A recent study aimed at distinguishing between these different modes of miRNA-mediated regulation of transcript levels, by simultaneously quantifying the levels of mRNA and ribosome-covered mRNA sequences in the process of being translated (Guo et al., 2010). This analysis, which was performed in mammalian cultured cells, suggests that miRNAs in most cases lead to a decrease in mRNA levels and thereby lower the amount of protein product. Interaction between miRNAs and the 3′UTR of target mRNAs leads to decreased stability of the mRNA, possibly by mediating transcript deadenylation.

Target mRNA recognition by miRNAs occurs by the direct interaction of short complementary sequences. Within the miRNA, sequences required for pairing with the mRNA usually include the "seed sequence" in nucleotides 2–7, but additional nucleotides can modulate the interaction. Complementary miRNA recognition sequences are generally encoded in the 3′UTR of the target mRNA. Even though the interaction between miRNA and mRNA relies on nucleotide sequence complementarity, several difficulties are encountered in the computational prediction of miRNA target genes (reviewed in Bartel, 2009). The RNA sequences recognized by miRNAs are generally short and do not necessarily depend on perfect complementarity (Pasquinelli, 2012). Thus computational predictions of miRNA/mRNA interactions indicate that many miRNAs target several hundred different transcripts. Overall, about 60% of protein-coding genes are predicted to be subject to miRNA regulation (Friedman et al., 2009). Overexpression of miRNAs in cultured human cells has supported the idea that a single miRNA can regulate dozens or even hundreds of mRNAs (Lim et al., 2005), but whether under normal conditions all these predicted target mRNAs are directly regulated by miRNAs remains to be experimentally validated. The difficulty in predicting functional miRNA/mRNA interactions is shown by the following example, in which the putative regulatory function of three predicted miRNA/mRNA pairs was addressed experimentally (Lemons et al., 2012). The three miRNAs are encoded in the *Drosophila hox* cluster and their putative target mRNAs are all *hox* gene transcripts possessing high confidence 3′ UTR miRNA target sites. But none of the three predicted interactions is actually used to regulate *hox* gene expression during embryogenesis, and it is clear that the presence of potential miRNA recognition sites alone is not sufficient to predict a regulatory interaction (reviewed in Pasquinelli, 2012). Even if such sites can be bound by a cognate miRNA, and even if the mRNA is downregulated in the context of miRNA overexpression, it still has to be determined whether such an interaction is biologically functional in development, physiology, or cell biology. The problem of computationally predicting potential target mRNAs is reminiscent of the difficulties in predicting transcription factor target sites. In general, computational predictions based only on short and imprecise DNA or RNA recognition sequences identify a large number of potential target sites, of which only a subset will be functional. Thus, just as with transcription factor binding site recognition, particular proposed interactions between miRNAs and their targets require specific experimental validation.

Regulation by any given miRNA usually has only a mild effect on target mRNA levels. For example, in HeLa cells transfected with individual miRNAs, only a small fraction of genes containing one or more miRNA recognition sites displays change in expression levels exceeding a factor of two (Guo et al., 2010). Similarly, in zebrafish embryos lacking miRNA function, about 75% of those genes displaying increased expression have only 1.2–2 fold change in transcript levels, and only about 25% of genes with increased expression levels are affected more than twofold (Giraldez et al., 2006). This relatively subtle effect on the expression of many target genes is consistent with the view that miRNAs predominantly act by slightly decreasing the stability of their target mRNAs. Thus, while not affecting mRNA synthesis, miRNAs result only in subtle reductions of target mRNA levels. We review in Chapter 2 how the synthesis and degradation rates of RNA and protein affect the levels of the protein product, and what the consequence would be on the kinetics of regulatory interactions were this protein product a transcription factor. Lowering the levels of transcription factors by a factor of two will in most cases not affect their function in embryonic development, but in other biological contexts, reducing the levels of a protein by a factor of two may have considerable biological consequences and some examples of such miRNA functions are discussed below. The effect of miRNAs on gene expression might be much more prominent when the gene encoding the target mRNA is no longer being transcribed, as the following examples illustrate.

3.2   General roles of miRNAs in development

General insights into the role of miRNAs during embryonic development can be obtained experimentally by inhibition of miRNA processing. Several experiments of this kind have been performed. For instance, *dicer* mutant mice, in which the maturation of most small ncRNAs including miRNAs is evidently inhibited, die during early development (Bernstein et al., 2003). On the other hand, zebrafish embryos lacking zygotic expression of Dicer develop without apparent defects, apart from a developmental delay at a later embryonic stage (Wienholds et al., 2003). However, due to the presence of maternal Dicer activity, miRNA processing is not completely blocked in these mutant embryos. Simultaneous mutation of maternal and zygotic Dicer (MZdicer) results in a more severe phenotype (Giraldez et al., 2005). In these embryos, most miRNAs which are expressed in wild-type embryos are not detectable. MZdicer mutants display defects in morphogenetic processes, somitogenesis and in the development of heart and brain, and in addition their development is delayed. Yet, axis formation and regional specification of the embryo was not affected in MZdicer embryos, and neither was the formation of most cell types. Thus almost all of the system controlling spatial organization and cell type specification is impervious to the simultaneous elimination of most processed miRNAs. But what then is the cause of those phenotypes observed? To answer this question, the prevalence of individual miRNAs during embryogenesis was analyzed. One particular family of miRNAs, the miR-430 family, is highly expressed in early zebrafish development. Amazingly, injection of individual members of this family, such as miR-430a, into MZdicer mutant embryos resulted either in complete (brain development) or partial (somitogenesis, gastrulation) rescue of many of the observed phenotypes as illustrated in Fig. 1.4(B) (Giraldez et al., 2005). Specific miRNAs therefore need not be restricted to particular embryonic regions in order to execute their specific functions, and systemic injection of one type of miRNA is sufficient to rescue a broad variety of phenotypes. These results suggest that spatial organization during early zebrafish development operates independently of miRNAs, but that miRNAs are nonetheless required for system-wide functions during embryogenesis. Indeed, as was later discovered, miR-430 directly regulates hundreds of maternal transcripts, inducing the clearance of these mRNAs in the early zebrafish embryo (Giraldez et al., 2006). The function of the miR-430 family therefore is to mediate the transition from maternal to zygotic gene expression and this to some extent affects the entire embryo.

Similar results were obtained in sea urchin embryos. Dicer transcripts are present maternally in the egg, whereas Dicer protein is produced in the early stages of embryogenesis (Song et al., 2012). Blocking translation of Dicer protein by injection of a gene-specific morpholino does not affect early development of sea urchin embryos, but later development is delayed and the formation of the gut, in this embryo the major morphogenetic process during gastrulation, is impaired. Similarly, interfering with Drosha protein expression causes developmental delay at gastrula stages. However, just as in zebrafish, injection of only

Figure 1.4

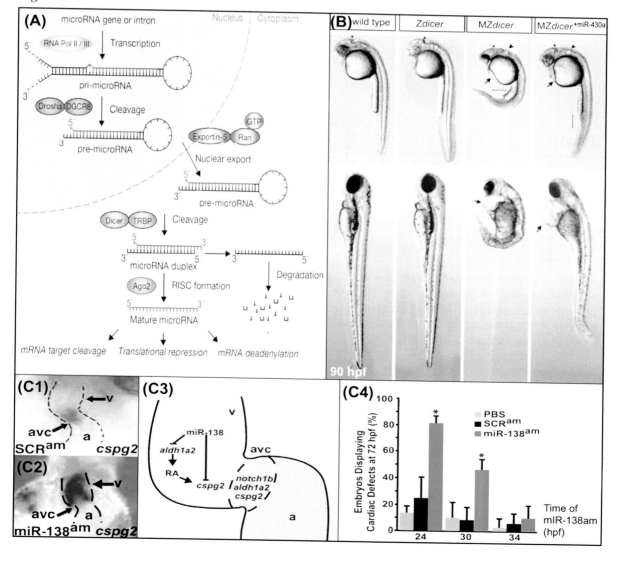

Figure 1.4

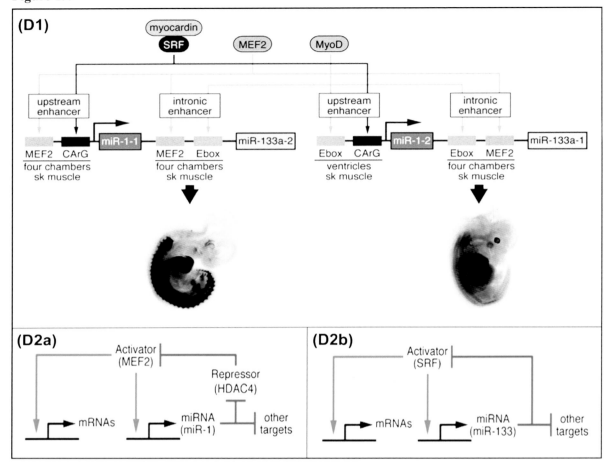

(D1)

myocardin
SRF   MEF2   MyoD

upstream enhancer   intronic enhancer   upstream enhancer   intronic enhancer

MEF2 CArG — miR-1-1   MEF2 Ebox — miR-133a-2   Ebox CArG — miR-1-2   Ebox MEF2 — miR-133a-1
four chambers sk muscle   four chambers sk muscle   ventricles sk muscle   four chambers sk muscle

(D2a)

Activator (MEF2)   Repressor (HDAC4)

mRNAs   miRNA (miR-1)   other targets

(D2b)

Activator (SRF)

mRNAs   miRNA (miR-133)   other targets

Figure 1.4 MicroRNA function and regulation in development. (A), Biogenesis and canonical sequence of miRNA processing steps *(from Winter et al. (2009))*. After intranuclear cleavage of the primary miRNA (pri-microRNA) and export to the cytoplasm by the Exportin 5-Ran-GTP complex, the pre-miRNA hairpin interacts with the double stranded RNA binding protein TRBP and is cleaved to its final length and denatured. Together with the Argonaut (Ago2) proteins the mature miRNA strand forms the "RNA-induced silencing complex" (RISC), which interacts with target mRNAs to produce the outcomes indicated at the bottom of the Figure. (B), Rescue of zebrafish embryos lacking all miRNAs by replacement of a single miRNA species *(from Giraldez et al. (2005))*. Embryos are shown in lateral view at 36 and 90 h post fertilization (hpf). Morphological effects of mutational inactivation of zygotic *dicer* gene expression (Z*dicer*), and of both maternal and zygotic *dicer* mutations (MZ*dicer*), can be seen in comparison with wild-type larvae at top. Injection into single-celled zygote of miR-430a alone can be seen in bottom panel to rescue the most severe defects caused by the double mutation, viz. brain morphogenesis (white bracket), midbrain/hindbrain boundary (asterisk), and general aspects of trunk and somite morphology (black bar), though heart edema (arrows) and defective ear formation (arrowheads) are not rescued. (C), miR-138 effects on heart development in zebrafish *(from Morton et al. (2008))*. (C1), Transcripts of the gene *cspg2* (encoding the Versican proteoglycan) expressed normally in valve development, here specifically in the atrioventricular canal, avc, as revealed by in situ hybridization: v, ventricle; a, atrium. A scrambled control antagomir (SCR^am) was present. (C2), Result of introduction of antagomir targeting miR-138 (miR-138^{am}): *cspg2* expression now extends throughout the ventricle. (C3), Function of miR-138: *aldh1a2* (the gene encoding Raldh2), and *cspg2* are both direct targets of miR-138 in the ventricle, and both genes are expressed in the avc in the absence of miR-138 function. (C4), Application of miR-138^{am} at 24 but not 34 h of development causes heart defects, including edema (ordinate): PBS (saline) and SCR^{am}, controls. (D), miR-1/133 transcription control, and their function in heart regulatory circuitry *(from Liu and Olson (2010))*. (D1), *Cis*-regulatory inputs to bicistronic mouse miR genes active in developing heart (not to scale). miR-1-1 and miR-133a-2 are encoded in one gene, miR-1-2 and miR-133a-1 in a second similarly organized gene. The relevant transcription factors are shown in ovals above the diagram, *cis*-regulatory elements are indicated by colored boxes, and below these are given the locations where the respective enhancers are active: four chambers, heart; sk, skeletal muscle. LacZ constructs containing these two regulatory systems drive expression as shown in the transgenic mice (e11.5) beneath each gene. (D2), miR-1 and miR-133 regulatory circuitry: (D2a), miR-1, (D2b), miR-133. In (D2a) miR-1 RNAs participate in a double negative gate circuit, repressing the HDAC4 repressor and thus stimulating expression of the Mef2 transcription factor, which in turn leads to increased miR transcription (D1). In (D2b), in contrast, miR-133 RNAs repress expression of the Srf transcription factor, thus limiting their own transcription.

the four most abundantly expressed miRNAs, in this case miR-1, miR-31, miR-2012, and miR-71, into Drosha knockdown embryos rescues their development and leads to the formation of feeding larvae (Song et al., 2012). Thus in sea urchin embryos as in zebrafish embryos, a few miRNAs are required in the entire embryo for normal development to proceed, again suggesting that these miRNAs regulate general processes that function in many cells of the embryo. However, many miRNAs are expressed in specific cell types. Nonetheless, systematic analyses of the function of individual miRNAs or miRNA families in *Caenorhabditis elegans* revealed that only very few miRNA knockouts resulted in developmental or morphological defects (Miska et al., 2007; Alvarez-Saavedra and Horvitz, 2010). Therefore, miRNAs are in general not responsible for the spatial organization of the embryo or for the specification of particular cell types, functions that are mainly controlled by transcription factors. Despite the large number of miRNAs encoded in the genome and the large number of predicted mRNAs targets, miRNAs do not significantly contribute to the generation of developmental complexity, though they have more important functions in post-developmental processes.

3.3 miRNA functions in specific developmental processes

In certain developmental processes, miRNAs have been supposed to affect the spatial distribution of mRNA in a particular way. For instance, there are several miRNAs encoded in the *hox* clusters in flies and vertebrates. One miRNA family, miR-10, is conserved from flies to mammals both in sequence and in its genomic location upstream of *hox4* genes. miRNAs are also encoded upstream of *hox9* genes in both flies and vertebrates, although their sequences differ. The *Drosophila* miRNA *iab4* locus encodes two miRNAs, miR-iab-4 and miR-iab-8 (also termed miR-iab-4AS or antisense) on opposite DNA strands (Ronshaugen et al., 2005; Stark et al., 2008). Potential target sites for both miRNAs are contained in the 3′UTR sequences of neighboring *hox* genes, including *Ubx*, and many of these sites are conserved across 12 *Drosophila* species. In vitro assays demonstrate that indeed these miRNAs are capable of repressing the expression of a reporter gene containing the *Ubx* 3′UTR. In the developing fly, *Ubx* is expressed in the haltere imaginal disc and represses genes which function in the formation of wings. Mutation of *Ubx* causes a homeotic transformation, producing wings in place of halteres (see Chapter 7). Consistently, overexpression of miR-iab-4 or miR-iab-8 in haltere imaginal discs leads to the transformation of halteres into wings (Ronshaugen et al., 2005; Stark et al., 2008). These experiments strongly suggest that in normal

Drosophila larvae, the two miRNAs function to repress *Ubx* expression. Furthermore, the expression of miR-iab-4 and miR-iab-8 is mostly exclusive with respect to distribution of *Ubx* transcripts. But despite these overexpression results, embryos lacking both miRNAs showed almost no change in *Ubx* expression and mutant adult flies showed no segmental transformation (Bender, 2008). Therefore, miR-iab-4 and miR-iab-8 are not normally required to control spatial expression of *Ubx* in the dorsal imaginal discs, although they can be made to do so by gross overexpression. Instead, as much evidence shows, the spatial expression of *Ubx* is regulated transcriptionally in the dorsal imaginal discs. Nevertheless, the conservation of seed sequences in the *Ubx* 3′ UTR suggests that these sequences have a biological function. It may be relevant that a recent observation indicates that *Ubx* and other *hox* genes use alternative 3′UTR sequences (Thomsen et al., 2010). Only a few miRNA seed sites are included in the short *Ubx* 3′UTR, the form which is predominantly expressed during embryonic germ band extension. On the other hand, most *Ubx* transcripts expressed in the developing CNS contain the longer 3′UTR which includes all predicted miRNA seed sequences, and in this location, expression of *Ubx* overlaps with miR-iab-4/8.

In the vertebrate *hox* cluster, a different family of miRNAs, the miR-196 family, is encoded in the same position as is miR-iab-4 in *Drosophila*, upstream of *hox9*. The function of the miR-196 family has been tested in the anterior–posterior patterning of the axial skeleton in chicken embryos (McGlinn et al., 2009). Since miR-196 family members are encoded in the A, B, and C *hox* clusters, perturbation of miR-196 activity was achieved in *trans*, by AntagomiR injection. Blocking miR-196 function leads to slight anterior expansion of *hox8b* expression and to a small increase in the frequency with which the last cervical vertebra is transformed into a thoracic vertebra. A revealing aspect of this result is that it represents a quantitatively subtle increase in a phenotype that occurs naturally at a low frequency. The implication is that the failure to clear *hox8b* transcript in the absence of miR-196 results in perturbation of a developmental process which is not robustly hardwired, since some flexibility is normally observed. The spatial role of the miRNA is best described as minor fine tuning.

A more decisive example of miRNA function in developmental patterning has been elaborated in zebrafish embryos. In the developing zebrafish heart miR-138 is transcribed in ventricular cardiomyocytes, where it represses genes that are expressed in the adjacent atrioventricular canal (AVC) (Morton et al., 2008). In the absence of miR-138 function, the expression of AVC-specific genes expands into the ventricle, as can be seen for a normally AVC-specific proteoglycan gene in Fig. 1.4(C1,2) (Morton et al., 2008). One of the mRNAs directly regulated by miR-138 encodes a protein that functions in the retinoic acid (RA) synthesis pathway, RA dehydrogenase (Raldh2). This enzyme converts retinaldehyde to RA, and is normally expressed in the AVC region. But in miR-138 knockdown embryos, its expression extends into the ventricle. Thus the function of miR-138 in this developmental context is to restrict the production of the RA signaling molecule to the AVC region and thereby contribute to the patterning of the heart (Fig. 1.4(C3)). Consistent with this interpretation, treating embryos with RA leads to a similar expansion of gene expression from the AVC into the ventricle (Morton et al., 2008). An interesting aspect of this system is that the function of miR-138 is very time sensitive. MiR-138 is specifically required between 24 and 34 h of zebrafish development, but is not necessary after that (Fig. 1.4(C4)). This suggests that while miR-138 functions to establish a boundary of gene expression between the AVC and the ventricle, it is not required for the long-term maintenance of this boundary. As heart development progresses, AVC-specific gene transcription evidently becomes sufficient to restrict gene expression to the AVC. Various other miR-NAs have been found to function in cardiovascular development (Liu and Olson, 2010). Their expression, like the expression of protein-coding genes, is driven by the specific set of transcription factors present in cardiac muscle cells. For example, as shown explicitly in the *cis*-regulatory diagrams of Fig. 1.4(D1), the transcription of miR1/133 is regulated by multiple enhancers, utilizing SRF, MEF2, and MyoD, and these miRNAs participate in the regulatory circuitry (Fig. 1.4(D2)). While one might imagine that the functions of miR1/133 would be required to avoid particular phenotypic errors in heart development, specific regulatory interactions of these miRNAs required for discrete developmental processes, as in the zebrafish miR138 case, are not evident. Rather, a major contribution of miR1/133 to biological process is in the control of stress response and other adult heart functions mediated by the subtle regulation of multiple factors, which cooperate to fine tune the outcome.

Thus, miRNAs function primarily by controlling the levels of target mRNAs, usually to a modest degree. The paucity of clear cases where miRNAs affect developmental processes illuminates the minor role of subtle level control in most such processes. Perhaps control of level kinetics is most important during regulatory state transitions, where it is useful to ensure rapid clearance of pre-existing transcripts, and it is interesting to see that those cases that display specific developmental miRNA functions occur at such transient junctures.

3.4   Common and distinct roles of transcription factors and miRNAs

The essential question is how transcription factors and miRNAs cooperate in the regulation of gene expression, and what their specific functions are in the complex process of development. There are several similarities in gene regulation by transcription factors and by miRNAs (Hobert, 2008). It is clear that both are capable of reading nucleic sequence by direct interaction. This interaction can affect the level of transcripts prevalent in the cell, either by regulating RNA synthesis or by regulating RNA decay. Interestingly, both types of molecule recognize relatively short sequences with a certain redundancy. Transcription factors recognize DNA-binding sites displaying a preference for certain nucleotides at certain positions, while allowing some variation of the preferred binding motif. Similarly, miRNAs interact with their target mRNAs preferably through the conserved seed sequence, although certain mismatches may be compensated by interactions of additional nucleotides. In other words, a variety of related sequences may be bound by individual regulatory molecules of both types. On the other hand, the presence of a binding motif is not necessarily sufficient for gene regulation. Transcription factors always work in combination with other transcription factors, but miRNAs also have been shown to function combinatorially together with other miRNAs. The similarities between gene regulation by transcription factors and miRNAs also extend to the fact that they both regulate multiple target genes.

Yet there are some profound differences in the way miRNAs and transcription factors regulate gene expression. Transcription factors may control gene expression either by activating or repressing the synthesis of a transcript, while miRNAs can further regulate either the stability of the transcript or the translation of the protein product, both resulting in reduction of protein expression. Thus, the function of miRNAs cannot explain why genes are expressed in given spatial domains, but rather explain why these genes are not expressed at higher levels. The formation of spatial boundaries during embryonic development often relies on repression mechanisms which lead to the Boolean exclusion of alternative cell fates. While transcriptional repression often acts dominantly, effectively turning gene expression completely off, miRNAs only lead to a reduction of mRNA levels. The formation of exclusive spatial compartments in embryogenesis requires expression of entirely different sets of genes and not simply different relative levels of gene expression. Spatial control of gene expression in development therefore clearly depends on on/off transcriptional regulation, as demonstrated by the numerous examples of CRMs with specifically confined spatial activity. Repression mediated by transcription factors and miRNAs may have quite different kinetic consequences, and regulation of a gene by both transcription factors and miRNAs may be required. For example, consider the process by which the expression of an active gene has to be turned off. A transcription factor which represses this gene must be expressed within the same cells at sufficient levels. The target gene therefore continues to be expressed while the repressor is being transcribed, transcripts are being transferred to the cytoplasm, proteins are being produced, and then transported back to the nucleus. Once sufficient levels of repressor are achieved, it will indeed bind the respective CRM of the target gene, recruit additional cofactors and decisively turn off transcription. But depending on the stability of the mRNA, it may require a considerable amount of time until this mRNA can be cleared from the cytoplasm exclusively by means of transcriptional control, easily several hours in typical animal systems. Were this same gene also regulated by miRNAs, clearance of the target mRNA could occur much faster. Rapid clearance of a transcript can be achieved if the transcript is intrinsically unstable and has a high turnover rate, but then this transcript would be unstable wherever it is expressed. In contrast, stable transcripts can be cleared in given conditions by the advent of a miRNA which targets it for destruction. Thus regulation by miRNA enables context-dependent control of turnover kinetics. It is not surprising in this regard, that miRNAs are frequently found to be involved in the regulation of physiological responses, e.g., in stress response, and for the regulation of cell cycle. Each of these processes depends on rapid and reversible changes of gene expression levels. In addition, developmental transitions that require the simultaneous clearance of many mRNAs might involve the function of miRNAs.

Thus, just as do transcription factors, miRNAs regulate gene expression by directly recognizing and binding to specific sequence motifs. Yet the mechanisms by which miRNAs regulate gene expression differ critically from transcriptional regulation, resulting in the distinct roles that transcription factors and miRNAs play in the spatial control of gene expression during development. In general transcriptional regulation generates the spatial gene expression patterns that fundamentally underlie the developmental process, while miRNAs kinetically and occasionally otherwise refine these patterns and modulate their downstream functional performance.

Farther downstream with respect to gene regulation lie additional layers of control. These include controls at the levels of translational output, the regulated sequestration and topological distribution of proteins within the cell, and beyond that, enzymatic alterations of protein properties. Ultimately such mechanisms pertain to much of cell biology, to the cytoplasmic processes by which proteins interact with one another and assemble in the structures of the cell. But the fundamental point remains that the proteins affected at these levels of control are present in given cells only because the genes encoding them are expressed in those cells. Our focus in this book is on the regulatory genome and its system level, sequence-dependent functions; the outer universe of cell biology lies beyond these matters.

4. Levels of Control of Gene Expression: Histone Modifications

We turn now to the regulatory significance of modifications of those nuclear components that, other than transcription factors, are in the most intimate contact with the DNA genome, chromatin proteins. The approximately six feet of linear DNA molecules present in each of our cell nuclei are structurally organized in chromatin, consisting of the genomic DNA and the proteins immediately associated with it. Nucleosomes provide the major structural unit of chromatin. The core nucleosome consists of 147 bp of DNA ($1\frac{3}{4}$ superhelical turns) wrapped around tetrameric protein complexes of the core histones H2A, H2B, H3, and H4. The status of the chromatin differs depending on the genomic region and the state of the cell, with direct consequences for the accessibility of DNA (Cruickshank et al., 2010; Bell et al., 2011). One way in which DNA accessibility is regulated involves the posttranslational modification of histone proteins. A huge body of recent literature reveals the prevalence of specific histone modifications at specific genomic regions and in specific developmental contexts, indicating correlations between histone modifications and genomic activity. Yet causal interrelations remain to be established among gene-specific insertion of chromatin modifications, their dynamic regulation during development, their inheritance through subsequent cell cycles, and their effect on different DNA-dependent processes, including regulation of transcription.

4.1   Usages of "epigenetics"

The field of biology concerned with the modification of histone proteins is today called epigenetics, descendant of a term that was redefined by C.H. Waddington in the 1940s. However, the use of this term has evolved dramatically over these last decades (Felsenfeld, 2014). The initial concept defined by epigenetics is that there is no simple relationship between individual genes and the emergence of developmental properties, rather there must exist a more complex system of interactions which determines how individual components affect the outcome of biological processes. Waddington defined epigenetics as "the branch of biology which studies the causal interactions between genes and their products, which bring the phenotype into being" (Waddington, 1942). Hence the early use of the term epigenetics reflects concepts not so different from what we are concerned with here. But the modern use of "epigenetics" is quite distinct from this. The definition of epigenetics as the study of "mitotically and/or meiotically heritable changes in gene function that cannot be explained by changes in DNA sequence" (Russo et al., 1996) reflects a shift in perception from how genes are used at the system level to how gene use is inherited at the cellular level. The early definition clearly includes the problem of how the increase in complexity and therefore the acquisition of different cellular states is achieved during development, though it was not clear at that time that all cells of an organism contain the same genetic information. However, the newer definition is much more oriented toward the problem of inheriting different obtained cellular states.

Unfortunately, this newer definition of "epigenetics" is interpreted by some to imply that differences in gene activity which are not due to differences in DNA sequence occur by mechanisms independent of genomic sequence. Throughout development, gene expression indeed varies dramatically though the DNA sequence is invariant, but these changes in expression occur neither randomly nor under the control of environmental (extra-organismic) influences. The changes in gene function are the product of DNA sequence-dependent regulatory processes, as is proved by the numerous examples throughout this book. Epigenetics, as currently used purports to show how transcriptional status, once obtained, is inherited through cell divisions. However, the maintenance of stable cellular states of gene expression matters only once the developmental specification of cell fate is achieved, and cells are terminally differentiated. At this stage, most cells divide only occasionally if at all, and the problem of how these chromatin modification marks are inherited through mitotic divisions is often not even relevant for such cells. Perhaps because of the various concepts associated with the term "epigenetics", or for other reasons (Maderspacher, 2010), the field of epigenetics has given rise to a problematic assumption, despite absence of direct causal relationships between chromatin modifications and gene regulation. This has drawn much attention, especially in public science debate. The assumption is that the different patterns of gene expression in cells of an organism have to be caused by cell type-specific chromatin modifications, which cover the otherwise identical genomic sequence and determine its activity. But in this assumption the true causality is reversed. It is the genomic sequence which determines the patterns of chromatin modifications, and these are installed as a result of developmental cell fate specification.

Considering the mechanisms associated with histone modification, perhaps a much more useful translation of the Greek word "epi" in this context is "attached to". "Epigenetic modification" describes covalent chemical alterations of histone proteins which are therefore not part of, but attached to the genetic information. These modifications are inserted at specific genomic locations in consequence of the binding of transcription factors that recognize specific DNA sequences, while histones and histone-modifying enzymes cannot read DNA sequence. The further inheritance of such covalent histone modifications could indeed be an important contribution to the maintenance of the state of gene expression throughout subsequent cell generations without requiring the continuous presence of DNA-binding transcriptional regulators. At least in the case of repression of genes associated with alternative cell fates, this mechanism seems to be frequently applied. Another mechanism of maintaining the silent states of genes which have been subjected to transcriptional repression is DNA methylation. The methylation of CpG residues has been associated with silencing of transposons, repeated sequence elements, inactive X chromosomes, and condensed heterochromatin. DNA methylation ensures transcriptional silence by interfering with binding of certain transcription factors or by attracting additional transcriptional silencers. However, DNA methylation generally occurs long after and far downstream of the repression of gene transcription by sequence-specific regulators and lies outside the scope of this discussion, which is focused on informational transactions in gene regulation (for reviews, see Bird, 2002; Deaton and Bird, 2011; Ghirlando et al., 2012).

The potential contribution of histone modifications in the regulation of gene expression during development is in maintenance, and not in change of gene expression states. Many of the genes encoded in the genome are expressed in specific cellular domains or specific cell types. These genes are all subject to transcriptional control, which determines both their active expression in given cells and their nonexpression in all other cells. When genes are activated, transcription factors that bind to the CRMs servicing them can be identified, and these factors are required for transcription. There are two alternative possibilities for why these same genes are not expressed in all other cells. Either the other cells simply do not possess the required activating transcription factors, or they express repressors which specifically block expression of these genes. Much evidence indicates that in fact specific repression is very frequently utilized to restrict gene expression during development. Thus once cells have been specified to acquire particular fates, they may need to permanently silence genes that will never again be expressed, for instance liver differentiation genes in pancreatic progenitor cells. While transcriptional silence is first introduced by specific repressors, it may later be maintained by the covalent modification of associated histone molecules and/or installation of other kinds of repressive chromatin protein complexes, for example the ubiquitous polycomb repressive protein complex. Thus, rather than requiring the permanent presence of high levels of negatively acting sequence-specific transcription factors, repressive chromatin modifications may be used to protect specific genes from future activation, and in addition these modifications contribute quantitatively to complete transcriptional silence.

The logical requirements for a role of histone modifications in gene regulation would be the following. First, that the chromatin modifications must be introduced in a DNA sequence-specific manner, at the right time and in the right cells, under the control of transcription factors that institute the transcriptional status of the gene. Second, that the presence of given chromatin configurations contributes in some manner to the specific transcriptional status of the associated locus. Third, that specific chromatin modifications be able to survive or be accurately regenerated over long developmental time periods, and sometimes through mitotic cell division. Current research focuses on the underlying mechanisms, even though at this point the evidence frequently consists of statistical correlations, and causal relationships remain scarce.

4.2   Histone modifications and transcriptional regulation

Recent genome-wide analyses have attempted to relate histone modifications to specific states of gene expression (Cruickshank et al., 2010; Beisel and Paro, 2011). The terminal polypeptide chains of histone proteins that protrude from the nucleosome DNA–protein complex, the histone "tails", are the major sites of covalent modification. Chemically, these modifications are primarily addition of acetyl- or methyl-groups on lysine (K) amino residues. For instance, it is claimed that acetylation of H3 and H4 histone tails contributes to a permissive environment for gene transcription and that acetylation of histone lysine residues weakens the DNA–nucleosome interaction thus increasing DNA accessibility. Genes that are specifically expressed are associated with H3K9/14 acetylated nucleosomes at the transcriptional start site. Histones with H3K4 trimethylation appear more frequently associated with promoters of genes that are transcriptionally active, whereas H3K4 monomethylation is often associated with enhancers. On the other hand, H3K9 and H3K27 methylation appear to be associated with repressed chromatin regions. Heterochromatin regions are marked by histone hypoacetylation, by H3K9 methylation, and by the presence of heterochromatin protein 1 (HP1). Genes that are repressed by polycomb group (PcG) proteins are associated with H3K27 methylation.

Yet despite observations that the distribution of histone modification correlates with specific functions in the associated DNA sequence, considering these modifications as a regulatory "histone code" is misleading (Henikoff and Shilatifard, 2011). Such a code would have to be predictive and causally connected to the function of the associated genomic site, but neither has been established in the many years of research in this field. Nor in principle could it be predictive: histones cannot read DNA sequence. Most purported relationships between histone modifications and gene expression states are based on correlations between genome-wide gene expression data and ChIP-seq data identifying given modifications (ChIP-seq, chromatin immunoprecipitation followed by sequencing to identify the genomic location of the modification identified by the antibody). Correlations cannot provide causal relationships between particular chromatin states and particular functional genomic outcomes. For example, methylation of H3K9 and H3K27 associates most frequently with silent genes and these are often termed "repressive marks". However, these histone marks are not predictive. Neither are they present on all silent genes, nor does their presence preclude gene transcription. They can also be present at genomic locations which at the same time harbor activating histone marks and which are actively expressed, as for example in the mouse β-globin locus (Vakoc et al., 2005). One problem in the causal study of histone modifications is that given types of modification can usually be inserted by multiple enzymes, and vice versa, individual histone-modifying enzymes may catalyze more than one type of histone modification. Furthermore, enzymes that catalyze the post-translational modification of histones may in addition modify other proteins. The functional contribution of individual modifications is therefore difficult to study, and their importance is not easy to demonstrate. Yeast in which the tails of either H3 or H4 histones are completely deleted are still viable, indicating that at least some modifications of these tails are not essential for regulated gene transcription to occur, though combinations of H2A and H3 tails are required for normal growth (Ling et al., 1996; Henikoff and Shilatifard, 2011; Kim et al., 2012). In other words, the distribution of histone modifications fails to predict the regulatory status of given regions of the genome and therefore it cannot be the case that the distribution of histone modifications qualitatively controls the regulatory status. However, the ubiquitous presence of histone modifications and their various correlations with status of gene expression indicates that they are not functionless. Histone modifications might be best considered as among the many working components of the enormously complex machinery that executes transcription or repression of transcription. As is generally true for large heterogeneous complexes, removal of a given component can wreck the function, but this rarely means that the component uniquely causes the function.

In development, these chromatin modifications are anything but static. Accordingly, enzymes which catalyze either the insertion or removal of histone modifications often colocalize at given genomic locations, e.g., methyltransferases and demethylases, or acetylases and deacetylases. Observing change of chromatin modification patterns in respect to change in gene expression status at subsequent stages of a developmental process can therefore be much more informative than is a snapshot evaluation of chromatin marks (Wamstad et al., 2012; Zhang et al., 2012). For example, a recent study of changes in chromatin modifications during heart development revealed that H3K4me1 marks enhancers (these are sometimes referred to as "poised" enhancers) even though only some of the associated genes will later be transcriptionally active (Wamstad et al., 2012). The marked enhancers were specific to the cardiac lineage in contrast to enhancers active in other cell types. A subset of genes which were at first marked with H3K4me1 later on displayed enrichment of H3K4me3 marks, correlated with gene expression. However, most of the apparently poised enhancers did not acquire an active state either during cardiomyocyte differentiation from embryonic stem cells or during cardiac precursor cell differentiation from cells representing a general mesodermal state, as summarized in Fig. 1.5. These marked enhancers could become active in cellular progeny which give rise to other cell fates. Another study on T cell development over five consecutive developmental stages shows the accumulation of H3Ac mostly at promoters of expressed genes, whereas H3K4me2 was often present at promoters even before as well as after the genes are expressed (Zhang et al., 2012). Observations of these chromatin marks in the context of the developmental process display their dynamic nature, and reveal them to be correlated with developmental changes in gene expression. For example, the T cell-specific gene *cd3* is transcriptionally silent at early stages when it did not show H3K4me2 marks at the promoter, while the locus was covered with low levels of H3K27me3. Yet at this stage the *cd3* enhancers were indeed marked by H3K4me2 and were already bound by the GATA-3 transcription factor. At a later stage, when transcription of *cd3* is initiated, H3K4me2 and H3Ac appeared at the promoter (Zhang et al., 2012). Whatever the mechanistic functional significance of these dynamic changes in histone modifications at the cd3 gene, their correlation with transcriptional state is unequivocal.

Figure 1.5

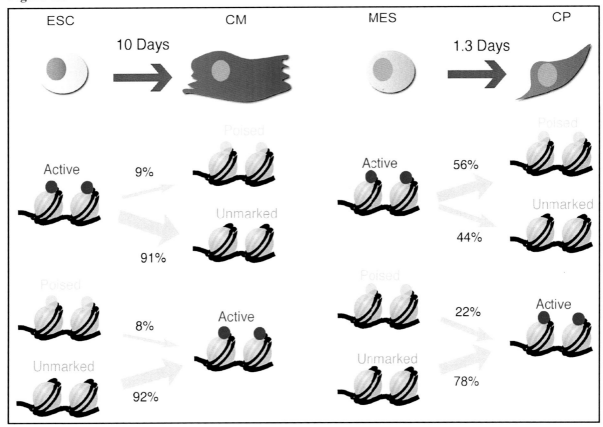

Figure 1.5 Transitions in chromatin state of specific enhancers during cardiac cell differentiation. The chart shows a summary of observations by ChIP-seq on enhancers of known genes active in various stages of cardiac cell differentiation in an embryonic stem cell (ESC) culture system *(from Wamstad et al. (2012))*. Four stages were compared: the starting ESC's marked by expression of the pluripotentiality genes *oct4* and *nanog*; a mesodermal precursor phase (MES) marked by expression of *brachyury* and *msp1*; a cardiac precursor (nonbeating) phase (CP) marked by expression of *nkx2.5*, *islet1*, and *tbx5*; and the final beating cardiomyocyte stage (CM) marked by expression of the cardiac specific *myosin heavy chain 6* and *7* genes. The enhancers were classified according to their chromatin states as active, i.e., positive for H3K27ac with or without H3K4me1; poised, i.e., H3K4me1 only; or inactive, without these marks. In the transition from ESC to CM, the large majority of enhancers active in ESCs became inactive, while only 8% of enhancers active at the CM stage had been poised at the ESC stage. Even in the relatively short transition between MES and CP stages, less than a fourth of the active CP enhancers had been in a poised state at the MES stage.

Certain chromatin modifications, particularly when observed in developmental context, have proved to be of great predictive value for the identification of active transcriptional regulatory elements such as promoters and distant enhancers. Thus H3K4me1, H3K4me2, H3K27Ac, H4Ac, and also the transcriptional coactivator protein p300 frequently accumulate on or around enhancers (Visel et al., 2009; Blow et al., 2010; Wamstad et al., 2012; Zhang et al., 2012).

4.3 Sequence- and cell type-specific histone modifications

Histone modification enzymes are not sequence specific and have to be recruited to particular genomic locations by interaction with transcription factors or other sequence-specific molecules. The developmental context or cell type in which the histones around a particular gene will be modified thus depends on expression of both the recruiting transcription factor and the modification enzyme. From what we now know the genome encodes many fewer modification enzymes than it does transcription factors. Thus the same enzymes are recruited by different regulatory factors, depending on the cellular context. For example, the lysine methyltransferase G9a catalyzes H3K9 dimethylation, which is often associated with transcriptional repression. In liver cells, G9a is recruited by the transcription factor E4bp4 to the gene encoding fibroblast growth factor 21 (FGF21) and is required for the repression of this gene (Tong et al., 2013). In undifferentiated myoblast cells, it is the homeodomain transcription factor Msx1which recruits G9a to genes associated with myogenic differentiation, for instance *MyoD*, resulting in the repression of these genes (Ling et al., 2012). Similarly, the bHLH transcription factor Sharp-1 inhibits the differentiation of skeletal muscle cells by mediating the recruitment of G9a to the *myogenin* promoter (Wang and Abate-Shen, 2012). Here, G9a catalyzes the methylation of both H3K9 and MyoD transcription factor. As these examples show, it is transcription factors which determine the set of genes to be repressed. However, the mechanism for gene repression does not uniquely depend on the transcription factors but also requires additional cofactors like G9a, which assist in the transcriptional silencing of the target genes. Sequence dependence of transcription factor-mediated histone modification provides the basis for gene- and cell type-specific chromatin modifications.

The addition of chromatin modifications can occur in particular developmental or physiological contexts. For example, the two genes *mdc* and *il-12b* are expressed specifically in dendritic cells and in macrophages upon inflammatory stimulation by lipopolysaccharide (LPS) (Zhu et al., 2012). *Cis*-regulatory sequences for both genes were identified based on the prevalence of H3K4me1 modifications in dendritic cells. The activating transcription factors are expressed in macrophages, dendritic cells and fibroblasts, but fibroblasts do not activate *mdc* and *il-12b* genes in response to LPS stimulation. An analysis of histone modifications revealed that while H3K27Ac and H4 acetylation modifications, usually associated with activation, were present at the endogenous enhancers in both dendritic cells and fibroblasts, there was a strong increase of repressive H3K9me3 modifications flanking the endogenous enhancers in fibroblasts upon LPS induction (Fig. 1.6(A1)). The proposed model for LPS-mediated induction of *mdc* and *il-12b* expression, normally restricted to dendritic cells and macrophages, can be summarized as follows: upon stimulation, sequences around the enhancers accumulate repressive histone modifications which can prevent these enhancers from activating transcription. The mechanism for insertion of H3K9me3 marks at these enhancers therefore must involve transcription factors which specifically recognize sequences in the enhancer. Since H3K9me3 accumulates upon LPS induction in fibroblasts, the recruitment or activity of the corresponding methyltransferase therefore depends on LPS induction. In dendritic cells, however, stimulation by LPS also results in an increase of expression of the H3K9me3 demethylase JmJD2d. This enzyme specifically accumulates at the enhancers of the *mdc* and *il-12b* genes (though not at enhancers of other LPS induced genes), and it prevents the accumulation of H3K9me3. Blocking the expression of JmjD2d results in an increase of H3K9me3 modifications around the enhancers and in a decrease in *mdc* and *il-12b* activation, as shown in Fig. 1.6(A2). Consistent with this result, ectopic expression of JmjD2d in fibroblasts resulted in a decrease of H3K9me3 marks and increased gene expression upon LPS induction. These experiments demonstrate a causal relationship. The mechanism for recruiting JmjD2d to the specific sites is not known so far, but again sequence-specific recognition must be the primary event. It is interesting that the insertion of H3K9me3 modifications occurs only in sequences flanking the enhancers and not within the regulatory sequence, so that interference with transcription factor binding to the enhancers in fibroblasts was not the cause of lack of inductive expression.

Figure 1.6

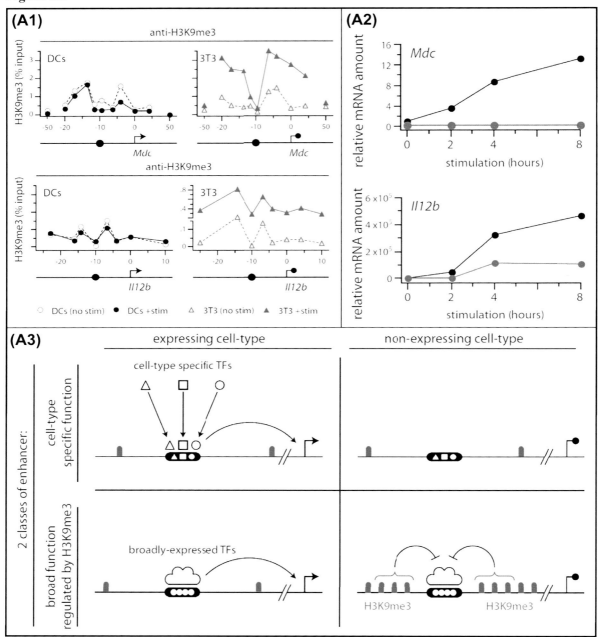

Figure 1.6 Functional correlations and maintenance of histone chromatin marks. (A), Involvement of repressive H3K9me3 marks in a cell type-specific induction response *(from Zhu et al. (2012))*. (A1), Induction of H3K9me3 marks in 3T3 fibroblast cells but not in dendritic cells (DC) by stimulation with LPS (or in 3T3 fibroblasts by stimulation with TNF-α). The amount of H3K9me3 marks as measured by ChIP (ordinates) is shown for *mdc*, a chemokine gene, and for the *il12b* cytokine gene, with and without stimulation. Only the DC cells normally respond to stimulation, and as discussed in the text the imperviousness of the fibroblasts to stimulation is correlated with the induction of H3K9me3 marks. (A2),

Figure 1.6

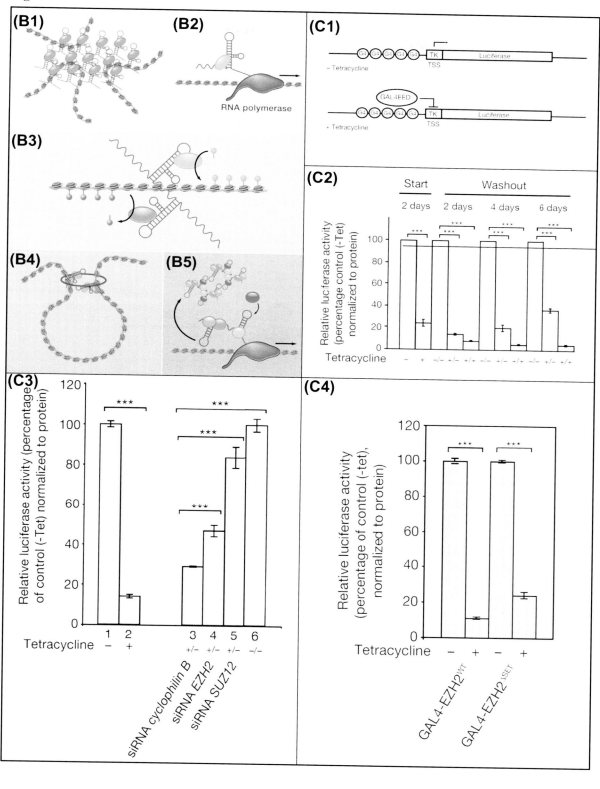

Figure 1.6

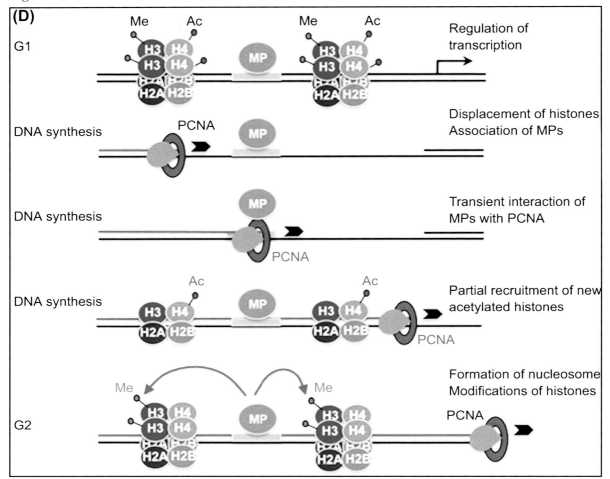

Effect of JmjD2d knockdown (red) on transcriptional response to LPS stimulation in DC's, compared to controls (black). The implication is that removal of H3K9me3 marks is required for the response. (A3), Proposed role of H3K9me3 marks in repression of certain classes of enhancer: activity of some enhancers depends directly on developmentally presented cell type-specific transcription factors while others respond to broadly present activators whereupon cell type specificity of expression is regulated by the specific addition of H3K9me3 modifications. (B), Various modes of lncRNA function that locally mobilize chromatin modifications *(from Nagano and Fraser (2011))*. (B1), Nucleation of chromatin in trans to form a "hub" where chromatin modifiers may act. (B2), Recruitment of chromatin modifiers to given loci. (B3), Recruitment of diverse modifiers by one multivalent lncRNA. (B4), Participation of lncRNA in cohesion mediated chromatin loops (cf Chapter 2). (B5), Generation of protein complexes within the nucleoplasm mediated by lncRNAs. (C), Role of a polycomb repressor complex (PRC2) in maintenance of histone modifications *(from Hansen et al. (2008))*. (C1), Diagram of tetracycline inducible luciferase (luc) reporter construct used to cause inducible expression of a Gal4-EED repressor (EED is a protein component of PRC2). G4, Gal4 site; TK, thymidine kinase promoter; TSS, transcription start site. (C2), Apparent persistence of repression after transient EED expression through subsequent cell divisions. "Start" samples were cells from a line bearing the construct in (C1), either stimulated or not with tetracycline for two days, then assayed for luc expression; the cells expressing

Gal4-EED efficiently repress construct activity. In the "Washout" experiments, cells previously treated with tetracycline were replated in tetracycline medium (+/+) or in medium lacking tetracycline (+/−), and cells never exposed to tetracycline were also replated without tetracycline (−/−); luc activity was then assayed at the indicated times. The experiment demonstrates the near complete maintenance of repression in the +/− cells even in the absence of tetracycline following washout. (C3), Further washout experiments illustrating necessity of the enzymatic PRC2 protein Ezh2 (see text) for maintenance of repression in absence of tetracycline. Lanes 1,2, luc expression after 2 days in presence or absence of tetracycline; lanes 3,4,5, show luc expression 4 days after washout and transfection with siRNAs targeting cyclophyllin as a control; Ezh2; and another PRC2 protein, Suz12. Lane 6 is a tetracycline − control. Three astericks indicates highly significant comparative results. (C4), Repression of construct by tetracycline dependent expression of Gal4-Ezh2 wild type (wt) but also by Gal4-Ezh2 lacking the SET domain (ΔSET). Observations were made after 2 days of tetracycline induction and they show that the Set domain is not required for the initial installation of repression. (D), Model for reinstallation of histone modifications mediated by Polycomb and Trithorax proteins, following DNA replication *(from Petruk et al. (2012))*. Successive stages are shown top to bottom. Nucleosomes are represented as histone octomers at G1 and G2 stages; MP, Polycomb, and Trithorax "maintenance proteins". These complexes remain attached to the DNA throughout the replication process, before and after the passage of the PCNA (proliferating cell nuclear antigen) which acts as a polymerase processivity factor associated with a DNA polymerase subunit. When nucleosomes reform in G2 their histone tails are remethylated by enzymatic functions executed by the Polycomb and Trithorax complexes.

Taken together, the results suggest that the activity of the *mdc* and *il-12b* enhancers is restricted by regulatory mechanisms involving the accumulation of repressive histone modifications specifically in sequences flanking the enhancers. Thus the "epigenetic" regulation of gene expression which is depicted here can be viewed as one of the mechanisms by which transcription factors achieve the regulation of gene expression from distantly located enhancers. The recruitment of histone-modifying enzymes to specific genomic sites depends on the regulatory state in specific cell types. In the case analyzed here, the sequence-specific recruitment of histone-modifying enzymes to *mdc* and *il-12b* enhancers occurs via transcription factors present in both dendritic and fibroblast cells. This results in the default repression of these enhancers. The activation of the *mdc* and *il-12b* enhancers in dendritic cells requires the (transcriptional) induction of an H3K9me3 demethylase, which is expressed specifically in this cell type upon LPS stimulation. Thus instead of direct activation by cell type-specific transcription factors, as is predominant for developmental enhancers, enhancers responding to physiological stimuli might operate on broadly expressed transcription factors in addition to conditional expression of derepressive cofactors such as histone-modifying enzymes. This idea is summarized in Fig. 1.6(A3) (Zhu et al., 2012).

Another well-elaborated example of the specific insertion of histone modifications and their consequence for gene expression is the mouse *β-globin* locus (Li et al., 2010). The *β-globin* gene is expressed in fetal liver cells where it is specifically enriched in H4R3me. This same modification is not enriched, however, at the silent *β-globin* gene in fetal brain cells. *β-globin* gene expression is regulated by the transcription factor Usf1, which binds both the promoter and the distant locus control region (LCR). Usf1 interacts with protein arginine methyltransferase 1 (Prmt1) and recruits this enzyme specifically to the *β-globin* locus where it methylates H4R3. Knockdown of Prmt1 not only resulted in reduced H4R3 methylation marks, but also in reduced H3K9/K14 acetylation at the same promoter and at LCR sites. Loss of H4R3me modifications also interfered with the looping between the LCR and the promoter and lead to a reduction in *β-globin* transcript levels. Thus specific histone modifications in the *β-globin* locus are required for the recruitment of additional modifications, for the communication between distant chromosomal locations, and for the transcription of the gene and possibly for the recruitment of particular chromosomal regions to transcriptional hubs (Li et al., 2010).

Transcription factor-mediated recruitment of histone modification enzymes does not necessarily affect gene expression state. Transcription factors may bind to cognate DNA sequences that lie outside of regulatory modules, but without apparent function. ChIP assays have demonstrated that the number of transcription factor–DNA binding events greatly exceeds the number of actual target genes for the respective factor. However, the mere association of particular transcription factors with histone-modifying enzymes can be sufficient to cause the modification of the associated chromatin, even when the factor binds to nonfunctional DNA sites. For example, it was recently shown that MyoD binds to approximately 60,000 sites in the genome, exceeding by a factor of three the total number of genes in the genome, but that even in the absence of any apparent regulatory function, binding of MyoD results in increased H4 acetylation (Cao et al., 2010).

Another mechanism by which chromatin-modifying molecules are recruited to specific sites in the genome has become apparent very recently, and this involves lncRNAs. These RNAs are encoded in the genome much like protein-coding genes, in multiple exons. Recent systematic analyses of lncRNA expression during zebrafish embryogenesis revealed 500–1000 lncRNAs which are transcribed by RNA polymerase II, spliced and polyadenylated just as are mRNAs, but lncRNAs lack open reading frames (Ulitsky et al., 2011; Pauli et al., 2012). Few lncRNAs have so far been functionally analyzed, but several examples clearly demonstrate the various potentials of these molecules to bind to protein complexes that mediate histone and DNA modifications, some of which are summarized in Fig. 1.6(B) (Nagano and Fraser, 2011; Pauli et al., 2011). One example is the role of lncRNAs in the imprinting of one parental genomic locus. In mice, the silencing of the paternal *igfr2* locus involves the expression of two lncRNAs from the paternal allele, which recruit chromatin modifiers such as the H3K9 histone methyltransferase G9, and which are required for the silencing of gene expression at this locus (Sleutels et al., 2002; Nagano et al., 2008; Pauli et al., 2011). Similarly, the expression of multiple lncRNAs is required for the inactivation of one of the two X chromosomes in placental mammalian females. Xist RNA expression occurs in the chromosome to be inactivated, but in mice (though not in humans) is inhibited on the active X chromosome by the expression of an antisense lncRNA, Tsix, transcribed from the opposite strand of the *xist* gene. Tsix RNA recruits DNA methyltransferases to the *xist* promoter. The inactive X chromosome retains expression of Xist, and this lncRNA mediates in *cis* the silencing of the entire chromosome. Though in most cases studied so far, lncRNAs seem to operate in *cis*, they have also been shown to function in *trans*, regulating gene expression from a different chromosome. The HOTAIR lncRNA, which is expressed in the intergenic region of the *hoxC* locus in mice, for example, is required in certain cells for the silencing of the *hoxD* locus on a separate chromosome (Rinn et al., 2007; Nagano and Fraser, 2011). HOTAIR has been shown to interact with two histone modification enzyme complexes, both of which mediate the formation of repressive chromatin modifications.

Consistent with the view that lncRNAs may function in *cis* as assembly scaffolds or concentration devices for protein complexes which affect the expression of protein-coding genes, mammalian lncRNAs are mostly encoded within <10 kb of protein-coding genes. In zebrafish, about 95% of sequences encoding lncRNAs are located in intergenic regions (Ulitsky et al., 2011). Even though the sequence of these lncRNAs is generally not conserved when compared to mouse or human lncRNAs, their position in respect to protein-coding genes is conserved. Thus, of about 900 protein-coding genes in the zebrafish genome which have conserved orthologs in mouse or human and which in zebrafish have flanking lncRNAs, about a third display similarly positioned lncRNAs in mammalian genomes.

4.4   Inheritance of histone modifications through mitotic cell divisions

A proposed role for histone modifications is that they contribute to transmission of information on previous regulatory activity at specific genomic locations through subsequent rounds of cell division. Yet, even though we can easily imagine how covalent modifications of specific histone residues can be maintained through several cell generations, it is less clear how modified nucleosomes can remain associated with a genomic locus as the replication fork dissociates the DNA strands, and how these modified nucleosomes become distributed to the two daughter loci. Even in the absence of DNA synthesis, a considerable turnover of nucleosomes has been observed within actively expressed genes (Deal et al., 2010; Beisel and Paro, 2011).

Perhaps not surprisingly, most evidence for the stable transmission of chromatin modifications derives from cases of gene repression, as the following illuminating examples show. Evidence for the stable inheritance of a repressive chromatin modification was generated in an experiment with cultured human fibroblast cells (Hansen et al., 2008). The experimental system consists of a reporter gene (Fig. 1.6(C1)), which is expressed under the control of GAL4-binding sites, and a conditionally expressed fusion protein, GAL4-EED (EED is a Polycomb protein), which causes the recruitment of the Polycomb Repressive Complex 2 (PRC2) to the reporter gene. Expression of the fusion protein GAL4-EED leads to the accumulation of H3K27me3 modifications and repression of reporter gene transcription. After transcription of GAL4-EED was turned off, H3K27me3 modifications were maintained over at least 4 days (in which time the cells divided about four times) as shown in Fig. 1.6(C2). The effects of residual levels of GAL4-EED, however, cannot be excluded. Endogenous PRC2 was shown to be required not just for initial addition but also for maintenance of the histone modification, implying that this histone modification depends on the enzymatic function of PRC2 for its persistence (Fig. 1.6(C3)). PRC2 was found to bind to H3K27me3 and is thus capable of recognizing its own product (Hansen et al., 2008). This interaction is proposed to result in a positive feedback loop that could function as a mechanism by which histone modifications can be stably transmitted. Even if the nucleosomes become redistributed and diluted by the two daughter strands after DNA replication, the presence of some nucleosomes harboring the H3K27me3 modification could, according to this idea, serve as a seed and recruit PRC2 to propagate the modification. The trimethylation of H3K27 is mainly mediated by the enzyme Enhancer of Zeste, Ezh2, which is the catalytic subunit of PRC2. This work shows that the catalytic SET domain of Ezh2, even though not required for the initial repression of gene transcription (see experiment reproduced in Fig. 1.6(C4)), is necessary for generating the H3K27me3 modification and for maintenance of gene repression (Hansen et al., 2008). Thus even though the histone modification H3K27me3 does not cause initial gene repression, it is involved in the feedback mechanism necessary for its maintenance.

Such feedback mechanisms may account for the spreading and maintenance of repressive complexes, both in heterochromatin and in polycomb-mediated repression. For example, the HP1 interacts with an H3K9 methyltransferase (SUV39H1) and also binds to di- and trimethylated H3K9 residues. While the initial targeting of HP1 to heterochromatic regions has to depend on sequence-specific interactions with the DNA mediated by other molecules, binding to methylated H3K9 residues might additionally enhance the recruitment of further HP1 to chromatin domains that already carry these modifications (for review, see Beisel and Paro, 2011).

A different conclusion was drawn from observations made in early *Drosophila* embryos (Petruk et al., 2012). At sites of DNA replication, the assembly of nucleosomes and the association of proteins of the Trithorax group (TrxG) and Polycomb group (PcG) were detectable. Yet, the histone modifications which are introduced by the TrxG complex, H3K4me3, and by PcG, H3K27me3, did not occur in proximity to the replication fork. In contrast to the nucleosome seed model proposed above, histone modifications might not be directly inherited through DNA replication. Rather these modifications are reestablished following replication through the actions of TrxG and PcG which remain associated with the DNA through replication, and not via the inheritance of histone-modified nucleosomes per se. This model is pictured in Fig. 1.6(D) (Abmayr and Workman, 2012; Petruk et al., 2012).

Whatever the precise mechanism might be by which histone modifications are propagated through cell divisions, it is important to see that they may persist for long periods of time. In the case that the cells divide, these modifications may either be maintained through cell division or reestablished thereafter. The maintenance of these chromatin modifications does not seem to rely on the DNA-binding transcription factors that are required for initial recruitment of histone-modifying enzymes to specific genomic regions. This is important from the point of view of the entire regulatory system. If transcription factors were required for the continuous control of gene expression for all genes that are either permanently turned ON (such as housekeeping genes in all cells or differentiation genes in terminally differentiated cells) or OFF (such as genes associated with alternative cell fates), a huge part of the regulatory system would have to be devoted to maintenance, since multiple transcription factors regulate each gene. The redeployment of transcription factors and signaling molecules in different developmental contexts could be problematic as the activation of unrelated and contextually irrelevant target genes would also have to specifically be prevented. There is also a quantitative argument for long-term repression of genes independent of the continuing presence of transcription factors: considering the short time for which many transcription factors remain bound to DNA, high levels of repressor proteins would be needed in the nucleus just to maintain the regulatory status quo, in order to keep expression of already silent genes OFF at all times. The existence of regulatory memory mechanisms which persist independently of transcription factors thus in principle has enormous advantages of economy in the control system. The developmental regulation of individual genes is tightly controlled by gene-specific *cis*-regulatory systems, each interacting with combinations of a few of the hundreds or thousands of encoded transcription factors. Thus the regulatory informational transactions mediated by transcription factors are required for programmed change in expression status, while in contrast, the chromatin modification system which is much more general and consists of many fewer components, may be used to maintain the states of gene expression developmentally instituted by transcription factor–DNA interactions.

4.5   Successive mechanisms controlling gene expression

All genes use as their primary logic device the *cis*-regulatory sequences that control their expression. As we further consider in the next chapter, the particular expression patterns for each gene depend directly on the composition of transcription factor binding sites in these regulatory modules. But only if the required transcription factors are present does the gene even have the potential to be expressed. However, not all of these factors have to bind to a given CRM at the same time. The observation that the binding of certain transcription factors leads to the "priming" of *cis*-regulatory elements, allowing or precluding the later binding of additional factors, indicates stepwise regulatory processes. Thus, the set of transcription factors interacting with a particular CRM might operate in a temporal order which requires additional mechanisms such as modification of associated histone molecules or recruitment of other cofactors to maintain memory of past binding events. One issue that is frequently confused in this context is what comes first. Is it transcription factor binding which leads to the modification of histones or does the modification of histones modulate the accessibility of DNA sequences to transcription factor binding? It is evident that histone modifications are not randomly distributed but rather occur at specific genes in particular developmental contexts. Their position depends on the recruitment of specific histone-modifying enzymes by sequence-specific transcription factors that are expressed in particular spatial and temporal domains of the developing organism. Every type of modification that is introduced nonrandomly has to be targeted by transcription factors or sequence-specific RNA molecules. Yet, some of these modifications will in turn affect the configuration of the chromatin and possibly alter the subsequent transcription factor binding capacity of certain sites and for certain factors. Many detailed aspects of each *cis*-regulatory design will affect the requirements for both transcription factors and chromatin modifications in the regulation of gene expression, including how tightly or cooperatively the cognate transcription factors bind, and previous regulatory history that might have affected the gene.

Ultimately, the production of a protein is the integral outcome of all levels of gene regulation. And even though all these levels might be required to produce specific amounts of gene products in a particular time and space, the relative contribution of each is very different. Many genes in the genome are not only regulated transcriptionally, but are also associated with dynamically changing chromatin modification signatures and produce transcripts that are further regulated by miRNAs and other posttranscriptional and posttranslational mechanisms. Thus the question of which type of mechanism regulates a given gene should be replaced by inquiry into how each of the various levels of regulation contributes to biological process. For the control of spatial gene deployment during animal development it has become clear that transcriptional regulation is of primary and ultimate importance.

5. The Regulatory Genome

5.1 Genomic sequence and regulatory genomic sequence

A remarkable feature of animal genomes first observed over 40 years ago (Avery et al., 1969) is the existence of discrete DNA sequence elements that stand out sharply from the rest when genomes of related animals are compared, because they are shared between species to an unusual extent. These blocks of sequence are evolutionarily conserved. Of course, as we now know, sequence conservation is a slippery slope. Some protein-coding sequences display obvious extreme conservation across the whole of the Bilateria, for example histones, many cell biology structural proteins, domains found in many proteins that display particular enzymatic functions, and DNA-binding domains of many transcription factors. There exist as well unusually conserved CRMs little changed in sequence since divergence from common ancestors of Cambrian vintage. But there is a continuum of conservation within every class of conserved sequence. For example, in well-studied echinoderm genes, CRMs can be recognized reliably as patches of relative sequence conservation in species about 50 million years apart, but as the evolutionary distance becomes greater, the conservation signal progressively fades out (Cameron and Davidson, 2009). The baseline limit is the largely sequence-independent majority component of the genome in which no significant conservation can be detected. Such sequence evolves at as high rates as do third bases in the degenerate codons of open reading frames, and it diverges as rapidly as does the majority of intronic and intergenic sequence. The best known functional components of the genome all display some sequence conservation relative to the surrounding nonconserved sequence: these functional components include protein-coding genes and sequences encoding miRNAs, small noncoding RNAs, ribosomal RNAs, and some other RNAs that execute various regulatory functions. Long ago the reasonable deduction was made that local evolutionary conservation means selective constraints on change, and that hence conservation per se indicates function. But there followed a classic logical error, for it is not true that lack of conservation means lack of function. The assertion that nonconserved intergenic and most intronic sequence is "junk DNA" (Orgel and Crick, 1980) produced one of the less felicitous moments in the history of molecular biology. The vast majority of the genomic DNA accounting for variation in genome size among animals, particularly among animals of given clades (Britten and Davidson, 1971) is repetitive sequence. These sequences, which originate as inserted families of mobile elements (Britten and Davidson, 1971; Moore et al., 1978; Jurka et al., 2007), are in general the most variable and least conserved of all genomic components, even between closely related species. About half of our genome, for example, consists of repetitive sequence. Most individual repetitive sequences are indeed likely to be functionless in the process of development, but not all are: for instance repeats are essential for X-chromosome inactivation by Xist RNA mentioned in the foregoing, as well as for escape from X-chromosome inactivation (Hall and Lawrence, 2010). And in evolution, mobilization of repeats is clearly one mechanism of producing novel associations between CRMs and genes (reviewed by Peter and Davidson, 2011). A large amount of generally nonconserved sequence is also represented in the hundreds of lncRNAs recently discovered (though some lncRNAs are conserved), and lncRNAs in turn have been implicated in a number of quite diverse functions, in addition to the emplacement of chromatin-modifying complexes discussed above. These functions include assembly of intranuclear structures, alternative splicing, genomic imprinting, and provision of structural links in multiprotein complexes (for reviews, Wang and Chang, 2011; Moran et al., 2012). Short specific sequences and secondary structures within lncRNAs, which are not easily detected by the metrics of sequence conservation, might provide intermolecular recognition signals that could be required for their functions. Similarly, many CRMs have been discovered which are not generally conserved in their sequence although their short specific transcription factor target sites persist during evolution and execute conserved regulatory functions. There is another very general and very necessary kind of function to be associated with nonconserved genomic sequence space. Genomic space per se is essential for a common feature of regulatory design in animals. It allows the multiple, distant CRMs servicing given genes to alternatively loop from their often distant positions around genes to the basal transcription apparatus, so as to mediate activation of given genes in different regulatory contexts, as discussed in the next chapter.

Here we are concerned with that portion of the detectably conserved genomic sequence which executes primary encoded gene regulatory functions. In the following we confine the term "regulatory genome" to DNA sequences constituting the CRMs discussed briefly above and in more detail in the next chapter. Functionally, CRMs are defined by the transcription factor target site sequences they contain. As we have seen, transcription factors are polyfunctional: they recognize and chemically interact with their target DNA sequences, and they also perform effector functions that in one way or another affect gene expression. Their DNA-binding domains execute the essential informational task of specifying to which gene these effector functions are to be applied. Directly or indirectly, these additional functions result causally in transcriptional expression or transcriptional silencing of the target genes. There are numerous biochemical mechanisms both of activation and repression of target genes, usually mediated by cofactors that interact with the transcription factors (for reviews of biochemical mechanisms of gene repression and activation, see respectively Beisel and Paro, 2011; Spitz and Furlong, 2012). Some proteins of regulatory significance which also recognize specific DNA sequences affect gene expression in other ways. For example, proteins that bind insulator sequences affect the choice of which CRM will interact with a target gene, and in well-studied cases it is clear that these proteins too are sequence specific (for example, CTCF, Ghirlando et al., 2012). CTCF also binds to nuclear lamina proteins, positioning genes to this transcriptionally quiescent peripheral region of the nucleus (Handoko et al., 2011), as do other sequence-specific transcription factors that control intranuclear position with indirect but potent effects on gene expression (for example, Zullo et al., 2012). Thus we arrive at a very simple conclusion. Beyond chromatin modifications, molecular biology is revealing a vast and continuously increasing repertoire of biochemical and cell biological functions which mechanistically affect gene expression, positively or negatively. But the all-important deployment of these functions to the right genes depends ultimately on the inherited regulatory DNA sequence code, and the means by which this occurs for all these functions is ultimately recognition of the relevant regulatory genomic sequence.

What is the complexity of the regulatory genome, that is, how much of the whole genome consists of sequences performing regulatory functions that include DNA sequence-specific recognition interactions? A crude a priori estimate extrapolated from known genes whose CRMs have been characterized suggested that there could be equal or twice the amount of regulatory DNA sequence, as there is of protein-coding sequence, if not more (Davidson, 2006). This was based on the number of CRMs found per gene in well-studied cases. CRMs are usually several hundred base pairs in length (for review of developmental *cis*-regulatory structure/function relationships, see Davidson, 2006). The sequence length of such modules includes both the actual transcription factor target site nucleotides and the inter site sequences. The sequences between functional binding sites are also though not always relatively conserved, either because in reality they too contain (unknown) target site sequences, or because for other reasons their deletion or insertion would affect function (Cameron and Davidson, 2009). The number of CRMs per gene in well-studied cases can range from 1 to 15 or more, often ~2–4. A similar number of modules per gene emerged from a recent "blind" scan for active CRMs that included almost 40 previously unstudied sea urchin regulatory genes using a high-throughput detection method (Nam et al., 2010). Extrapolation of these values to all protein-coding genes would then indicate about $1–5 \times 10^5$ individual CRMs in the whole animal genome. Now, with an array of high-resolution whole-genome assays in hand, it is possible to do very much better than such rough extrapolations. For example a study was carried out on 19 different mouse tissues, some such as brain regions very complex in terms of cell types and functions, using chromatin immune precipitation to detect regulatory regions (Shen et al., 2012). Putative active promoters were identified by mapping PolII and histone H3K4me3 marks, and active CRMs were indicated by mapping H3K4me1 and H3K27Ac marks, coincident with the presence of the very common coactivator p300. An example from these data is reproduced in Fig. 1.7(A). Validation controls using large sets of known promoters and CRMs yielded ~80% detection scores. The results

Figure 1.7

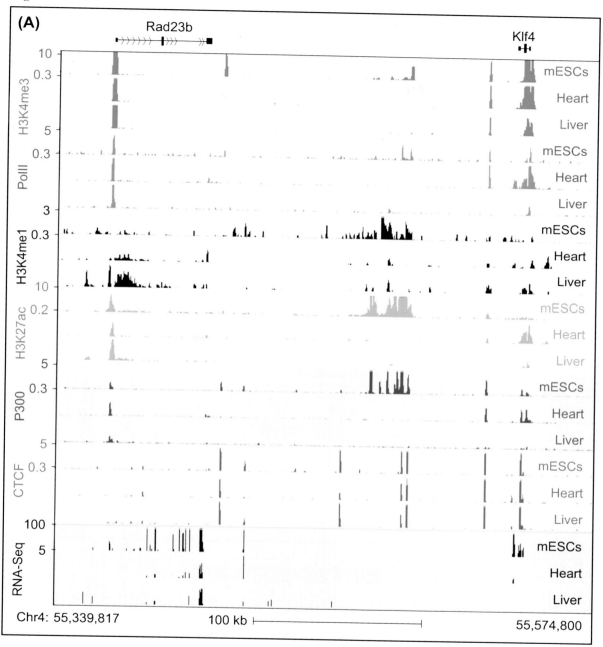

Figure 1.7 Detection of functional CRM *cis*-regulatory modules and active promoters by whole-genome methods. (A), A representative portion of the genome-wide analysis, from mouse chromosome 4, is illustrated for three of the 19 tissues analyzed in this study, ESC, heart, and liver *(from Shen et al. (2012))*. The mapped features are shown superimposed on replicate tracks from the UCSC genome browser. The signals shown indicate normalized intensity from chromatin immunoprecipitation, located by sequencing the bound DNA fragments ("ChIP-seq"). The top two sets of measurements identify the location of trimethylated lysine 4 of histone H3 and of

Figure 1.7

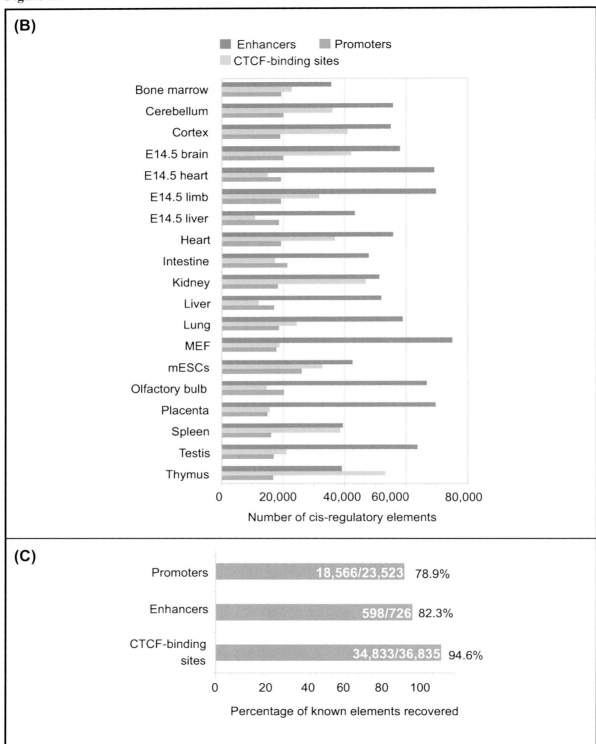

PolII in this portion of the genome, both indicative of active promoters. The following three sets of tracks display marks correlated with the presence of active CRM (enhancers): these are monomethylated lysine 4 of H3 histone, acetylated lysine 27 of the same histone, and the presence of P300, a coactivator utilized by diverse transcription factors. Other data in this study show that only a minority of the active CRM genome wide are actually associated with P300, however. The following tracks identify CTCF binding sites; CTCF is an insulator binding protein and also plays other roles in locating given genes to functionally important regions of the nucleus. The bottom tracks indicate the positions of nascent RNA transcripts in this region of the genome in the three tissues. (B), The numbers of nonoverlapping active promoters and CRM in each of the complete sets of tissues investigated is tallied in the bar graph. (C), The fractions of independently known validated promoters, of enhancers and of CTCF sites that were identified in this study are shown; these are underestimates of the actual recovery rate since some of the validated sets are expressed elsewhere than in the tissues studied.

are revealing: there were evidently on the average about 2–3 active promoters per active gene; and 2–3 times as many CRMs as active promoters (Fig. 1.7(B)). In all 19 tissues taken together, about 2.4×10^5 nonoverlapping CRMs were indicated genome wide, or, if the 80% validation (Fig. 1.7(C)) were taken literally, $\sim 3 \times 10^5$. It is also interesting to note that at a cutoff where only 10% of random mouse and human DNA appears conserved in sequence, about 50% of CRM sequences and about 75% of active promoter sequences are detectably conserved at this evolutionary distance. These numbers suggest that in an animal genome roughly $5–10 \times 10^7$ bp of sequence is included in the regulatory genome, similar to the complexity of the total protein-coding sequence.

So we may conclude that there are indeed several hundred thousand, and perhaps a million or more functional, independent *cis*-regulatory units in the regulatory genome of a mammal, as well as other sequences such as insulators. Each CRM is to be thought of as a little information processing, computational logic device. The information contained in the CRM sequence is processed by means of the multiple regulatory factors that the CRM responds to. The way CRMs controlling regulatory genes are linked to one another functionally is what determines the outcome of development, as we shall see in the next chapter.

5.2 Conclusion

We began this chapter with an external consideration of the complexity of developmental process, and the a priori requirement that it be encoded in the animal genome. Of all the aspects of gene expression control that we have traversed in this discussion, there is only one way by which genomic complexity can be utilized to build a program encoding development: by direct recognition of DNA sequence. Only transcription factors (and possibly occasional lncRNAs) have the capacity to read DNA sequence and initiate the cascade of processes resulting in patterns of gene expression. Many biochemical mechanisms are required for this cascade, which generates developmental gene expression, including the regulation of transcript levels by miRNAs and the mediation of prolonged states of gene repression by chromatin modification. But the decision of where these mechanisms are to be deployed, and which gene functions are to be applied in which developmental context, depends almost entirely on the sequence-specific interaction of transcription factors with the regulatory genome. The direct consequence of this fundamental conclusion is that the causal process in development is the regulatory control of the genes which encode transcription factors, in time and space. Indeed, much of the remainder of this book is focused on how the networks of regulatory interactions control function of these genes in development.

REFERENCES

Abmayr, S.M., Workman, J.L., 2012. Holding on through DNA replication: histone modification or modifier? Cell 150, 875–877.

Alvarez-Saavedra, E., Horvitz, H.R., 2010. Many families of *C. elegans* microRNAs are not essential for development or viability. Curr. Biol. 20, 367–373.

Armit, C., Venkataraman, S., Richardson, L., Stevenson, P., Moss, J., Graham, L., Ross, A., Yang, Y., Burton, N., Rao, J., Hill, B., Rannie, D., Wicks, M., Davidson, D., Baldock, R., 2012. eMouseAtlas, EMAGE, and the spatial dimension of the transcriptome. Mamm. Genome 23, 514–524.

Arnone, M.I., Rizzo, F., Annunciata, R., Cameron, R.A., Peterson, K.J., Martinez, P., 2006. Genetic organization and embryonic expression of the ParaHox genes in the sea urchin *S. purpuratus*: insights into the relationship between clustering and colinearity. Dev. Biol. 300, 63–73.

Avery, R.J., Bolton, E.T., Britten, R.J., Cowie, D.B., Hoyer, B.H., Kohne, D.E., Rice, N.J., Roberts, R.B., Shleser, R., 1969. Biophysics, Carnegie Institution of Washington Yearbook.

Bartel, D.P., 2009. MicroRNAs: target recognition and regulatory functions. Cell 136, 215–233.

Beisel, C., Paro, R., 2011. Silencing chromatin: comparing modes and mechanisms. Nat. Rev. Genet. 12, 123–135.

Bell, O., Tiwari, V.K., Thomä, N.H., Schübeler, D., 2011. Determinants and dynamics of genome accessibility. Nat. Rev. Genet. 12, 554–564.

Bender, W., 2008. MicroRNAs in the *Drosophila* bithorax complex. Genes Dev. 22, 14–19.

Bernstein, E., Kim, S.Y., Carmell, M.A., Murchison, E.P., Alcorn, H., Li, M.Z., Mills, A.A., Elledge, S.J., Anderson, K.V., Hannon, G.J., 2003. Dicer is essential for mouse development. Nat. Genet. 35, 215–217.

Bird, A., 2002. DNA methylation patterns and epigenetic memory. Genes Dev. 16, 6–21.

Blow, M.J., McCulley, D.J., Li, Z., Zhang, T., Akiyama, J.A., Holt, A., Plajzer-Frick, I., Shoukry, M., Wright, C., Chen, F., Afzal, V., Bristow, J., Ren, B., Black, B.L., Rubin, E.M., Visel, A., Pennacchio, L.A., 2010. ChIP-Seq identification of weakly conserved heart enhancers. Nat. Genet. 42.

Britten, R.J., Davidson, E.H., 1971. Repetitive and non-repetitive DNA sequences and a speculation on the origins of evolutionary novelty. Q. Rev. Biol. 46, 111–138.

Cameron, R.A., Davidson, E.H., 2009. Flexibility of transcription factor target site position in conserved *cis*-regulatory modules. Dev. Biol. 336, 122–135.

Cao, Y., Yao, Z., Sarkar, D., Lawrence, M., Sanchez, G.J., Parker, M.H., MacQuarrie, K.L., Davison, J., Morgan, M.T., Ruzzo, W.L., Gentleman, R.C., Tapscott, S.J., 2010. Genome-wide MyoD binding in skeletal muscle cells: a potential for broad cellular reprogramming. Dev. Cell 18, 662–674.

Cruickshank, M.N., Besant, P., Ulgiati, D., 2010. The impact of histone post-translational modifications on developmental gene regulation. Amino Acids 39, 1087–1105.

Damle, S.S., Davidson, E.H., 2012. Synthetic in vivo validation of gene network circuitry. Proc. Natl. Acad. Sci. U.S.A. 109, 1548–1553.

Davidson, E.H., 2006. The Regulatory Genome. Gene Regulatory Networks in Development and Evolution. Academic Press/Elsevier, San Diego, CA.

Deal, R.B., Henikoff, J.G., Henikoff, S., 2010. Genome-wide kinetics of nucleosome turnover determined by metabolic labeling of histones. Science 328, 1161–1164.

Deaton, A.M., Bird, A., 2011. CpG islands and the regulation of transcription. Genes Dev. 25, 1010–1022.

Felsenfeld, G., 2014. A brief history of epigenetics. Cold Spring Harb. Perspect. Biol. 6.

Friedman, R.C., Farh, K.K., Burge, C.B., Bartel, D.P., 2009. Most mammalian mRNAs are conserved targets of microRNAs. Genome Res. 19, 92–105.

Frise, E., Hammonds, A.S., Celniker, S.E., 2010. Systematic image-driven analysis of the spatial *Drosophila* embryonic expression landscape. Mol. Syst. Biol. 6.

Ghirlando, R., Giles, K., Gowher, H., Xiao, T., Xu, Z., Yao, H., Felsenfeld, G., 2012. Chromatin domains, insulators, and the regulation of gene expression. Biochim. Biophys. Acta 1819, 644–651.

Giraldez, A.J., Cinalli, R.M., Glasner, M.E., Enright, A.J., Thomson, J.M., Baskerville, S., Hammond, S.M., Bartel, D.P., Schier, A.F., 2005. MicroRNAs regulate brain morphogenesis in zebrafish. Science 308, 833–838.

Giraldez, A.J., Mishima, Y., Rihel, J., Grocock, R.J., Van Dongen, S., Inoue, K., Enright, A.J., Schier, A.F., 2006. Zebrafish MiR-430 promotes deadenylation and clearance of maternal mRNAs. Science 312, 75–79.

Guo, H., Ingolia, N.T., Weissman, J.S., Bartel, D.P., 2010. Mammalian microRNAs predominantly act to decrease target mRNA levels. Nature 466, 835–840.

Hall, L.L., Lawrence, J.B., 2010. XIST RNA and architecture of the inactive X chromosome: implications for the repeat genome. Cold Spring Harb. Symp. Quant. Biol. 75, 345–356.

Handoko, L., Xu, H., Li, G., Ngan, C.Y., Chew, E., Schnapp, M., Lee, C.W., Ye, C., Ping, J.L., Mulawadi, F., Wong, E., Sheng, J., Zhang, Y., Poh, T., Chan, C.S., Kunarso, G., Shahab, A., Bourque, G., Cacheux-Rataboul, V., Sung, W.K., Ruan, Y., Wei, C.L., 2011. CTCF-mediated functional chromatin interactome in pluripotent cells. Nat. Genet. 43, 630–638.

Hansen, K.H., Bracken, A.P., Pasini, D., Dietrich, N., Gehani, S.S., Monrad, A., Rappsilber, J., Lerdrup, M., Helin, K., 2008. A model for transmission of the H3K27me3 epigenetic mark. Nat. Cell Biol. 10, 1291–1300.

Hardison, R.C., Taylor, J., 2012. Genomic approaches towards finding *cis*-regulatory modules in animals. Nat. Rev. Genet. 13.

Henikoff, S., Shilatifard, A., 2011. Histone modification: cause or cog? Trends Genet. 27, 389–396.

Hobert, O., 2008. Gene regulation by transcription factors and microRNAs. Science 319, 1785–1786.

Howard-Ashby, M., Materna, S.C., Brown, C.T., Chen, L., Cameron, R.A., Davidson, E.H., 2006a. Gene families encoding transcription factors expressed in early development of *Strongylocentrotus purpuratus*. Dev. Biol. 300, 90–107.

Howard-Ashby, M., Materna, S.C., Brown, C.T., Chen, L., Cameron, R.A., Davidson, E.H., 2006b. Identification and characterization of homeobox transcription factor genes in *Strongylocentrotus purpuratus*, and their expression in embryonic development. Dev. Biol. 300, 74–89.

Howard-Ashby, M., Materna, S.C., Brown, C.T., Tu, Q., Oliveri, P., Cameron, R.A., Davidson, E.H., 2006c. High regulatory gene use in sea urchin embryogenesis: Implications for bilaterian development and evolution. Dev. Biol. 300, 27–34.

Howe, D.G., Bradford, Y.M., Conlin, T., Eagle, A.E., Fashena, D., Frazer, K., Knight, J., Mani, P., Martin, R., Moxon, S.A., Paddock, H., Pich, C., Ramachandran, S., Ruef, B.J., Ruzicka, L., Schaper, K., Shao, X., Singer, A., Sprunger, B., Van Slyke, C.E., Westerfield, M., 2013. ZFIN, the zebrafish model organism database: increased support for mutants and transgenics. Nucleic Acids Res. 41, D854–D860.

Jurka, J., Kapitonov, V., Kohany, O., Jurka, M., 2007. Repetitive sequences in complex genomes: structure and evolution. Annual Review of Human Genetics 8, 241–259.

Kim, J.A., Hsu, J.Y., Smith, M.M., Allis, C.D., 2012. Mutagenesis of pairwise combinations of histone amino-terminal tails reveals functional redundancy in budding yeast. Proc. Natl. Acad. Sci. U.S.A. 109, 5779–5784.

Kvon, E.Z., Kazmar, T., Stampfel, G., Yáñez-Cuna, J.O., Pagani, M., Schernhuber, K., Dickson, B.J., Stark, A., 2014. Genome-scale functional characterization of *Drosophila* developmental enhancers in vivo. Nature 512, 91–95.

Lemons, D., Paré, A., McGinnis, W., 2012. Three *Drosophila* Hox complex microRNAs do not have major effects on expression of evolutionarily conserved Hox gene targets during embryogenesis. PLoS One 7, e31365.

Li, X., Hu, X., Patel, B., Zhou, Z., Liang, S., Ybarra, R., Qiu, Y., Felsenfeld, G., Bungert, J., Huang, S., 2010. H4R3 methylation facilitates beta-globin transcription by regulating histone acetyltransferase binding and H3 acetylation. Blood 115, 2028–2037.

Lim, L.P., Lau, N.C., Garrett-Engele, P., Grimson, A., Schelter, J.M., Castle, J., Bartel, D.P., Linsley, P.S., Johnson, J.M., 2005. Microarray analysis shows that some microRNAs downregulate large numbers of target mRNAs. Nature 433, 769–773.

Ling, B.M., Gopinadhan, S., Kok, W.K., Shankar, S.R., Gopal, P., Bharathy, N., Wang, Y., Taneja, R., 2012. G9a mediates Sharp-1-dependent inhibition of skeletal muscle differentiation. Mol. Biol. Cell 23, 4778–4785.

Ling, X., Harkness, T.A., Schultz, M.C., Fisher-Adams, G., Grunstein, M., 1996. Yeast histone H3 and H4 amino termini are important for nucleosome assembly in vivo and in vitro: redundant and position-independent functions in assembly but not in gene regulation. Genes Dev. 10, 686–699.

Liu, N., Olson, E.N., 2010. MicroRNA regulatory networks in cardiovascular development. Dev. Cell 18, 510–525.

Maderspacher, F., 2010. Lysenko rising. Curr. Biol. 20, R835–R837.

Malpighi, M., 1673. Dissertatione Epistologica de Formatione Pulli in Ovo.

Malpighi, M., 1686. Opera Omnia.

Materna, S.C., Howard-Ashby, M., Gray, R.F., Davidson, E.H., 2006. The C_2H_2 zinc finger genes of *Strongylocentrotus purpuratus* and their expression in embryonic development. Dev. Biol. 300, 108–120.

McGlinn, E., Yekta, S., Mansfield, J.H., Soutschek, J., Bartel, D.P., Tabin, C.J., 2009. In ovo application of antagomiRs indicates a role for miR-196 in patterning the chick axial skeleton through Hox gene regulation. Proc. Natl. Acad. Sci. U.S.A. 106, 18610–18615.

Miska, E.A., Alvarez-Saavedra, E., Abbott, A.L., Lau, N.C., Hellman, A.B., McGonagle, S.M., Bartel, D.P., Ambros, V.R., Horvitz, H.R., 2007. Most *Caenorhabditis elegans* microRNAs are individually not essential for development or viability. PLoS Genet. 3, e215.

Moore, G.P., Scheller, R.H., Davidson, E.H., Britten, R.J., 1978. Evolutionary change in the repetition frequency of sea urchin DNA sequences. Cell 15, 649–660.

Moran, V.A., Perera, R.J., Khalil, A.M., 2012. Emerging functional and mechanistic paradigms of mammalian long non-coding RNAs. Nucleic Acids Res. 40, 6391–6400.

Morton, S.U., Scherz, P.J., Cordes, K.R., Ivey, K.N., Stainier, D.Y., Srivastava, D., 2008. microRNA-138 modulates cardiac patterning during embryonic development. Proc. Natl. Acad. Sci. U.S.A. 105, 17830–17835.

Nagano, T., Fraser, P., 2011. No-nonsense functions for long noncoding RNAs. Cell 145, 178–181.

Nagano, T., Mitchell, J.A., Sanz, L.A., Pauler, F.M., Ferguson-Smith, A.C., Feil, R., Fraser, P., 2008. The air noncoding RNA epigenetically silences transcription by targeting G9a to chromatin. Science 322, 1717–1720.

Nam, J., Dong, P., Tarpine, R., Istrail, S., Davidson, E.H., 2010. Functional *cis*-regulatory genomics for systems biology. Proc. Natl. Acad. Sci. U.S.A. 107, 3930–3935.

Orgel, L., Crick, F.H.C., 1980. Selfish DNA: the ultimate parasite. Nature 284, 604–607.

Pasquinelli, A.E., 2012. MicroRNAs and their targets: recognition, regulation and an emerging reciprocal relationship. Nat. Rev. Genet. 13, 271–282.

Pauli, A., Rinn, J.L., Schier, A.F., 2011. Non-coding RNAs as regulators of embryogenesis. Nat. Rev. Genet. 12, 136–149.

Pauli, A., Valen, E., Lin, M.F., Garber, M., Vastenhouw, N.L., Levin, J.Z., Fan, L., Sandelin, A., Rinn, J.L., Regev, A., Schier, A.F., 2012. Systematic identification of long noncoding RNAs expressed during zebrafish embryogenesis. Genome Res. 22, 577–591.

Peter, I.S., Davidson, E.H., 2011. Evolution of gene regulatory networks controlling body plan development. Cell 144, 970–985.

Petruk, S., Sedkov, Y., Johnston, D.M., Hodgson, J.W., Black, K.L., Kovermann, S.K., Beck, S., Canaani, E., Brock, H.W., Mazo, A., 2012. TrxG and PcG proteins but not methylated histones remain associated with DNA through replication. Cell 150, 922–933.

Richardson, L., Venkataraman, S., Stevenson, P., Yang, Y., Burton, N., Rao, J., Fisher, M., Baldock, R.A., Davidson, D.R., Christiansen, J.H., 2010. EMAGE mouse embryo spatial gene expression database: 2010 update. Nucleic Acids Res. 38.

Rinn, J.L., Kertesz, M., Wang, J.K., Squazzo, S.L., Xu, X., Brugmann, S.A., Goodnough, L.H., Helms, J.A., Farnham, P.J., Segal, E., Chang, H.Y., 2007. Functional demarcation of active and silent chromatin domains in human HOX loci by noncoding RNAs. Cell 129, 1311–1323.

Rizzo, F., Fernandez-Serra, M., Squarzoni, P., Archimandritis, A., Arnone, M.I., 2006. Identification and developmental expression of the ets gene family in the sea urchin (*Strongylocentrotus purpuratus*). Dev. Biol. 300, 35–48.

Ronshaugen, M., Biemar, F., Piel, J., Levine, M., Lai, E.C., 2005. The *Drosophila* microRNA iab-4 causes a dominant homeotic transformation of halteres to wings. Genes Dev. 19, 2947–2952.

Russo, V.E.A., Martienssen, R.A., Riggs, A.D., 1996. Epigenetic Mechanisms of Gene Regulation. CSHL Press.

Shen, Y., Yue, F., McCleary, D.F., Ye, Z., Edsall, L., Kuan, S., Wagner, U., Dixon, J., Lee, L., Lobanenkov, V.V., Ren, B., 2012. A map of the *cis*-regulatory sequences in the mouse genome. Nature 488, 116–120.

Sleutels, F., Zwart, R., Barlow, D.P., 2002. The non-coding air RNA is required for silencing autosomal imprinted genes. Nature 415, 810–813.

Small, S., Blair, A., Levine, M., 1996. Regulation of two pair-rule stripes by a single enhancer in the *Drosophila* embryo. Dev. Biol. 175, 314–324.

Song, J.L., Stoeckius, M., Maaskola, J., Friedländer, M., Stepicheva, N., Juliano, C., Lebedeva, S., Thompson, W., Rajewsky, N., Wessel, G.M., 2012. Select microRNAs are essential for early development in the sea urchin. Dev. Biol. 362, 104–113.

Spitz, F., Furlong, E.E., 2012. Transcription factors: from enhancer binding to developmental control. Nat. Rev. Genet. 13, 613–626.

Stark, A., Bushati, N., Jan, C.H., Kheradpour, P., Hodges, E., Brennecke, J., Bartel, D.P., Cohen, S.M., Kellis, M., 2008. A single Hox locus in *Drosophila* produces functional microRNAs from opposite DNA strands. Genes Dev. 22, 8–13.

Thomsen, S., Azzam, G., Kaschula, R., Williams, L.S., Alonso, C.R., 2010. Developmental RNA processing of 3′UTRs in Hox mRNAs as a context-dependent mechanism modulating visibility to microRNAs. Development 137, 2951–2960.

Tong, X., Zhang, D., Buelow, K., Guha, A., Arthurs, B., Brady, H.J., Yin, L., 2013. Recruitment of histone methyltransferase G9a mediates transcriptional repression of Fgf21 gene by E4BP4 protein. J. Biol. Chem. 288, 5417–5425.

Tu, Q., Brown, C.T., Davidson, E.H., Oliveri, P., 2006. Sea urchin Forkhead gene family: phylogeny and embryonic expression. Dev. Biol. 300, 49–62.

Ulitsky, I., Shkumatava, A., Jan, C.H., Sive, H., Bartel, D.P., 2011. Conserved function of lincRNAs in vertebrate embryonic development despite rapid sequence evolution. Cell 147, 1537–1550.

Vakoc, C.R., Mandat, S.A., Olenchock, B.A., Blobel, G.A., 2005. Histone H3 lysine 9 methylation and HP1gamma are associated with transcription elongation through mammalian chromatin. Mol. Cell 19, 381–391.

Visel, A., Blow, M.J., Li, Z., Zhang, T., Akiyama, J.A., Holt, A., Plajzer-Frick, I., Shoukry, M., Wright, C., Chen, F., Afzal, V., Ren, B., Rubin, E.M., Pennacchio, L.A., 2009. ChIP-seq accurately predicts tissue-specific activity of enhancers. Nature 457.

Visel, A., Prabhakar, S., Akiyama, J.A., Shoukry, M., Lewis, K.D., Holt, A., Plajzer-Frick, I., Afzal, V., Rubin, E.M., Pennacchio, L.A., 2008. Ultraconservation identifies a small subset of extremely constrained developmental enhancers. Nat. Genet. 40.

Waddington, C.H., 1942. The epigenotype. Endeavour 1, 18–20.

Wamstad, J.A., Alexander, J.M., Truty, R.M., Shrikumar, A., Li, F., Eilertson, K.E., Ding, H., Wylie, J.N., Pico, A.R., Capra, J.A., Erwin, G., Kattman, S.J., Keller, G.M., Srivastava, D., Levine, S.S., Pollard, K.S., Holloway, A.K., Boyer, L.A., Bruneau, B.G., 2012. Dynamic and coordinated epigenetic regulation of developmental transitions in the cardiac lineage. Cell 151, 206–220.

Wang, J., Abate-Shen, C., 2012. The MSX1 homeoprotein recruits G9a methyltransferase to repressed target genes in myoblast cells. PLoS One 7, e37647.

Wang, K.C., Chang, H.Y., 2011. Molecular mechanisms of long noncoding RNAs. Mol. Cell 43, 904–914.

Wienholds, E., Koudijs, M.J., van Eeden, F.J., Cuppen, E., Plasterk, R.H., 2003. The microRNA-producing enzyme Dicer1 is essential for zebrafish development. Nat. Genet. 35, 217–218.

Winter, J., Jung, S., Keller, S., Gregory, R.I., Diederichs, S., 2009. Many roads to maturity: microRNA biogenesis pathways and their regulation. Nat. Cell Biol. 11, 228–234.

Zhang, J.A., Mortazavi, A., Williams, B.A., Wold, B.J., Rothenberg, E.V., 2012. Dynamic transformations of genome-wide epigenetic marking and transcriptional control establish T cell identity. Cell 149, 467–482.

Zhu, Y., van Essen, D., Saccani, S., 2012. Cell-type-specific control of enhancer activity by H3K9 trimethylation. Mol. Cell 46, 408–423.

Zullo, J.M., Demarco, I.A., Piqué-Regi, R., Gaffney, D.J., Epstein, C.B., Spooner, C.J., Luperchio, T.R., Bernstein, B.E., Pritchard, J.K., Reddy, K.L., Singh, H., 2012. DNA sequence-dependent compartmentalization and silencing of chromatin at the nuclear lamina. Cell 149, 1474–1487.

Gene Regulatory Networks

"Gene regulatory network" (GRN) is shorthand for the system of regulatory genes and their encoded interactions that determines the genetic functions to be expressed in cells of each spatial domain in the organism, at every stage of development. This includes the expression of regulatory genes (i.e., genes encoding transcription factors), genes that encode intercellular signaling functions, and genes that participate in downstream differentiation and morphogenesis functions. Since everything the cells do depends on the genes they express, the GRN control system essentially operates the developmental process. By linking the expression of every gene to its upstream transcriptional regulators, GRNs determine the coexpression of particular cellular functions exclusively in different cells. During development, the spatial expression of transcriptional regulatory genes is the driving force for the formation of the discrete structures of organisms, such as organs, body parts, and cell types. Therefore the most important function of GRNs, that is most directly causal for development of the specific body plan, is organizing the spatial allocation of regulatory gene expression. This determines the developmental fate and ultimately the differentiated cellular activity of the descendants of embryonic cells giving rise to each portion of the organism. GRNs are thus the "brains" of each act of the developmental process.

GRN theory emerges from diverse insights gained over the last quarter century regarding mechanisms of transcriptional control in animal development. This includes understanding of the expression of the regulatory apparatus itself, the structure and function of the DNA sequences that process regulatory inputs, the dimensions of the regulatory system operative in given contexts, and experimentally acquired knowledge of GRN structure/function relationships. We begin this chapter with a brief introduction to GRN theory, and then go on to discuss particular foundational principles of this concept. These include the Boolean nature of spatial gene expression in development, the combinatorial utilization of the transcriptional regulatory apparatus, the sequence-specific control of gene transcription by *cis*-regulatory modules (CRM), and the mechanisms of choice among alternative modules. Finally, this chapter concludes with a discussion of transcriptional dynamics and regulatory cascade behavior, which provides a first principles approach to thinking quantitatively about these regulatory processes. We revisit this dynamic treatment in the context of a discussion of GRN models in Chapter 6.

1.   Introductory Overview of Developmental GRNs

1.1   GRN function and how GRNs are encoded in the genome

Every bilaterian animal consists of tissues, organs, and cell types, the function of which depends ultimately and completely on what genes each cell expresses. Yet as we all know, each of these cells contains an identical genome, with a few specialized exceptions. The relation between DNA sequence and gene expression activity is therefore not linear, in that DNA sequence per se does not directly predict the developmental process in the way that it does predict the proteins that can be synthesized. Instead, genomic control of development is encoded in a complex program which is active throughout the process and which functions to define the different parts of the organism by sequential specification. This program is distributed in many parts of the genome, those referred to in Chapter 1 as the regulatory genome.

GRNs provide the fundamental control mechanism directing developmental process. As briefly summarized in Chapter 1 (see also below), gene expression is regulated sequence-specifically by the interaction of transcription factors with *cis*-regulatory DNA modules. Thus, the control operations which assign diverse cellular functions are those determining when and where transcription factor encoding genes will be expressed. By encoding the *cis*-regulatory inputs of every regulatory gene, GRNs specify the interactions among regulatory genes that are responsible for the expression of particular sets of transcription factors. These transcription factors in turn also control cohorts of genes encoding many other kinds of protein, here referred to as effector genes, that is, differentiation genes and morphogenesis genes. Cells manifest their fates in development by the programmed activation of distinct suites of effector genes, directly determining their biological properties, the final specific readout of developmental GRNs. Thus ultimately the expression of all genes in the genome is linked by interactions within GRNs.

Regulatory genes have the special feature that they play dual roles in the GRN, in that their expression is at once the output of the upstream regulatory genes which provide their transcriptional inputs, and at the same time they provide inputs to other target genes within the same network. Thus, the set of transcription factors present in a given time and place determines the new set of transcription factors to be expressed, which then in turn establishes the expression of another regulatory condition. The continuous changes of regulatory gene expression in developmental time can be regarded as the major driver of developmental progression. Development is powered by changes in states of regulatory gene activity and as a consequence of these changes, new cell fates are established in the construction of the body plan. Development is ultimately controlled by GRNs, and these constitute the primary machinery of control in Metazoa.

The genomic components of developmental GRNs are on the one hand the genes encoding transcription factors, signaling components, and effector functions, and on the other hand, the CRMs controlling the expression of all of these genes. "GRNs" denote the physical and functional relationships among regulatory genes. The term "network" in this context is not metaphorical but literal: the network structure grows directly out of two fundamental physical facts of regulatory life in animal systems. First, the CRMs which control spatial developmental gene expression function combinatorially in that they always require qualitatively multiple inputs, as we discuss below (i.e., inputs encoded by several different regulatory genes). Second, the outputs of regulatory genes, the targets of the transcription factors that they encode, are always qualitatively multiple as well (i.e., each regulatory gene has multiple targets). Therefore, each regulatory gene operates at the node of a network. Unlike certain other kinds of "interaction networks" such as protein:protein networks, these are directed, oriented networks in which information flows in only one direction, from transcription factor to *cis*-regulatory target sites and from regulatory genes to transcription factor production. Therefore GRNs are intrinsically hierarchical. Much of the ultimate functional import of GRNs in development and in evolution devolves from the hierarchical structure of these networks. The main point to be retained in this introductory discussion is that all of the general characteristics of developmental GRNs have as their physical basis the way gene regulation works in multicellular animal systems.

1.2 GRN hierarchy and modular organization

The basic elements of GRNs are the genes themselves, the modular *cis*-regulatory control systems that regulate gene expression at every level of the genetic hierarchy, and the causal directional relationships between regulatory and target genes. Regulatory interactions are denoted by statements such as "Gene 1 directly activates Gene 2", meaning that the transcription factor encoded by Gene 1 binds to and functions as an activator at a *cis*-regulatory target site in a regulatory module controlling the expression of Gene 2. We refer to this relationship as a "regulatory linkage" from Gene 1 to Gene 2 (in graph theory what we term "linkages" are referred to as "edges"). GRN linkages function in a strictly unidirectional manner, and irreversibility is a definitive property of animal development at every stage and phase, distinguishing development from other processes such as physiological response (Amit et al., 2009). The fundamental mechanistic basis for this irreversibility is the unidirectionality of the regulatory interaction between an upstream transcription factor and the *cis*-regulatory sequences of its target gene. This is intrinsically a one-way street. Even though the transcription factor–target site interaction is a reversible equilibrium reaction, the upstream gene controls the downstream gene.

As we consider more extensively in Chapter 6, the function of developmental GRNs emerge not just from the individual CRMs at its nodes, but from the structural features of the assemblages of genes that execute most of the developmental operations that GRNs perform. Individual CRMs set the conditions upon which each given gene is expressed or repressed, accounting causally (and in a causal sense entirely) for the activity of each gene. However, individual events of development never depend only on single genes, but always on expression of multiple regulatory genes and eventually on the expression of very large numbers of downstream effector genes. This requires that appropriate linkages among these genes be encoded in the GRN. Furthermore, there are many canonical developmental "jobs" that, as we shall see in the next three chapters of this book, every developmental process entails, each of which requires participation of multiple genes "wired" together and often including both activators and repressors. Examples of subprocesses or jobs that constitute the developmental process as a whole include formation of boundaries between spatial domains of gene expression; lockdown or stabilization of states of gene expression originally installed by transient inputs; reception of signals and organization of downstream regulatory consequences; mediation of cell fate choice mechanisms; interpretation of initial inputs; and so forth. Such functions are executed by GRN "subcircuits". These subcircuits consist of small sets of genes (typically 3–8) which together execute a particular "job" according to a specific architecture or topology of their regulatory linkages. An important insight that has emerged from structure/function studies of such sub-circuits is that the type of developmental function executed by a network subcircuit is determined by the subcircuit architecture, and that similar developmental functions are often executed by subcircuits of simi-lar architecture, regardless of the particular genes they are composed of. We address subcircuit structure/function relationships in detail in Chapter 6. Suffice it to say here that the architecture, and hence function, of every GRN subcircuit is hardwired in the regulatory genome. A given developmental GRN will include several separate subcircuits joined by encoded regulatory linkages. Thus, considered from the perspective of the structural elements that perform its overall control functions, the developmental GRN has a modular character.

The two salient features of developmental GRN structure are its internal subcircuit composition and its strongly hierarchical organization. Both features directly reflect the intrinsic nature of the developmental process. Development from egg to adult body plan requires many phases, each phase depending on the preceding phase, and these phases occupy different successive spatial and temporal domains. Since each of these phases is controlled by a regional GRN, as a whole the GRN has de facto a deep hierarchical struc-ture. That is, the earliest phases at the top of the GRN hierarchy give rise to the subdivisions of embryonic space controlled by the next hierarchical level. Subsequently, embryonic precursor domains undergo fur-ther subdivisions in the development of the body parts of the adult form, an increasingly complex process controlled by additional levels of GRN hierarchy. The GRN hierarchy terminates with the allocation of differentiation gene batteries and morphogenesis gene cassettes. The expression of these effector genes is directly controlled by the developmental GRNs, but since their gene products do not have gene regulatory functionality, these genes provide no further input into the regulatory circuitry. Therefore, they define the downstream periphery of the network. In contrast, at the upper levels of hierarchy, the network consists essentially of cross-regulatory circuitry.

1.3   Models of GRN topology

GRNs are statically encoded in the linear DNA sequence and this code is invariant in every cell of the organism throughout life. In every phase of development, given portions of the overall GRN interactions are utilized to direct the progressive creation of local regulatory states. Our problem is how to convey the network of regulatory interactions in play in any given phase of development in a meaningful way. The key objective is to represent graphically the origin and destination of every input into the *cis*-regulatory control loci of each gene in the system, and the destinations of the outputs of each gene, if it encodes a transcription factor. Network models that graphically represent genes as nodes and display regulatory interactions as input/output linkages are referred to as topological network models. In this book we deal essentially with one standard representation of topological models, although we also discuss mathematical models of different kinds in Chapter 6. Our standard mode of presentation of topological GRN models is by use of the BioTapestry computational and graphical platform (Longabaugh et al., 2005, 2009). In these BioTapestry models, individual spatial domains or phases of development are depicted in separate frames (for example, see Fig. 3.5I). The frames can be used to indicate the linkages which are in operation at each point in time and in each spatial domain in order to activate or repress target genes. All participating regulatory genes in each frame are linked by validated interactions, such that the inputs into each relevant *cis*-regulatory module and the outputs of the genes controlling these modules are explicit. This presenta-tion contrasts with the standard epistasis diagrams traditionally used in genetic presentations which do not distinguish between direct and indirect interactions. BioTapestry models are genomically centered in that they explicitly represent the driver regulatory inputs into the genes occupying each node of the network.

That is, these models indicate the existence of specific binding sites for input transcription factors in the CRMs where the inputs terminate. In addition, these models clearly distinguish regulatory genes, signaling interactions, effector genes and off the DNA functions. Developmental GRN models are focused on control of spatial gene expression and do not in general include either ubiquitous biochemical functions or functions that affect transcript, levels only to a minor extent. For these reasons, except where they are shown to directly cause spatially specific gene expression, neither miRNAs nor epigenetic chromatin modifications are included (see Chapter 1).

The developmental GRN models with which we are here concerned are based on standards of evidence implied by the foregoing definitions. Direct causal evidence is required to demonstrate the existence of a functional GRN linkage. *Trans*-perturbations, in which a transcription factor input is removed from the system (by genetic mutation, treatment with morpholino antisense oligonucleotides or siRNAs, for instance) can reveal the requirement of this factor in the control of given target genes. *Cis*-regulatory analyses, including experimental assessment of the requirement of target sites for given factors within the regulatory sequences of the target gene, distinguishes whether observed *trans*-regulatory effects are direct or indirect. Additional evidence of many kinds can be useful for making this distinction (for example, Oliveri et al., 2008; Peter and Davidson, 2011). Many networks can be found in the literature constructed solely on the basis of statistical analysis of gene expression information, such as clustering of transcriptome data, or genome-wide physical interaction data such as ChIP-seq observations. But such network models are not considered in this volume, because the contribution of the proposed linkages to gene regulation and to the causality of the developmental process has not been experimentally assessed.

2. Boolean Spatial Output

Animal bodies display a fundamentally Boolean character at every level of organization. The body parts of which they are composed are discrete and their structure and function as well as location within the body are deterministically programmed and spatially bounded. Heads are not partly legs, eyes do not blend into mouths, and the pelvis does not grade into the spine. Similarly, at a microscopic level, differentiated cell types are discrete. Neurons, muscle cells, keratinocytes, and gland cells each express discrete sets of effector genes which determine their respective functions, and their spatial location within each body part is also discretely determined in development. This discrete organization of the body plan is the outcome of a sequential series of regulatory definitions of space in the developing organism. In molecular terms, such definitions are achieved by the spatially discrete expression of sets of regulatory genes. From the point at which the cells of an embryo begin to express their genes, the patterns of expression are discrete and Boolean. That is, cells in every domain of the embryo express some genes that are not expressed in other domains. The expression of genes in some cells but not in others requires that their regulators be present only in some but not in other cells. How this occurs right from the beginning of embryogenesis is taken up in the following chapter.

Thus the fundamental explanatory challenge is to understand how the genomic code that underlies the developmental process gives rise to spatially Boolean patterns of gene expression. Illustrations of such patterns are now available for every developmental system that has been studied at the molecular level, including both embryogenesis and subsequent processes of body part formation. As an illustration, in Fig. 2.1(A) we show a collection of in situ hybridization images in developing sea urchin embryos, which at this stage consist of several hundred cells differentially expressing their genomes in geometrically organized patterns. These patterns indicate the domains of the embryo in which the gene is expressed and also the domains in which it is not expressed. Most genes in Fig. 2.1(A) are regulatory genes, and the absence of detectable in situ hybridization signal in sea urchin embryos indicates a level of transcripts too low to produce a functional level of transcription factor (see below for quantitative basis). There is nothing particular to the sea urchin embryos shown in Fig. 2.1(A). Similar

Figure 2.1

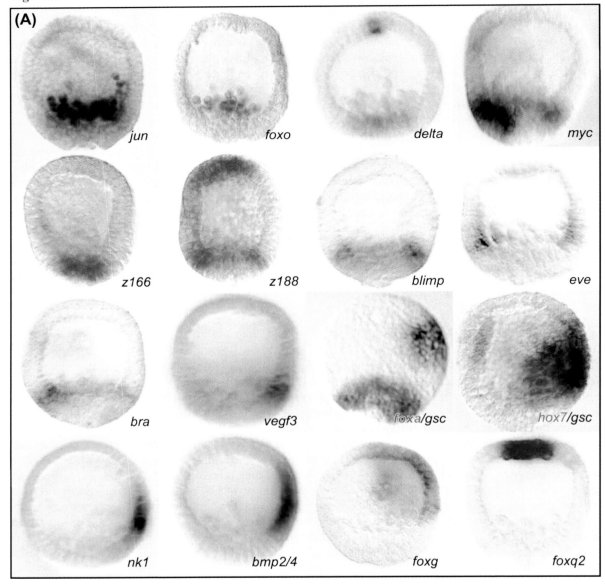

Figure 2.1 Boolean spatial patterns in development. (A), Typical examples of discrete spatial gene expression patterns in sea urchin (*Strongylocentrotus purpuratus*) embryos visualized by whole mount in situ hybridization. Genes shown all encode transcription factors or signaling ligands. All embryos shown are lateral views at 24 h of development. (B), Combined digitalized display of expression patterns of regulatory and signaling genes in the regulatory state territories of this embryo up to gastrulation at 30 h after fertilization. Dark red: active expression; gray: no detectable expression; white: no data. Genes are listed alphabetically on left. Each cell in the vertical columns of the chart represents a 3 h interval and the spatial domains of the embryo are shown across the top of the diagram; micr, micromeres; skel, skeletogenic; MesoA and MesoO, aboral and oral non-skeletogenic mesoderm respectively; EndoA and EndoO, aboral and oral endoderm; EctoA and EctoO, aboral and oral ectoderm; Mesom, mesomeres. *(From Peter and Davidson (2011)).*

Figure 2.1

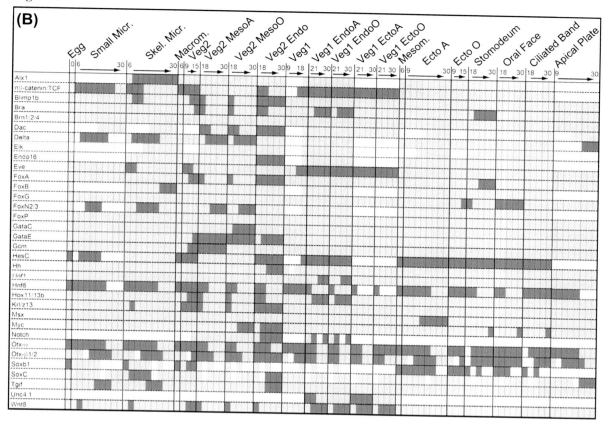

demonstrations of discrete spatial expression patterns fill thousands of pages of literature on mouse, worm, frog, and fly development. Where expression of virtually all of the regulatory genes involved in a specific phase of a developmental process has been studied, the patterns of gene expression can be formalized in a global Boolean matrix, as shown in Fig. 2.1(B). This chart includes all regulatory genes specifically expressed in the endoderm and mesoderm of the developing sea urchin embryo up to gastrulation, when this embryo (*Strongylocentrotus purpuratus*) consists of approximately 600 cells. Temporal and spatial expression information is summarized across the whole period from cleavage to gastrulation for dozens of genes encoding transcription factors. Just as in any Boolean system, there is as much information content in the cells of the matrix showing absence of expression as there is in cells showing presence of expression. Thus, a mechanistic explanation of the expression pattern must provide causes for both states. The essential difference between Fig. 2.1(A) and Fig. 2.1(B) is that in the latter the developmental output of the system has been converted to a digital form which enables the computational processing of large-scale spatial expression data.

Underlying the Boolean spatial expression of genes in development are the discrete functions of the array of spatially active enhancers that service each gene. Just as genes are expressed in only a fraction of the cells of an embryo or body part, each of their enhancers is expressed in only a fraction of the cells expressing each gene. A classic example was shown in Fig. 1.3A2 which demonstrated the modularity of the genomic regulatory system that controls the expression of the *even skipped* gene in the *Drosophila* embryo. This gene is expressed in seven discrete stripes, but each of the enhancers controlling this function specifically causes expression in only one or two of the stripes. Thus these regulatory modules display canonical Boolean behavior, as illustrated for the *eve* stripe 3 + 7 module in Fig. 1.3A2. This is the general rule for developmentally active enhancers. The elemental unit in animal gene regulation is the CRM and the Boolean spatial output of developmentally active CRMs is fundamentally the source of the Boolean character of animal body plan organization.

3.   Regulatory States

Regulatory states are the collective sum of specifically coexpressed transcriptional regulators in a given place at a given time. Unlike individual transcription factors, which are used repeatedly in many developmental contexts, the regulatory state uniquely defines a developmental condition and phase. Regulatory states determine all developmentally specific functions, and later in development cell type specific functions. For this reason, considering GRN output in terms of regulatory states provides the key to the causal relation between genome and developmental function.

Returning to Fig. 2.1(B), each column shows the regulatory genes expressed at a given time in a given domain of the embryo, that is, each column as a whole specifies the particular regulatory state for that domain and that time. Moving to the right within each domain in this figure we can perceive the progressive change in regulatory state with time. Comparison of regulatory states among different domains allows the identification of spatially specific regulatory signatures. Here we consider the output of regulatory genes at the mRNA level as a proxy for the transcription factors that they will give rise to. As just noted, a significant or functional level of transcription factor protein can be generated from the levels of regulatory gene transcripts which are detectable by in situ hybridization (about 10 copies of message per cell). Therefore, assessment of regulatory gene transcript matrices as in Fig. 2.1(B) is directly informative of both regulatory state per se and of downstream transcriptional functions.

Among the genes which are expressed under the control of a given regulatory state are the regulatory genes which will compose the next regulatory state. Regulatory states are generated by the networks of interacting regulatory genes. Regulatory states are therefore both the output and the input to GRNs and they represent the individual active states the network can assume.

4.   Regulation in *Cis*

Controlled transcription of animal genes in development and physiology depends on the specific dedicated DNA sequences for which we use the term "*cis*-regulatory modules" (Chapter 1). In general, the role of CRMs is to recruit specific sets of transcription factors which in turn control the expression of the genes to which the CRMs are dedicated. The specificity of CRM function devolves from their content of the particular DNA sequence elements that individual transcription factors recognize and physically bind to. Within a short stretch of DNA sequence, CRMs contain target sites for multiple transcription factors, and combinatoriality in the function of inputs in CRMs is a fundamental principle of gene regulation in animal cells. Often the transcription factors binding in a given CRM interact with one another cooperatively (see below) or with third-party cofactors which do not themselves bind DNA but are recruited to specific CRMs by the transcription factors bound to their specific target sites.

4.1   CRM genomics

An initial concept immediately illuminates the meaning of the word "module" in the term CRM. This concept highlights the difference between a transcription factor-binding site located within a CRM and anywhere else in the genome; by elemental probability such sites should and do appear randomly many thousands of times per genome. For example, a six base pair sequence would occur in a random DNA sequence on average once every 4000 bp or almost a million times in a mammalian-sized genome. The difference between the target site(s) within the CRM and the much larger number of these sites occurring randomly is not necessarily in the recruitment of their cognate transcription factor, as shown by many ChIP-seq results, but rather in their regulatory function. The organization of CRMs ensures that multiple functions can be deployed when the CRM is fully loaded. These functions depend on (1) integration of multiple diverse transcription factor inputs; (2) recruitment of cofactors which in turn determine the transcriptional activity of the gene, including particular coactivators (such as p300), or corepressors (such as Groucho) or nucleosome modification enzymes (see Chapter 1); (3) genomic looping which brings the CRM and its associated protein complex into immediate contact with the basal transcription apparatus at the start site of the gene; (4) topological positioning of the CRM and the gene in specific intranuclear compartments where transcription is favored or disfavored. In the absence of this multiplicity of events, regulatory function cannot occur. This is why random orphan-binding sites do not contribute to specific regulation of gene expression. While individual orphan sites occur frequently, clusters of particular binding sites within a few hundred base pairs of sequence are uniquely improbable. The functional regulatory apparatus of the genome is thereby uniquely encoded in its CRMs.

At the next level of genomic regulatory organization, a dominant feature of animal gene control systems is the multiplicity of CRMs per gene. As discussed briefly in Chapter 1, there are on average about 5–10 CRMs per gene. Some genes have many more CRMs than this, like for example the *Ubx* and *string* genes in *Drosophila*, and the *mrf4, myf5, sox2*, and *otx2* genes in mouse (Davidson, 2006). In general, each of the multiple CRMs that service a given gene contains a unique set of binding sites for transcription factors that in concert are present at particular times and places in development. Therefore most CRMs act only in particular developmental contexts. This is the genomic basis for the polyfunctionality of many genes. Thus, typically, if a gene is expressed in many domains of a developing organism, its control system will be mosaic in that the same gene will be controlled by a separate CRM for each phase of its expression. A great example discussed in Chapter 5 is the aforementioned *mrf4* and *myf5* cis-regulatory system, which operates in muscles all over the body, while expression in each particular part is driven by different individual CRMs (Hadchouel et al., 2003; see Chapter 5). The essential basic principle is that developmental gene regulation is a modular process and the structural basis underlying this fundamental feature is the modularity of the *cis*-regulatory apparatus.

4.2 CRM logic functions

CRMs have the potential to affect the basal transcription apparatus in multiple ways. Functions mediated by CRMs ultimately control activation, repression, and also rate of expression. The many different types of biochemical interactions between CRM-bound transcription factors and cofactors on the one hand and the transcription complex on the other, lie beyond the scope of this chapter (for reviews see Ma, 2011; Marsman and Horsfield, 2012; Spitz and Furlong, 2012). Most of the functions of the basal transcription apparatus, which consists of an enormous complex of over 50 polypeptides, are not specific to particular genes but instead are generally deployed for gene transcription. The answers to the question why any particular gene is expressed in a specific developmental context therefore usually do not lie within the structure of the basal transcription apparatus. Rather, developmental specificity resides in the encoded combination of transcription factor target sites that constitute the working sequences of CRMs.

Active CRMs bind many different sequence-specific transcription factors, as well studied cases demonstrate (Thanos and Maniatis, 1995; Yuh et al., 1998, 2001; Swanson et al., 2010). These factors contribute in various ways to the function of the CRM and interference with the binding of any of them affects the regulatory output. Factors which perform positive spatial and temporal control of gene expression are commonly referred to as "drivers". In the case of *endo16*, a gene which encodes a secreted protein of the midgut in the sea urchin embryo, only three out of thirteen sequence-specific transcription factors bound to the *endo16* CRMs, function as drivers. The other proteins have a variety of functions including quantitative amplification of the regulatory output, spatial repression, internal communication among different modules constituting the overall *cis*-regulatory system and regulation of the use of driver inputs (Davidson, 2006). From an informational point of view, the determinants of developmental gene expression are the drivers which animate each CRM plus any repressors that restrict CRM activity from given domains.

It is useful to consider CRMs in terms of their input and output functions. The inputs are defined by the ambient transcription factors which convey spatial regulatory information to the CRM (or in other words by the transcription factors which are expressed in a non-ubiquitous manner). The regulatory output of a CRM, which it conveys to the basal transcriptional apparatus when it is in play, is the result of informational processing of these inputs. CRMs mandate logic functions by which these driver inputs will be combinatorially processed (Istrail and Davidson, 2005). For example, a gene responds to input A, which is expressed in a given set of cells, and also to input B, expressed in a second set of cells. If the CRM utilizes these inputs by AND logic, the consequence is that it will produce a positive output only in the unique spatial subset of cells where A and B are both present, or else it produces no output at all. Thus the structure/function relations encoded in the information processing system in this CRM require that it runs only when both A and B are present simultaneously. Or else, the same inputs might be utilized by OR logic, which would result in a positive CRM output wherever either A or B are expressed. Other CRMs respond to specific repressors, which if present execute NOT logic functions. Often it is the case that whenever such repressive inputs are received by the CRM, the CRM is prevented from producing a positive output, irrespective of the presence of activating inputs. In life, given CRMs may process multiple inputs with the use of multiple logic functions. The consequence is that the output of genes driven by such information processing systems is different from any one of their individual inputs in time and/or space, and in development this property of CRMs lies at the heart of their ability to generate novel patterns of gene expression. CRM logic functions can be conveniently modeled using Boolean logic operators and this provides a way of predictively treating very large developmental GRNs, as we discuss in Chapter 6.

4.3   Flexibility and constraint in CRM sequence

CRMs may consist of densely packed functionally important sequences. Recent studies in which the significance of virtually every nucleotide has been assessed for several known enhancers demonstrate that the majority of sites may contribute to CRM function (Kwasnieski et al., 2012; Melnikov et al., 2012), although this is not always the case (Patwardhan et al., 2012). A detailed analysis of the *Drosophila sparkling* enhancer demonstrated that sequences between the known driver target sites were also essential for CRM spatial activity (Swanson et al., 2010). Mutation of these sequences showed that most of the *sparkling* enhancer contains critical regulatory information (Fig. 2.2(A)). The significance of such observations is to remind us that CRMs are far more complex machines than simply target sites for one or two driver transcription factors. When we consider the complexity of interactions in enhancers such as *sparkling*, it is clear that we have only begun to assess the actual information processing functions of CRMs. Integrating over the many thousands of developmentally active CRMs that exist in animal genomes (see Chapter 1) directly implies the informational complexity of the developmental process.

Despite the dense informational content of CRM sequence, comparison of orthologous enhancers with conserved function among more and less closely related species reveals an amazing flexibility at the sequence level. When *eve* stripe 2 regulatory sequences were isolated from the genomes of even distantly related flies, associated with reporter gene expression constructs and introduced into *Drosophila melanogaster*, these constructs generated accurate stripe 2 expression (Fig. 2.2(B1,2);

Figure 2.2

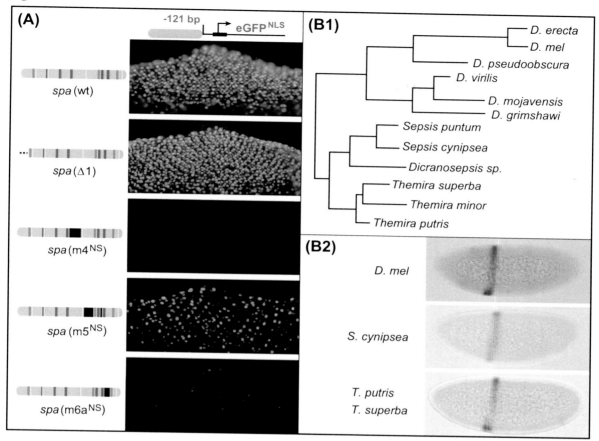

Figure 2.2 Sequence requirements for conservation of *cis*-regulatory function. (A), *Drosophila sparkling* enhancer of the *pax2* gene, active in cone cells of the eye *(from Swanson et al. (2010).)* Target sites are color coded in diagrams of the 362 bp enhancer to left. This enhancer is known to respond to Notch (Su(H), red), EGFR via Ets (yellow), and the runt family factor Lozenge (blue). Here in a construct diagrammed at top the enhancer is placed only 121 bp from the GFP reporter transcriptional start site, and expression of constructs in which regions other than the Su(H), Ets and Lz sites were systematically deleted is shown; wt, wild type. The 5′ sequence deleted in Δ1 is required only for interaction with the promoter from its normal distal position but not when located only 121bp away as in this construct; deletions m4^NS and m6a^NS abolish expression, and m5^NS severely weakens expression indicating the necessity of additional inputs. (B), Constraint and flexibility in sequence requirements for *eve* stripe 2 enhancer function, demonstrated by evolutionary comparisons in dipteran flies *(from Hare et al. (2008b).)* (B1), Phylogenetic tree displaying relationships of flies included in following analyses; note comparisons are within Drosophilidae and from *Drosophila melanogaster* (D. mel) to the distant clades Sepsidae and Themidiridae. (B2), Demonstration by transgenesis in *D. mel* that *eve* stripe two enhancers isolated from genomes of sepsid and themirid flies respond to the *D. mel* regulatory state exactly as does the *D. mel* enhancer, producing a normal stripe 2 pattern. (B3), Sequence organization of stripe 2 enhancers from the species included in (B1). Site identities are color coded as at bottom of figure; the only site arrangements conserved either inside or outside of the Drosophilidae are closely adjacent boxed site pairs. The figure demonstrates that it is the presence of sites for all requisite

Figure 2.2

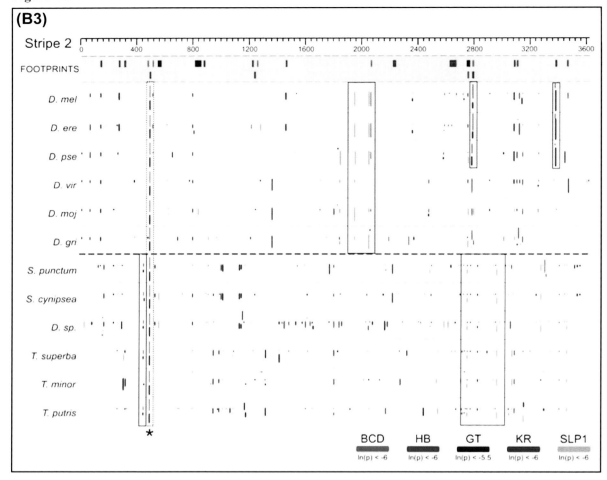

factors that is required for similar function. Neither similar site arrangement nor inter-site sequence need be conserved. (C), Interclade transgenesis of an extremely conserved *cis*-regulatory module of a *soxb* gene *(from Royo et al. (2011).)* (C1), top tier, almost identical expression of enhancer constructs from orthologous human, amphioxus and sea urchin *soxb* genes in developing zebrafish brain; e, eye; h, hindbrain; m, midbrain; s, spinal cord; t, telencephalon. Bottom tier, from left to right, endogenous expression of sea urchin (*S. purpuratus*) *soxb2* gene in neurogenic ciliated band, oral perimeter, and midgut wall of feeding larva, detected by WMISH; GFP expression of sea urchin (Sp), human (HS), and *Nematostella* (Nv) enhancers (CNR) in individual neurons of these regions in sea urchin larvae of same stage. (D), Organizational requirements for spatial function in the *sparkling* enhancer (color coding as in (A); *from Swanson et al. (2010))*. (D1), Interactions within module. Green arrows indicate short-range, i.e., position dependent, intra-modular interactions among diverse factors bound at indicated sites. These interactions are necessary for expression in cones ("patterning circuitry"). (D2), Alteration of relative positions of target sites changes spatial specificity of enhancer, even though same set of sites is present (i.e., same "combinatorial code"); alternative cell types are cones and photoreceptors (R1 and R6).

Figure 2.2

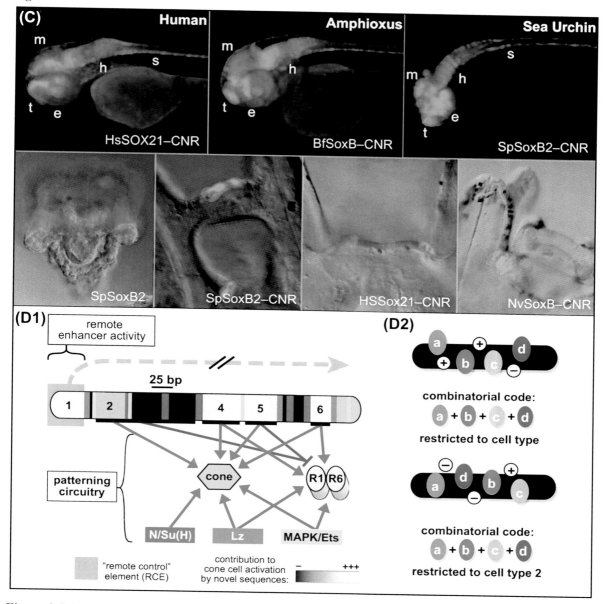

Figure 2.2 (continued...)

Hare et al., 2008a,b). Thus the function of these enhancers is perfectly conserved. Yet as shown in Fig. 2.2(B3), the arrangement of the well characterized target sites in the *eve* stripe 2 enhancers is clearly different. Most of those features that are conserved appear to consist of target sites for two different factors which directly abut one another or even overlap. In general, the binding sites for all the known crucial drivers of the *eve* stripe 2 enhancer are present in the CRMs of every species. However, their relative positions and their location within the enhancer have changed since divergence, and the sequences of these CRMs have turned over extensively, including both the driver sites and the sequences separating them. This suggests that with the exception of the closely contiguous conserved site pairs, what is important for function is essentially just the presence of the driver sites, and not their relative positions. A similar import derives from an experimental study on a *Drosophila sog* CRM (Liberman and Stathopoulos, 2009). Comparison of the regulatory architecture of this CRM among *Drosophila* species again displays significant variation of the organization of target sites for the transcription factors that spatially regulate *sog*. Nonetheless when functionally tested in *D. melanogaster* embryos, these various CRM architectures are all capable of generating similar lateral stripes of gene expression. Furthermore, experiments using synthetic *cis*-regulatory elements demonstrated that various arrangements of the necessary target sites suffice for driving gene expression specifically in the lateral stripes.

Observations on the naturally occurring flexibility of CRMs of similar spatial function led to the proposition of the "billboard model" of CRM structure/function relations (Kulkarni and Arnosti, 2003; Arnosti and Kulkarni, 2005). According to this idea, the essential information within CRMs is comprised by the identity of the target sites within it, but the relative organization of these target sites is not usually significant. However, as pointed also out by these authors, some types of CRMs are known that are organized in an entirely different manner (Kulkarni and Arnosti, 2003; Arnosti and Kulkarni, 2005). The canonical example is the *interferon-β* "enhanceosome", which consists of tightly apposed sites for six different transcription factors and which operates only when all six factors are in place. Here the architecture of the binding sites is required to enable the essential cooperative interactions among bound transcription factors, and almost every nucleotide in this 44-bp-long enhancer is required for function (Melnikov et al., 2012). In vertebrates, that small fraction of developmentally active CRMs in which the nucleotide sequence is conserved across the whole length of the CRM, from fish to mammals, may belong to this CRM structural class (Siepel et al., 2005; Woolfe et al., 2005; Katzman et al., 2007; Elgar and Vavouri, 2008). An extreme example illustrated in Fig. 2.2(C), is an enhancer for a *soxb* class regulatory gene, which at the sequence level is conserved from *Nematostella* (starlet sea anemone) to human genomes (Royo et al., 2011). Remarkably, when tested by transgenesis in sea urchins, zebrafish, and mice, this CRM functions almost identically, irrespective of the genome of origin, promoting expression in developing or mature neurons. Thus, despite the vast evolutionary distance, these CRMs are capable of responding to similar regulatory states expressed in neurogenic tissues.

An alternative example of structure/function relations in a developmental CRM is provided by the *sparkling* enhancer referred to above. This enhancer is very poorly conserved in sequence, even within the Drosophilidae; yet it contains tightly packed regulatory sequence for many factors. As shown in the diagram of Fig. 2.2(D1), multiple factor interactions are required for the cell type specific expression that it mediates. When the arrangement of the target sites is altered experimentally, the spatial function of the CRM changes, so that it promotes expression in a different cell type (Fig. 2.2(D2)). The implication is that functional tightly organized *cis*-regulatory architecture can turn over rapidly in evolution, and sequence conservation is not a required indication of functionality in the regulatory genome.

5. Module Choice

As we have seen in Chapter 1, animal genes are controlled by multiple CRMs located in different regions within and surrounding the gene, and frequently these genes utilize alternative promoters as well. With the advent of elegant methods for cross-linking chromatin in its native configuration in living cells ("chromatin capture"; 3C, hiC), it has now become clear that the mechanism of transcriptional control by distantly located CRMs involves the formation of specific loops which bring the CRM and the promoter into direct contact. Loops are not an intrinsic structural property of the genome sequence but rather their formation is determined by physiological or developmental context. Thus given loops have been shown to be cell type specific and their formation to be tightly correlated with transcriptional activity. In some circumstances there are multiple enhancers looping to one promoter or individual enhancers may service an array of multiple promoters, with the formation of alternative loops (Guo et al., 2012). A number of important issues devolve, among which are: what is the mechanism by which these loops form; what controls the specificity of loop formation during development; and what determines the choice of enhancers and promoters participating in any given loop.

5.1 General looping mechanisms

The physical stability of transcriptional looping configurations is dependent on special protein complexes. A major component of these is the polyfunctional multiprotein complex known as cohesin. Cohesin forms rings which assist in maintaining the physical contiguity of enhancer and promoter components (see Fig. 2.3(A) for several such configurations; Ong and Corces, 2011). Cohesin often interacts with the mediator protein complex, an enormous polypeptide entity associated with a large fraction of promoters in yeast and animals. Furthermore, cohesin interacts with CTCF, which was initially discovered as a major component of insulator complexes, where it specifically binds the CCCTC sequence motif. CTCF has now been associated with many chromatin looping functions because of its capacity to interact homotypically once bound to DNA (Bell et al., 1999; Yang and Corces, 2012). Some loops are facilitated by the interaction between insulator CTCF and CTCF bound to sites within or near enhancers, which assist in generating productive transcriptional loops (Krivega and Dean, 2012). Numerous observations have shown that

Figure 2.3

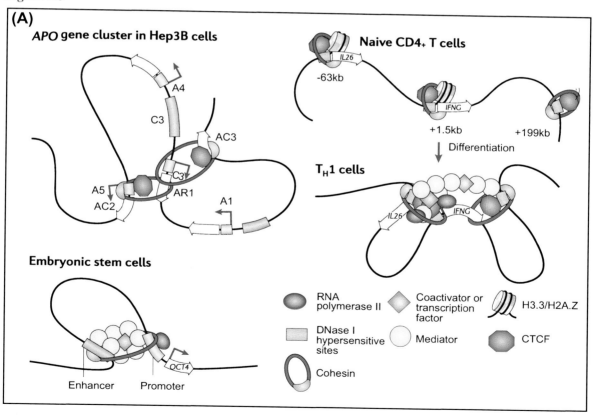

Figure 2.3 Looping interactions between distant enhancers and promoters. (A), Specific examples of cohesin loop stabilization structures *(from Ong and Corces (2011))*. Components are shown in key at bottom of figure; of these, transcription factors and CTCF factor interact with DNA sequence

Figure 2.3

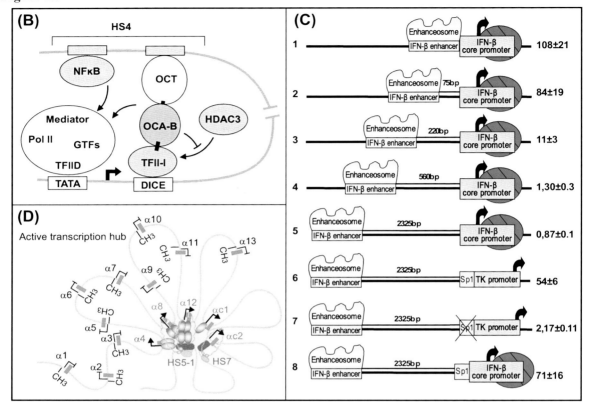

specifically. Promoter sequences are shown as blue cylinders. In the APO gene, abbreviations are: C3 denotes an enhancer, which is present in the same loop as the C3, A4, and A5 promoters; A1 promoter is present in a different loop. AC3, AC2, and AR1 are insulator sites. (B), The transcription factors NFκB and OCT interact with the enhancer HS4. NFκB in addition interacts with Mediator complex while OCT interacts with the cofactor OCA-B, through which it is tethered to the initiation site by TFII-I. The result is the direct physical interaction between proteins bound to the immunoglobulin heavy chain enhancer and proteins bound to the promoter of this gene, while the intervening DNA sequence (break) forms a loop *(from Ren et al. (2011))*. (C), Synthetic experiments demonstrating that transcriptional induction of the *interferon-β* gene enhancer requires the transcription factor Sp1 to be bound near the promoter when the enhancer is distantly located though not if proximity makes looping unnecessary *(from Nolis et al. (2009))*. The experiment shows decreasing levels of inductive gene expression with increasing distance between promoter and enhancer sequences in the absence of SP1 binding sites (Exp. 1–5). But even at the normal distance of 2325bp, presence of an SP1 site restores activity of either heterologous (Exp. 6) or endogenous promoters (Exp. 8) to levels comparable to the construct where enhancer and promoter are juxtaposed (Exp. 1). Numbers at right indicate fold induction of gene expression upon addition of stimulatory reagent to which the enhanceosome responds. (D), Enhancer-promoter looping in the protocadherin gene cluster *(from Guo et al. (2012))*. Promoters are indicated as short blue bars. Promoters in contact with enhancers (HS5-1 and HS7, red) cause transcription of protocadherin gene segments. Those active here are α4, α12, α8, αc1, and αc2, which have looped to the enhancers forming a transcription hub held together by CTCF (green)/cohesin (red circles) complexes.

CTCF, cohesin, and mediator are present in the transcriptional complexes detected by chromatin capture, and that they are required for loop stability and for continued transcription. Thus removal of CTCF or knockdown of cohesin disrupts the chromatin loop and reduces gene transcription (Mishiro et al., 2009; Guo et al., 2012). But these generally utilized structural facilitators of looping are present in all cells and thus cannot explain the differences in chromatin loops observed when different cell types are compared.

5.2 Specificity of looping

The multiple CRMs controlling expression of a given gene generally respond to distinct sets of transcription factors, endowing the gene with the capacity to be expressed in various developmental contexts. This is particularly true of the classes of genes controlling development, those encoding signaling molecules and transcription factors. For the gene to be regulated in each particular context by the appropriate enhancer requires the formation of a loop that brings this enhancer into immediate contact with a promoter. What mechanisms ensure specificity of interaction between promoter and the particular CRM, thus regulating enhancer choice? A number of recent studies demonstrate that transcriptional loops occur only when the enhancers are loaded with their specific transcription factors. Thus the sequence specificity of loop formation with respect to which enhancer is to be engaged depends directly on the binding of these transcription factors just as do the other functions of the enhancer. A particularly clear case concerns the regulation of the interaction between an immunoglobulin heavy chain enhancer and promoter (Ren et al., 2011). In B cells, Oct2 is a sequence-specific activator of this enhancer and the same factor participates in looping of the enhancer to the promoter, as shown in Figure 2.3B (Ren et al., 2011). The loop is formed by an interaction between the Oct2 cofactor Oca-b and the basal transcription factor TFII-1, which binds in the promoter region. Additionally there is an interaction between another transcription factor bound in the enhancer, NFκB, and the mediator/PolII/TFIID complex at the transcription start site. Thus the sequence-specific binding of Oct2 and NFκB to their *cis*-regulatory target sites in the CRM directs the enhancer/promoter interaction. Another illuminating case concerns the *interferon-β* (*ifn-β*) enhancer discussed above (Nolis et al., 2009). Here the loop is formed by the interaction between transcription factors of the enhanceosome and the transcription factor Sp1, which binds directly upstream of the promoter. The necessity of Sp1 for interaction between the enhanceosome and the promoter is demonstrated in the experiments shown in Fig. 2.3(C) (Nolis et al., 2009). In the absence of the Sp1 site, the enhanceosome is capable of activating gene expression only when located immediately adjacent to the promoter. When located distantly, a productive looping interaction occurs only in the presence of the promoter-proximal Sp1. Specific transcription factors and their cofactors have been shown to be required for enhancer/promoter looping in many other circumstances. Many clear examples of such transcription factor-dependent mechanisms of enhancer/promoter interactions have appeared in the literature. Among these are: class-switch recombination, in which enhancer-bound Pax5 and its cofactor Ptip are required for long-range interactions (Schwab et al., 2011); inflammatory gene expression in response to interleukin-1 (IL-1), in which transcriptional loop formation is cooperatively regulated by sequence-specific transcription factors Lef1, RelA, and c-Jun (Yun et al., 2009); the *β-globin* gene cluster in which a distant locus control element loops to the promoter of the gene in a process requiring binding of the transcription factors Klf1 and Gata-1, as well as the cofactor Fog (Drissen et al., 2004; Vakoc et al., 2005); *opsin* loci, where the transcription factors Crx, Nrl, and Nr2e3 bind specifically in rods and cone cells and are required for the formation of transcriptional loops (Peng and Chen, 2011; see also the examples in Fig. 2.3(A)).

Looking at the process from the perspective of the actively loaded CRM, the looping mechanism must be able to define the promoter sequence to which the CRM should loop. This includes recognition of a promoter sequence from which transcription may be initiated, or choice among multiple promoter sequences. Recognition of promoter sequences as opposed to other sequences is mediated by the general factors of the transcription complex, which bind at promoter sequences and may include mediator and cohesin tethers. However, there are many examples in which given promoters must be distinguished from other promoters to ensure specific enhancer/promoter interaction. These cases include genes with multiple promoters, each serviced by different CRMs, and genes in which the relevant CRM is located at a great distance from the promoter to which it must loop, often over an array of other intervening promoters. Specific promoter proximal tethering sequences have been identified in *Drosophila* that mediate particular interactions to distantly located CRMs. In such cases, specific *cis*-regulatory sequences located directly at the promoter serve as target sites for tethering factors, the binding of which is required for looping with distant CRMs (examples are reviewed in Davidson, 2006; Ho et al., 2011). An interesting contrast to the use of individual promoter proximal transcription factors (as in the Sp1 case above) is the absence of any promoter-specific proximal tethering sequences where choice amongst multiple promoters is required to be stochastic. This mechanism is utilized in *α-protocadherin* gene clusters, where two enhancers loop to and activate a large number of alternative promoters (Fig. 2.3(D); Guo et al., 2012).

In summary, looping is generally facilitated by a particular set of looping proteins that bind in most promoters such as mediator/cohesion, and by CTCF which binds specific DNA sequences located in appropriately positioned flanking regions of the gene. But the specificity of transcriptional looping depends on sequence-specific transcription factors bound to the CRMs and often, but not always, specific promoter proximal transcription factors as well. Thus, the potential for the formation of particular loops is directly encoded in the genomic sequence, and its context-specific realization occurs according to the developmental regulatory state.

6.   Transcriptional Dynamics

GRNs consist mainly of interactions between transcription factors, the products of regulatory genes, and target sites in the CRMs controlling other regulatory genes. In development, these networks can be large and complex, and many individual interactions must be taken into account. It soon becomes impossible to think clearly about these interaction systems without considering their dynamics. This subject includes many often encountered conundrums: What is the time interval between successive gene activations in regulatory gene cascades? How do the real time kinetics of transcriptional systems in different animals living at different temperatures compare? How long does it take for the cell to make given amounts of mRNAs and what do you have to know to compute that? If we measure the approximate number of molecules of a regulatory gene mRNA in a cell, how do we determine if that number will suffice to generate enough transcription factor protein to affect downstream genes? What is the relation between transcription factor concentration and effective occupancy of a *cis*-regulatory target site, and how can we think about this occupancy with respect to the rate of transcription of the target gene? These are matters arising continuously in network regulatory molecular biology. Different mathematical approaches have been used to treat such problems, some more and some less closely tied to the actual nature of the physical processes of transcription control. Here we briefly summarize an elemental set of approaches to transcriptional dynamics, using as a guide the principle that we would like each step in the analysis to be related transparently to the basic processes of gene expression. We have tried to steer between Scylla and Charybdis; between on the one hand overly abstract formulations that behave mathematically as desired but do not directly and literally represent the real processes occurring in the cell, and on the other overly detailed mechanistic biochemical models which quickly become useless because they include too many microscopic constants which are hard to establish for most animal systems.

6.1   Transcriptional initiation and the life and death of mRNA

Productive gene transcription is biochemically an immensely complex process. Transcription is mediated by an assembly of dozens of proteins, only a few of which are well understood. The overall reason for the biochemical complexity is the number of diverse jobs that must be done for transcription to occur: the polymerase must be recruited to the transcriptional start site; it must be released from this site; if it is paused in situ it must be released from its tether; the polymerase complex must ratchet along the DNA; the DNA must be rendered accessible and transiently melted so that one strand can be copied; etc. Nonetheless, two precepts enable a practical treatment of the rate-limiting features of the process that are relevant to GRN dynamics. First, in processes such as embryonic development, the microscopic stochasticity of transcriptional processes and their bumpy instant by instant local rates in individual cells affect only insignificantly the accumulation dynamics of given transcripts over relatively long periods of real time. Thus we can deal in average rates and processes. Below we illustrate the fact that for a typical gene expressed in a field of similarly functioning cells, such cell by cell stochasticity is of little real consequence anyway. Secondly, one major dynamic variable among the multiple microscopic processes involved in transcription largely determines the rate of output of transcriptional systems, and that is the rate of initiation. This is defined as the frequency of the events by which the productive traverse of the gene by the RNA polymerase complex begins at the transcriptional start site of the gene. The rate of initiation per gene over time, which we shall term I (molecules of transcript initiated/minute) is a variable dependent probabilistically on the concentration of the relevant transcription factors, as we discuss in the next section.

I varies between 0 and the maximum initiation rate, I_{max}, which is a temperature-adjusted constant for each animal system: if we imagine a maximally loaded gene being transcribed at the maximum rate, the polymerases will be tightly packed, abutting one another as they traverse the gene (Fig. 2.4(A)), and the maximum rate of initiation is simply the inverse of the amount of time it takes for the first polymerase to move far enough down the gene so another polymerase can load on and initiate a new transcription event. The time required to clear the transcriptional start site depends on how fast the polymerase transits along the DNA, average values for which have been measured in many animal systems (Davidson, 1986; Bolouri and Davidson, 2003). The polymerase transit rate is among the basic numerical "facts of life", and it may seem surprisingly slow. For example, in sea urchin embryos at 15 °C the polymerase transit rate is only 6–9 base pairs per second, and since a polymerase occupies about 100 base pairs, it takes 11 s or more for the initial polymerase to move out of the way so that another may move in. Thus I_{max} is no more than about 5.5 molecules of RNA initiated per minute and gene, on average. For animals which operate at different temperatures, comparison of polymerase transit rates showed that this parameter roughly obeys the same "Q10" rule as do most chemical reactions, i.e., it, and I_{max}, increase about twofold for each 10 °C rise in temperature. Thus for example in mice, the transit rate is ~30 base pairs per second.

Since the polymerase transit rate is more or less constant for all genes in a given animal at a given temperature, the initiation rate averaged over time directly predicts the average length of DNA separating adjacent polymerases and nascent transcripts on the gene. As a vast amount of direct and indirect measurement demonstrates, in a typical developing embryo such as a sea urchin or frog embryo, about 90% of the genes are being transcribed at rates far below I_{max}, and the expected spaced appearance of transcription arrays on such genes can be seen in Figs 2.4(B)1,2,3. An important conceptual point is that the rate of completion of transcription and release of finished transcripts is about equal to the rate of initiation (except in some special circumstances where there is attrition and fewer polymerases complete than initiate transcription). Thus, given continuing initiation, the length of the gene does not affect the ongoing rate of transcript release although at the low polymerase transit rates several hours may elapse between initiation and release of the first transcript. This interval is merely a time lag (dependent on gene length) and the completion and release rate in molecules per minute and gene is generally equal to I.

The primary RNA transcript is processed in a biochemically complex chain of events that includes intron removal, splicing, capping, polyadenylation, and interaction with proteins that chaperone and transit the mature mRNA out of the nucleus. But here again, in terms of dynamics, we are looking only at a lag, since in most (though not all) circumstances processing is 100% efficient. Thus in molecules per minute and gene, the rate of pre-mRNA synthesis is also the rate of mRNA entrance into the cytoplasm, again equal to I. The average processing time lag is 20–30 min in 15 °C sea urchin embryos, shorter at higher temperatures, following the Q10 rule.

Figure 2.4

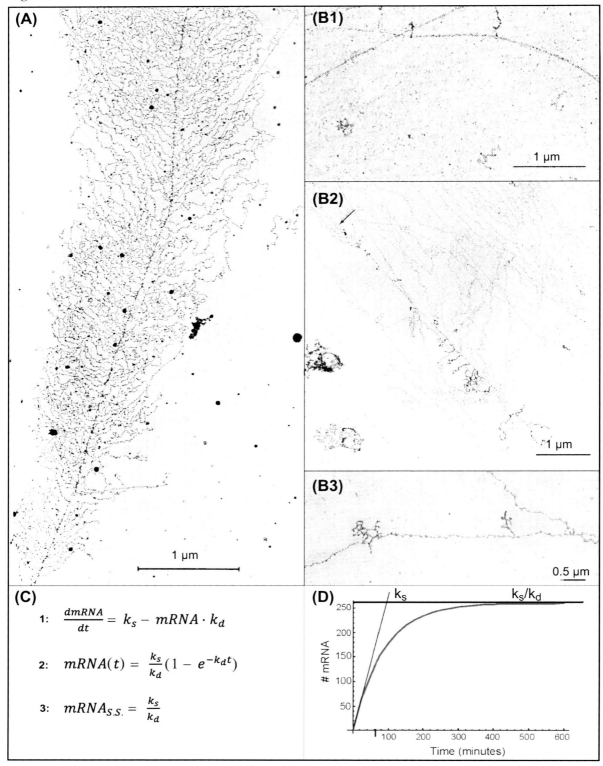

(A)

(B1)   1 µm

(B2)   1 µm

(B3)   0.5 µm

(C)

1: $\dfrac{dmRNA}{dt} = k_s - mRNA \cdot k_d$

2: $mRNA(t) = \dfrac{k_s}{k_d}\left(1 - e^{-k_d t}\right)$

3: $mRNA_{S.S.} = \dfrac{k_s}{k_d}$

(D)

Figure 2.4 Transcription dynamics. (A) and (B), "Miller chromosome spreads", electron micrographic displays of individual transcription units with their nascent transcripts, after treatment with weak detergent; the technology was developed by the late Oscar Miller. (A), Densely packed transcription unit from *Triturus* (newt) oocyte lampbrush chromosome. Calculations show that this transcription unit is being transcribed at near the maximum possible rate at ambient temperature (Davidson, 1986), and transcripts are spaced only about 100 nucleotides apart. Some polymerase molecules can be seen on the chromosomal DNA at the base of individual nascent transcripts. The initiation point for this gene is just beyond the lower left corner of the image *(from Miller and Bakken (1972))*. (B), Miller spreads of genes transcribed at much lower rates. These are far more typical (see text), and in the majority of transcription units visualized in these tissues only single nascent transcripts are observed by this method at any one time (Davidson, 1986). Because the intervals between successive initiations are long, in these cases several minutes, the nascent transcripts are much farther apart. (B1, B2), Sea urchin embryo *(Strongylocentrotus purpuratus)* transcription units. Scale bars represent 1 µm (i.e., ~3000 bp of DNA). The transcription unit in (B2) is transcribed with a slightly higher initiation rate than that in (B1), but compare (A); the arrow marks the point of initiation *(from Busby and Bakken (1979))*. (B3), Rabbit embryo transcription unit, also displaying infrequent initiations. Scale bar is 0.5 µm *(from Cotton et al. (1980))*. (C), Fundamental equations for mRNA synthesis dynamics, as discussed in text. The O.D.E. in Eqn (1) gives the rate of mRNA synthesis; Eqn (2) gives the solution of Eqn (1); and Eqn (3) defines the steady state amount of mRNA (mRNA$_{s.s.}$): parameters are k_s, the mRNA synthesis rate constant (number of newly synthesized mRNA molecules appearing in cytoplasm per minute); k_d, mRNA decay rate constant (fraction of mRNA pool decaying per minute, i.e., t^{-1}). (D), Graphical evaluation of Eqn (2), illustrating the eyeball estimation of these same constants: k_s is the initial slope, and mRNA$_{s.s.}$ is k_s/k_d, the plateau value, thus knowing k_s, k_d can be obtained. In this measurement, $k_s = 3$ molecules/min and $k_d = 0.012$ min^{-1}, i.e., the half-life of the mRNA $ln2/k_d$ (tick on abscissa), is 60 min. *(From Ben-Tabou de-Leon and Davidson (2009).)*

A most useful quantitative relationship for those working with transcriptional processes or cascades thereof, is that which relates mRNA synthesis rate, mRNA accumulation over time, and mRNA turnover rate. This is Eqn (1) in Fig. 2.4(C), which we will consider to represent events on an average per cell basis. It is straightforward to think about dynamic transcriptional problems as processes that can be represented literally by differential equations, and Eqn (1) illustrates this perfectly. Equation (1) states the obvious: the rate of change in quantity of an mRNA per unit time *(dmRNA/dt)*, is given by the rate of its synthesis (k_s) minus the rate of its decay by turnover ($k_d \cdot mRNA$); that is to say by the difference between the rate of entry of the mRNA into the cytoplasm and the rate of its disappearance. Each of the three terms in Eqn (1) has the same units, molecules/min. Let us dwell momentarily on each of these terms: If the differential equation is solved, the left side becomes the absolute number of mRNA molecules present at each given point in time (this solution is Eqn (2) of Fig. 2.4(C), a plot of which is seen in Fig. 2.4(D)). Considering that there are two of each gene in the diploid cell we see that k_s is just $2 \times I$, which directly relates the flow rate of the mRNA into the cytoplasm to the rate of transcriptional initiation per gene encoding the message. In most circumstances decay of any individual molecule in a pool of molecules occurs independently of how long it has been present, i.e., all mRNAs have an equal probability of being targets of the decay mechanism (nuclease attack) whether they have just arrived or are many hours old. However, the mRNA cannot be subject to these decay processes until it is present in the cytoplasm. The probability of decay (k_d) is expressed as the fraction of the mRNA pool that is likely to decay per time interval, for example 1% per minute (here $k_d = 0.01$ min^{-1}). Thus the net number of molecules decaying per minute is the number of mRNA molecules present in the pool at a given time multiplied by the probability of decay and the decay term would be written *mRNA* (molecules) $\times$ **0.01** per minute; the whole term is again molecules/min.

Here is why Eqns (1) and (2) are so useful. It has become experimentally straightforward to measure quantitatively the time course of accumulation of any specific mRNA or set of mRNAs by QPCR or other means. Consider the situation at early times after a gene is activated: the amount of mRNA yet made is small and so the decay term is insignificant, and thus the initial slope of the output of Eqn (2) directly gives the transcription initiation rate, a valuable parameter. When the decay term and synthesis term balance out and become equal, the left side of Eqn (1) becomes 0 since there is now no net change. The mRNA accumulation process for that gene is said to have attained steady state. This is seen graphically in Fig. 2.4(D) as the ultimate plateau. From Eqn (1), when $dmRNA/dt = 0$ we have Eqn (3), which defines the amount of mRNA (molecules) at steady state as k_s/k_d. This is again very useful for it provides a way of estimating either one of these constants if the other and the easily measured steady state quantity are known. The decay rate constant k_d is not always simple to measure directly, but it is easily extracted by estimation of the mRNA half-life ($t_{1/2}$). Imagine that the gene has turned off and whatever mRNA is present is decaying. From Eqn (1), when $k_s = 0$ we can derive k_d from the time required for 50% of the mRNA to decay: $k_d = ln2/t_{1/2}$.

The logical simplicity of Eqn (1) allows us to relate directly measurements of mRNA at given times to the dynamic parameters of transcriptional synthesis and mRNA turnover, and to define computationally the mRNA half-life and steady state. Thus we are enabled to rationalize the amounts of message present through real time in the quantitative terms of mRNA life and death processes.

6.2 Productive *cis*-regulatory occupancy by transcription factors in the nucleus

In a GRN, cascades of transcriptional events occur in which given regulatory genes are transcribed, producing transcription factors that activate other regulatory genes immediately downstream in the network, and by a similar process these in turn activate further regulatory genes. To evaluate such cascade dynamics, we require a means by which the real time intervals between the steps in the cascade can be computationally resolved. The steps of this computation are as follows: (1) consideration of the probability of *cis*-regulatory occupancy in terms of transcription factor concentration and of the intrinsic affinity of the factor for its specific target site sequence; (2) consideration of the nonspecific interaction of transcription factors with DNA sequences as well as the specific interactions at their target sites; (3) generalization, for occupancy of a *cis*-regulatory module by more than one transcription factor; (4) relation between occupancy of the *cis*-regulatory module and the initiation rate of the gene it controls; (5) computation of the accumulation of the transcription factor protein based on the computed initiation rate; (6) computation of (1)–(5) for the next step of the cascade. By this means we may arrive at what turns out to be an extremely useful parameter in considering network dynamics: the step time. This is the time period that elapses for a given system running at a given temperature between the transcriptional activation of a gene encoding an upstream transcription factor and the transcriptional activation of its target gene in a transcriptional cascade.

Transcription factor—DNA interaction

In life, sequence-specific interaction of transcription factors with their *cis*-regulatory target sites is considered productive if such interaction contributes to regulatory function, such as transcriptional initiation, or repressor-dependent transcriptional silencing. Sequence-specific DNA–protein interactions have been studied in vitro for decades, although conditions in the animal cell nucleus are in many ways dissimilar from the conditions controlling interactions of these same factors with oligonucleotides bearing their target sites in vitro. The main purpose of most current in vitro studies is to determine the target site sequence preferences ("position weight matrices, PWMs") for given factors. Recent high-throughput methods have provided enormous, statistically supported databases that provide PWMs for a large number of known transcription factors (Berger and Bulyk, 2009; Stormo and Zhao, 2010; Christensen et al., 2011; Robasky and Bulyk, 2011; Zhao and Stormo, 2011). These databases prove to be of very great value in qualitative analyses of regulatory DNA sequence, since target site specificity is an intrinsic property of the amino acid sequence of the DNA-binding domain of the transcription factor, and similar DNA sequences are bound by individual factors in vivo and in vitro. Exceptions may occur if protein–protein interactions affect the binding behavior of given factors in ways that cannot be accessed easily in purified in vitro systems. In order to calculate the real life occupancy of a DNA-binding site within a nucleus, we have to be concerned with the concentration of the factor as well as the qualitative characteristics of the binding interaction. We shall treat the amount of binding as the fractional occupancy, Y, where $Y=1$ denotes complete occupancy of a site, or of required multiple sites, and $Y=0$ denotes no binding or zero occupancy.

Considering network dynamics for developing systems in terms of real amounts and real times, an initial project is to try to understand what range of occupancies is being deployed in vivo. For any given genes, are the initiation rates often near maximal, indicating high values of Y, or on the other hand are they usually expressed at modest or low levels? Not surprisingly the answer is different depending on the biological context. This and the next three chapters concern GRNs for various phases of embryonic development, and the vast majority of the regulatory genes at the nodes of these networks are expressed at relatively low levels. One exception is the syncytial fourteenth cleavage cycle *Drosophila* embryo, where it appears that prominent regulatory genes are being transcribed near maximum possible rates, suggesting a close to maximal occupancy of the relevant CRMs. Similarly, in Chapter 5 we deal with cell type specification, including terminal differentiation, and in that context some downstream differentiation genes also run near maximum rates of initiation. Here we focus on more typical developmental GRN dynamics. Many kinds of evidence converge on the conclusion that developmentally active regulatory genes are typically transcribed at relatively low rates of initiation. This is demonstrated in the sea urchin embryo as an example, where many relevant measurements are available (Davidson, 1986; Bolouri and Davidson, 2003). Direct time course measurements of transcript accumulation revealed a "default" mRNA half-life of about 3–5 h. This conclusion is substantiated in dozens of individual regulatory gene mRNA time courses that include decay phases following turn off of transcription (Materna et al., 2010). The same data show directly the low synthesis rates of regulatory gene transcripts. In addition, extensive quantitative transcriptome data confirm the low levels of regulatory gene mRNA transcripts (Tu et al., 2014). These data demonstrate that most active regulatory genes in this embryo are represented by only about 10–50 molecules of mRNA per cell; these mRNA's are generated at rates within a factor of two of about 100 molecules/h-embryo (Peter et al., 2012). Considering the number of cells in the respective domains expressing these genes, this means the initiation rates per gene and minute are for virtually all regulatory genes being expressed in the embryo only a few percent of the I_{max} calculated above, i.e., per active gene initiation occurs no more frequently than once every several minutes. The implication of these low I values is that there are low average levels of occupancy of the relevant CRMs by the driver factors controlling the expression of these genes. The significance of this for occupancy calculations is that we require a treatment that deals appropriately with intermittent low occupancy rather than preferentially with saturation occupancies.

In Fig. 2.5, Eqns (1) and (2), we see the equilibrium constant for transcription factor–DNA target site interaction from several vantage points. In a bimolecular reaction between transcription factor A and DNA target site D, equilibrium is the point at which the rate of formation of the product, here factor A-DNA site complex, or AD, is equal to its rate of dissociation. Equation (1) shows the bimolecular reaction kinetics. Here k_{as} is the rate constant for formation of the complex; given this constant, the actual rate of formation of course depends on the concentrations of factor A and site D. Once formed, the tightness of binding (affinity) determines the rate of complex dissociation, k_{ds}, an intrinsic chemical feature of each DNA–protein sequence interaction, i.e., the tighter the binding (the more energy released) the longer the complex will last. For a given k_{ds}, the actual amount of complex dissociating per unit time in Eqn (1) depends on the amount of complex there is. The factor finds its target site by diffusion, hence the rate at which this occurs, k_{as}, depends essentially on the rate of diffusion. Since most transcription factors fall within a range of a factor of two or so in protein mass, on which the diffusion rate depends, k_{as} is about the same for almost all transcription factors. However, k_{ds} reflects the intrinsic energetics of the factor–DNA interaction, and since transcription

Figure 2.5

(A)

Parameters and Symbols

1: $\dfrac{dAD}{dt} = k_{as}A \cdot D - k_{ds}AD$

At equilibrium: $\dfrac{dAD}{dt} = 0$ and now

2: $\dfrac{AD}{A \cdot D} = \dfrac{k_{as}}{k_{ds}} = K_{eq} \; [\text{M}^{-1}]$

3: $(S)K_{eq} = e^{-\Delta G/RT}$

4: $Y = \dfrac{e^{-\Delta G/RT} \cdot A}{1 + e^{-\Delta G/RT} \cdot A} = \dfrac{K_{eq} \cdot A}{1 + K_{eq} \cdot A}$

5: $Y_{AB} = \dfrac{K_{eqA} \cdot A \cdot K_{eqB} \cdot B \cdot K_q}{1 + A \cdot K_{eqA} + B \cdot K_{eqB} + K_{eqA} \cdot A \cdot K_{eqB} \cdot B \cdot K_q}$

6: $K_r = \dfrac{K_{eq}(specific)}{K_{eq}(non\text{-}specific)}$

7: $Y_{AB} = \dfrac{A \cdot K_{rA} \cdot B \cdot K_{rB} \cdot K_q}{D_N^2 + A \cdot D_N + B \cdot D_N + A \cdot K_{rA} \cdot D_N + B \cdot K_{rB} \cdot D_N + A \cdot K_{rA} \cdot B \cdot K_{rB} \cdot K_q}$

8: $I = I_{max} \cdot (1 - e^{-k_b \cdot Y_{AB} \, I_{max}})$

$mRNA$	molecules of mRNA
k_s	synthesis rate constant
k_d	mRNA decay rate constant
A, B	concentration of transcription factors A and B
D	concentration of specific target sites
AD, BD	concentration of [factor:DNA] complex
k_{as}	association rate constant
k_{ds}	dissociation rate constant
K_{eq}	equilibrium constant
S	standard medium activity coefficient
T	temperature
ΔG	change in free energy per mole
R	natural gas constant
Y	cis–regulatory occupancy
Y_{AB}	cis–regulatory occupancy for two factors, A and B
K_q	cooperativity constant
K_r	relative equilibrium constant
D_N	concentration of non-specific DNA target sites
I	initiation rate
k_b	initiation efficiency factor
k_T	translation rate constant
k_{dP}	protein decay rate constant
P	molecules of protein per cell

(B)

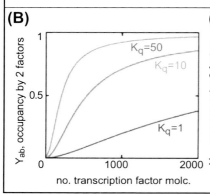

(C)

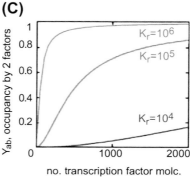

(D)

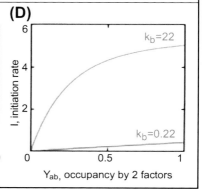

Figure 2.5 Kinetic treatment of DNA-protein interaction, *cis*-regulatory occupancy, and transcriptional initiation. (A), Equations for quantitation of DNA–protein interactions and transcriptional initiation; see list of terms at right for designations of symbols. These equations have been reviewed earlier (Emerson et al., 1985; Bolouri and Davidson, 2003; Ben-Tabou de-Leon and Davidson, 2009; Phillips et al., 2009). Equations (1–3), Definitions of equilibrium constant for formation of complex between transcription factor at concentration A and its specific DNA target site, at concentration D (a convenient use of concentration is to consider the number of molecules of factor per nuclear volume). Equations (1), O.D.E. giving rate of formation of the complex AD in a second order reaction. When this rate is 0 the reaction is in a state of equilibrium. Eq. (2), The equilibrium constant K_{eq} defined kinetically from Eqn (1) at $dAD/dt=0$, in terms of the association and dissociation rate constants k_{as} and k_{ds}; and stoichiometrically as the ratio of the molecules of complex to molecules of site D and factor A at equilibrium. Eq. (3), Thermodynamic definition of K_{eq} in terms of molar free energy change ΔG occurring in the reaction $A+D \leftrightarrow AD$; note that for this reaction the units of K_{eq} are M^{-1} while the right side of Eqn (3) is unit-less. This is because of an implicit term S [M], the activity coefficient, which accounts for the activity of the relatively high ionic strength of the solution in which these reactions take place, and is conventionally set at 1. Eq. (4), Fractional occupancy, Y, of a single target site when its factor is present at concentration A, and it binds the site with the equilibrium constant K_{eqA}; Y is the fraction of total site in complex, or the occupancy. Eq. (5), Double occupancy, Y_{AB}, for a *cis*-regulatory system with sites for factor A and also factor B when these factors are present at concentrations A and B, respectively; K_q, cooperativity constant for interaction between factor A and factor B (see text for discussion of this equation). Eq. (6), Definition of relative equilibrium constant K_r, i.e., ratio of K_{eq} for specific to K_{eq} for nonspecific DNA–protein interaction. Eq. (7), Y_{AB}, computed using K_r rather than K_{eq} for both A and B factors, and including K_q terms for cooperative interactions between A and B. Eq. (8), Relation between I and Y_{AB}; the higher the value of Y_{AB} the closer I will be to the maximum rate of initiation I_{max} (Eqns (7) and (8) from Bolouri and Davidson (2003)). (B–D), Simulations using typical sea urchin parameters *(from Bolouri and Davidson (2003))*. (B), Effect of cooperativity on Y_{AB}. (C), Effect of K_r on Y_{AB}. (D), Effect of initiation activation efficiency of given factors, k_b, on I, initiations per minute and gene.

factors vary over orders of magnitude in how strongly they bind their DNA target sites, k_{ds} varies likewise. The equilibrium constant K_{eq} is defined in Figure 2.5 Eqn (2) for the condition when $dAD/dt=0$ in Eqn (1). We see that the equilibrium constant is the ratio of the two rate constants, k_{as}/k_{ds}. Variations in the equilibrium constants for different transcription factor–target site interactions, depend almost entirely on k_{ds} since the values of k_{as} are usually similar. K_{eq} also gives the stoichiometry of the ratio of complex to unbound factor and site: thus from Eqn (2) we also see that the more factor there is the more complex will be formed, and the higher the equilibrium constant the more complex will form per amount of factor. This is the essential behavior which basically explains gene regulation: binding of a transcription factor to its target site is controlled by factor concentration, that is, by the activity of the upstream regulatory gene encoding the factor.

Occupancy

A classic approach is to treat occupancy as the physical manifestation of the probability that the specific sites in a regulatory DNA sequence will be in the productive bound state. The probability of productive occupancy of a CRM consisting of multiple binding sites is given by the likelihood with which the productive state is obtained compared to that of all possible states of binding. Thus the probability is calculated by normalizing the likelihood of occurrence of the desired bound state under given conditions to the sum of the likelihoods of occurrence of all the possible states of binding of the sites on the sequence. Here the normalization factor is the sum of the probabilities of all sites remaining unbound, of all combinations of some sites bound and others not, plus the probability of the functional bound state. Transcription factor interaction with DNA is an energetic transaction, since the amino acid side chains of the DNA recognition domain interact chemically with residues of the nucleotides constituting the DNA target site. It follows that the probability of a particular state of transcription factor occupancy for a given DNA sequence, is actually the ratio of the particular interaction energy of the productive bound complex to the sum of all the interaction energies of all the complexes the sequence could generate ("Boltzmann partition function"; for modern derivation and discussion in a biophysical context see Phillips et al., 2009). It is worth spending a few lines to explore the parameters and behavior of this representation of *cis*-regulatory occupancy as a probabilistic function.

We need first to be clear about what quantitatively controls individual transcription factor–DNA target site interactions: intuitively, this has to depend on the amount of the factor in the nucleus and on its innate affinity for its target site(s). The first is conveyed by the factor concentration (molar), and the second by the equilibrium constant for this particular factor–site interaction. Since the equilibrium constant value depends on k_{ds}, which depends on the energetic exchange at binding, K_{eq} can also be defined in terms of energy released (per mole) upon binding: for reactions in ionic solutions we can evaluate the equilibrium constant as in Eqn (3) of Fig. 2.5(A) (see note in caption). We can now use the principle of the Boltzmann partition function to see how occupancy, Y, will change for the maximally simple system of a single site and a single-factor species as a function of its concentration, A: this is given in Eqn (4) of Fig. 2.5(A). Here the likelihood of complex formation is calculated by the ratio of bound state to bound plus unbound state. Thus, the energetic definition of the equilibrium constant for factor A multiplied by its concentration is used to compute the bound state, and in the denominator the same term appears plus that for the alternative that the site is not bound (for naked DNA site, ΔG is 0, hence the 1 in the denominator, see Eqn (3)). We can see that at low A concentrations, Y will rise linearly with A, and with increasing concentrations of A the value of Y in Eqn (4) approximates 1.

Now, to become a step more realistic, consider a system where a second factor at concentration B must also bind its site in the sequence to generate an occupancy configuration required for function. By the same algorithm as used in Eqn (4), we obtain Eqn (5) (Ackers et al., 1983). This equation includes an additional term, K_q which captures the possibility that the two factors A and B interact cooperatively on the DNA, contributing a further energetic term that additionally stabilizes the complex. The definition of K_q could be extended to include any mechanism by which binding of one factor energetically facilitates the binding of a second factor, for example by introducing torsion of the DNA, or by increasing the accessibility of binding sites for the second factor so that more complex is formed for a given concentration of factor.

One further step is required to transit closer to the real regulatory world of the animal cell nucleus. It was realized early on that because of the enormous amount of genomic DNA in animal cell nuclei (relative to bacteria, let alone phage), and because all transcription factors display some affinity for any DNA sequence, factors not bound to specific target sites will in general be transiently bound nonspecifically to the DNA. This must be taken into account when computing specific DNA–protein interactions (Von Hippel et al., 1974; Lin and Riggs, 1975; Emerson et al., 1985). The nonspecific binding of transcription factors to DNA is due to the general property of this class of proteins that they include basic domains which interact transiently with the acid phosphate bridges of the DNA helix as the proteins swivel along the genome. The average $t_{1/2}$ for these nonspecific interactions is very brief, on the order of one or a few milliseconds (e.g., Emerson et al., 1985; Calzone et al., 1988), but on the other hand there is a very large number of sites since every base pair in internucleosomal sequence throughout the genome begins a new binding site for nonspecific interactions. So what is really required for in vivo considerations of site occupancy is the relative equilibrium constant, K_r, as defined in Eqn (6) of Fig. 2.5(A). Exhaustive studies of DNA–protein interactions in complex sea urchin *cis*-regulatory systems revealed that all those displaying $K_r > \sim 5 \times 10^4$ as measured in vitro will have regulatory function in vivo (Calzone et al., 1988; Kirchhamer and Davidson, 1996; Yuh et al., 1998, 2001). Consistent with such K_r values, recent measurements of transcription factor/DNA interaction dynamics in mammalian cells indicate that these complexes last typically from a few seconds to several minutes, as might be expected from comparison with in vitro binding results, i.e., taking into account the effects of high endogenous salt concentration in the nucleus (e.g., Poorey et al., 2013).

Incorporation of K_r into Eqn (5) of Fig. 2.5(A) results in Eqn (7) (Bolouri and Davidson, 2003). Plots of occupancy as a function of factor concentration and of cooperative interaction between the bound factors using Eqn (7) are also shown in Fig. 2.5. Note that cooperativity, which is probably the rule rather than the exception in loaded CRMs, has a strong effect on levels of occupancy (Fig. 2.5(B)). More importantly, significant occupancy is attained even at relatively low numbers of transcription factor molecules, given a very typical K_r of 10^5 (Fig. 2.5(C)). Here we see also that the computation supports the observation quoted above, namely that $K_r > \sim 5 \times 10^4$ means functional interaction; much less than this requires relatively large numbers of protein molecules to attain any occupancy.

Finally, it remains to relate occupancy to initiation rate, I. The probability of an initiation occurring in a time window might be expected to rise linearly at low occupancies, but to increase less efficiently with increasing occupancy at high levels because of interference by polymerase molecules that have not yet moved out of the way as the system operates closer and closer to I_{max}. The probability of an initiation is (1−the Poisson probability of no initiation). These simple ideas are captured in Eqn (8) of Fig. 2.5(A) (Bolouri and Davidson, 2003), where the rate of initiation as a fraction of I_{max} is related to the value of Y, computed as in Eqn (7); the higher the value of Y the lower the probability of no initiation, and at low Y, the initiation rate is essentially linear with Y. This result is also shown graphically in the simulations of Fig. 2.5(D). We are now equipped to deal with cascades of regulatory gene expression, that is, with network dynamics of the nature of those occurring in developmental GRNs.

6.3   Regulatory cascade dynamics

Familiar issues that arise continuously in working with GRNs include the following: if the transcription factor encoded by Gene 1 directly activates Gene 2, how much time will elapse between the onset of transcription of Gene 1 and the onset of transcription of Gene 2, i.e., the step time? About how many molecules of Gene 1 mRNA are likely to be required to make enough transcription factors to produce sufficient occupancy of the regulatory system of Gene 2 so that this gene can in turn generate enough molecules of its product to affect the next gene downstream? What do gene cascade kinetics look like? How important in a developing embryonic system is control of activity level? Here we apply real time measurements of kinetic constants using the computational apparatus in Figs. 2.4 and 2.5, and direct observations on regulatory gene cascades, in order to arrive at the answers to an interconnected set of important questions. It is relatively easy to measure accumulation time courses for multiple mRNAs simultaneously, and if we know enough to interpret the behavior of regulatory gene cascades, the measured kinetics can provide evidence that is useful in discriminating between alternative network architectures. Our examples will again be taken from the sea urchin embryo because both GRN architecture and the gene expression kinetics it generates are available. However, the conclusions are unlikely to be grossly different for other embryonic systems once a Q10 correction for temperature effects is made.

Before we can deal with these issues the rate of protein synthesis per mRNA needs to be taken into account; this has been measured at 2 molecules/min-mRNA for sea urchin embryos at 15 °C, a constant (Davidson, 1986). Application of Eqn (1) of Fig. 2.6(A), where this constant appears as k_T, now allows us to estimate the kinetics of transcription factor synthesis in terms of the kinetics of regulatory gene mRNA generation, computed as above.

A series of simulations was carried out (Bolouri and Davidson, 2003), using the occupancy, initiation, synthesis, and turnover treatments in Figs. 2.4, 2.5, and 2.6, and applying average values taken from a large number of prior measurements for the relative equilibrium and decay constants called for in these equations. Fig. 2.6(B) shows the kinetics of mRNA synthesis from a gene controlled by two transcription factors present at various concentrations per cell: note the predicted insensitivity of the

Figure 2.6

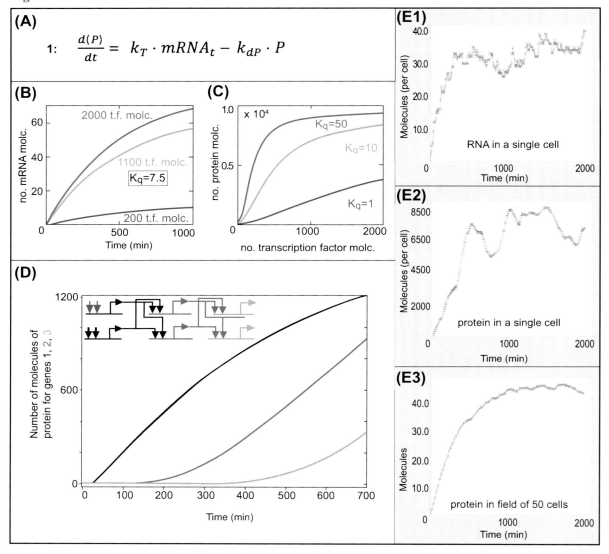

Figure 2.6 Protein output dynamics. (A), O.D.E giving rate of protein synthesis as a function of mRNA concentration (per cell); see Fig. 2.5 for terms. (B), mRNA output from a gene requiring two transcription factors as in Fig. 2.5 Eqn (7), as a function of the level of transcription factors present per cell (the concentrations of the two factors are set to be equal). (C), Protein output using the expression in (A), given 1100 transcription factor molecules, as in (B, green curve), shown for different degrees of cooperativity between the two transcription factors. (D), Gene cascade diagrammed at top, and kinetics with which the protein output of the second (red) and third (green) gene appear following activation of the first (black) gene (B–D from Bolouri and Davidson (2003)). (E), Effect of microscopic stochasticity in the intervals between successive initiations of transcription on mRNA and protein output (unpublished simulations of Bolouri and Davidson). Intervals were distributed around a mean by Poisson statistics and the RNA and protein output dynamics reported. (E1), mRNA output from a single cell. (E2), Protein output from a single cell, computed from the expression in (A). (E3), Protein output from a field of 50 uncoupled cells, all expressing the same gene.

response to very low factor concentrations. In this light it is interesting to see in Fig. 2.6(C) how quickly effective numbers of transcription factor molecules can be generated by a regulatory gene operating on two driver inputs present at intermediate levels, even in the absence of cooperativity ($K_q = 1$). The most important of these predictions is seen in Fig. 2.6(D). Here a cascade of three gene pairs in a tandem causal array is considered, such that double occupancy of the regulatory system of each gene by its immediately upstream transcription factors is required, and the output of the first pair of genes (black) activates the second two genes, the output of which (red) activates the third pair of genes (green). The plot shows, for typical constants including $K_q = 7.5$ and $K_r = 10^5$, that the interval between activation of an upstream Gene 1 and the activation of its downstream target Gene 2 is about 3 h, and this predicts the step time for this 15 °C system. There are two additional very important lessons, combining the import of Figs. 2.6(B) and 2.6(D). First, about 500–1000 transcription factor molecules per cell suffice to trigger activation of a direct target gene in a regulatory cascade (Fig. 2.6(D)), but this level of input factor leads to the accumulation of only <60 mRNA molecules per cell (Fig. 2.6(B)). Note that a similar level of mRNA is also produced if the level of upstream transcription factor is doubled to 2000 molecules, illustrating a profound insensitivity to transcription factor level once this rises beyond the effective threshold indicated in Fig. 2.6(B). This is why typically <60 regulatory gene mRNA molecules are to be found per cell in this system, and why the transcriptional initiation rates of regulatory genes are so leisurely. Second, from the shape of the curves in Fig. 2.6(D), we see that the downstream genes in the cascade are activated long before the gene products of the upstream genes ever attain steady state. Thus as soon as mRNA encoding transcription factors appears, it gets translated, and as soon as even modest levels of this transcription factor have accumulated it is used to activate target genes. A similar result, that is, activation of target genes long in advance of steady state, was obtained in a kinetic study of Ftz target gene transcription in *Drosophila* (Nasiadka et al., 2002). The direct implication is that control of transcription factor concentration at steady state is not relevant for target gene activation in a developing embryo, even though the only general way expression levels can ever be controlled closely is at steady state. Thus it is no surprise that when specific regulatory gene mRNA levels are measured in diverse batches of embryos, they often differ by 30–50% or more (Materna et al., 2010). As opposed to the finely tuned balanced homeostatic systems of physiology, developing embryos are forward drive systems and relatively level insensitive.

The insensitivity of cascade kinetic behavior to factor level in these systems provides a way of estimating the significance of microscopic transcriptional stochasticity. In Fig. 2.6(E) we see the stochasticity of gene output modeled under the assumption that transcriptional initiation occurs stochastically in time according to a Poisson distribution. The gene output is shown for a single cell as mRNA level over time (Fig. 2.6(E1)) and given canonical turnover rates, as protein level over time (Fig. 2.6(E2)). Even though the accumulation curves are noisy, the actual amount of variation will have little effect on cascade kinetics. In a field of 50 similarly functioning cells, even this amount of noise is effectively averaged out (Fig. 2.6(E3)).

The above computation, built using the kinetics of basic molecular processes of gene expression, predicts a step time of 3 h for this sea urchin embryo (Bolouri and Davidson, 2003). A remarkable confirmation that this step time is indeed the metric by which the sea urchin GRN operates was obtained 10 years after these a priori simulations were generated. In a dynamic model analysis of the sea urchin embryo GRN (Peter et al., 2012), which we discuss later in this book (Chapter 6), the 3 h step time was applied across the whole GRN to the large number of known direct regulatory gene inputs, over a 30 h period. The result was to produce a computed sequence of expression patterns extremely close to those observed, while imposition of 2 h or 4 h step times created disastrous nonconcordance with observation. This general result is also in agreement with many individual experimental observations made during study of the sea urchin embryo GRN that repeatedly revealed approximately 3 h intervals between activation of known regulatory genes and of their immediate target genes.

In summary, the quantitative arguments given here provide a basis on which the dynamics of gene expression and regulation can be rationalized, in terms of the rate limiting molecular functions that actually control the dynamics. A major conclusion that speaks to the general character of regulatory system dynamics in the embryo is that the regulatory state changes progressively in real time, in a way not dependent on steady state levels. We have here ignored the intermediate dynamics of nuclear pre-mRNA processing, since essentially it contributes only a time lag (for this see Bolouri and Davidson, 2003). Nor have we dealt with repression, for the reason that repression is often a multistep process in which the role of the transcription factors that initiate the process is transient. Transcriptional repressors may recruit chromatin repression complexes that maintain the silenced state in time (Chapter 1), and which cannot be encompassed in a literal transcription process model such as we have dealt with here. Both repression and activation are conventionally modeled mathematically by expressions taken from (approximate) kinetic analysis of enzymatic reactions, although mechanistically neither transcriptional activation nor transcriptional repression actually resembles such reactions. These mathematical approaches give the required external behavior, however, and they have enabled very interesting and useful analyses of small regulatory circuits, as we review later in this book (Chapter 6). Here we have structured the discussion so as to preserve as direct as possible a relation between the mathematical terms and the physical processes of gene regulation.

7. Historical Origins and Antecedents of GRN Theory

In this chapter we have taken up the different aspects of regulatory system mechanism which, when combined, produce the powerful synthetic concept that animal development is controlled by GRNs. These mechanistic aspects are now well and deeply supported by multiple kinds of experimental result. Our current concepts grew from separate ideological roots of diverse origin; some are of modern vintage, others descend from insights that go back to the initial period of causal developmental biology. To remind ourselves of the breadth and depth of the conceptual terrain in which developmental GRN theory is embedded is to enrich our appreciation for it.

It is interesting to consider how many different scientific trajectories have contributed to the overall concept of developmental GRNs. First there are the basic concepts that the process of development is encoded in the genome, that genomic information is equivalent in all cells of the embryo, and that therefore the differential readout of this information in different cells requires the existence of a regulatory system which is itself encoded in the genome. Second is the fundamental idea that developmental spatial gene expression differs from place to place and time to time because of combinatorial transcriptional control at the *cis*-regulatory sequences of each individual gene. Third are the concepts of systems biology, in particular the idea that a complex process like development must be the output of a large number of individual control functions and that the framework explanation must include all or most of these in order to arrive at a satisfactory explanation. Furthermore, this explanation cannot be derived by adding up observations on individual corners or beams without knowledge of the architecture of the edifice as a whole. Finally there is what could be described as the biological regulatory theory of GRN structure/function relations per se. This includes the spatial logic outputs of network subcircuits according to their architecture, the significance of the innate hierarchy of developmental GRNs, and the means by which GRNs encompass spatial changes of state by interdomain signaling. We leave for later in this book (Chapter 7) the large additional body of concept that arises when evolution of the body plan is considered in terms of change in GRN architecture.

Each of these areas of concept and advance could of course profitably be the subject of a treatise in the history of modern molecular biology, and what follows are only very brief sketches. Some aspects are only of recent origin: for example, the systems biology arguments are of essentially modern vintage. The antecedents of most of the strains of thought comprising our current synthetic concept lie further back, however. A brief reminder of some of the twentieth century way stations along each of these scientific pathways shows how wide and deep are these antecedents.

7.1 Organism-wide genomic regulatory system required for development

The concept that there is resident in the chromosomes of every cell a genomic program that controls cellular functions during development arose explicitly in the first decades of the twentieth century. A landmark along this pathway was the famous polyspermy experiment of Boveri (Boveri, 1905, 1907; Laubichler and Davidson, 2008). Using a protocol that engendered aneuploidy in early blastomeres of sea urchin embryos, Boveri showed quantitatively that only those blastomeres containing complete chromosome sets are capable of giving rise to morphologically complete larvae. He concluded that every chromosome contains unique genetic determinants and that all chromosomes (i.e., the whole genome) are required to be present in every cell for embryogenesis to be completed: ergo, the genomic program for development. Contemplating Boveri's work together with the accumulating evidence of chromosome behavior in mitosis and meiosis, interspecific nuclear transfer experiments, and much else that was qualitatively understood about development in the era before molecular biology, E.B. Wilson enunciated clearly the proposition that "hereditary" information in the chromosomes encodes the developmental process (to use our modern words; Wilson, 1925). He then followed a similar trail of logic as in the introductory section of Chapter 1 of this book, deducing from the increase of complexity during the process of development (which he refers to as "epigenesis") that: "…heredity is effected by the transmission of a nuclear preformation which in the course of development finds its expression in a process of cytoplasmic epigenesis" (Wilson, 1925; Peter and Davidson, 2013). Thus there had already coalesced the fundamental concept of a genomic program for development operative in all cells. For about the next half century there was no great conceptual advance in the state of this completely accurate idea per se. However, the fundamentally important point that all cells of the developing organism contain the same DNA genome was established in a particularly incontrovertible manner by John Gurdon's nuclear transplantation experiments of the 1960s (Gurdon and Uehlinger, 1966; Laskey and Gurdon, 1970). Evidence accumulated rapidly after the mid-1980s from what we have called the fundamental experiment in developmental gene regulation, i.e., gene transfer into eggs, demonstrating how spatially localized transcription is controlled by *cis*-regulatory interactions. Only then could come into focus what the regulatory system controlling development mechanistically entails, but its majestic dimensions emerged only with genomics.

7.2 Combinatorial transcriptional regulation of spatial gene expression

As we now know, *cis*-regulatory logic processing of multiple positive and negative inputs is the basic mechanism of spatial gene expression control, operating at all the nodes of developmental GRNs. But in terms of the overall history of our field, among the insights essential to our current concepts this was one of the last to coalesce. Until the late 1940s and early 1950s, what genes are in material terms remained very unclear, and the problem of how gene expression is regulated was inaccessible. This is not to say that no one earlier than this had figured out from first principles the outlines, if not the molecular biology, of how things actually do work: consider the remarkable prescience of the following quote, again from E.B. Wilson, but this from back in 1896: "If chromatin be the idioplasm (genetic material) in which inheres the sum total of hereditary forces, and if it be equally distributed at every cell division, how can its mode of action so vary in different cells as to cause diversity of structure, i.e., *differentiation*?… My own conception…is as follows. All the nuclei are equivalent, and all contain the same idioplasm… Through the influence of this idioplasm the cytoplasm of the egg or of the blastomeres derived from it undergoes specific progressive changes, each change then reacting upon the nucleus and then initiating a new change. These changes differ in different regions of the egg because of pre-existing differences…such as the distribution of different substances in the egg cytoplasm" (Wilson, 1896). If for "change" we substitute "new transcription factor", we have essentially how spatially differential gene expression does in fact get going in the early embryo (see next chapter). But this was not a dominant view, particularly among those who actually worked on genes, i.e., geneticists. As T.H. Morgan complained of dominant opinion in the area of what we would call developmental genetics in 1934, "the implication in most genetic interpretations is that all the genes are acting all the time in the same way." He then proceeds to argue in the opposite direction: "…an alternative view would be that different batteries of genes come into action as development proceeds" (Morgan, 1934). This anticipates the correct view that genes are differentially expressed. Morgan went on in this passage to adduce exactly the same egg cytoplasmic mechanism as in Wilson's 1896 quote (without attribution). Thus, regulation of gene expression reentered the arena, this time to stay. By the early 1950s the theory of "variable gene activity" had been explicitly developed as a general explanation of differentiation (Brachet, 1949; Sonneborn, 1950; Stedman and Stedman, 1950; Mirsky, 1951, 1953). Overwhelming evidence accumulated in the next three decades from molecular genetics and molecular developmental biology that this theory is correct, and that during development regulatory control of gene activity is specifically what accounts for differential gene expression, which is in turn the driver of development.

The weakness of this body of mechanistic knowledge was that it was all about how individual genes behave with little direct illumination of the holistic processes of development. On the other hand, an entirely unrelated theoretical trajectory oriented toward these holistic aspects was based on evidence of inductive gradients. This gave rise to attempts to interpret the spatial patterns formed in development from a completely different starting point, as the global output of "reaction diffusion" gradient systems. The relevance of these ideas to developmental pattern formation was challenged by accumulating molecular evidence regarding the mechanisms for control of spatial gene expression. A particularly illuminating step forward was the identification of multiple, specific CRMs which respectively generate the stripes of pair rule gene expression in *Drosophila*, utilizing combinatorial inputs from activators and repressors which at the DNA level delimit transcriptional activity and hence form the sharp stripe boundaries (Stanojevic et al., 1991; Small et al., 1996). By now, insights have accumulated from thousands of structure/function experiments on *cis*-regulatory expression constructs inserted into eggs, which show exactly where the explanation for developmental spatial gene expression actually lies: in the encoded *cis*-regulatory DNA sequence (for review of spatial information processing in specific developmental CRMs from many different embryonic systems, see Davidson, 2006).

7.3   Hierarchical structure of GRNs

An early attempt at constructing a model for gene regulation in development was formulated in 1969 (Britten and Davidson, 1969). This was a model for gene regulation which proposed many large-scale structural and logical features that much later turned out to be intrinsic to the organization of actual developmental GRNs. These features include the following. First, this model assumed a very large-scale system of interacting regulatory genes. Second, it was a hierarchical model in which upper level regulatory genes control other regulatory genes at the top of the hierarchy in each developmental process. Third, it was a network model in which individual genes respond to multiple *trans*-regulatory inputs produced by other regulatory genes, while each such *trans*-regulator has multiple targets. Fourth, it treated inductive signals as activators of special response elements which cause transcription of upper level hierarchy regulatory genes. Fifth and most generally, its presumption was that development is indeed determined by a distributed genomic sequence code. Virtually nothing was then known of the molecular biology of transcriptional regulation of the gene in animals. This model assumed that RNA rather than protein is the active product of the regulatory genes. In retrospect it is interesting to see that this largely erroneous assumption had so very little effect on the structure of its underlying logic. The arguments of the 1969 model soon led to the explicit proposition that change in the genomic sequence code for development that results in change in the architecture of regulatory networks, would turn out to be the engine of evolutionary change in body plans (Britten and Davidson, 1971). The spectacular technical advance of gene cloning soon ensued, and for many years thereafter the novel experimental opportunities to examine expression, function, and regulation of individual genes dominated research. As an indirect consequence, system-scale thought on the subject of developmental gene regulation was almost buried, and so were system-scale measurements, except for studies on a few specific large gene families, and some measurements of global parameters of genome organization and expression. However, by the end of the 1990s, in the light of all that had been learned about gene regulation, it became possible to perceive the shape of developmental gene regulatory networks the existence of which had been proposed by Britten and Davidson in 1969. The subject of developmental gene networks was rejuvenated first in principle (Arnone and Davidson, 1997), and then in respect to the initial experimental attempt at large-scale GRN analysis in the sea urchin embryo (Davidson et al., 2002). This has now led to the first relatively complete experimentally obtained GRN model for a large-scale developmental process for which the general explanatory value has been computationally demonstrated (Peter et al., 2012).

In the next three chapters we see all of the conceptual aspects of transcriptional regulatory networks that we have reviewed here in operation at different levels of the process of development. In Chapter 3, the GRNs considered control the formation of the embryo, in Chapter 4 they control the formation of adult body parts, and in Chapter 5 they control the specification of cell types from multipotential precursors. Throughout, our focus has been on developmental processes for which experimental analysis has revealed the genomic basis of the underlying regulatory logic. This leads inexorably to consideration of development through the lens of GRN structure and function.

REFERENCES

Ackers, G.K., Shea, M.A., Smith, F.R., 1983. Free energy coupling within macromolecules. The chemical work of ligand binding at the individual sites in co-operative systems. J. Mol. Biol. 170, 223–242.

Amit, I., Garber, M., Chevrier, N., Leite, A.P., Donner, Y., Eisenhaure, T., Guttman, M., Grenier, J.K., Li, W., Zuk, O., Schubert, L.A., Birditt, B., Shay, T., Goren, A., Zhang, X., Smith, Z., Deering, R., McDonald, R.C., Cabili, M., Bernstein, B.E., Rinn, J.L., Meissner, A., Root, D.E., Hacohen, N., Regev, A., 2009. Unbiased reconstruction of a mammalian transcriptional network mediating pathogen responses. Science 326, 257–263.

Arnone, M.I., Davidson, E.H., 1997. The hardwiring of development: organization and function of genomic regulatory systems. Development 124, 1851–1864.

Arnosti, D.N., Kulkarni, M.M., 2005. Transcriptional enhancers: intelligent enhanceosomes or flexible billboards? J. Cell. Biochem. 94, 890–898.

Bell, A.C., West, A.G., Felsenfeld, G., 1999. The protein CTCF is required for the enhancer blocking activity of vertebrate insulators. Cell 98, 387–396.

Ben-Tabou de-Leon, S., Davidson, E.H., 2009. Modeling the dynamics of transcriptional gene regulatory networks for animal development. Dev. Biol. 325, 317–328.

Berger, M.F., Bulyk, M.L., 2009. Universal protein-binding microarrays for the comprehensive characterization of the DNA-binding specificities of transcription factors. Nat. Protoc. 4, 393–411.

Bolouri, H., Davidson, E.H., 2003. Transcriptional regulatory cascades in development: initial rates, not steady state, determine network kinetics. Proc. Natl. Acad. Sci. U.S.A. 100, 9371–9376.

Boveri, T., 1905. Über die Abhängigkeit der Kerngrösse und Zellenzahl bei Seeigellarven von der Chromosomenzahl der Ausgangszellen. Zellenstudien, Jena.

Boveri, T., 1907. Zellenstudien VI. Die Entwicklung dispermer Seeigeleier. Ein Beitrag zur Befruchtungslehre und zur Theorie des Kerns. Gustav Fischer, Jena.

Brachet, J., 1949. L'hypothese des plasmagenesdans le developpement et la differentiation. Colloq. Int. C.N.R.S 8.

Britten, R.J., Davidson, E.H., 1969. Gene regulation for higher cells: a theory. Science 165, 349–357.

Britten, R.J., Davidson, E.H., 1971. Repetitive and non-repetitive DNA sequences and a speculation on the origins of evolutionary novelty. Q. Rev. Biol. 46, 111–138.

Busby, S., Bakken, A., 1979. A quantitative electron microscopic analysis of transcription in sea urchin embryos. Chromosoma 71, 249–262.

Calzone, F.J., Thézé, N., Thiebaud, P., Hill, R.L., Britten, R.J., Davidson, E.H., 1988. Developmental appearance of factors that bind specifically to cis-regulatory sequences of a gene expressed in the sea urchin embryo. Genes. Dev. 2, 1074–1088.

Christensen, R.G., Gupta, A., Zuo, Z., Schriefer, L.A., Wolfe, S.A., Stormo, G.D., 2011. A modified bacterial one-hybrid system yields improved quantitative models of transcription factor specificity. Nucleic Acids Res. 39, e83.

Cotton, R.W., Manes, C., Hamkalo, B.A., 1980. Electron microscopic analysis of RNA transcription in preimplantation rabbit embryos. Chromosoma 79, 169–178.

Davidson, E.H., 1986. Gene Activity in Early Development, third ed. Academic Press/Elsevier, San Diego, CA.

Davidson, E.H., 2006. The Regulatory Genome. Gene Regulatory Networks in Development and Evolution. Academic Press/Elsevier, San Diego, CA.

Davidson, E.H., Rast, J.P., Oliveri, P., Ransick, A., Calestani, C., Yuh, C.H., Minokawa, T., Amore, G., Hinman, V., Arenas-Mena, C., Otim, O., Brown, C.T., Livi, C.B., Lee, P.Y., Revilla, R., Schilstra, M.J., Clarke, P.J., Rust, A.G., Pan, Z., Arnone, M.I., Rowen, L., Cameron, R.A., McClay, D.R., Hood, L., Bolouri, H., 2002. A provisional regulatory gene network for specification of endomesoderm in the sea urchin embryo. Dev. Biol. 246, 162–190.

Drissen, R., Palstra, R.J., Gillemans, N., Splinter, E., Grosveld, F., Philipsen, S., de Laat, W., 2004. The active spatial organization of the beta-globin locus requires the transcription factor EKLF. Genes. Dev. 18, 2485–2490.

Elgar, G., Vavouri, T., 2008. Tuning in to the signals: noncoding sequence conservation in vertebrate genomes. Trends Genet. 24, 344–352.

Emerson, B.M., Lewis, C.D., Felsenfeld, G., 1985. Interaction of specific nuclear factors with the nuclease-hypersensitive region of the chicken adult beta-globin gene: nature of the binding domain. Cell 41, 21–30.

Guo, Y., Monahan, K., Wu, H., Gertz, J., Varley, K.E., Li, W., Myers, R.M., Maniatis, T., Wu, Q., 2012. CTCF/cohesin-mediated DNA looping is required for protocadherin α promoter choice. Proc. Natl. Acad. Sci. U.S.A. 109, 21081–21086.

Gurdon, J.B., Uehlinger, V., 1966. "Fertile" intestine nuclei. Nature 210, 1240–1241.

Hadchouel, J., Carvajal, J.J., Daubas, P., Bajard, L., Chang, T., Rocancourt, D., Cox, D., Summerbell, D., Tajbakhsh, S., Rigby, P.W., Buckingham, M., 2003. Analysis of a key regulatory region upstream of the Myf5 gene reveals multiple phases of myogenesis, orchestrated at each site by a combination of elements dispersed throughout the locus. Development 130, 3415–3426.

Hare, E.E., Peterson, B.K., Eisen, M.B., 2008a. A careful look at binding site reorganization in the even-skipped enhancers of Drosophila and sepsids. PLoS Genet. 4, e1000268.

Hare, E.E., Peterson, B.K., Iyer, V.N., Meier, R., Eisen, M.B., 2008b. Sepsid even-skipped enhancers are functionally conserved in Drosophila despite lack of sequence conservation. PLoS Genet. 4, e1000106.

Ho, M.C., Schiller, B.J., Akbari, O.S., Bae, E., Drewell, R.A., 2011. Disruption of the abdominal-B promoter tethering element results in a loss of long-range enhancer-directed Hox gene expression in Drosophila. PLoS One 6, e16283.

Istrail, S., Davidson, E.H., 2005. Logic functions of the genomic cis-regulatory code. Proc. Natl. Acad. Sci. U.S.A. 102, 4954–4959.

Katzman, S., Kern, A.D., Bejerano, G., Fewell, G., Fulton, L., Wilson, R.K., Salama, S.R., Haussler, D., 2007. Human genome ultraconserved elements are ultraselected. Science 317, 915.

Kirchhamer, C.V., Davidson, E.H., 1996. Spatial and temporal information processing in the sea urchin embryo: modular and intramodular organization of the CyIIIa gene cis-regulatory system. Development 122, 333–348.

Krivega, I., Dean, A., 2012. Enhancer and promoter interactions-long distance calls. Curr. Opin. Genet. Dev. 22, 79–85.

Kulkarni, M.M., Arnosti, D.N., 2003. Information display by transcriptional enhancers. Development 130, 6569–6575.

Kwasnieski, J.C., Mogno, I., Myers, C.A., Corbo, J.C., Cohen, B.A., 2012. Complex effects of nucleotide variants in a mammalian *cis*-regulatory element. Proc. Natl. Acad. Sci. U.S.A. 109, 19498–19503.

Laskey, R.A., Gurdon, J.B., 1970. Genetic content of adult somatic cells tested by nuclear transplantation from cultured cells. Nature 228, 1332–1334.

Laubichler, M.D., Davidson, E.H., 2008. Boveri's long experiment: sea urchin merogones and the establishment of the role of nuclear chromosomes in development. Dev. Biol. 314, 1–11.

Liberman, L.M., Stathopoulos, A., 2009. Design flexibility in *cis*-regulatory control of gene expression: synthetic and comparative evidence. Dev. Biol. 327, 578–589.

Lin, S., Riggs, A.D., 1975. The general affinity of lac repressor for *E. coli* DNA: implications for gene regulation in procaryotes and eucaryotes. Cell 4, 107–111.

Longabaugh, W.J., Davidson, E.H., Bolouri, H., 2005. Computational representation of developmental genetic regulatory networks. Dev. Biol. 283, 1–16.

Longabaugh, W.J., Davidson, E.H., Bolouri, H., 2009. Visualization, documentation, analysis, and communication of large-scale gene regulatory networks. Biochim. Biophys. Acta 1789, 363–374.

Ma, J., 2011. Transcriptional activators and activation mechanisms. Protein Cell 2, 879–888.

Marsman, J., Horsfield, J.A., 2012. Long distance relationships: enhancer-promoter communication and dynamic gene transcription. Biochim. Biophys. Acta 1819, 1217–1227.

Materna, S.C., Nam, J., Davidson, E.H., 2010. High accuracy, high-resolution prevalence measurement for the majority of locally expressed regulatory genes in early sea urchin development. Gene Expr. Patterns 10, 177–184.

Melnikov, A., Murugan, A., Zhang, X., Tesileanu, T., Wang, L., Rogov, P., Feizi, S., Gnirke, A., Callan, C.G., Kinney, J.B., Kellis, M., Lander, E.S., Mikkelsen, T.S., 2012. Systematic dissection and optimization of inducible enhancers in human cells using a massively parallel reporter assay. Nat. Biotechnol. 30, 271–277.

Miller, O.L., Bakken, A.H., 1972. Morphological studies of transcription. Acta Endocrinol. Suppl. (Copenh) 168, 155–177.

Mirsky, A.E., 1951. Some chemical aspects of the cell nucleus. In: Dunn, L.C. (Ed.), Genetics of the 20th Century. Macmillan, New York.

Mirsky, A.E., 1953. The chemistry of heredity. Sci. Am. 188, 47–57.

Mishiro, T., Ishihara, K., Hino, S., Tsutsumi, S., Aburatani, H., Shirahige, K., Kinoshita, Y., Nakao, M., 2009. Architectural roles of multiple chromatin insulators at the human apolipoprotein gene cluster. EMBO J. 28, 1234–1245.

Morgan, T.H., 1934. Embryology and Genetics. Columbia Univ. Press, New York.

Nasiadka, A., Dietrich, B.H., Krause, H.M., 2002. Advances in Developmental Biology and Biochemistry. Elsevier.

Nolis, I.K., McKay, D.J., Mantouvalou, E., Lomvardas, S., Merika, M., Thanos, D., 2009. Transcription factors mediate long-range enhancer-promoter interactions. Proc. Natl. Acad. Sci. U.S.A. 106, 20222–20227.

Oliveri, P., Tu, Q., Davidson, E.H., 2008. Global regulatory logic for specification of an embryonic cell lineage. Proc. Natl. Acad. Sci. U.S.A. 105, 5955–5962.

Ong, C.T., Corces, V.G., 2011. Enhancer function: new insights into the regulation of tissue-specific gene expression. Nat. Rev. Genet. 12, 283–293.

Patwardhan, R.P., Hiatt, J.B., Witten, D.M., Kim, M.J., Smith, R.P., May, D., Lee, C., Andrie, J.M., Lee, S.I., Cooper, G.M., Ahituv, N., Pennacchio, L.A., Shendure, J., 2012. Massively parallel functional dissection of mammalian enhancers in vivo. Nat. Biotechnol. 30, 265–270.

Peng, G.H., Chen, S., 2011. Active opsin loci adopt intrachromosomal loops that depend on the photoreceptor transcription factor network. Proc. Natl. Acad. Sci. U.S.A. 108, 17821–17826.

Peter, I.S., Davidson, E.H., 2011. A gene regulatory network controlling the embryonic specification of endoderm. Nature 474, 635–639.

Peter, I.S., Davidson, E.H., 2013. Transcriptional network logic: the systems biology of development. In: Walhout, A.J.M., Vidal, M., Dekker, J. (Eds.), Handbook of Systems Biology. Concepts and Insights. Academic Press/Elsevier, pp. 211–228.

Peter, I.S., Faure, E., Davidson, E.H., 2012. Feature Article: predictive computation of genomic logic processing functions in embryonic development. Proc. Natl. Acad. Sci. U.S.A. 109, 16434–16442.

Phillips, R., Kondev, J., Theriot, J., 2009. Physical Biology of the Cell. Garland Science, New York.

Poorey, K., Viswanathan, R., Carver, M.N., Karpova, T.S., Cirimotich, S.M., McNally, J.G., Bekiranov, S., Auble, D.T., 2013. Measuring chromatin interaction dynamics on the second time scale at single-copy genes. Science 342, 369–372.

Ren, X., Siegel, R., Kim, U., Roeder, R.G., 2011. Direct interactions of OCA-B and TFII-I regulate immunoglobulin heavy-chain gene transcription by facilitating enhancer-promoter communication. Mol. Cell. 42, 342–355.

Robasky, K., Bulyk, M.L., 2011. UniPROBE, update 2011: expanded content and search tools in the online database of protein-binding microarray data on protein-DNA interactions. Nucleic Acids Res. 39, D124–D128.

Royo, J.L., Maeso, I., Irimia, M., Gao, F., Peter, I.S., Lopes, C.S., D'Aniello, S., Casares, F., Davidson, E.H., Garcia-Fernandez, J., Gomez-Skarmeta, J.L., 2011. Transphyletic conservation of developmental regulatory state in animal evolution. Proc. Natl. Acad. Sci. U.S.A. 108, 14186–14191.

Schwab, K.R., Patel, S.R., Dressler, G.R., 2011. Role of PTIP in class switch recombination and long-range chromatin interactions at the immunoglobulin heavy chain locus. Mol. Cell. Biol. 31, 1503–1511.

Siepel, A., Bejerano, G., Pedersen, J.S., Hinrichs, A.S., Hou, M., Rosenbloom, K., Clawson, H., Spieth, J., Hillier, L.W., Richards, S., Weinstock, G.M., Wilson, R.K., Gibbs, R.A., Kent, W.J., Miller, W., Haussler, D., 2005. Evolutionarily conserved elements in vertebrate, insect, worm, and yeast genomes. Genome Res. 15, 1034–1050.

Small, S., Blair, A., Levine, M., 1996. Regulation of two pair-rule stripes by a single enhancer in the *Drosophila* embryo. Dev. Biol. 175, 314–324.

Sonneborn, T.M., 1950. The cytoplasm in heredity. Hered. (Edinb) 4, 11–36.

Spitz, F., Furlong, E.E., 2012. Transcription factors: from enhancer binding to developmental control. Nat. Rev. Genet. 13, 613–626.

Stanojevic, D., Small, S., Levine, M., 1991. Regulation of a segmentation stripe by overlapping activators and repressors in the *Drosophila* embryo. Science 254, 1385–1387.

Stedman, E., Stedman, E., 1950. Cell specificity of histones. Nature 166, 780–781.

Stormo, G.D., Zhao, Y., 2010. Determining the specificity of protein-DNA interactions. Nat. Rev. Genet. 11, 751–760.

Swanson, C.I., Evans, N.C., Barolo, S., 2010. Structural rules and complex regulatory circuitry constrain expression of a Notch- and EGFR-regulated eye enhancer. Dev. Cell. 18, 359–370.

Thanos, D., Maniatis, T., 1995. Virus induction of human IFN beta gene expression requires the assembly of an enhanceosome. Cell 83, 1091–1100.

Tu, Q., Cameron, R.A., Davidson, E.H., 2014. Quantitative developmental transcriptomes of the sea urchin *Strongylocentrotus purpuratus*. Dev. Biol. 385, 160–167.

Vakoc, C.R., Letting, D.L., Gheldof, N., Sawado, T., Bender, M.A., Groudine, M., Weiss, M.J., Dekker, J., Blobel, G.A., 2005. Proximity among distant regulatory elements at the beta-globin locus requires GATA-1 and FOG-1. Mol. Cell. 17, 453–462.

Von Hippel, P., Revzin, A., Wang, A., 1974. Non-specific DNA binding of genome regulating proteins as a biological control mechanism. Proc. Nat. Acad. Sci. U.S.A. 71, 4808–4812.

Wilson, E.B., 1896. The Cell in Development and Heredity, first ed. Macmillan, New York.

Wilson, E.B., 1925. The Cell in Development and Heredity, third ed. Macmillan, New York.

Woolfe, A., Goodson, M., Goode, D.K., Snell, P., McEwen, G.K., Vavouri, T., Smith, S.F., North, P., Callaway, H., Kelly, K., Walter, K., Abnizova, I., Gilks, W., Edwards, Y.J., Cooke, J.E., Elgar, G., 2005. Highly conserved non-coding sequences are associated with vertebrate development. PLoS Biol. 3, e7.

Yang, J., Corces, V.G., 2012. Insulators, long-range interactions, and genome function. Curr. Opin. Genet. Dev. 22, 86–92.

Yuh, C.H., Bolouri, H., Davidson, E.H., 1998. Genomic *cis*-regulatory logic: experimental and computational analysis of a sea urchin gene. Science 279, 1896–1902.

Yuh, C.H., Bolouri, H., Davidson, E.H., 2001. *Cis*-regulatory logic in the endo16 gene: switching from a specification to a differentiation mode of control. Development 128, 617–629.

Yun, K., So, J.S., Jash, A., Im, S.H., 2009. Lymphoid enhancer binding factor 1 regulates transcription through gene looping. J. Immunol. 183, 5129–5137.

Zhao, Y., Stormo, G.D., 2011. Quantitative analysis demonstrates most transcription factors require only simple models of specificity. Nat. Biotechnol. 29, 480–483.

Genomic Strategies for Embryonic Development

The embryos of Bilateria display astoundingly diverse morphologies. They differ not only in appearance but in apparent developmental strategies, so different that for a century, and even in recent texts, it conventionally went without saying that the developmental process had to be presented separately for each species considered. Yet, since it is clear that the Bilateria descend from a common ancestor, we know intuitively that there has to be something fundamentally wrong with this picture. Should we not just focus on the basic developmental mechanisms that all bilaterian embryos utilize? But, on the other hand, it is inescapable that some real and significant differences exist among modes of embryogenesis: to take an extreme case, it is by now (at last) generally realized that the particular mechanisms of spatial specification

Genomic Control Process
http://dx.doi.org/10.1016/B978-0-12-404729-7.00003-4

in the syncytial *Drosophila* embryo cannot be taken as a model for how spatially defined gene expression is determined in sea urchin or mouse or frog embryos. To achieve a global view of the control systems for bilaterian embryogenesis we must begin by recognizing what are the fundamental, universally shared elements of mechanism for development that were inherited from the common bilaterian ancestor, and that can be observed in all branches of its modern descendants. But we also need to understand those differences in control mechanism among embryonic processes that are real and profound, rather than superficial and illusory. Thus we may enjoy a rich comparison of the genomic regulatory programs that account for the special features of the various ways in which animals develop, and that show us what is essential and general, while also illuminating mechanisms underlying the endless variety of bilaterian life.

1. Common Principles of Embryonic Development

Early in the embryonic development of all bilaterians, specific regulatory states, sets of expressed transcription factors, are installed in the appropriate spatial domains of the multicellular embryo. Generally, this is the period in which zygotic gene expression is initiated in response to maternal inputs, and where spatial domains of regulatory gene expression are first formed all over the embryo with respect to the axial coordinates of the future body plan.

1.1 Specification in embryogenesis

The informational requirement is the same in all bilaterian embryos: what are the mechanisms which initially specify the diverse territories of the embryo, arranged according to bilateral axes of symmetry? Here "territory" means (with rare exceptions of secondary simplification) a multicellular domain, the cells of which all express a given regulatory state, and this regulatory state mediates their descendants' fate. From such specified territories given parts of the embryo will uniquely arise. "Specification" is a typically fuzzy term from the older days of embryology, but for us it has a sharp and mechanistic definition. The initiation of specification means neither more nor less than the initial acquisition of a particular transcriptionally controlled regulatory state, i.e., execution of a unique program of regulatory gene expression. Thus the fundamental project of starting an embryo off is installing the correct territorial specifications. Or more precisely, it is the project of activating defined sets of regulatory genes in particular sets of cells located in particular spatial domains of the embryo. A prerequisite is of course generation of multiple embryonic cells ("blastomeres") which are formed by division of the pre-formed mass of egg cytoplasm by cell membranes as the nuclei divide ("cleavage"). In all but amniote eggs (which absorb nutrients from their environment) there is no net growth during this period, and the number of ribosomes for example is the same in a postgastrular embryo that consists of hundreds or thousands of cells as in the newly fertilized egg of the same species (Davidson, 1986). Certain mechanisms of embryogenesis are utilized by all Bilateria, and these mechanisms can be regarded as universal principles of early bilaterian development (this is not to say that some such mechanisms are not used outside the Bilateria as well). In the following sections we consider these common Bilaterian regulatory strategies from the standpoint of the underlying genomic control mechanisms, which constitute a legacy from the distant Precambrian ancestor.

1.2 Properties of the egg

Animal eggs have four major functional properties necessary for embryonic development. In order of their historical discovery, these are: first, their genetic function, that they convey a pronucleus containing a complete haploid genome to the future zygote (i.e., the egg following fertilization and pronuclear fusion); second, their logistic function, that in addition to the genome they carry an immense store of the molecular requirements for life to be utilized by the embryo; third, their activation function, that they biochemically respond to fertilization or the events immediately preceding fertilization by dramatically revving up protein synthesis and metabolism; and fourth, their spatial regulatory function, that they provide asymmetrically localized regulatory molecules which are directly utilized by the early embryo in its axial specification processes. Their genetic function was discovered in the late 1870s and early 1880s by careful observation of meiosis, fertilization, pronuclear fusion, and mitosis. Their logistic function was suspected from measurements of their huge RNA and protein content in the 1930s, 1940s, and 1950s (Brachet, 1933), and proved by the discovery of maternal mRNA and its utilization for embryonic protein synthesis in the 1960s (Brachet et al., 1963; Monroy and Tyler, 1963; Denny and Tyler, 1964; Gross and Cousineau, 1964). Their activation function was first indicated by observations on oxidative metabolism following fertilization in the premolecular biology era, and the biochemical sequence of events leading to activation of protein synthesis was uncovered in the decades after 1970. Their spatial regulatory function, though long suspected (cf. the 1896 quote from E.B. Wilson in Chapter 2), was correctly predicted to indicate cytoplasmic sequestration of specific gene regulatory factors in eggs decades ago (Davidson, 1968; Davidson and Britten, 1971). But only recently, as we discuss in the next section, have abundant examples been authenticated by modern molecular biology.

Here we are focused on the unusual features of genomic regulatory control in oogenesis that, after fertilization has taken place, enable the egg to execute these three major classes of function besides genetic transmission. During oogenesis the oocyte must accumulate the large and complex storehouse of RNAs and proteins needed for its logistic function, as well as generate the latent cytoskeletal and enzymatic machinery needed for its fertilization response function. And somehow it must acquire spatial polarity, and with respect to the future embryonic axes localize cytoplasmic molecules that directly or indirectly will serve to differentially activate genes in those blastomere nuclei inheriting these special cytoplasmic domains.

The oocyte genomes are always active in the accumulation of macromolecular maternal products (more of which below), but they never work alone. For example, yolk protein, the major source of amino acids used for protein synthesis during embryogenesis, which is also the main protein constituent of the mature oocyte, is always made by differentiated cells elsewhere and is the consequence of their own transcriptional control programs. Yolk gene expression is usually hormonally responsive. The yolk protein enters the oocyte by endocytic processes. The extreme example of oogenetic reliance on non-oocyte genomes is what is termed "meroistic" oogenesis, in which during growth the oocyte is syncytially connected through open canals to multiple "nurse cells" or "trophic cells" which are also of germ line origin (for review see Davidson, 1986). The nuclei of these cells generate most of the mRNAs and ribosomal RNAs the oocyte contains. Meroistic oogenesis is found in many insect orders including beetles, bugs, bees, butterflies, and flies, and also is known in some annelids, urochordates, and elsewhere; it has evidently arisen multiple times. For an idea of the logistic power of this transcriptional strategy, in *Drosophila*, each nucleus of the 15 nurse cells contains roughly a thousand copies of every transcribed gene, the products of which are all pumped into the oocyte. By contrast, oocytes of species lacking nurse cells make all of their own RNA. Oocyte growth takes place during a prolonged meiotic prophase, and so they contain four copies of their genomes, all transcriptionally active. It is clear that "logistic" is the right word for oocyte transcription: the rates of transcription are quantitatively just what is needed to accumulate the numbers of molecules required to fill up the oocyte in the time allowed for oogenesis. Large eggs, such as those of amphibians, sharks and cephalopods, which lack nurse cells, produce a high multiplicity of each stored maternal mRNA and their maximally loaded transcriptional complexes constitute lampbrush chromosome structures (see Fig. 2.4(A)). In small oocytes, such as those of sea urchins and mammals, transcription of maternal products may occur at quite leisurely rates and lampbrush chromosomes are not found (for review and quantitative interpretation, see Davidson, 1986).

The transcriptional control problem during oogenesis is different than anywhere in the embryonic development of a multicellular animal. Each oocyte is only one cell and the problem of spatially differential gene expression, as in the future embryo, does not arise. During the growth period of oogenesis some gene products continuously accumulate, while others only appear at certain stages; and for the logistic reasons above genes may run at high or low rates; but once the *cis*-regulatory module accounting for the expression of a gene in the oocyte is activated, there is no further transcriptional choice of cell type to be made. In this light, a particularly interesting result has been obtained in *Caenorhabditis elegans*, where the gene products accumulating at the sequential stages of oogenesis have been closely monitored and the regulatory cause analyzed. While the transcriptional regulatory apparatus operates continuously throughout oogenesis, different maternal transcripts accumulate specifically at different stages. The stage specificity of the transcript accumulation program turns out to be determined posttranscriptionally, by means of the 3′ trailer sequences of the mRNAs (Merritt et al., 2008). In direct contrast, once development begins, in *C. elegans* as in every other bilaterian embryo, spatially differential control of blastomere gene expression occurs at the transcriptional level (Murray et al., 2012).

Among the most important functional attributes of oogenesis is asymmetric distribution of regulatory materials within the oocyte that will affect the spatial organization of the embryo, and that is often used to determine at least one of the major embryonic axes. While maternal anisotropies of regulatory significance occur very generally in bilaterians, the means by which they are installed vary greatly. In oocytes that develop solo, lacking either nurse cells or follicle cells, transcriptional control cannot be responsible for localizing the intracellular spatial functions that determine the axial polarity of the egg, where this is fixed before fertilization. In these cases, another mechanism is used to cause the deposition of factors of later regulatory significance in given positions in the cortex of the mature oocyte. Such oocytes, for example sea urchin or sea star oocytes, probably use their intrinsic apical/basal polarity as epithelial germ cells in the ovary to develop a polarized cytoskeleton during oogenesis. This polarity is ultimately reflected in the animal/vegetal (A/V) polarity of the early embryo and the distribution of regulatory molecules therein. Oocytes of most species, however, are associated with follicular accessory cells. Here, the example of *Drosophila* shows how transcriptionally regulated intercellular interactions provide complex possibilities for generating molecular pre-pattern in the mature oocyte, which will later be of regulatory significance.

In *Drosophila*, both axes of the future embryo are determined during oogenesis, which is atypical for animal eggs in general. The positioning of the oocyte at the posterior end of the oocyte–nurse cell complex, and the resulting location of the specialized canals from nurse cell into oocyte mark the anterior end, and lead to internal cytoarchitectural reorganization. The oocyte nucleus later migrates to what will be the dorsal-anterior position of the egg. Much of the subsequent patterning depends on regional transcriptional spatial specification in the sheath of follicle cells surrounding the oocyte. A fascinating and unique feature is the influence of the oocyte on this: the dorsal anterior oocyte nucleus locally generates a Tgf-α ligand which affects the specification of the adjacent dorsal anterior follicle cells. These cells signal to their more ventral neighbors and the ultimate result is differential follicular expression of an extracellular matrix protein on the opposite (ventral) 40% of the egg chamber. At the beginning of development this protein is utilized to trigger a cascade of enzymatic events ultimately resulting in the ventral-only activation of Toll receptor in the egg membrane. The eventual consequence is to establish a ventral to dorsal nuclearization gradient of the Dorsal transcription factor within the egg, the fundamental outcome of which we see below (for detailed review of these and other signaling interactions in the patterning of the *Drosophila* oocyte, see López-Schier and St Johnston, 2001; Roth, 2003; Huynh and St Johnston, 2004; Roth and Lynch, 2009; Cheung et al., 2011).

Thus, as we see here, where spatial oocyte patterning depends on intercellular interactions with accessory cells, it is controlled by transcriptional regulatory processes just as in embryogenesis or elsewhere in development. But stockpiling of transcripts within the oocyte is not a spatial control function, and at least in the well-studied *C. elegans* case (Merritt et al., 2008), the temporal pattern of transcript accumulation is regulated posttranscriptionally.

1.3 Regulatory anisotropy in eggs/very early cleavage embryos and the initiation of spatial specification

The initial inputs used to organize gene expression spatially in the early embryo are always at least in part localized maternal molecules that determine regional zygotic regulatory gene expression. The universal property of embryological development is that it operates by the principles of an automaton-like logic processing system. Regulatory gene expression can generate such automaton-like behavior, by causing the transcription of other regulatory genes which then can do likewise by means of their networks of interactions. What makes the embryo's spatial control system get up and go, and then keep on going, is the differential transcription of new regulatory genes in space. This elemental point divides the immensely complex phenomenon of localization of maternal RNAs in eggs into two categories: those localizations that causally affect blastomere regulatory gene transcription, and all the rest. For example, a great variety of maternal mRNA species have been identified in early cleavage molluscan embryos which are dramatically localized to given sets of blastomeres, but what they encode are centrosomal proteins, RNA-binding proteins, metabolic proteins, cytoskeletal proteins of many kinds (Kingsley et al., 2007; Henry et al., 2010; Rabinowitz and Lambert, 2010), plus, a small minority of transcription factors. The individual cell biological properties of given early blastomeres might indeed depend on these various localized cytoplasmic mRNAs. But only those molecular species that result in differential activation of zygotic regulatory genes can operate as components of the developmental automaton. These species include localized mRNAs encoding transcription factors, localized already translated transcription factors, localized necessary cofactors of transcription factors and localized signaling ligands that when received by other blastomeres will alter patterns of regulatory gene expression. So in what follows we focus only on asymmetrically sequestered maternal regulatory molecules inherited regionally in the embryo, which clearly act to cause expression of specific regulatory genes uniquely in those regions. Regional regulatory state initiation by this mechanism is a universal principle of bilaterian embryogenesis, except in mammals as we see below (Davidson, 1990, 2006). Reliance on localized maternal inputs, however, can only yield relatively simple, initial patterns of regionally distinct regulatory state, but that is all it takes to begin the definition of the axes of the overall embryonic transcriptional system.

What has been traditionally referred to as axial determination (the process by which the primordial orthogonal polarities in the embryo are defined) happens at different times in different ways in different animals. As we have seen, in *Drosophila* both anterior/posterior (A/P) and dorsal/ventral (D/V) axes are set during oogenesis, essentially by intercellular interactions with accessory cells. In *C. elegans*, both axes are set only after fertilization, the A/P axis a result of cleavage asymmetry and the D/V axis with respect to the point of sperm entry. In sea urchins, the A/V axis is set during oogenesis while the future second axis of the embryo, the oral/aboral axis, forms after fertilization by an asymmetric distribution of mitochondria resulting in a redox gradient. In *Xenopus*, the A/V axis is formed during oogenesis and the A/P axis is oriented after fertilization due to a cortical rotation that is oriented with respect to the point of sperm entry, and is required for specification of the future posterior end (for references and review, see Davidson, 2006). But the concept of "axial determination" per se has a metaphysical flavor, as if there were orthogonal magnetic fields established throughout the fertilized egg. Less mysteriously, what happens is that whenever the polar orientations are established, in oogenesis or following fertilization, one or both of the poles of each future embryonic axis becomes the site of differentially localized or activated maternal molecules that directly or indirectly control zygotic transcription. A number of examples of anisotropic localization of maternal components that will affect spatial zygotic regulatory gene expression are shown in Fig. 3.1. In all these examples the result is to define in regulatory terms polar embryonic territories, thereby beginning the functional zygotic development of the future embryonic axes.

In the eggs of ascidians both embryonic axes are physically specified by localization of molecules of regulatory significance by the time first cleavage occurs, due to complex cytoskeletal rearrangements. In this embryo, it is possible to follow an unbroken chain of evidence from cytoplasmic localization of

Figure 3.1

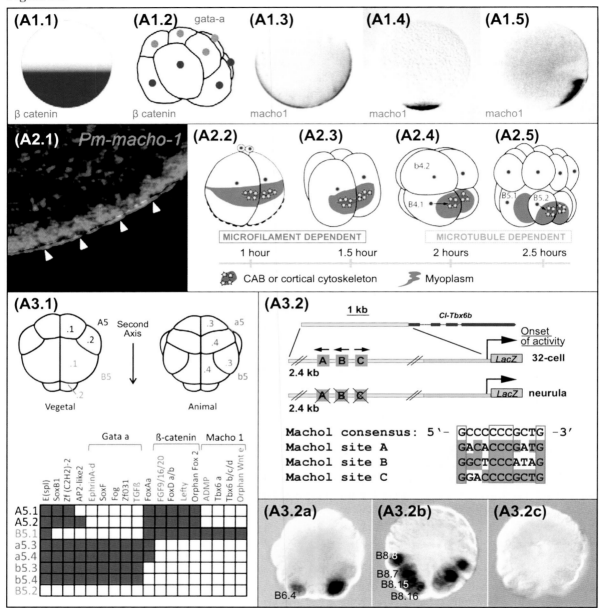

Figure 3.1 Anisotropic localization in bilaterian eggs and early embryos of maternal regulatory factors causing differential zygotic transcription. (A) *macho1* mRNA and other maternal regulatory factors in ascidian eggs and embryos. (A1.1 and A1.2), Diagrams showing animal versus vegetal localizations *(from Lemaire (2009))*. (A1.1), Maternal β-catenin in vegetal half of uncleaved egg. (A1.2), Resulting distribution of *gata-a* mRNA (red dots) in animal blastomeres and β-catenin (blue dots) in vegetal blastomeres of 16-cell embryo; though *gata-a* mRNA is initially ubiquitous, β-catenin:Tcf restricts vegetal transcription of the *gata-a* gene, lateral view. (A1.3–A1.5), Maternal *macho1* mRNA, visualized in *Halocynthia* eggs by in situ hybridization *(from Nishida (2002))*. (A1.3), Unfertilized egg; (A1.4), first stage of localization following fertilization; (A1.5), second

Figure 3.1

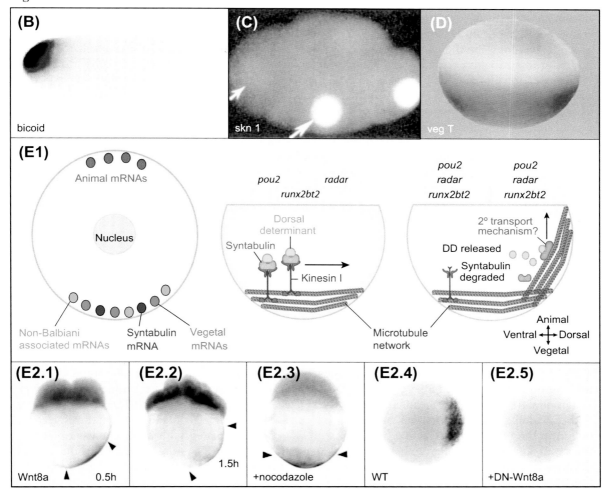

stage of localization to future posterior region of embryo, all prior to first cleavage. (A2), Distribution of *macho1* mRNA to early posterior blastomeres. (A2.1), High-magnification demonstration by fluorescent in situ hybridization of cortically localized *macho1* mRNA in unfertilized *Phallusia* egg (green, white arrows), present in a thin layer under the egg membrane and outside the mitochondria (purple immunostain) enriched in the vegetal cytoplasm *(from Paix et al. (2009))*. (A2.2–A2.5), first, second, third, and fourth cleavages respectively, drawings from lateral-posterior view, displaying association of localized posterior maternal mRNAs including *macho1* mRNA (yellow stars), associated initially with cortical cytoskeletal elements (red, A2.2 and A2.3) and from eight-cell stage with a compacted organelle, the centromere attracting body (CAB; red, A2.4 and A2.5) which lies within the mitochondria-rich "myoplasm" (green). The posterior mRNAs are partitioned between the first two blastomeres, and the discrete association of *macho1* mRNA with the CAB then ensures its distribution to the right and left B4.1 muscle founder cells and their B5.2 descendants *(from Sardet et al. (2007))*. (A3), Spatially differential transcriptional response to the initial, anisotropic localizations of regulatory factors in mid-cleavage *Ciona* embryos. (A3.1), Establishment of regulatory states in blastomeres of the 16-cell embryo *(from Lemaire (2009))*. At the top the embryo is shown diagrammatically from the vegetal (left) and animal (right) aspects, with blastomeres labeled on one side: anterior vegetal blastomeres (A lineage; black labels); posterior vegetal blastomeres

(B lineage; orange labels); anterior animal blastomeres (a lineage; red labels); posterior animal blastomeres (b lineage; green labels). Eleven regulatory genes known to be expressed at the 16-cell stage (Satou et al., 2009) are shown in violet, together with some signaling ligand genes shown in green listed along the top of the expression matrix. The blastomeres labeled above are listed on the matrix ordinate, color coded as in the embryo diagrams, and the observed Boolean expression pattern is indicated in color (red cells denote expression). Most of these genes fall into spatial sets by virtue of receiving transcriptional inputs from the three localized regulators considered in (A1) and (A2), as indicated. These expression patterns define six regulatory states (all but the epidermal b5.4 giving rise to more than one later tissue type). For example, the B5.1 blastomere gives rise to endoderm, mesenchyme, and tail muscle, the latter due to the Macho1 input. (A3.2), Cis-regulatory function of Macho1 in fifth and seventh cleavages B lineage muscle cell progenitors *(from Kugler et al. (2010))*. Key early zygotic regulatory genes of the muscle GRN are *tbx6b* and *tbx6c*, of which a map of the first is shown here at top, blue rectangles denoting exons. Cis-regulatory constructs containing wild type or mutated Macho1 target sites (sequences shown below) were electroporated into eggs. In panel A expression of the wild type construct is shown in the B6.4 cells of the fifth cleavage embryo and in panel B the wild type construct can be seen expressing in multiple B8 cells, which are solely destined to produce tail muscle. But if as in panel C the construct lacks Macho1 sites, no expression is obtained in an embryo of the same stage as in panel B. (B), Localization of maternal *bicoid* mRNA at anterior end of *Drosophila* egg *(from Ochoa-Espinosa et al. (2005))*. The mRNA is the product of the 15 polyploid nurse cells and during oogenesis is pumped into the oocyte through ring canals at the anterior end, where it is attached to the cytoskeleton until translation after fertilization. (C), Maternal Skn1 protein in EMS and P2 cell nuclei of 4-cell *C. elegans* embryo *(from Bowerman et al. (1993))*. Initially present in all four blastomeres, the protein is post translationally localized. (D), Localization of maternal *vegt* mRNA to vegetal portion of fertilized *Xenopus* egg *(from Zhang and King (1996))*. Localization occurs during oogenesis and occurs by transit along microtubules driven by kinesin motors (King et al., 2005). (E), Dorsal determinants of maternal origin in zebrafish. (E1), Diagrammatic summary of localization processes *(from Langdon and Mullins (2011))*. From left, later oocyte, 1-cell and 2-cell embryos; animal and vegetal mRNAs have segregated. Vegetal mRNAs are bound in the oocyte cortex, some having previously been associated with a perinuclear organelle, the "Balbiani body". In the embryos several regulatory gene mRNAs are indicated in the animal blastomeres, and maternal syntabulin, a kinesin motor protein linker is translated and facilitates the transport of dorsal determinants (yellow balls, DD) to the future dorsal side (right in 2-cell embryo). (E2), Wnt8 function as the dorsal determinant in zebrafish *(from Lu et al. (2011))*. During oogenesis maternal *wnt8a* mRNA is localized at the vegetal pole of the oocyte. (E2.1,2), In early embryos *wnt8a* mRNAs are initially localized on one side of the yolk cell and are then transported to an animal position. (E2.3), This transport is microtubule dependent, since it fails in nocodazole treatment. (E2.4), A gene expressed downstream of Wnt8a signaling, *chordin*, is normally expressed on the dorsal side. (E2.5), Introduction of a dominant negative Wnt8a abolishes this expression. Other experiments show that Wnt8a is confined to the dorsal side by Wnt antagonists, and that both Wnt8a gain and loss of function produce the phenotypic effects expected if Wnt8a is a dorsal determinant.

specific maternal regulatory molecules to their *cis*-regulatory function in particular cell lineages of the early embryo, as shown in Fig. 3.1(A). Thus, in Fig. 3.1(A1.1) and (A1.2) can be seen the vegetal localization of maternal β-catenin just before first cleavage, and at the eight-cell stage the resulting exclusive localization of *gata-a* mRNA in the animal blastomeres, where it contributes to activation of the initial animal and neural regulatory genes; β-catenin:Tcf, which contributes to endomesodermal fates, also restricts *gata-a* activity to the animal blastomeres (Rothbächer et al., 2007). Fig. 3.1(A1.3)–(A1.5) show the progressive localization of maternal *macho1* mRNA toward the future posterior pole where it will serve as a regulatory determinant of muscle fate. This relocalization is driven by cortical microfilaments and sperm aster microtubules (Sardet et al., 2007; Paix et al., 2009). Initially, before fertilization, *macho1* mRNA is associated specifically with the egg cortex (Fig. 3.1(A2.1)). *Macho1* and other maternal mRNAs are then associated with an organelle (CAB) and distributed in an all or nothing way to certain blastomeres during early cleavage (Fig. 3.1(A2.2)–(A2.5)). Fig. 3.1(A3.1) shows how these three maternal localized regulatory species, *gata-a* mRNA, β-catenin, and *macho1* mRNA contribute to the combinatorial Boolean gene regulatory states which specify individual embryonic blastomeres (Lemaire, 2009). Fig. 3.1(A3.2) completes the chain of causality with the *cis*-regulatory demonstration that the Macho1 transcription factor is required to activate a zygotic muscle regulatory gene, *tbx6* (Kugler et al., 2010).

Other examples are shown in Fig. 3.1(B)–(D). Respectively, these panels illustrate localization of *bicoid* mRNA in *Drosophila*, *skn1* transcription factor in *C. elegans*, and *vegt* mRNA in *Xenopus*. For each of these particular cases, detailed evidence reviewed later in this chapter demonstrates that the localized maternal regulatory factor participates directly in the installation of regional zygotic regulatory states once development begins. We encounter additional examples in sea urchin embryos, *C. elegans* and *Drosophila* below. Even in the yolky, incompletely cleaving egg of the zebrafish, maternal regulatory factors are anisotropically segregated by a microtubule network at the beginning of cleavage, whence they control early zygotic gene expression, as summarized in Fig. 3.1(E) (Langdon and Mullins, 2011).

So we see that anisotropic localization of gene regulatory factors is indeed in Bilateria a general and apparently universal mechanism of initiating differential zygotic gene expression. The means of localization are different in each case; the biochemical nature of the maternally encoded factors varies also, and the identities of the target genes are unique. But such are all details, compared to the universality of the use of localized regulatory molecules of maternal origin to install in the embryo polar transcriptional regulatory states.

1.4   Signaling, and its causal developmental consequences

The remaining common principles of embryonic development that we treat here are common not only to all manner of bilaterians but also to all stages of development, not just the beginning. They are the subject of great literatures on their own, and in the following we focus only on the framework of the regulatory control logic.

Inductive signaling is a stereospecific interaction between a ligand expressed specifically by cells of one spatial territory and receptors on cells of another spatial territory, and ultimately it causes new regulatory gene transcription in the latter cells. When considered globally, in space and time, inductive signaling is required for the increase in spatial complexity during development, that fundamental property observed so long ago (Fig. 1.1). The general consequence of any directional signaling interaction, whether in the context of a pregastrular embryo or a developing body part, is subdivision of developmental space into more and more specific regulatory state domains. This is because of the basic aspects of inductive signaling: the only cells able to respond to a signal are those located within the range of the signaling ligand which also express a cognate receptor. The signal ligand may be tethered so it acts only at short range, such that the immediately adjacent cells are exclusively affected, differentiating them from the next row of cells. Or it may be allowed to be active only in a given subregion because of dedicated extracellular "anti-signals" expressed by other cells which in one way or another sequester or inactivate it. Or it may be present at different concentrations in spatial domains located at different distances from its source, interpreted differently in the receiving cells by their transcriptional gene regulatory networks (GRNs). The fundamental result in those cells that respond to the signal is always to activate new regulatory gene expression, thus affecting the regulatory state. Consequentially, an originally homogenous field of cells becomes subdivided, such that cells receiving the signal will express one regulatory state and hence developmental fate and cells not receiving the signal will express a different regulatory state and a different developmental fate.

As with so much else in biology, developmental signal transduction biochemistry includes a daunting mass of detailed interactions of large numbers of cytoplasmic components. From the nuclear perspective, however, the different developmental inductive signaling systems most prominent in bilaterian embryogenesis all amount to the same way of doing a certain kind of business. The universal outlines of signaling regulatory logic are as follows. Reception of the extracellular signal by the transmembrane receptor causes an intracellular response cascade which results in the activation of a dedicated transcription factor ("immediate response factor"). This factor binds target gene *cis*-regulatory sites for which it has sequence specificity. Binding of the receptor by the ligand results in activation of the immediate response factor in many different manners. There are not so many different signaling systems, as defined by the signaling ligand families, used commonly in animal development. Most prominent developmental inductive processes in the embryo utilize Notch ligands, Wnt, TGFβ, EGF, FGF, retinoic acid, or Hedgehog (HH) signaling ligands, though we encounter a number of more rarely used signal transduction systems later in this book. However, particularly in mammals, some of the ligand gene families are large and the cognate receptors are also multiple, and in addition, there exist many different extracellular antagonists to some ligands, increasing the potential complexity of spatial signaling interactions. The immediate response factor may be complexed with a fragment cleaved in a signal-dependent manner from the receptor (Notch signaling); it may depend on an activating cofactor which is saved from destruction by signal-dependent interference with a default degradation enzyme (Wnt signaling); its activity may depend on signal-dependent phosphorylation (TGF-β signaling); it may be deployed due to a signal-dependent kinase cascade (FGF, EGF signaling); its activity may be directly regulated by the signaling ligand (retinoic acid signaling); or, it may be proteolytically processed by signal-dependent mechanisms (HH signaling). But the result is the same: a new transcriptional activator appears in the nucleus in signal receiving cells. The signaling effects are encoded in the genomic regulatory sequence, for only those genes of which the *cis*-regulatory modules bear the target sites for the activated response factor are turned on inductively. In many signaling systems these same genes are actively repressed by the early response factor in cells which do not receive the signaling input. This is because in the absence of the activating input, the early response factor instead binds a default repressive cofactor (such as Groucho). Thus, inductive signaling has a Janus-like quality, in that the signal response factor activates genes in the signal-receiving subdomain and actively represses these same genes elsewhere (as first emphasized by Barolo and Posakony, 2002).

An example is shown in Fig. 3.2, where the ligand is Delta, the receptor is Notch, the response factor is Suppressor of Hairless (Su(H)), the target gene is *gcm*, and the cells receiving the signal are presumptive sea urchin embryo mesoderm (Ransick and Davidson, 2006). Delta is expressed in cells immediately adjacent to mesodermal cells, under the control of the specification GRN operating in these cells. Its receptor, Notch, is at this stage expressed on the membrane of all cells, as it is provided maternally. Because the Delta signaling ligand is embedded in the cell membranes, the only cells within reach of this molecule are cells in direct contact to the signaling source. Fig. 3.2(A) shows a single ring of cells responding to the Delta signal, which is emanating from the unlabeled cells in the center of the image. The target of this interaction is the *gcm* gene, the transcripts of which are stained blue in Fig. 3.2(A). In its *cis*-regulatory sequence, *gcm* contains the binding sites for Su(H), the early response factor of Delta/Notch signaling (Fig. 3.2(B)). Reception of the signal indeed regulates spatial expression of *gcm* via Su(H), as mutation of *cis*-regulatory target sites proves (Fig. 3.2 (B) and (C)). Thus, beyond the single cell deep ring receiving the Delta signal, Su(H)-binding site mutations cause ectopic expression because they disrupt the default repression function of the same response factor in the absence of signal. This can be seen in the increased ectopic expression in Fig. 3.2(C) and (D) when either the Su(H) site is mutated or a dominant negative form of Su(H) is used. The Janus-like behavior of these systems makes inductive signaling a global spatial control system in the embryo (Davidson, 2010). It indeed causes specific regulatory gene activation in target cells different than those that emit the signal, but in the absence of the signal it also ensures silence of these same genes wherever the usually widely distributed response factor is present. In the case portrayed in Fig. 3.2 this is everywhere in the embryo.

Figure 3.2

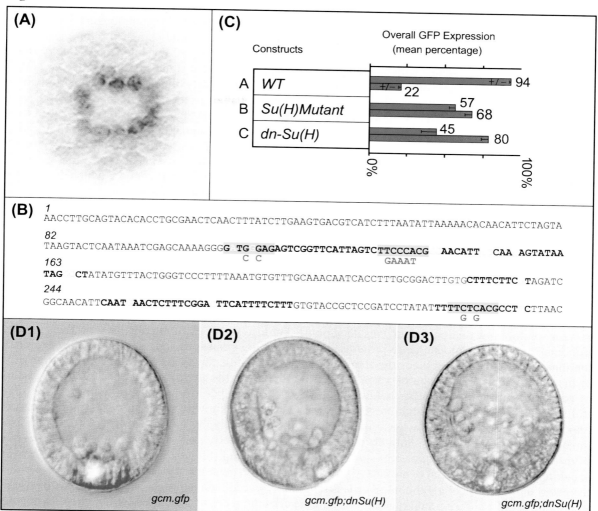

Figure 3.2 "Janus" behavior in inductive signaling. This example is from the sea urchin embryo *(from Ransick and Davidson (2006))*. (A) Ring of veg2 cells expressing *gcm*, as seen by whole mount in situ hybridization. Unlabeled skeletogenic cells in center express the Delta ligand, and responding cells do so by the Notch signaling pathway, for which the immediate response factor is Su(H). (B) *Cis*-regulatory module of *gcm* gene with Su(H) target sites highlighted, and nucleotide changes made to mutate these binding sites (red). Bold indicates sequence conserved between the species of sea urchin used in this experiment, *Strongylocentrotus purpuratus*, and *Lytechinus variegatus* (about 50 my apart), except for gray residues within these regions. (C) *Cis*-regulatory results of mutation of Su(H) sites and of introduction of mRNA encoding a dominant negative form of Su(H) (dn-Su(H)), which traps the intracellular Notch fragment but cannot bind DNA. Results of both are equivalent: less correct expression and greatly more ectopic expression are obtained. Green bars show the percentage of embryos expressing a *gcm:GFP* construct (normal in A,C, mutant in B) correctly in mesodermal precursors only; red bars show percentage with additional ectopic expression. (D) Images of embryos expressing injected *gcm*:GFP constructs. Fluorescent GFP reporter expression is shown in green. (D1), Correct expression in mesodermal cells of blastula stage embryos seen laterally. (D2,3), Ectopic expression in endoderm and ectodermal regions in the presence of dominant negative Su(H). Su(H); suppressor of hairless.

Inductive signaling thus is a powerful means of establishing new spatial regulatory states where they did not exist before. Indeed it is the major mechanism of increase in spatial complexity during development. Indirectly, this was why signaling became an obsessive concern of developmental biology in the twentieth century: interference with signaling or ectopic presentation of signals has cascading and therefore dramatic effects because of the direct alterations in spatial regulatory organization these perturbations generate. Unfortunately however, looking at developmental signaling effects without concern for the regulatory system in which these interactions operate, is simply to obfuscate the mechanism. The focus of large swaths of research in this field thus became the phenomenology of signal effects in place of the causal genomic regulatory code. As we now know, each signaling system is used in multiple developmental contexts, in which it has multiple molecular consequences. The specific consequences of signal reception are thus determined by the particular GRNs operating in each developmental context, which execute the regulatory signal response. Expression of the signal ligand and designation of the signal target genes are both one to one outputs of the encoded genomic control system for development.

1.5 Differentiation

Differentiation gene batteries are sets of more or less coordinately expressed effector genes that work together to produce a given specialized cell phenotype or function, including their immediate regulatory drivers. For example, a muscle gene battery would contain not only the genes encoding the contractile proteins required of that particular muscle subtype, but also those encoding ancillary proteins such as troponins and muscle bioenergetic proteins such as creatine kinase. In addition to the effector genes, the muscle gene battery includes the transcription factors immediately responsible for the expression of these particular downstream genes in the muscle fiber cells (de Joussineau et al., 2012). Expression of particular differentiation gene batteries defines each cell type. The general principles of the control systems directing differentiation gene battery function are known from many examples. The most important is that every effector gene is controlled independently, by its own *cis*-regulatory transcriptional apparatus. Though most unlikely a priori, it has been directly excluded that coordinate expression of differentiation genes requires any sort of genomic clustering (Shoguchi et al., 2011). Therefore the absolutely coordinate spatial expression of the differentiation gene battery (i.e., expression of the effector genes in the same cells), and more or less temporally coordinate expression, is caused by the use of a common set of transcriptional regulators by all the effector genes of the battery. When these regulators are expressed, the effector genes are expressed. The *cis*-regulatory systems of multiple effector genes of given differentiation gene batteries have been studied, and the generality that emerges is that all genes of each battery respond to a subset of the battery driver regulators. From a set of say three cell type-specific drivers, an individual gene of the differentiation gene battery might require inputs from only two to get specific expression. Usually other factors also have inputs to individual genes, accounting for the inexactly coordinated amplitude and dynamics of expression of the various effector genes of each battery (Davidson, 2006). There is always, however, a small, specific set of regulatory inputs that defines each differentiation gene battery (we discuss transcriptional features of differentiation gene batteries in mechanistic detail in Chapter 5).

Fundamental conceptual consequences follow. First, it means that the terminal regulatory transactions in the GRNs that ultimately specify cell type are those that cause expression of the few transcriptional regulatory drivers for each differentiation gene battery. Thereby, the large numbers of effector genes defining each cell type are spatially deployed. This conclusion places the hard wiring that spatially deploys differentiation gene batteries specifically in the control circuitry of the drivers of each differentiation gene battery.

Differentiation, the installation of expression of cell type-specific effector gene batteries, occurs throughout the developmental process, beginning in some modes of embryogenesis even before gastrulation, in others after. It continues into post-embryonic development and throughout life, as stem cells contribute to the working populations of the adult body, and it recurs on a large scale in wound healing and regenerative processes. Differentiation is to be regarded as a modular regulatory package which is on call simply by activation of the definitive gene battery regulatory drivers. The crucial appearance of mechanisms for spatial deployment of differentiation gene battery drivers was among the regulatory innovations required for metazoan life.

1.6 Morphogenetic functions

Embryogenesis, and post-embryonic development as well, require a set of specialized, spatially deployed activities in addition to those implied by deployment of differentiation gene batteries in the strict sense. While only erythrocytes execute the terminal effector programs of the red blood cell branch of hematopoiesis, and only muscle cells express the muscle differentiation gene battery, cells of many kinds undergo programmed epithelial to mesenchyme transitions, or change their surface properties to enable ingression or invagination, or become migratory, or form sheets or tubes. These kinds of functions, loosely denoted as cellular morphogenesis, all involve complex cytoarchitectural changes. Those of which anything is known involve large numbers of protein-coding genes, of many diverse kinds. Thus, as for the effector genes of differentiation gene batteries, there is a dual problem of genomic control. Both the deployment of morphogenetic effector function to the right spatial component of the embryo and coordination of all the necessary protein functions therein must be organized in such a way that they can be called in appropriately by the genomic control program for development. Indications from the presently limited cases where control of a cellular morphogenetic process has been addressed in detail are that the structure of these control systems differs in a basic way from those of differentiation gene batteries.

The essential difference is that many or most of the effector genes involved in cellular morphogenetic systems are widely expressed, so their products are present before the onset of the morphogenetic process, while only some key genes are expressed under the control of the developmental GRN. Thus both aspects of the deployment control problem for morphogenetic effector functions resolve to control of a minority of linchpin effector genes, the products of which nucleate or trigger the process. This concept is explicit in the example shown in Fig. 3.3 (Christiaen et al., 2008). In *Ciona* embryos known GRN subcircuits specify the heart precursor cells, in part by inducing their migration, which exposes them to BMP signaling, resulting in further changes in their regulatory states, followed by cessation of migration and heart cell differentiation (Beh et al., 2007; Christiaen et al., 2009, 2010). The key initial regulator is encoded by the *mesp* gene, which according to *cis*-regulatory studies (Christiaen et al., 2009) is activated in only two cells where its two regulatory drivers overlap (Fig. 3.3(A) and (B)). The relevant portion of this sequence of events is the relation between the developmental genomic control apparatus, and the onset of migration, and the main point of interest here is summarized in Fig. 3.3(C) (Christiaen et al., 2008). The *mesp* regulatory gene is at the head of a feed-forward activation of the regulatory gene *foxf*, which participates in another feed-forward subcircuit to transcriptionally activate expression of the *rhoDF* gene, encoding a Rho GTPase. RhoDF together with a number of already expressed cytoskeletal effector genes (blue in Fig. 3.3(C)), causes dynamic membrane protrusion and motility. At least 130 different effector genes are involved in the cytoarchitectural changes that migration requires, only several of which, including the Rho GTPase, are directly under developmental transcriptional control.

Though it lies beyond the scope of our (genomic) concerns, this type of cellular morphogenesis mechanism implies extensive and complex self-assembly processes at the protein level. The transcriptional regulation of key nucleators for deployment control provides the logic levers by which cellular morphogenesis functions are integrated into the developmental process.

Figure 3.3

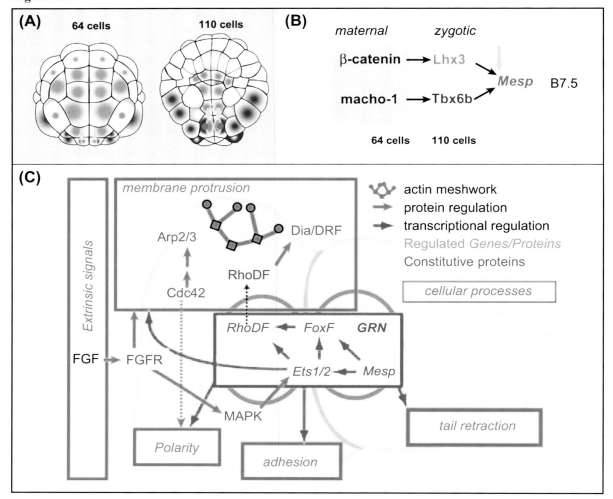

Figure 3.3 Transcriptional control of a morphogenetic function, migration of *Ciona* heart progenitor cells. (A), Diagrammatic ventral views of *Ciona* embryos at indicated stages, showing cells expressing *lhx3* in green and *tbx6b* in blue. The two heart founder cells are marked in red. These are the only cells which express both *tbx6b* and *lhx3* and which activate *mesp*, encoding a bHLH transcription factor required for heart cell migration. (B), Sequence of maternal and zygotic inputs resulting in *mesp* transcription. The *mesp* gene *cis*-regulatory module contains apposed *lhx3* and *tbx6* sites required for accurate expression *(A,B from Christiaen et al. (2009))*. (C) Network of interactions triggered by *mesp* expression, which results in migratory behavior. Transcriptional subcircuit in green box, cell biology functions indicated in red boxes, and the specific activities produced by RhoDF in upper red box labeled "membrane protrusion". Genes in blue are expressed anyway, not under Mesp transcriptional control; RhoDF acts as a nucleating driver of this cellular morphogenetic process *(from Christiaen et al. (2008))*. GRN, gene regulatory networks.

2. Phylogenetic Framework

Though the Bilateria share universal mechanisms used in development, as we have just seen, these cannot explain the differences in regulatory strategy that do distinguish diverse modes of embryogenesis. The remainder of this chapter is devoted to analysis of these differences at the deep level of GRN structure/ function and other aspects of the developmental gene regulatory program. But first, since we here apply a global perspective to the Bilateria, it is useful to pause for a moment to place this discussion in a phylogenetic context. Without reliable phylogeny, no mechanistic comparisons make sense, because only phylogeny can tell us whether a certain way of doing business is inherited from an ancestral mechanism, or has rather been invented anew in evolution. Phylogeny defines the polarity of evolutionary change in developmental mechanisms and provides a necessary organizational framework for the comparison of the diverse modes of embryogenesis that follow in this chapter.

2.1   Bilaterian phylogeny

Linnaean phylogeny, i.e., hierarchical ordering of animal relationships based on the distribution of shared features, dates back to the mid-eighteenth century (Linnaeus, 1758), though some of our current cladistic levels (Phylum, Class, Order, Family, Genus, Species) were inserted in the nineteenth century. The phylogenetic placement of animal groups within this branching hierarchical framework has gone through many phases. Its modern history begins in the late nineteenth and early twentieth centuries with attempts to classify the morphological forms of animal embryogenesis, which reveal many relationships otherwise buried in the evolution of secondarily diversified adult forms. Despite the subjective and often nonsensical quality of some of this work, it had its early successes, such as the astounding intuition on extremely thin morphological evidence that bilaterians could be divided into "deuterostomes" (the name given to embryos in which the initial opening into the interior of the embryo becomes the anus) and "protostomes" (in which the initial opening becomes the mouth) (Grobben, 1908). With the predominance of the logical paradigms of formal cladistics from the 1970s on, morphological animal phylogeny was put on a less subjective footing, but it still suffered fundamentally from the basic and inescapable inadequacy of anatomical character sets, so that the results remained far from what we have at present. It is scarcely surprising in retrospect that whole genomic sequences provide invaluable data for constructing correct phylogenies. What is surprising is how much, in different respects, each generation of molecular methods achieved in the pre-genomic era: applying strict cladistic logic, phylogenies were first based on inter-phylum and interclass comparison of sequences of particular universal gene families, particularly rRNAs, then augmented with comparisons of particular highly conserved protein sequences. Now they are derived from massive statistical computations based on hundreds of genomically predicted proteins ("phylogenomics"). But all these comparisons require extensive computational assessments of the pathways and likelihoods of processes of evolutionary sequence change, and competing algorithms are not always convergent. With the vast current expansion of genome sequencing and annotation, however, a new form of discontinuous phylogenetic metric is providing valuable additional evidence of a more Boolean nature. This kind of evidence concerns unit features, such as presence of introns at certain positions in a protein-coding sequence; presence of specific genes; and presence of given mobile sequence families. The evolutionary events creating each of these features occur very infrequently and thus they are shared only among descendants of a common ancestor which possessed such characters.

Fig. 3.4 shows a simplified phylogenetic tree for Bilateria. It is simplified only in that many animal groups which do not figure in the discussion in this book have been omitted (for the full phylogenies upon which this is based, see Bourlat et al., 2008; Philippe et al., 2011; Rota-Stabelli et al., 2011; Telford and Copley, 2011). The basic principle of modern phylogenies is that the descendants of any given node are

Figure 3.4

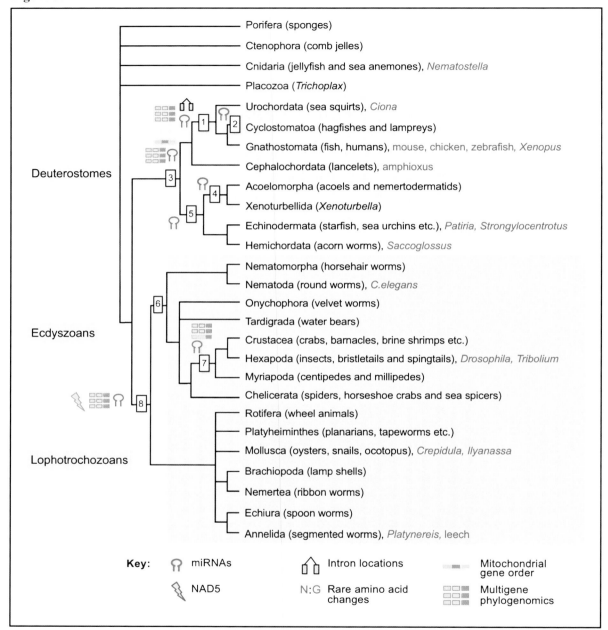

Key:
- miRNAs
- NAD5
- Intron locations
- N:G Rare amino acid changes
- Mitochondrial gene order
- Multigene phylogenomics

Figure 3.4 A simplified modern phylogeny of the Bilateria. The three great bilaterian superclades, are indicated by the background coloring according to labels at left *(figure slightly simplified from Telford and Copley (2011))*. Only some Phyla in each are included, particularly those of which members figure in the discussion in this chapter, the generic or common species names of which are indicated in red. The main mass of evidence upon which the phylogeny is based is phylogenomics. "Phylogenomics" denotes genome-based analyses, using elegant computational methods, to arrange very large datasets of encoded protein sequences from different animals in such a way as to reveal common ancestry and shared derived sequence features (for references, see text). Despite the general power of these methods, however, they leave occasional relationships contentious or unresolved, particularly in respect to clades in which the rate of evolutionary sequence change is atypically high. Thus, as the Figure indicates, a variety of other metrics of common ancestry have come into use (see key at bottom of Figure). These include intron location, possession of particular miRNAs, mitochondrial gene order, a particular form of a mitochondrial gene (nad5) present only in all protostomes, and some rare amino acid changes. These metrics of relationship are of immense value in locking down phylogeny because they are Boolean parameters rather than statistical conclusions based on continuous sequence divergence; they are discontinuous properties that are either present or absent, and are of such complexity that they could not occur the same way by accident. As indicated, they bolster most of the key relationships in the phylogeny. Each numeral indicates a (monophyletic) clade consisting of the Phyla shown to its right, and descendant from a common ancestor at the position of the numeral: 1, "Olfactores", i.e., urochordates (such as *Ciona*) plus all jawed and jawless vertebrates; 2, "Cyclostomata", i.e., lampreys and hagfish, and all extinct jawless vertebrates; 3, deuterostomes; 4, "Xenacoelomorpha", i.e., acoel worms and their relatives plus Xenoturbellids; 5, "Xenambulacraria", i.e., the clade that includes as sister groups echinoderms/hemichordates together with Xenacoelomorpha; 6, ecdysozoans; 7, "Mandibulata", i.e., arthropods, other than spiders and horseshoe crabs and their relatives (Chelicerata); 8, protostomes, i.e., Ecdysozoa plus Lophotrochozoa.

genetically, historically, and lineally more closely related to one another than is any one of them to any other animal, as revealed by their shared characters, and each node represents the evolutionary common ancestor of all its descendants. We see that the Bilateria all descend from a common ancestor. The closest major sister group to the Bilateria are the cnidarians (jellyfish, sea anemonies, and hydras). Fig. 3.4 shows that Bilateria is composed of three great super-phyletic assemblages, or "superclades", of which we ourselves belong to the deuterostomes. By the metrics of molecular phylogenomics the deuterostomes are a reproducibly resolved and coherent superclade. The deuterostomes include vertebrates and invertebrate chordate organisms such as *Ciona* and other ascidians, as well as amphioxus. The echinoderms (e.g., sea urchins) and their sister group, the hemichordates, are also deuterostomes, as well as are some evolutionarily simplified animals, the acoel worms, and several more obscure forms such as the recently discovered "Xenoturbellida" (Fig. 3.4). The remaining bilaterians, Grobben's "protostomes", consist of another robust superclade, "Ecdysozoa", plus a diverse set of phyla which have been united by the results of molecular phylogeny in a superclade termed "Lophotrochozoa". The ecdysozoans (the name refers to their common property of molting) include most prominently the arthropods and thus insects such as *Drosophila* as well as other arthropod classes including crustaceans, spiders and horseshoe crabs, centipedes, and so forth; but also nematodes, including *C. elegans*. The lophotrochozoans are comprised of two large subgroups for which the argument of common ancestry is perhaps more tenuous. One of these includes diverse tentaculate phyla ("lophophorates") of which we shall have little to say as there is almost no mechanistic information available on their development, but the other large assemblage ("trochozoa" or "spiralians") is more prominent, having been a focus of embryological attention since the end of the nineteenth century. This group includes molluscs, from clams and snails to octopus, as well as annelids, among which are some model systems of rising developmental and evolutionary interest such as the marine annelid *Platynereis*. The common ancestry of spiralians is indicated by their special "spiral cleavage" pattern known nowhere else, and by the form of their larvae.

The phylogeny provides important background illumination on many issues. For example, the universal principles of development just considered must descend from the bilaterian common ancestor, since they obtain in all three superclades. Another very important conclusion is that there are no living bilaterians basal to all other bilaterians. Contrary to earlier subjective interpretations, we see that nematode worms cannot be taken as representative ancestral bilaterians, rather they are descendants of the ecdysozoan common ancestor just as are arthropods; nor can acoel worms be regarded as basal to all other bilaterians, rather they are descendants of the deuterostome common ancestor. Furthermore, though intuitively flies may seem to share more body parts with us (hearts, paired eyes, legs, brains) than do say sea urchins or ascidians, we are actually much more closely allied by descent to the latter than we are to flies. We shall have occasion to refer to Fig. 3.4 many times in what follows, as it provides a portion of the logical underpinning for any comparison of developmental mechanisms.

2.2   Three modes of pregastrular regulatory development

Considered not by how the embryos of various bilaterians look, but by the mechanisms through which they install spatially differential regulatory gene expression at the very beginning of development, there are basically only three distinct types of early embryonic process (Davidson, 1990). They differ in their modes of pregastrular spatial specification, as a consequence of differences in the underlying genomic programs for development.

Mode 1 (*C. elegans*, echinoderms, molluscs, for example)

Spatial regulatory states are set up in Mode 1 embryos on the basis of their canonical cleavage planes, which position blastomeres with respect to the polar coordinates of the egg the same way in every member of the species. The nuclei are transcriptionally active from the very beginning of development. Thus each blastomere in the early embryo expresses a specific regulatory state and gives rise to a particular set of cell fates in every member of the species. That is, the lineage and fate of Mode 1 embryos is invariant. They use the three-dimensional geometrical spatial template generated by the cleavage process to produce clonal or polyclonal spatial regulatory state territories: the initially specified territories are those inheriting sequestered maternal regulatory components. The cells of these territories then transcriptionally express signals which install specific regulatory states in the neighboring blastomeres. Further subdivisions of regulatory state territories depend upon inter blastomere signaling as cleavage continues to generate further cellular spatial compartments. The process is distinguished by the generation of a mosaic of regulatory states by the end of the blastula stage which defines the future territorial domains of the whole embryo, before there is any cell migration. The corollaries of this mode of specification are (1) the embryo has a fixed, species-specific cleavage pattern and cell lineage; (2) the embryo nuclei become transcriptionally active beginning in early cleavage, or even at the pronuclear fusion stage, and directly control regulatory state from the beginning; (3) a constraint is that since canonical cleavage patterns cannot be maintained for more than perhaps 8–10 cleavages, often less, the eggs are relatively small and gastrulation occurs when there are only on the order of a few hundred or so cells; (4) differentiation, marked by expression of cell type-specific effector genes occurs precociously in some domains of the embryo even before gastrulation ("direct cell type specification"); (5) some early domains of the embryo, in which regulatory state is determined initially by inheritance of localized maternal factors, are specified autonomously, i.e., if these blastomeres are isolated they will continue to express their regulatory states absent any signal inputs from other cells.

Mode 2 (vertebrates, cephalopods)

Spatial regulatory states in Mode 2 embryos are set up very differently. In these large eggs there are no invariant canonical cleavage patterns, transcription is quiescent until hundreds or thousands of cells have formed, and in many or all parts of the embryo massive cell migration precedes completion of cell fate specification. Both localized maternal regulatory inputs, which provide geometrical information relative to the polar coordinates of the egg, and widely diffusing signals, are utilized once transcription resumes to instruct cells in the establishment of their regulatory states. Most cells of the embryo are specified not in respect to their lineage origins but in respect to the influences they are exposed to during and after their migration. Further subdivision of regulatory states in space all depends on intersection and integration of multiple signals from multiple sources. Corollaries of this mode of embryonic specification include the following: (1) this is a potent means of mobilizing tens of thousands of cells to install diverse territorial regulatory states by the end of gastrulation, and thus permits rapid construction of large, complex postgastrular embryos; (2) thus Mode 2 eggs are characteristically large and yolk-filled; (3) there are no autonomously specified cell lineages, and instead unspecified plastic cells abound until after gastrulation; (4) differentiation does not occur until after gastrulation; (5) Mode 2 embryos always develop directly and proceed immediately to formation of elements of the adult body plan.

Mode 3 (*Drosophila*, other arthropods)

Syncytial blastular development requires again a different strategy for embryonic regulatory state pattern formation. Though many kinds of very large eggs, for example some fish eggs, undergo what is called "meroblastic" cleavage in which the cleavage planes do not traverse the whole of the egg cytoplasm resulting in a transient syncytial structure, the noncellular portion is soon walled off and the rest of the embryo develops by Mode 2 processes. But most extremely in long germ band insect orders (meaning that the whole A/P length of the embryo is specified simultaneously), the embryo remains syncytial until just before gastrulation. Nuclei clothe the surface of the embryo in a dense two-dimensional monolayer within which complex spatial regulatory state patterns are set up. By the end of this stage these spatial regulatory states foreshadow the fates of the respective areas once cellularization occurs. The major mechanistic difference from Mode 1 and Mode 2 specification is that internuclear spatial communication occurs by diffusion of transcription factors to neighboring nuclei in the syncytium, and that there can be no use of intercellular signaling since there are no cell membranes enclosing the zygotic nuclei in which to mount receptors and within which to sequester the products of signal transduction. This has many consequences, distinguishing this from other modes of embryogenesis: (1) the nuclei are mainly transcriptionally quiescent until they reach the surface of the egg and all the regulatory state patterning in the syncytial stage takes place within a few nuclear cycles; (2) since signaling cannot serve to reinforce or set spatial boundaries of regulatory state domains, these are instead set exclusively by direct transcriptional repression; (3) dynamics of pattern change are unprecedentedly rapid, all of the syncytial blastoderm nuclei of the embryo are patterned simultaneously, and when cellularization does occur signaling is at once instituted; (4) since there are no cells until this point, cell migration cannot begin until after that time.

2.3   Phylogenetic distribution of modes of embryonic specification

Probably Mode 1 specification is the pleisiomorphic means of establishing embryonic regulatory states, that is, a character descendant from the common ancestor of the bilaterians. This is the most direct interpretation of its phylogenetic distribution in the modern world. Of model systems that we know a good bit about, this is the way *Ciona*, sea urchin, *C. elegans*, and annelid embryos develop. In other words, deuterostome, lophotrochozoan, and ecdysozoan embryos manifest this mechanism, the hallmark of an originally pan-bilaterian trait (cf. Fig. 3.4). An earlier appraisal based on the morphological processes of early embryogenesis also concluded that Mode 1 was the original bilaterian form (Davidson, 1990). This would imply that Modes 2 and 3 are derived. For Mode 3 this is an obvious conclusion. Syncytial cleavage divisions occur sporadically in various Arthropod groups, but they are thought to have evolved from cellularizing ancestral forms. The phylogenetic tree shown in Fig. 3.4 illuminates this point with respect to Mode 2 mechanisms: all vertebrates initiate embryogenesis in this way, but so do cephalopod molluscs. However, since in molluscs all the more basal classes oper-ate by Mode 1 embryogenesis, and in deuterostomes, more basal chordates (ascidians, amphioxus) as well as nonchordate deuterostomes (echinoderms, hemichordates) also utilize Mode 1 embryo-genesis, the Mode 2 developmental features in vertebrates and cephalopod molluscs are obviously convergent and derived.

This raises another question, that of indirect development. Many echinoderms, which display canon-ical Mode 1 embryogenesis, produce small, free living feeding larvae, within which the adult body plan develops in a second stage, by a lengthier, and more complex process than is required to complete embryogenesis. It is clear for echinoderms that this type of indirect development is pleisiomorphic (Raff, 1987; Wray and Bely, 1994; Davidson et al., 1995). But in many echinoderm clades indirect development has been wholly or partially lost, so that the embryo proceeds to form elements of the adult body plan directly. For this phylum direct development is derived, with respect to indirect develop-ment. The immediate outcome of indirect development is the self-supporting larva, a relatively simple bilaterian organism consisting initially of only a few thousand cells. But it is nonetheless equipped with feeding, sensory, immunological and motile, not to say further developmental, capacities. Because of this simple developmental outcome the study of Mode 1 specification processes in indirect developing embryos presents enormous advantages. So it is not surprising that, as we see in the following section, understanding of Mode 1 specification systems is more comprehensive than for other large-scale embry-onic processes.

How far toward complete development of an adult body plan can Mode 1 specification take an embryo? The answer is almost all the way if the adult is *C. elegans*. The adult hermaphrodite consists of 952 cells exclusive of the indeterminate germ line cells of the reproductive system. The somatic cells of which the organism is constructed have a canonical lineage right to the end, which develops by Mode 1 mechanisms much as do the embryos considered here. At the end, however, other events ensue; secondary sexual struc-tures form post-embryonically from set aside "blast cell" progeny, the skin becomes syncytial, molting occurs, etc. Nonetheless it is worth keeping in the back of one's mind that Mode 1 specification is poten-tially sufficiently potent to produce an adult animal body plan as long as it consists of a relatively small number of cells.

3.   Genomic Strategies of Control in Mode 1 Embryonic Processes

This and the following two sections outline the genomic bases for the special strategies of specification mobilized in each of the three embryonic modes, that is, our aim is to show by example exactly why mode-specific characters obtain and how they are encoded. We begin with Mode 1 specification, where our examples are drawn from the indirect early development of the sea urchin and the direct early development of *C. elegans*.

3.1   Mode 1 strategies in the sea urchin embryo GRNs

Appreciable parts of the pregastrular development of the sea urchin *Strongylocentrotus purpuratus* have been experimentally and conceptually reduced to the underlying genomic regulatory code. This embryo displays a typical Mode 1 specification (see above and Fig. 3.5(A)), and our project here is to use the GRN to show how these Mode 1 mechanisms are encoded in the genome. Four specific aspects are how maternal anisotropies give rise to an autonomously specified, polar cell lineage; how short range signaling installs new regulatory states; how the canonical cleavage pattern (i.e., spatial cell lineage/fate relationships) are used to set up dif-ferent spatial regulatory states; and how territorial fate decisions and territorial boundaries are made.

Maternal anisotropy: coding initial specification of the polar skeletogenic lineage

By fourth cleavage the maternal regulatory state of the four micromeres that arise at the vegetal pole of the embryo (Fig. 3.5(A) and (B)) is different from that of the other 12 cells with respect to several regulators: in particular, the nuclei of these cells contain the highest levels of both β-catenin and Otx transcription factor (Fig. 3.5(C)). β-catenin is a cofactor of the Tcf transcription factor required to convert Tcf from a complex with an obligate repressor to a factor that permits target gene activation. These two inputs, Otx and Tcf:β-catenin, activate a regulatory gene, *pmar1*, transforming a maternal spatial anisotropy into a zygotic *cis*-regulatory function with immediate downstream consequences (Oliveri et al., 2008; Smith and Davidson, 2009). This is the opening of a double-negative regulatory gate: *pmar1* encodes a repressor which, in the only cells in which it is active, transcriptionally represses a second gene which is otherwise globally active, also encoding a repressor, the h*esC* gene (Revilla-i-Domingo et al., 2007). The genetic regulatory circuitry is shown in Fig. 3.5(D). Here it can be seen that the HesC targets of repression include *ets*, *alx1*, *tel*, and *tbr*, the products of which together constitute the founding regulatory state of the skeletogenic lineage (Oliveri et al., 2008; Damle and Davidson, 2011). As we have pointed out (Peter and Davidson, 2009) this double-negative gate is actually an encoded global control device in that it allows expression of these genes only in the cells descendant from the four micromeres, but it also ensures the regulated silencing of these same genes by HesC everywhere else in the embryo. The way in which the initial skeletogenic regulatory state is set up provides a paradigmatic example of the strategy by which polar specification is installed in a Mode 1 embryonic process, and we see here exactly how it is encoded in the regulatory genome.

But that is not all, in that the same initiation system is used for additional purposes with effects that radiate outward to adjacent domains of the embryo (Fig. 3.5(A)). The precocious specification of the skeletogenic micromere lineage not only constitutes the skeletogenic regulatory state but also accounts for transcription of a gene encoding a signal ligand which affects the fates of surrounding cells. This is the *delta* gene, the product of which specifically activates the Notch signal transduction pathway, as discussed above. Like the skeletogenic regulatory genes, the *delta* gene is also under control of the double-negative gate (Fig. 3.5(D)).

Encoded transcriptional function of short-range signaling in spatial specification

We have already seen how in the sea urchin embryo the expression of the Delta ligand causes transcriptional activation of the mesodermal pioneer gene *gcm* in the ring of cells immediately positioned to receive this signal (Fig. 3.2). Delta/Notch signaling is confined to the cells adjacent to the Delta source because the Delta ligand is anchored in the membrane of the signaling cell, whereas the Notch receptor is maternal and ubiquitously present. The causality, as in all short-range signaling, depends simply on where the *cis*-regulatory target sites for the response factor are located in the genome. This determines which gene(s) will be activated and hence the identity of the new spatial regulatory state triggered by the signaling event. As noted above, Mode 1 embryos accomplish spatial specification with relatively small numbers of canonically positioned cells, the correlate of using invariantly positioned cell lineages to position their regulatory states. Here early specification processes typically depend on cell-to-cell signaling that affects only adjacent or almost adjacent cells, in contrast to the broad ranges of the graded signals and signal inhibitors ("morphogen gradients") common in the relatively huge embryonic systems of vertebrate eggs for example. In *C. elegans* embryos (for review, see Maduro, 2010) and *Ciona* embryos (Lemaire, 2009) short-range signaling processes are used to specify single individual blastomeres or pairs of blastomeres located in specific positions defined precisely by their adjacency to the signaling cells. In sea urchins the targets of a given short range signal more typically generate an incipient regulatory state territory of identically specified cells, such as the single cell deep ring of *gcm* positive cells shown in Fig. 3.2(A), all 16 of which at this stage constitute the veg2 lineage.

Cell lineage and cell fate in the sea urchin endomesoderm GRN

Several spatial fate decisions are made within the endomesodermal vegetal plate of the sea urchin embryo well in advance of gastrulation. The cell fates at this stage are morphologically invisible, but the discrete regulatory states these decisions generate are determinate for the developmental fates of the cellular descendants of the respective territories. These specification events occur initially in concentric domains located radially in respect to the central skeletogenic territory, as indicated diagrammatically in Fig. 3.5(E1) (Peter and Davidson, 2010), which, left to right, unfolds along a linear radial vector. We have already discussed the autonomous specification of the skeletogenic lineage. All other mesodermal cells as well as the endoderm derive from two embryonic cell lineages, veg2 and veg1. These separate at the horizontal sixth cleavage, as rings of eight cells each, veg1 lying atop veg2 (Fig. 3.5(A), 10h embryo). The cell lineage origins from which the endoderm and mesoderm arise are shown in Fig. 3.5(E2). In the sea urchin embryo all nonskeletogenic mesodermal cell types of the embryo derive from veg2, and all endoderm of the future gut derives from veg2 plus the sister lineage veg1. Cell fates in both lineages will be further subdivided. Within veg2, the mesodermal regulatory state must be separated spatially from what after gastrulation will become anterior endoderm; and within veg1 what will become posterior endoderm must be spatially separated from what will become the adjacent ectoderm (Fig. 3.5(A) and (E1)). The endomesoderm GRN model shows how the spatial domains set up by the canonical cleavage pattern become endowed with the regulatory states underlying these respective fate decisions.

The veg2 decision is made in the following way. Within the nuclei of the same veg2 ring, now 16 cells, expressing the mesodermal *gcm* gene under control of Delta/Notch signaling (Fig. 3.2(A)), a small set of endodermal genes are as well being expressed, activated by Tcf:β-catenin input. As Fig. 3.5(C) shows, veg2 cells contain nuclearized β-catenin of maternal origin, and their descendants in addition receive a boost of β-catenin nuclearization from the zygotic expression of Wnt signaling ligands in these cells (Cui et al., 2014). As we saw earlier, the Delta signal that activates Notch signal transduction in veg2 cells derives from the polar skeletogenic cells that lie within the veg2 ring (see Fig. 3.2(A)). But the incipient endodermal subcircuit, and the incipient mesodermal subcircuit which includes other regulatory genes activated in turn by *gcm*, operate independently in veg2 cells, without cross-regulation (Peter and Davidson, 2010). By about 15h, the ring of veg2 cells has divided radially to form two concentric rings, and what happens next is quite dramatic, as illustrated in Fig. 3.5(F) and (G). Here mesoderm subcircuit expression is represented by *gcm* (green), and endoderm subcircuit expression by *foxa* (red): the inner veg2 ring still in contact with the Delta expressing skeletogenic cells continues to express both subcircuits, visualized in the 16h embryo of Fig. 3.5(F) as yellow; the outer ring is no longer touched by the Delta ligand and *gcm* expression disappears while *foxa* expression continues, visualized as red. But by 18h endodermal subcircuit expression has been extinguished in the inner ring, effecting a permanent mesoderm/endoderm spatial regulatory state separation in the two rings of veg2 cells. The mechanism of extinction occurring specifically in these cells is interference with the Tcf:β-catenin feed into endodermal regulatory genes, as a further consequence of Delta/Notch signaling (Fig. 3.5(G); Peter and Davidson, 2010, 2011). Here again we see directly the causality of a Mode 1 spatial specification decision. The genomic code determines the activation of genes of the initial mesodermal and endodermal subcircuits, as well as the extinction of expression of the endodermal subcircuit in cells that continue to receive the Delta signal. These regulatory events capitalize on the radial eighth cleavage to effect final spatial separation of mesodermal and endodermal regulatory states.

The anterior endoderm/posterior endoderm choice utilizes the veg2/veg1 boundary originally generated at 6th cleavage. The mechanism is again dramatic. A regulatory gene, *eve*, is initially expressed in both veg1 and veg2 in response to maternally supplied Tcf:β-catenin. But by 15h its veg2 expression is extinguished transcriptionally, repressed by one of the early veg2 endoderm genes, *hox11/13b*, and by autorepression (Smith et al., 2008; Peter and Davidson, 2011). As shown in Fig. 3.5(H), the result is the separation of *eve* expressing veg1 cells, from which arises the future posterior endoderm, from *foxa* expressing veg2 cells, the future anterior endoderm (Peter and Davidson, 2011). The entire endomesoderm GRN model is shown in Fig. 3.5(I). This model, which represents the encoded *cis*-regulatory transactions, explains how the initial fate separation into spatial domains defined physically by cleavage planes (lineage) occurs.

A leitmotif of all these specification functions is that each begins with the spatially novel institution of transcription of genes thereby initiating the spatial specification function: in the skeletogenic domain *pmar1*, in the mesodermal domain *gcm*, in the anterior endoderm *hox11/13b* and *foxa*, in the posterior endoderm domain *eve*. It is in the *cis*-regulatory modules of these genes that the onset of each spatial activation event is encoded.

Progressive territorial boundary formation by the sea urchin embryonic ectoderm GRN

Until gastrulation (30h) the sea urchin embryo is essentially a single cell thick structure, spherical in form, with a large blastocoelar cavity which contains only the skeletogenic cells once they have ingressed (Fig. 2.1(A)). The fundamental process of cell fate specification occurs by successive subdivision of regulatory domains, as we have already seen in the endoderm and mesoderm. Such processes extend to the whole embryo and result in patterns of regulatory state domains of unexpected complexity. At least 18 spatial regulatory state domains have been defined at the molecular level just to the onset of gastrulation, some of which are visualized in the diagram in Fig. 3.5(E1), and additional domains can be seen in the ectoderm state maps shown in Fig. 3.5(J). The main import is that regulatory state subdivision occurs in a temporal sequence, such that larger domains of gene expression give rise to smaller ones, and that each subdivision is a binary process by which a new boundary is established (arrow pairs in Fig. 3.5(J1)). What begins as a simple animal–vegetal stratification (12h embryo) produces an orthogonal regulatory state grid (30h embryo; Fig. 3.5(J2)).

The encoded mechanisms by which many of the ectodermal domains are defined have been incorporated in a GRN model for pregastrular ectoderm development (Li et al., 2014). An example is afforded by the progressive definition of the boundary between the domain of expression of the neurogenic regulatory gene *foxq2* and the remainder of the ectoderm. *Foxq2* is the initial regulatory gene to be expressed specifically in the apical neurogenic region of the embryo (purple in Fig. 3.5(J)). The *foxq2* domain forms as follows (Li et al., 2014): The earliest transcriptional expression of *foxq2* (7h) extends over the entire animal half of the embryo, down to the original third cleavage boundary that separates the animal and vegetal half (that between mesomeres and macromeres in Fig. 3.5(E2)). Its driver is a pan-ectodermal transcriptional regulator, SoxB1. Elegantly, SoxB1 is also the driver of the regulatory gene *emx*, which encodes a repressor of *foxq2* and which confines *foxq2* expression to the apical domain shown in the 30h embryo. In turn, *emx* expression is excluded from the apical neurogenic region by Foxq2 repression.

In the sea urchin embryo, the second axis, or oral–aboral axis, is formed in response to a redox gradient generated by concentration of mitochondria on one side of the egg. Patterning along the oral-aboral axis occurs as follows. An early transcriptional response to the redox gradient is activation of redox-sensitive maternal transcription factors which initiate expression of the *nodal* signaling gene on the future oral side (Coffman et al., 2004, 2009; Nam et al., 2007; Range et al., 2007, Ben-Tabou de-Leon et al., 2013). In response to Nodal signaling, a very important transcriptional repressor encoded by the *not* gene is expressed throughout the oral ectoderm (Li et al., 2012). It plays several crucial roles in restricting the spatial boundaries of regulatory state domains adjacent to the oral ectoderm illustrated in Fig. 3.5(J2) for the 30h embryo (Li et al., 2014). First, Not acts synergistically with Emx to confine *foxq2* expression apically, but it is also responsible for restricting the transcription of certain regulatory genes to the lateral oral ectoderm regions (orange in Fig. 3.5(J2)). Furthermore, it excludes expression of genes transcribed in the aboral ectoderm from the oral side. The *not* gene thus executes boundary forming patterning functions by repression in both animal–vegetal and oral–aboral axes. To complete the picture, the boundaries between the aboral ectoderm and the lateral oral ectoderm domains are also determined by repression. The genes expressed in the lateral oral ectoderm domain are driven by broadly expressed activators but repressed on the aboral side by regulators generated within the aboral ectoderm specification GRN. A similar strategy accounts for the formation of the crucial endoderm–ectoderm boundary within the veg1 domain (see Fig. 3.5(J2)). Here, the initially expressed regulatory gene is *eve*, which is transcribed throughout veg1 and contributes to the expression of *hox11/13b* in the subset of cells giving rise to endoderm (Fig. 3.5(I); Peter and Davidson, 2011). Hox11/13b in turn represses transcription of ectodermal genes such as *lim1* which were previously expressed throughout the veg1 domain (Li et al., 2014). Thus, in the formation of these specific boundaries as well as the others indicated in Fig. 3.5(J), the general rule is that broadly expressed regulators first block out a prospective regulatory state territory, the final boundaries of which are established by transcriptional repression.

3.2   The sea urchin embryo GRNs and the code for territorial embryonic fate

Thus far we have focused on specific aspects of the sea urchin endomesoderm GRN which illuminate particular Mode 1 specification mechanisms. Each of these aspects is embedded in the overall 0–30h endomesodermal GRN model reproduced in Fig. 3.5(I), but the network contains a very large number of additional genes and linkages. Knowledge of this network is unusually complete, and indeed a computational automaton built upon the endomesoderm GRN model shows that it includes sufficient genomic regulatory information to permit a remarkably accurate de novo calculation of spatial and temporal gene expression that matches observation in all but a few particulars (as we discuss in Chapter 6; Peter et al., 2012). So there is more to be learned about the biology of embryogenesis from the GRN than so far mentioned, and here we briefly summarize some further insights that emerge from the network architecture.

From transient inputs to definitive spatial regulatory states

In all well-studied embryonic specification processes the initial specification inputs do not last, but function rather to begin the process of activating new spatial regulatory states. These initial inputs, for example maternal inputs or inductive signaling inputs, are always transient, and they are always soon followed by a particular kind of circuitry that is of deep functional consequence per se, namely positive feedback circuitry consisting of two or three regulatory genes locked in a mutually stimulatory embrace. For example, in the skeletogenic domain the *erg*, *hex*, and *tgif* genes become engaged in a skeletogenic positive feedback loop, downstream of the transiently expressed *pmar1* gene. In the aboral mesoderm the *gcm*, *gatae*, and *six1* genes interact positively in a feedback configuration, which persists long beyond the initiating Delta input. Such feedback loops are of such regular occurrence in specification GRNs that in explorations of unknown networks their occurrence can be confidently predicted. Their significance is functional: they set the output levels of the regulatory states anew, irrespective of the prior transient activities; and for given spatial domains, they provide the major stabilization and state lockdown devices of these control systems (see Chapter 6 for detailed discussion of feedback circuitry; Peter and Davidson, 2009).

Direct cell type specification

In Mode 1 embryos early specified lineages often begin to express effector genes of differentiation gene batteries long before gastrulation, and before most overt differentiation properties of the cells

Figure 3.5

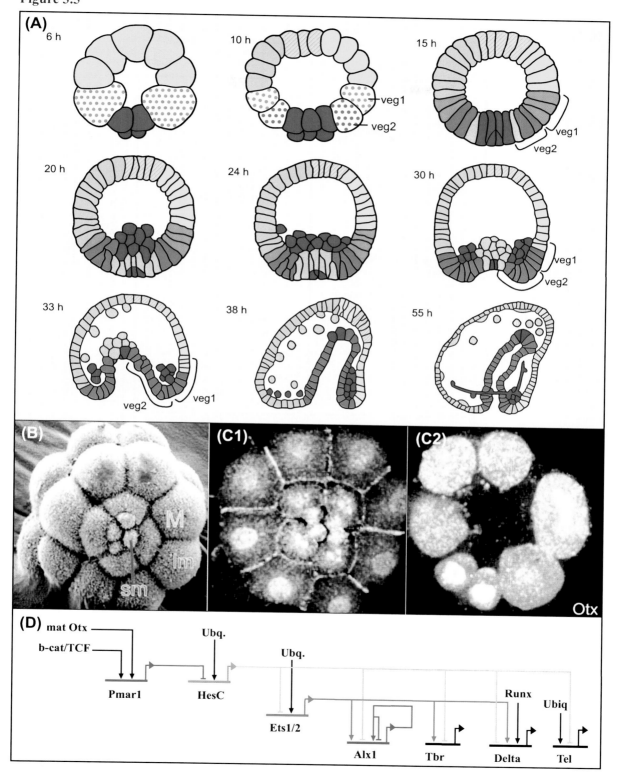

Figure 3.5

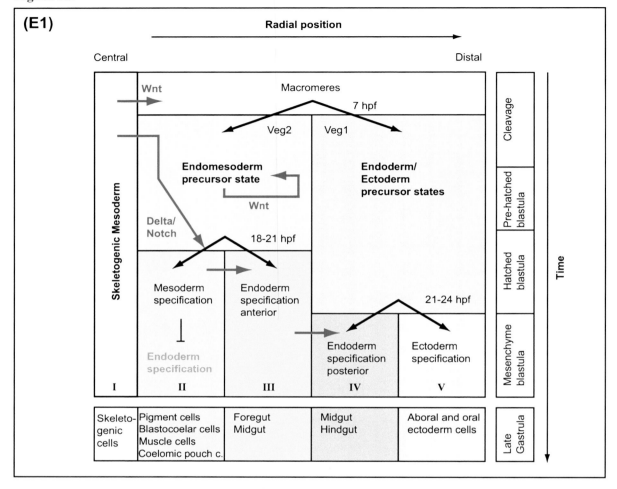

Figure 3.5 **Genomic control of regulatory domain formation in the sea urchin embryo.** (A), Disposition of early lineages, and cell fate ancestry in embryos of the sea urchin *Strongylocentrotus purpuratus.* Regulatory state domains are color coded up to 24h post fertilization. Spatial regulatory states become much more complex than indicated all over the embryo after late blastula stage, and the color coding in the diagrams of the later embryo represents only very general embryonic components: red, skeletogenic domain and skeleton of late embryo; purple, small micromere domain and coelomic pouches to which this lineage contributes; blue and red dots, "macromeres" of fifth cleavage embryo (6h); blue, veg2 lineage and orange, veg1 lineage. After 24h both the veg2 and veg1 endodermal descendants are colored blue though as discussed in text veg 1 gives rise to posterior endoderm and veg2 to anterior endoderm regulatory states. Following gastrulation many distinct regulatory state domains are set up in the developing archenteron, foreshadowing the foregut, midgut, hindgut, sphincters, and anus, not shown. Lavender, nonskeletogenic mesoderm domain also deriving from veg2, and mesenchymal immune and pigment cells to which this domain gives rise; orange stipple, apical neurogenic domain; green, aboral ectoderm domain; yellow, oral ectoderm domain, each of which also develops complex internal regulatory state sectors, not shown here, nor is the neurogenic ciliary band that separates oral and aboral ectodermal territories indicated. (B), SEM view of vegetal end of a fifth cleavage embryo in which the large skeletogenic micromeres (lm) and their small

Figure 3.5

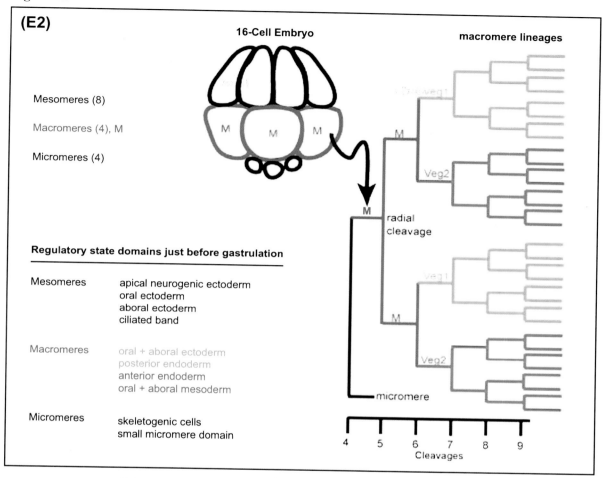

micromere sister cells (sm) can be seen; M, macromere *(image from J.B.Morrill, reproduced from Oliveri et al. (2008))*. (C), Maternal regulatory molecules localized in specific blastomeres of early embryos, revealed by immunocytology. (C1), Nuclear localization of β-catenin in the eight veg2 cells and large and small micromeres of a sixth-cleavage *Lytechinus variegatus* embryo, vegetal view *(from Logan et al. (1999))*. The anisotropy of β-catenin is initially the consequence of its association with vegetally localized Disheveled in the unfertilized egg (Weitzel et al., 2004). (C2), Nuclear localization of Otx transcription factor in micromeres of a fourth cleavage *S. purpuratus* embryo *(from Chuang et al. (1996))*. (D), Network architecture of the double-negative gate controlling initial zygotic skeletogenic regulatory state in the large micromeres (i.e., expression of *ets, alx1, tbr, tel,* and *delta*). As a result of the initial expression of *pmar1* beginning at the end of fourth cleavage, driven by the maternal Otx and Tcf:β-catenin activators in these cells (C1,C2), the encoded Pmar1 repressor blocks *hesc* transcription in skeletogenic micromeres, though this gene is expressed everywhere else. Thereby *ets1/2* transcription is permitted to be activated, and since HesC repression has been relieved on the other genes as well, by 8–10h Ets1/2 initiates transcription of *alx1, tbr, tel,* and *delta*; ubq, ubiquitous, probably maternal transcriptional activator. As the diagram indicates, *alx1* auto-activates, but after its product achieves a certain concentration it represses itself, resulting in a sharp peak of transcription *(modified from Damle and Davidson (2011))*. (E), Endomesoderm specification and the veg1 and veg2 lineages. (E1), Process diagram of endomesodermal specification events: ordinate, time, top to bottom, stages given at right;

Figure 3.5

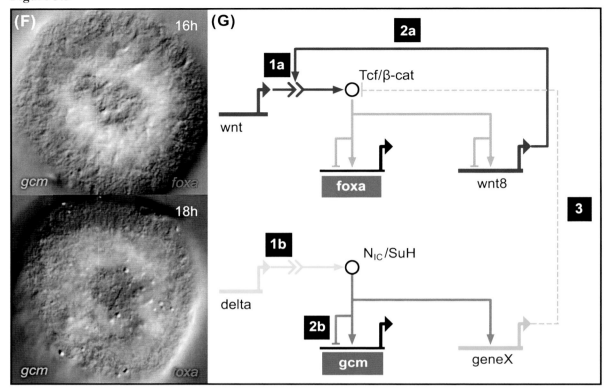

abscissa, radial position viewed from the vegetal pole, as indicated above the diagram. Territorial regulatory state domains are shown in large boxes; at bottom in small boxes are listed the portions of the later embryo to which these domains give rise. Five endomesodermal regulatory states are generated: from center outward the skeletogenic regulatory state (cf. D); the veg2 nonskeletogenic mesoderm regulatory state; the veg2 anterior endoderm regulatory state; the veg1 posterior endoderm regulatory state; the several veg1 ectoderm regulatory states *(from Peter and Davidson (2010))*. (E2), Origins of veg1 (orange) and veg2 (green) lineages, and cell types descendant from each *(from Peter and Davidson (2011))*. (F), Dramatic separation of mesodermal regulatory state from anterior endodermal regulatory state within 2 h, visualized by double fluorescent in situ hybridization: at 16h the inner ring of veg2 cells (yellow) expresses both *gcm* (green), a mesodermal regulatory gene, and *foxa* (red), a canonical endodermal gene; but by 18h expression of *foxa* (and all other endodermal regulatory genes) has been extinguished in the inner ring of cells, resulting in exclusive mesodermal and endodermal specification states *(from Peter and Davidson (2011))*. (G), Network subcircuit for endoderm/mesoderm cell fate decision. Endodermal regulatory gene expression depends on Tcf/β-catenin, as discussed in text, and mesodermal regulatory gene expression requires Delta/Notch signaling. After initial co-expression, endodermal regulatory genes are shut off in inner veg2 ring (cf. F) by means of an unknown gene (geneX) that probably causes clearance of β-catenin from the nuclei of these cells. GeneX expression is activated by Delta-Notch signaling and thus clearance of the essential input into the endodermal GRN such as *foxa*, occurs in the same cells where mesodermal regulatory genes are expressed. Absence of β-catenin causes instead repression of endodermal genes, mediated by the Janus function of the Wnt early response factor Tcf *(from Peter and Davidson (2011))*. (H), Separation of veg2 anterior from veg1

Figure 3.5

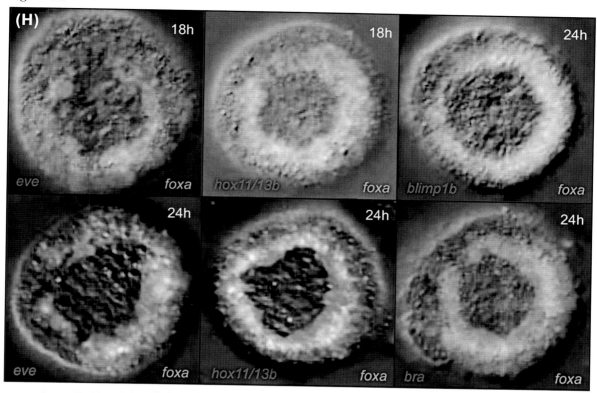

posterior endoderm regulatory states between 18 and 24h, visualized by double fluorescent in situ hybridization. The green probe is *foxa* throughout and the identity of the red probe is indicated. Left pair of images shows that *eve* and *foxa* mark veg1 and veg2 respectively and exclusively throughout this period. Middle pair: at 18h veg2 endodermal cells express both *hox11/13b* and *foxa* but by 24h *hox11/13b* expression is extinguished in veg2 and activated in veg1, marking the posterior endoderm regulatory state. Right pair, top: *blimp1b* gene continues to be co-expressed with *foxa* in veg2 endoderm at 24h; bottom, the veg1 endoderm domain is now also marked by *brachyury* expression which like *hox11/13b* is no longer expressed in veg2 *(from Peter and Davidson (2011))*. (I), A current version of the endomesodermal GRN, represented in the BioTapestry platform. For details, recent updates, temporal progression of regulatory states, regional views, and extensive underlying data, see Web site at http://sugp.caltech.edu/endomes/#EndomesNetwork. Results from several labs are included, as indicated there. The network is based on a global experimental dataset including perturbation data, *cis*-regulatory data, quantitative kinetic and spatial expression data. Linkages shown in bold lines are directly validated at the *cis*-regulatory level. The network demonstrates explicitly the direct inputs and outputs of the genes it contains. A few effector genes are shown in the boxes at the bottom; otherwise genes encode transcription factors and several signaling components. This GRN has been shown to suffice for a correct predictive recomputation of the observed spatial and temporal expression pattern of almost every individual gene (Peter et al., 2012). (J), Regulatory state domains of the pregastrular ectoderm *(from Li et al. (2014))*. (J1), Process diagram organized as in (E1) displaying regulatory state subdivisions during pregastrular development (timescale at right). Colored regions represent ectodermal domains formed during this period. As indicated, paired orthogonal arrows represent cell fate divergence:

Figure 3.5

(I)

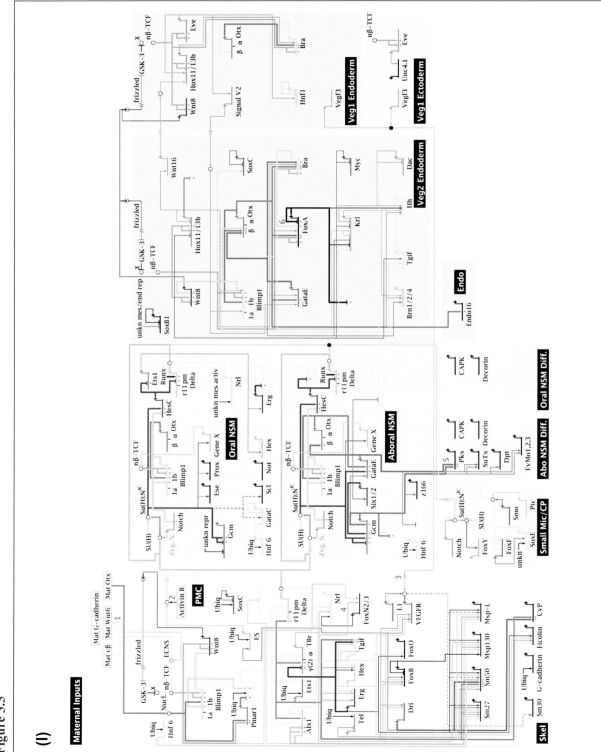

Figure 3.5

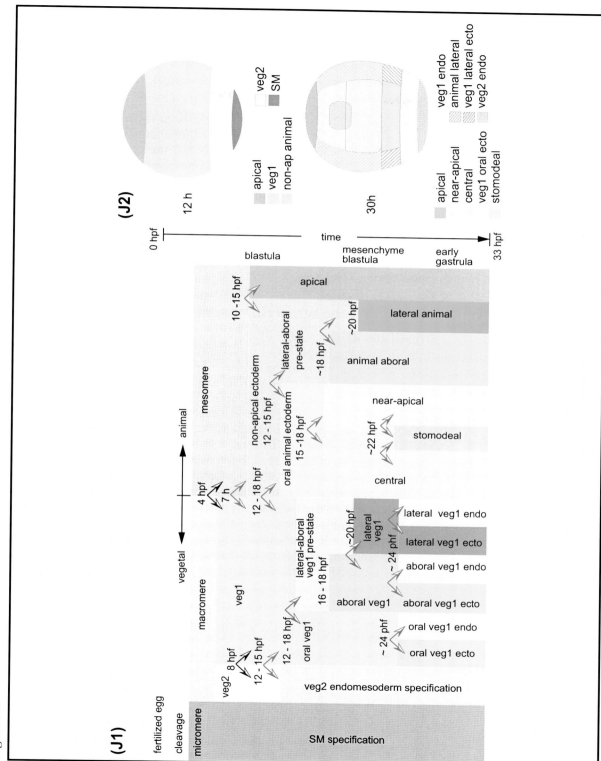

black arrows, divergence boundary corresponds to a canonical cleavage plane; green arrows represent boundaries perpendicular to the animal–vegetal axis; red arrows represent boundaries perpendicular to the oral-aboral axis. (J2), Diagrams of cleavage stage (12h; top) and late blastula stage (30h; bottom) embryos, domains color coded as in (J1). *For an always updated version of the ectoderm GRN model, see* http://sugp.caltech.edu/endomes/#EctodermNetwork.

descendant from them become manifest. This is because the circuitry in these lineages is relatively shallow, in that these systems transit directly from the generation of regulatory states to expression of downstream effector genes as soon as the regulatory states include the transcriptional drivers of differentiation genes. Thus in Fig. 3.5(I) we see that some but not all skeletogenic genes encoding biomineralization proteins of the skeletogenic matrix ("sm" genes) are activated in this pregastrular network, and this occurs even before ingression of these cells into the blastocoel and thus well before the appearance of the first skeletal elements. Similarly, long before pigment cells per se differentiate from their aboral mesoderm precursors, genes encoding pigment synthesis enzymes such as polyketide synthase (Pks) and other effector genes shown in the terminal region of this portion of the aboral mesoderm GRN model are expressed. The reason is simply that the drivers of these genes have already appeared, here Gcm and Gatae (Calestani and Rogers, 2010).

Summary: encoded functions of the sea urchin endomesoderm GRN

The GRN model in Fig. 3.5(I) includes a large number of encoded interactions among regulatory genes, almost all of evident function, in that the output of each gene directly affects the activity of other genes. In the foregoing we touch briefly on some of the biological phenomena of development that are explained by these interactions: thus we see exactly how an initial maternal anisotropy is turned into a lineage-specific regulatory state; how a short-range signal generated by the GRN of one territory results in installation of a new regulatory state in an adjacent territory; how lineage boundaries become spatial regulatory state boundaries. Space does not permit a comprehensive discussion of all the endomesoderm GRN functions here but we return to this network in Chapter 6 in context of a Boolean computational model built upon the interaction map shown in Fig. 3.5I. (Oliveri et al., 2008; Peter and Davidson, 2009, 2010, 2011, 2013; Peter et al., 2012; Materna et al., 2013; Cui et al., 2014; Li et al., 2014). The GRN controls the dynamic progression of spatial gene expression throughout (Peter et al., 2012). Ultimately the main function of the GRN is to erect in each spatial domain of the developing embryo a specific regulatory state which always includes multiple, cross-regulating control genes. Their cross-regulation endows the regulatory state with independence from external signals or inputs. As shown by the skeletogenic domain, many of the control genes that constitute the mature regulatory state also provide inputs into effector genes, and directly or indirectly most are required for effector gene function (Rafiq et al., 2012, 2014).

3.3 Endomesoderm specification in the *C. elegans* embryo

A GRN for specification of endomesoderm in the *C. elegans* embryo (Owraghi et al., 2010) illustrates the same Mode 1 strategies of establishing spatial regulatory states as seen in the sea urchin embryo, though in terms of interactions, on a more compact scale. In the initial stages the specification functions affect single individual blastomeres of the early embryo which become founder cells for particular lineages. It is interesting to note that the single cell or single pairs of cells per regulatory state seen in some early *Ciona* (Imai et al., 2009; Lemaire, 2009) and *C. elegans* specification processes are only found in Mode 1 embryos (considering here embryogenesis just up to postgastrular stages).

Maternal anisotropies, short-range signals, and initial spatial specifications

Mode 1 embryonic specification processes are all about assigning specific regulatory states to specific cell lineages, using the canonical positions of early lineage ancestors with respect to the egg axes and to one another for spatial organization of embryonic fates. This can be seen particularly clearly in *C. elegans*. In this embryo unique lineage founder cells are easily recognized because of the asymmetry of the cleavage pattern, and because the numbers of cells at the time of specification is small. As early as third cleavage (Fig. 3.6(A)) the lineages E, D, and P4 arise, which share the particular feature that all descendants of each lineage express the same regulatory state at a given time and give rise to only one differentiated fate. As in all Mode 1 systems, transcription is active by very early cleavage, and specific blastomeres right away begin to transcribe specific regulatory genes. This has been observed most dramatically using expression constructs driving fluorescent reporters, as well as by in situ hybridization (for a recent general review of early specification processes in this embryo, see Maduro, 2010).

The transcriptionally active blastomeres in second and third cleavage *C. elegans* embryos with which we are here concerned are ABa and EMS at second cleavage, E and MS (daughters of EMS) at third cleavage, and from fourth cleavage the ABa granddaughters and their successors (see Fig. 3.6(A)). The sequence of regulatory events leading to definitive specification of these cells is particularly simple: (1) one or two specific zygotic regulatory genes are activated in each of the blastomere lineage founder cells by a combination of very early interblastomere short-range signals and anisotropically localized maternal transcription factors. This anisotropy is dependent on the asymmetric cytoplasmic mobilization of maternal components, accomplished by posttranscriptional mechanisms prominently mediated by mRNA-binding proteins, and proteins that bind to and localize these (Huang et al., 2002; Farley et al., 2008; Farley and Ryder, 2008); (2) these initial zygotically transcribed factors activate dedicated additional regulatory genes in each lineage according to the encoded GRN linkages; (3) these genes engage in auto- or cross-regulatory circuitry which renders their respective regulatory states independent of the initial specification pioneers which all quickly cease to be expressed; (4) these same regulatory genes determine further cell fate decisions and the expression of next sets of regulatory genes, and also immediately activate terminal differentiation effector genes of the respective tissue types. We see these steps in detail in the GRN models that we discuss below.

Initial specification genes in endomesoderm lineages in *C. elegans*

In Fig. 3.6(A) the initial lineages of the *C. elegans* embryo are shown, with those endoderm and mesoderm components included in the following GRN analysis highlighted. The disposition of the founder blastomeres at the eight-cell stage, and the relevant endoderm and mesoderm components of the completed larva are indicated in Fig. 3.6(B). Similar to the four skeletogenic micromeres of the fifth cleavage sea urchin embryo which produce descendants that execute only one developmental function, skeletogenesis, the single E blastomere of the third cleavage *C. elegans* embryo gives rise only to gut endoderm cells. The larval intestine consists exclusively of all 20 descendants of the E blastomere. From this we see that once the regulatory state of the E blastomere is initiated, its autonomously operating GRN equates the E cell lineage with endodermal fate. The MS blastomere produces many kinds of progeny, most but not all mesodermal (Fig. 3.6(A)), and here we focus on two major products of its descendants, the muscular posterior pharynx and body wall muscle. AB descendants generate the muscle of the anterior pharynx, among many other cell types. The specification processes encompassed in the GRN model that we now briefly examine are those underlying development of the intestine, posterior pharynx, body wall muscle, and anterior pharynx. The canonical sequence of these zygotic specification processes is encoded in lineage-specific GRNs. An overall diagram indicating these steps in abbreviated form for the MS and E lineages is shown in Fig. 3.6(C) (Maduro, 2010), and the GRNs per se are reproduced in Fig. 3.6(D) (Owraghi et al., 2010).

Figure 3.6

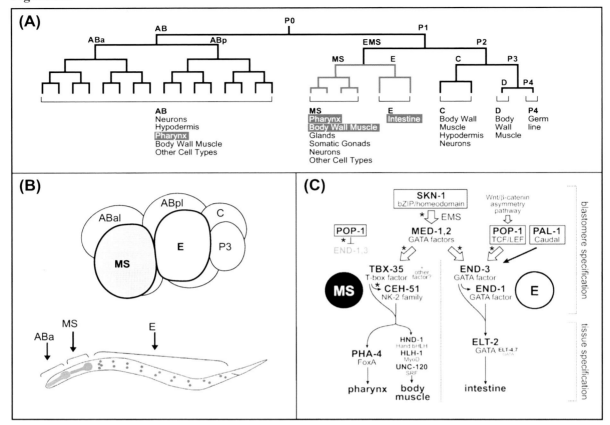

Figure 3.6 Genomic control of embryonic cell fate specification in pharyngeal muscle and endoderm in *C. elegans*. (A), Early lineage and some ultimate cell fates *(data from Sulston et al. (1983))*. The MS and E lineages (red) contribute posterior pharyngeal muscle and gut endoderm respectively, while anterior pharynx derives from AB descendants (red boxes). (B), Diagram of eight-cell embryo, ventral view. The E gut founder blastomere is shown in green and its sister cell MS in gray *(from Goldstein (1995))*. Below is a drawing of the larva with its 20 gut cells. Anterior to the larval intestine (lavender, purple nuclei), are the anterior and posterior pharynx (blue and red respectively). The anterior pharynx is formed from descendants of the ABa blastomere (anterior AB daughter), the sister cell of which, ABp, migrates to the rectum; the posterior pharynx derives from MS blastomere products *(diagram of larva from Maduro and Rothman (2002))*. (C), Outline of initial specification network that distinguishes E and MS regulatory states. Spatially localized maternal inputs are boxed. In the parent cell, EMS, Skn1 transcription factor and other factors activate zygotic *med1* and *med2*, as described in text. In E, these factors plus Tcf:β-catenin, activated by Wnt signaling from the adjacent P2 cell at the EMS stage (cf. B) in turn activate the *end3* gene, triggering a feed-forward chain of different *gata* gene expressions (*end1* to *elt2*); *elt2* thence plays a major role in E lineage specification. In MS the same Tcf factor lacking β-catenin, instead represses *end1* and *end3*. Activation of *med1,2* in MS causes activation of the MS-specific *tbx35* regulatory gene. Downstream of *tbx35* and its target *ceh51*, the pharynx specification factor *pha-4* is turned on in some MS descendants, while in others the body muscle subcircuit including the canonical muscle regulators *hand*, *myod*, and *srf* is brought into action. The initial interactions are known to be direct (*) and the genes expressed at the stage labeled "blastomere specification" are

Figure 3.6

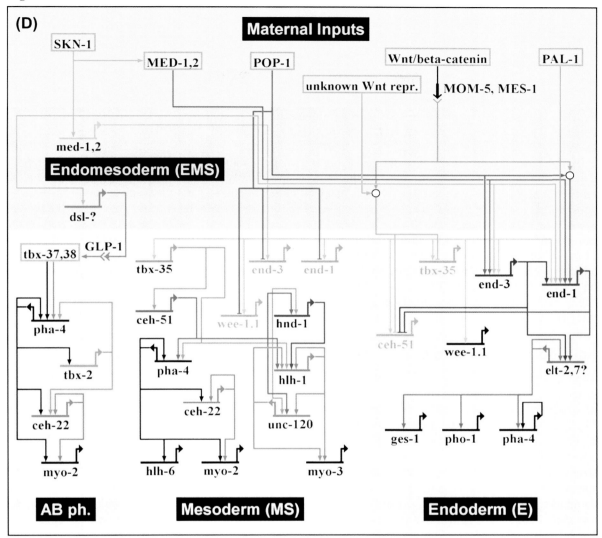

expected only very transiently *(from Maduro (2010))*. (D), The pharyngeal and endodermal GRN model; see text for discussion and interpretation. Here the regulatory interactions summarized in (C) are explicit, and the further steps toward direct cell type specification are shown. The network is portrayed in BioTapestry (cf Fig. 3.5(I)). Note that the same pharyngeal muscle regulators *pha4* *(foxa)*, *myo2*, and *ceh22* *(nkx2.5)* are turned on by different routes in anterior and posterior pharynx. *(Slightly modified from Owraghi et al. (2010))*.

The inputs responsible for activation of the earliest genes expressed in each of the four lineage/cell fate elements here considered, posterior to anterior, are as follows. At the four-cell stage, EMS contains three relevant maternal transcription factors that are all differentially localized to the posterior cells of the embryo, the b-zip-homeodomain factor Skn-1 (Fig. 3.1(C)), a Caudal ortholog (Pal-1), and a Tcf ortholog (Pop-1). This last factor is the "Janus" response factor for Wnt signaling; i.e., in the presence of Wnt signaling Tcf forms a complex with β-catenin and becomes permissive for activation, while in the absence of the Wnt signal, Tcf acts as a repressor for the same target genes (see Section 1.4, this Chapter). As Fig. 3.6(C) and in more detail Fig. 3.6(D) shows, in EMS a pair of zygotic genes encoding divergent Gata factors, *med1,2* are activated by Skn-1. Meanwhile the immediately posterior sister cell to EMS, P2, signals to EMS by both Wnt and MAPK pathways, polarizing EMS so that when it divides to form E and MS only the E nucleus receives a Tcf-β-catenin input. The initial regulatory linch-pin of E specification is the *cis*-regulatory apparatus of an essential pair of genes encoding Gata factors, *end-3* and *end-1*, which are directly activated by a combinatorially unique input set: the *end-3* gene is activated by Med-1,2 and Skn-1 inputs, plus Tcf-β-catenin; and the *end-1* gene by inputs from *end-3* and *med-1,2* genes, the maternal Skn-1 and Pal-1 factors, plus Tcf-β-catenin (Maduro, 2010; Owraghi et al., 2010). In MS, which does not experience the Wnt-driven Tcf-β-catenin input, the *end-1,3* genes are specifically repressed by the same Tcf factor, while the initial specification gene *tbx-35* is activated by *med-1,2* gene products. In contrast, in E, *tbx-35* expression is specifically excluded by Wnt signaling-dependent repression. In the AB lineage, two separate Notch signaling events in early cleavage account for the activation of the initial specification genes in the founder cells from which the anterior pharynx will develop. These specification genes are *tbx-37* and *tbx-38*. At second cleavage EMS sends a Delta signal to the immediately adjacent ABp blastomere (Fig. 3.6(A)), and the result of this is to exclude the transcription of *tbx-37* and *tbx-38* in the posterior AB descendants. Two cleavages later MS (Fig. 3.6(B)) expresses a Notch ligand which now impinges on the anterior AB descendants adjacent to it, causing *tbx-37,38* to be activated in these cells (Good et al., 2004). Thus the zygotic pioneer genes at the top of the GRNs in Fig. 3.6(D) are now all specifically and exclusively activated: *end-1,3* in E, *tbx-35* in MS, *tbx-37,38* in the ABa granddaughters.

Specification GRNs in *C. elegans* endoderm and mesoderm lineages

The remaining genomically encoded steps of the sequence are illustrated in Fig. 3.6(D) for each of four different pathways. In the E blastomere, the *end-1,3* genes activate one or more downstream *gata* genes of which the most important is *elt-2*, and this in turn activates the *C. elegans foxa* gene (*pha-4*). Note in Fig. 3.6(D) that while *end-1,3* expression is transient, *elt-2* and *foxa* are both equipped with positive autoregulatory devices, and their expression is hence permanent. By as early as fourth cleavage when there are now two E cells, an effector gene which encodes a gut esterase (*ges-1*) has been activated as a direct *elt-2* target, as are many other gut effector genes (Fukushige et al., 1998; McGhee et al., 2009). In MS the initial specific zygotic regulatory gene *tbx-35* activates in turn an *nk-2* gene (*ceh-51* in Fig. 3.6(D)). In MS granddaughters and their descendants, two different pathways are then activated. In the body wall muscle progenitors *tbx-35* and *nk-2* together, in a feed-forward circuit, apparently turn on the *hand-1* muscle gene regulator and other muscle regulatory genes. In the pharynx progenitors they turn on the *foxa* gene, apparently by similar wiring (Broitman-Maduro et al., 2009; Owraghi et al., 2010). In the body wall muscle GRN downstream from *hand-1* a lockdown feedback permanently ensures activation of *myod* (*hlh-1*) and *srf* (*unc-120*), canonical muscle regulators, and thereupon effector genes such as muscle myosin (*myo-3*) are activated. In the pharynx pathway *foxa* locks itself on by autoregulation and activates cell type-specific regulators including *nkx2.5* (*ceh-22*), which also drives its own expression, and these two together activate the pharynx muscle myosin gene *myo-2*. Direct cell type specification thus occurs in both pathways. In the ABa anterior pharynx lineage where *foxa* is again the upstream specification gene, it activates *tbx-2*, and together these feed forward to activate *nkx2.5* and then the *myo-2* gene (Fig. 3.6(D); Owraghi et al., 2010).

Thus we see in detail how this Mode 1 specification process works. In the first step transient regulators are assigned to each lineage element by combinations of very early intercell signaling and asymmetrically inherited maternal transcriptional regulators. Of course these processes are directly encoded at the genomic level in that both the targets of signal reception and of the maternal transcription factors are "written" in the *cis*-regulatory sequences of the pioneer genes. Following this initial step, the same thing happens in each domain. GRN wiring ensures the installation of stable, dynamic intra- or intergene feedbacks among a small number of cell type-specific regulators, and these proceed directly to activate terminal effector genes. The clarity with which we can see step by step how the genomic code mediates Mode 1 spatial specification processes in early *Ciona*, sea urchin and *C. elegans* embryos can at this point not be matched for any vertebrate embryo, the subject to which we now turn.

4. Genomic Strategies of Control in Mode 2 Embryonic Processes

Vertebrate embryos contain thousands of cells at gastrulation, and in contrast with Mode 1 embryos they do not rely on the canonical positions of cleavage stage blastomeres to spatially organize their regulatory states. Nor do they in general utilize strictly fixed cell lineages, and the exact ancestry of given postgastrular regulatory state domains will differ in detail individual to individual within the species. This can be seen in the interdigitation of specific blastomere descendants in the color-coded maps of *Xenopus* embryos in Fig. 3.7(A)–(E). Vertebrate embryos are endowed with maternal anisotropies of crucial regulatory significance, organized with respect to the primordial axes of the embryo (Fig. 3.1(D) and (E)). These maternal factors provide initial spatial specification instructions for the definition of regional zygotic regulatory state domains. The subsequent specification processes of vertebrate embryos, here referred to in shorthand terminology as Mode 2 specification, are dominated by regional clines of long range signals, and by negative modulators of both the signals and the signal responses. The combinations of dynamically changing signals and their antagonists are received by masses of migratory, dividing cells. The intense focus in vertebrate embryology over the past half century on the signals, their modulators and antagonists, has resulted in a large body of data describing stage-dependent signal interactions and their consequences. However, there remains unresolved a fundamental mechanistic aspect of signaling in these systems, which is how combinations of graded signals and their antagonists are used by the developmental GRNs to set up precisely positioned spatial regulatory state domains within each of which all cells perform similarly and yet differently from the cells of adjacent domains. Nor is it clear in general in these embryos how regulatory state domain boundaries are precisely defined. We return to the subject of interpretation of signal gradients later in this book (Chapter 6). Large-scale embryogenesis GRN models up to postgastrular stages have begun to be constructed for *Xenopus* (Loose and Patient, 2004; Koide et al., 2005; Swiers et al., 2010) and zebrafish (Chan et al., 2009; Tseng et al., 2011), based on an extensive literature on *cis*-regulation of individual genes of interest, and on both gain and loss of function perturbations. These GRN models illuminate the interactions that for example lead to mesodermal or endodermal regulatory states, and they display interesting aspects of circuitry (some of these models are explicitly reviewed in Chapter 4). As yet however, they mostly illuminate the acquisition of cell type-specific functions rather than the cell fate specification processes which determine the spatial regulatory organization of the embryo with which we are here concerned.

Here we consider two aspects of vertebrate embryogenesis that are particular to their mode of development, for which some mechanistic explanation at the transcription control level has been discovered. These are the global temporal control of embryonic gene expression in Mode 2 developmental processes; and the *cis*-regulatory integration of diverse inputs at key genes in the Spemann organizer.

4.1   Global temporal control of transcription in *Xenopus* and zebrafish embryos

In direct contrast to Mode 1 embryos, vertebrate embryos produce thousands of cells before generally activating their transcriptional apparatus. Furthermore, pregastrular expression of differentiation effector genes is explicitly precluded in vertebrate embryos, while as we have seen, precocious expression of such genes is a hallmark of Mode 1 embryogenesis. These temporal aspects are special features of Mode 2 embryos and their mechanisms are of interest in this light.

The initial transcriptional silence in *Xenopus* lasts until the mid-blastula transition when all classes of nuclear transcription suddenly open up (Newport and Kirschner, 1982; Davidson, 1986). This occurs at around 7h when there are about 4000 cells. Prior to this most of the transcriptional apparatus is apparently subjected to global repression, though a few specific genes are transcribed earlier (Collart et al., 2014). At least one essential component of the global repression apparatus is a prevalent, polyfunctional maintenance methyltransferase, xDnmt1, but it is another activity of the protein than its enzymatic methyltransferase function that is responsible for transcriptional repression (Dunican et al., 2008). Premature transcriptional activation occurs if translation of the maternal mRNA encoding this protein is blocked, and transcriptional repression can be rescued by introduction of a morpholino-insensitive Dnmt1; but this result is obtained just as well if a mutant form of the protein lacking the enzymatic methyltransferase activity is introduced. Further observations showed that the global repression is due to binding of the xDnmt1 protein by a noncatalytic domain to promoters of target genes. Since the amount of maternal mRNA encoding this protein is not augmented by new transcription during early development, general transcriptional reactivation would occur when the level of the protein becomes insufficient to bind the number of promoter targets present in the embryo at twelfth cleavage. This is undoubtedly not the only mechanism at work here, however. For example, the time of transcriptional reactivation is also just when the amount of the basal transcription protein TBP (TATA-binding protein) reaches its normal level, as a result of continuous accumulation by translation from maternal mRNA during the period of relative transcriptional quiescence (Akhtar and Veenstra, 2011). None of these mechanisms, nor the others that have been adduced, such as excess histones or lack of histone modifications (Akhtar and Veenstra, 2011) operate by gene-specific protein–DNA sequence recognition.

In zebrafish however, pregastrular differentiation gene expression is specifically suppressed by a sequence-specific regulatory mechanism. Here transcription resumes at the 512-cell stage at 3h of development. The earliest expressed cohort of genes is primed by binding of the transcription factor Pou5f1 (Oct4) in advance of transcription, which functions synergistically with ubiquitous SoxB1 factors in the activation of these genes at the mid-blastula transition (Leichsenring et al., 2013). These genes are marked by compound *cis*-regulatory Sox-Pou binding sites. However, many tissue-specific differentiation genes and the regulatory genes that control them are activated only following gastrulation at 6–8h. The timing of expression of hundreds of such differentiation genes is dependent on the maternally and zygotically expressed zebrafish *oct4* regulatory gene (*pou5f1*), which for some large gene sets works together with *sox2* (Onichtchouk et al., 2010; Iwafuchi-Doi et al., 2011). Genome-wide microarray studies of the zygotic transcriptome in *pou5f1* mutants, *cis*-regulatory analysis, and much other experimental data demonstrate that *pou5f1* directly controls tissue-specific regulatory repressors, and these in turn keep differentiation genes off until the appropriate time. Thus in *pou5f1* mutants these genes turn on several hours early. An interesting aspect is the implied network wiring: single repressive regulatory genes under *pou5f1* control, such as *foxd3,* control many mesoderm and endoderm differentiation genes; the *pou5f1* targets *klf2* and *klf4* repress many ectoderm genes; *pou5f1* and *sox2* activate *hes* class repressors, and thereby indirectly prevent premature expression of many neuroectoderm differentiation genes. The choice of which of these pou5f1 target genes is activated where depends on additional *cis*-regulatory inputs from regionally expressed transcription factors (Kotkamp et al., 2014). This temporal control system suggests that very large sets of genes expressed in given major domains of the embryo, each probably including many diverse differentiation gene batteries, share relatively noncomplex, overriding systems of repressive *cis*-regulatory controls. Almost the same temporal network operates in mouse embryos (Onichtchouk et al., 2010), and in fact hundreds of the same genes are thus controlled. Furthermore, mouse *oct4* can rescue zebrafish *pou5f1* mutants. This places in a new light the role of *oct4* as a pluripotency gene in stem cells and ES cells, where the same repressors of differentiation genes, such as *foxd3, klf,* etc., are also operative. The implication is that the *oct4* pluripotency subnetwork of mammals originated as a lower vertebrate temporal control device for pre-gastrular suppression of embryonic differentiation gene transcription. The evolutionary insight illuminates a special encoded regulatory function intrinsic to the Mode 2 specification processes of early vertebrate embryos.

4.2 *Cis*-regulatory signal integration at a key control gene of the Spemann organizer

The Spemann organizer is a signaling center directly and indirectly required for formation of the major dorsal and anterior structures of the postgastrular embryo. It is specified at the gene regulatory level beginning immediately after the mid-blastular resumption of transcription, in a confined group of cells on the dorsal side of the posterior pole of the embryo (Fig. 3.7(A)–(F)). Its specification has been intensively studied in *Xenopus* embryos, and offers an interesting window on the spatial allocation of decisive regulatory functions in a Mode 2 specification process, beginning with anisotropically localized maternal determinants. Specification of the organizer, that is, installation of its spatially confined regulatory state, begins with the transient expression of two regulatory genes, *siamois* and its paralog *twin*, which in turn provide inputs to a definitive organizer gene *goosecoid* (*gsc*). This gene affects the expression of genes encoding many downstream organizer functions, such as producing extracellular antagonists of BMP signaling, stimulating cell motility, and repressing genes encoding ventral fates (Sander et al., 2007). By early gastrula stage the cells expressing *gsc* and the remaining suite of organizer genes comprise most of the prospective mesoderm and the adjacent dorsal endoderm (Fig. 3.7(F)). Our problem is how this regulatory state domain is formulated and positioned.

The process begins with the initial regulatory organization of the egg. As illustrated in Fig. 3.1(D), the maternal transcription factor VegT is asymmetrically distributed to the vegetal region of the egg. In addition, a cytoskeletal reorganization, oriented by the events of sperm entry, results in a cortical rotation in the direction of the future posterior pole. Oriented microtubule tracks form, along which kinesin molecular motors transport toward the posterior pole a complex of proteins which cause the local accumulation of β-catenin there (Weaver and Kimelman, 2004). The functions of β-catenin in the embryo begin long before the transcriptional activation at mid-blastula of the earliest zygotic regulatory genes, *siamois* and *twin* that are directly necessary and sufficient for organizer specification (Blythe et al., 2010; Bae et al., 2011). As early as the 256-cell stage, six cleavages before general resumption of transcription, the β-catenin/Tcf (here β-catenin/Lef3) complex binds at regulatory sites in several Nodal-like ligand genes (*xnr* genes), which are activated due to this and maternal VegT inputs (Yang et al., 2002). β-catenin/Tcf also binds in the *siamois cis*-regulatory sequence in advance of its transcription (Blythe et al., 2010). *Cis*-regulatory studies confirm that activation of both *twin* and *siamois* depend on their β-catenin/Tcf target sites (Brannon et al., 1997; Laurent et al., 1997; Crease et al., 1998; Fan et al., 1998; Nishita et al., 2000). Activation of *siamois* and *twin* then provides an essential spatial activating input to the *gsc* gene. However, the *gsc cis*-regulatory system acts as an integrator of two different spatial inputs. Via *twin* and *siamois*, one of these is the β-catenin/Tcf input, while the other is a vegetal to dorsal/posterior Nodal signal input (Reid et al., 2012; Sudou et al., 2012). This is shown in the diagrams of Fig. 3.7(G) and (H), where an activated Smad complex is the transducer of the Nodal signal input. Soon after, by gastrula stage, additional factors, viz. zygotic VegT, and Mix, Otx and Lim, all of zygotic origin, all Nodal targets (Koide et al., 2005), appear and bind at target sites in the *gsc cis*-regulatory region (Fig. 3.7(I) and (J)). As a result, the domain of *gsc* expression extends to the dorsal endoderm component of the organizer, as well as to the mesodermal component (Fig. 3.7(E) and (F)). Fig. 3.7(G)–(J) shows similarly how the regulatory state assembled in both the endodermal and mesodermal domains of the organizer cause the transcription of the *cerberus* (*cer*) gene, which encodes an extracellular BMP antagonist required for neuroectoderm induction in dorsal ectoderm.

Figure 3.7

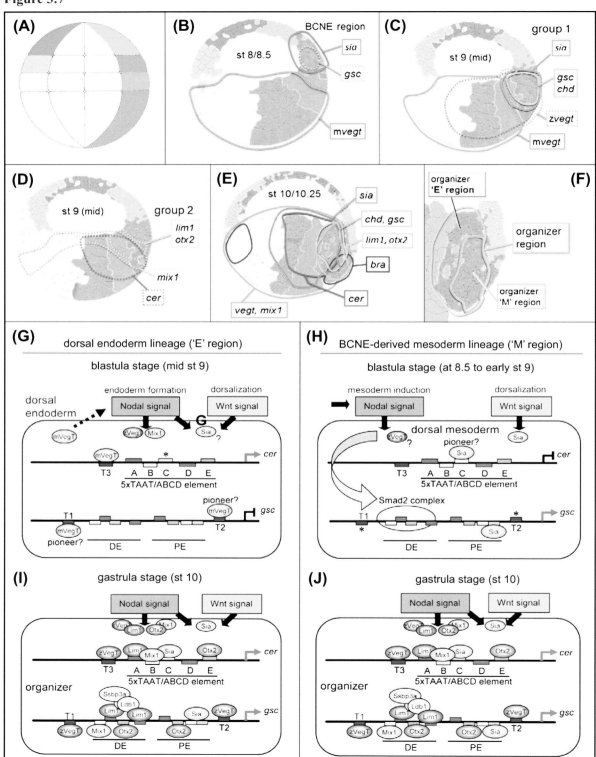

Figure 3.7 Progressive transcriptional specification of the organizer in *Xenopus*. (A–F) Detailed zygotic gene expression domains of the organizer, superimposed on color-coded maps of cellular descendants of fifth cleavage blastomeres, animal pole toward top, posterior to right (Bauer et al., 1994). (A), Canonical blastomere positions of fifth cleavage *Xenopus* embryo, color-coded map *(from Bauer et al. (1994))*. (B), Map of blastomere descendants at mid-blastula transition when transcription resumes, and location of some transcriptional regulators. The BCNE region which overlaps the domain populated mainly by descendants of the second tier dorsal fifth cleavage blastomere (red), will express genes encoding the secreted BMP antagonists Noggin and Chordin, downstream of the initial organizer regulator that is zygotically activated, *siamois* (*sia*). Within the *sia* domain *gsc* is activated. A yellow line shows on this map the location of maternally encoded VegT transcription factor. (C), Slightly later blastula map, zygotic *vegt* transcription has begun; note expanded *sia* expression domain. Group1 refers to genes expressed in BCNE (blastula *chordin-* and *noggin*-expressing) region. (D), Same stage, showing zygotically expressed Group 2 organizer genes expressed outside BCNE region. These include *cer* and the homeodomain genes *otx2*, *lim1*, and *mix1*. (E), Early gastrula map of same genes plus *brachyury*. (F), The organizer, defined here by transcription of *chordin*, *gsc*, *lim1*, and *otx2* genes: this derives mainly from two 5th-cleavage blastomeres (orange and blue). It includes both dorsal mesoderm (green line, "M region") and the remaining endodermal tissue (E region) within the yellow perimeter outlining the organizer. (G–J), *Cis*-regulatory occupancy of relevant control regions of *gsc* and *cer* genes, showing occupancy by organizer factors and more generally distributed factors, for which see (B–F). Evidence is from literature, including extensive *cis*-regulatory experiments, and ChIP studies (Sudou et al., 2012). (G) and (I), E region of organizer at mid-blastula and gastrula respectively; (H) and (J), M region of organizer just after transcriptional activation and at gastrula respectively *(B–J from Sudou et al. (2012))*.

The import of this sequence of events is that it shows in mechanistic detail how the encoded *gsc* and *cer* *cis*-regulatory systems utilize the earliest localized signal and transcriptional regulatory inputs to produce organizer specific functions. It is dramatically evident how the combinatorial input integration at the *gsc* and *cer* *cis*-regulatory systems accounts for an expression domain which is more confined and distinct than that of any of its inputs because of its AND logic function. However, as Fig. 3.7(E) and (F) shows in detail, the exact boundaries of *gsc* expression can still not be explained. More broadly, we see how the spatial regulatory organization of the egg and early embryo is translated progressively into the first steps of the conclusively transcriptional control system that ultimately defines the organizer.

4.3 A brief note on early mammalian embryogenesis

Eutherian mammals (placental mammals) develop indirectly, in that the pre-implantation embryo consists for an extended period of time of an extra-embryonic life support system surrounding multiplying totipotent cells. This is in contrast to embryogenesis in all nonmammalian vertebrates, in which ab initio complex regional patterning processes rapidly give rise to the various body parts of the postgastrular embryo, such as neural tube, gut, dorsal axis, etc. Nonplacental mammals differ: monotremes such as the platypus produce eggs which do not implant, and marsupial embryos implant only after gastrulation and germ layer specification. In eutherian development, only upon uterine implantation do the descendants of the totipotent cells of the embryo (epiblast) begin to execute the embryonic processes by which the vertebrate body plan is typically built. From this time on, the homology of the developmental process with other gastrular and postgastrular vertebrate embryos at every level, from morphology to signal deployment and gene regulatory transactions, is obvious. At the regulatory level, the preimplantation development of the eutherian embryo also deviates almost entirely from any other vertebrate process of embryogenesis (Stern, 2004).

Its regulatory state specification processes are extremely simple relative to that of any other developing embryo: it first produces a specialized external wall (trophectoderm) surrounding the embryonic inner cell mass. After implantation the trophectoderm does not contribute to the embryo proper, but only forms extraembryonic ectoderm. The inner cell mass then, by a process that initially appears stochastic, generates epiblast cells that retain totipotency, plus future "primitive endoderm" cells which do not ultimately contribute to the embryo proper. Sheltered within the pre-implantation trophectodermal wall, these two cell types express distinct transcriptional profiles (Chazaud et al., 2006). By implantation they have sorted out into a peripheral layer of primitive endoderm and the underlying epiblast cell mass. In other words, the whole blastocyst essentially expresses three regulatory states. There is no mid-blastula transition, and thus transcription is already active in the two-cell mouse embryo. Instead of operating on a totally closed nutritive system reliant only on stored yolk and other components, as do all free-living vertebrate embryos, even pre-implantation eutherian embryos take up nutrients from their fallopian environment. Most fundamentally, there is no obvious relation between localized regulatory components of the fertilized egg and transcriptional specification of definitive embryonic domains of special functional state, such as the organizer, the future gut endoderm, etc., as there is in frogs or fish (Fig. 3.1 and preceding Section). Thus development of the eutherian cleavage and blastula stage embryo is an evolutionary add-on with its own ways of doing business. The insight that the Oct4 based pluripotency system of its epiblast cells might derive from the ancestral global control system for temporal gene expression in vertebrate embryos (Onichtchouk et al., 2010) suggests one of the evolutionary routes toward invention of pre-implantation eutherian development. True Mode 2 specification processes are installed only as the polarity of the embryo is established on implantation. But then, in place of maternal anisotropies, the embryo instead positions its body axes with respect to the uterine wall by using signal inputs from its asymmetrically located extraembryonic tissues, the extraembryonic ectoderm, and the visceral endoderm, a descendant of the primitive endoderm (Rossant and Tam, 2009).

5. Global Aspects of A/P Spatial Regulatory Patterning in the Syncytial *Drosophila* Blastoderm

At ninth cleavage the dividing nuclei of the syncytial *Drosophila* embryo migrate to the surface of the egg, where they are embedded in a peripheral layer of yolk-free cytoplasm. By tenth cleavage zygotic expression of the earliest gap genes begins in broad patterns within the two-dimensional syncytial blastoderm. The spatial mosaic of gene expression rapidly increases in complexity, and within a period of about 70 min between the thirteenth cleavage cycle and the onset of cellularization in mid-fourteenth cleavage cycle, sharply bounded spatial domains of nuclear gene expression are set up throughout the syncytial blastoderm, forming an orthogonal grid of both A/P and D/V bands and stripes of regulatory gene transcription. These syncytial regulatory state patterns foreshadow many aspects of the postgastrular body plan, including, for the A/P patterns, its segmental metamerism, the initiation of head regionalization, and the initial regionalization of thorax and abdomen; and for the D/V stripes, the specification of future mesodermal, neuroectodermal, and dorsal epidermal domains and the initial template for the later mediolateral patterning of the CNS. Immediately after cellularization, gastrulation ensues; the embryo is little more than 3 h post fertilization.

Formation of these complex regulatory state patterns in a two-dimensional syncytial system presents unique conceptual challenges which have attracted a great deal of intellectual attention. Unlike any other of the early developmental processes discussed in this chapter, the early *Drosophila* embryo cannot rely on intercellular signaling and in addition it has to form sharp boundaries even though transcription factors are free to diffuse to adjacent nuclei. In order to specify positional regulatory states, this system utilizes two different initial maternal anisotropies, causing anterior localization of *bicoid* mRNA and ventral nuclearization of Dorsal transcription factor, in order to generate orthogonal gradients of transcription factor proteins. The issue is how these graded inputs give rise to sharply bounded discrete regulatory state domains. Many individual enhancers active in this process have been beautifully analyzed. These regulatory modules read the dynamically changing ambient regulatory landscape so that when attached to reporters and introduced into the *Drosophila* germ line, they faithfully recreate the specific A/P or D/V patterns of expression of their particular parent genes, and a large literature on these continues to accumulate (for reviews see Schroeder et al., 2004; Davidson, 2006; Fujioka et al., 2012). The performance of these enhancers is in itself of enormous importance, because it demonstrates that the genomic *cis*-regulatory code contains all the necessary spatial information. For many years it was popularly thought that the instructive spatial information determining the locations of the stripes of expression is resident in the values of the concentrations of the "Bicoid (Bcd) morphogen" along the A/P axis, and of the "Dorsal morphogen" along the D/V axis. The presumption was that these concentrations are directly read by the respective *cis*-regulatory modules of downstream genes, and that spatial transcriptional activity is specifically determined by the morphogen concentration acting on the *cis*-regulatory designs at each position. We discuss the actual roles of these and of other graded morphogens per se in Chapter 6 of this book. But, it has become increasingly clear that the formation of regulatory state patterns based on graded inputs depends instead on gene network circuitry, particularly institution of networks of repressive interactions among spatially expressed zygotic regulatory genes. Thus, understanding of the progressive mechanisms by which these complex patterns are set up in the syncytial blastoderm during those 70 min requires a system level analysis. That is, we need a sufficiently complete GRN model which explains how the regulatory transactions at each step give rise to the next stage in the generation of the landscape of spatial regulatory states. Most is known about interactions in the A/P gene network. In addition to a vast amount of experimental information, understanding of the A/P network has been enriched by diverse predictive mathematical models, as also reviewed in Chapter 6.

The first point to be made expands to a global perspective the importance of the individual *cis*-regulatory analyses alluded to above. In one study, a computer program was trained on a set of known A/P *cis*-regulatory target site sequences to derive predictive parameters in order to discover novel A/P *cis*-regulatory modules, many of which were indeed experimentally validated (Schroeder et al., 2004). Another analysis predicted expression patterns of *cis*-regulatory modules active in A/P patterning with remarkable fidelity on the basis of the expression of input transcription factors, their DNA-binding specificities, and the *cis*-regulatory site configurations (Segal et al., 2008). In a third study 32 new Bcd-dependent enhancers were found by use of Bcd ChIP and site clustering data (Fig. 3.8(A)), bringing to 66 the number of known validated A/P enhancers, all utilizing Bcd inputs (Chen et al., 2012). The main point here is that aspects of the genomic *cis*-regulatory sequence suffice to predictively identify A/P modules active in early cycle 14, and also to predict their spatial activities. Some generalizations emerge from these system-wide analyses of A/P *cis*-regulatory modules: the maternal inputs are positive, while the products of gap genes often function as repressors; usually given enhancers operate on a small variety of activators but respond to multiple repressive inputs; there is no evidence that the strength of the Bcd input, i.e., the local concentration of Bcd, determines the spatial boundaries of the expression domains of Bicoid-responsive genes (Jaeger et al., 2004; Schroeder et al., 2004; Ochoa-Espinosa et al., 2005; Segal et al., 2008; Jaeger, 2011; Chen et al., 2012).

The most intensively explored GRN of cycle 14 is the trunk and abdominal gap gene system, including *hunchback* (*hb*), *Kruppel* (*Kr*), *knirps* (*kni*), *giant* (*gt*), *huckebein* (*hkb*), and *tailless* (*tll*). The very broadly distributed initial maternal inputs are indicated in the drawings of Fig. 3.8(B). A large amount of experimental information has been accumulating for 20 years on most if not all of these genes, including very detailed and quantitative high-resolution spatial and temporal expression measurements, mutational interaction data, some *cis*-regulatory analyses, and computational analyses of target site sequences. The completeness of our understanding has been materially advanced by large-scale modeling efforts (Sánchez and Thieffry, 2001; Perkins et al., 2006; Jaeger, 2011; Jaeger et al., 2013). We discuss these models in detail in Chapter 6. As summarized in Fig. 3.8(C) (Jaeger, 2011), gap gene pattern formation is the output

Figure 3.8

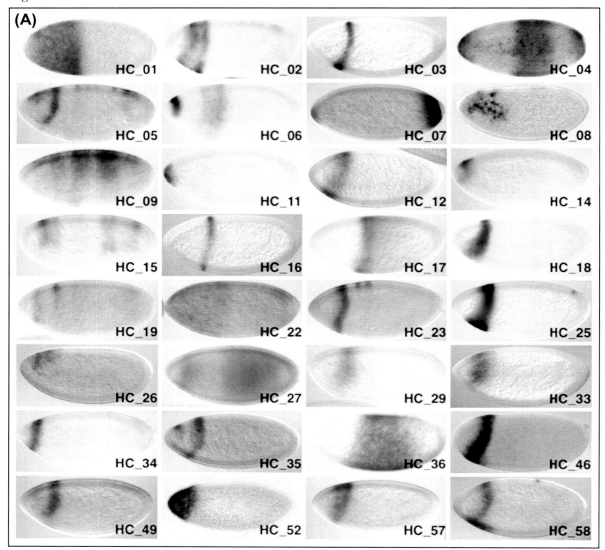

Figure 3.8 Transcriptional control of anterior/posterior regulatory gene expression in the syncytial *Drosophila* embryo. (A), Demonstration that Bicoid (Bcd) DNA-binding sites suffice to reveal Bcd-dependent enhancers active in early *Drosophila* embryos. In this study putative enhancers were associated with lacZ reporters (cf. Fig.1.3) and tested by gene transfer; expression patterns in cycle 14 embryos are shown anterior to left, dorsal up. Constructs 1–33 contain fragments that bound Bcd in a ChIP study (+/−500bp from peak); these were from all of the top 50 peaks not already known to be Bcd target genes. Of 33 fragments tested 28 were active, of which the 24 shown here were demonstrated to be Bcd dependent by crossing them into *bcd* mutant flies. The eight additional Bcd-dependent enhancers shown, constructs 34–58, were from 25 putative enhancers tested which had lower ChIP signals but also computationally identified Bcd site clusters; computational predictions which lacked ChIP signal failed to identify any active enhancers (*from Chen et al. (2012)*). (B), Broadly positioned maternal factors which provide inputs into the gap gene GRN. Data (except for Nanos) were acquired by digital imaging. Top, the approximate Bcd and Caudal (Cad) transcription factor gradients; for Bcd see Fig. 3.1(B). Bcd

Figure 3.8

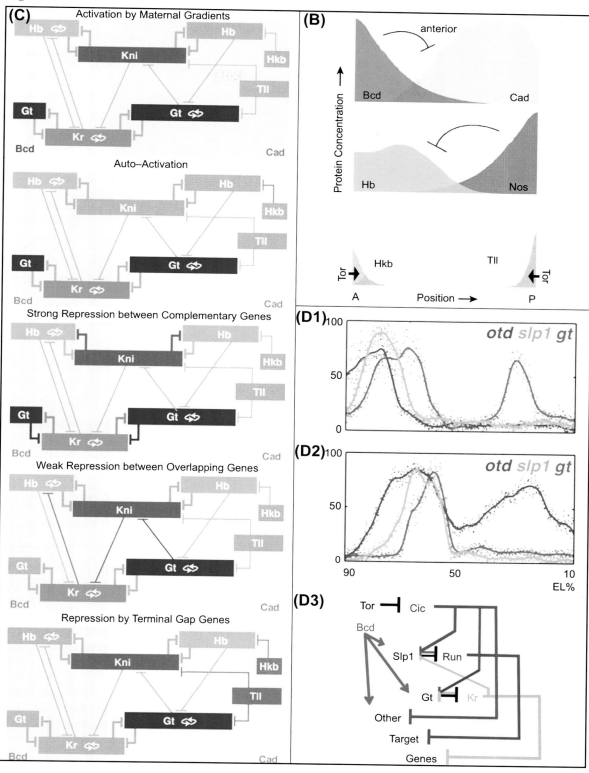

(C) Activation by Maternal Gradients

Auto–Activation

Strong Repression between Complementary Genes

Weak Repression between Overlapping Genes

Repression by Terminal Gap Genes

Figure 3.8

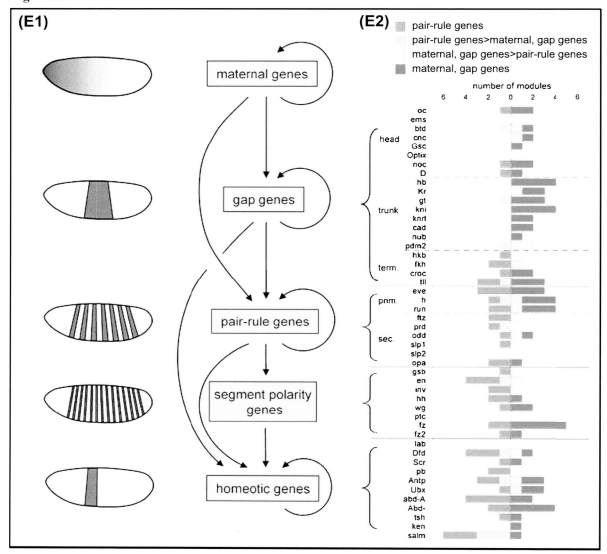

represses translation of maternal *cad* mRNA, binding to a 3′ mRNA sequence element, and resulting in a graded anterior absence of Cad. Middle, ubiquitously distributed *hunchback* (*hb*) maternal mRNA is not translated in the posterior region due to by a posteriorly localized maternal translational repressor, Nanos (Nos). Bottom, maternally encoded Torso tyrosine kinase receptor (Tor) is activated at both ends of the egg by an extracellular peptide, leading to transcriptional derepression of *huckebein* (*hkb*) at the anterior end and *tailless* (*tll*) at posterior end. This occurs by Torso-mediated phosphorylation of a maternal repressor of these genes, Capicua (cic) (Jiménez et al., 2012). (C), Gap gene network circuitry features, fourteenth cycle. Color indicates genes and linkages affected by the respective mechanisms, from top to bottom activation by maternal factors, auto-activation, exclusionary repression, weak repression in overlap domains, and effect of repression by Hkb and Tll *((B) and (C) from Jaeger (2011))*. (D), Demonstration that repressors set the posterior expression boundaries of Bcd-dependent regulatory genes and repression circuitry *(from Chen et al. (2012))*. (D1, 2), Expression data from triple fluorescent in situ hybridizations detecting expression of the head gap gene *orthodenticle* (*otd*), the trunk gap gene *giant* (*gt*) and the *sloppy paired* gene (*slp*) which encodes an anterior transcriptional repressor. (D1), Wild type spatial patterns; (D2), patterns observed in embryos mutant for two repressors which were experimentally demonstrated to control many Bcd target genes, the expression of which is confined to the anterior regions of the embryo. Gross ectopic posterior expression is observed for *otd* in this mutant. The repressors missing from these embryos are encoded by the *runt* gene, at this stage normally expressed throughout the middle of the embryo, and the maternal effect gene *cic* which, as above, is normally downregulated by terminal Torso activity. (D3), Network circuit indicating the repressive mechanism by which Bcd responsive gene expression is spatially controlled. Multiple repressors set these boundaries combinatorially for each gene, i.e., depending on the *cis*-regulatory target sites present in the control system of each gene (including those encoding these repressors). Note that the *runt* and *cic* genes are not activated by Bcd. However, Bcd does activate *slp* and *gt* expression and these two repressors provide the anterior boundaries for *run* and for *Kr* transcription. The *slp* and *run* genes mutually repress each other and their expression domain boundaries are positioned independently of Bcd concentration; similarly *gt* and *Kr* repress each other (cf. C). These relations set the positions of the Cic, Runt, and Kr repressors that control the expression boundaries of this set of Bcd target genes. (E), Input sources in the hierarchical anterior/posterior GRN *(from Schroeder et al. (2004))*. (E1), Hierarchy of the zygotic GRN (i.e., from the gap genes down), displaying sources of the regulatory states to which the genes at each level respond. (E2), *Cis*-regulatory modules were predicted computationally, by use of an algorithm that successfully identified known and novel enhancers of gap genes and pair rule genes. These modules were examined for target sites and classified (color coded) according to their inputs. The genes are arranged according to the hierarchy in (E1), as indicated by brackets: term., terminal regions; prim., primary pair rule genes; sec., secondary pair rule genes.

of five different classes of interactions, supported by experimental observation and dynamic modeling: (1) activation by maternal Bcd and Caudal factors; (2) positive auto-activation of three of the gap genes once they are turned on; (3) powerful repression between pairs of genes whose domains of expression directly abut, whereby the boundaries are mutually set; (4) mild repression amongst genes expressed in overlapping patterns, which is in every case stronger toward the posterior domain boundaries and which accounts for dynamic boundary shifts such that the posterior genes are dominant; (5) repression at the posterior end of the egg by terminal gap genes which sets the posterior boundaries of *hb*, *gt*, and *kni* expression.

The head gap genes *orthodenticle* (*otd*), *ems*, *buttonhead* (*btd*), anterior *gt*, and (initially) *sloppy paired* (*slp*) are subject to a different regimen (Chen et al., 2012). These are Bcd-dependent genes, the posterior expression boundaries of which are determined by two other genes encoding repressors, *runt* which is expressed in a broad domain in the middle of the embryo in early cycle 14, and *capicua*, which is of maternal origin and functions throughout the syncytial embryo except at the terminal regions, where it is down regulated by Torso. The effects of mutational loss of *runt* and *capicua* expression on the posterior boundaries of *otd*, *slp*, and *gt* expression domains can be seen in Fig. 3.8(D1),(D2). The gap gene interactions are summarized in Fig. 3.8(D3), where Bcd inputs and mutual repression circuits are indicated.

Finally, downstream of the gap gene network, activation of pair rule genes (still in the syncytial period of cycle 14), and then segment polarity and *hox* genes follows a logic that is indicated on a system level by the type of input the relevant *cis*-regulatory modules receive. Thus as Fig. 3.8(E) (Schroeder et al., 2004) shows, the gap gene *cis*-regulatory modules receive inputs from maternal genes and gap genes; the primary pair rule genes also receive inputs from these sources; the secondary pair rule genes receive inputs mainly from the other pair rule genes; and so forth, though as can be seen in the flow chart, the gap genes continue (briefly) to provide inputs at all subsequent levels (Fig. 3.8(E2)). In sum, although the 70 min process of A/P syncytial pattern formation in *Drosophila* is anything but paradigmatic of developmental processes in the vast majority of animal groups, it is a dramatic example of complex and precise spatial regulatory gene expression in an early embryo. And for some portions of the process, particularly the gap gene network, we have an unusually extensive causal understanding at the system level which reveals the regulatory basis of syncytial pattern formation in a bald way. This is a process stripped of signaling, cell movement, or cell lineage: the causal lines of encoded genomic regulation are thus clearly shown to be the fundamental, necessary, and sufficient mechanism by which novel programs of spatial gene expression are dynamically installed in the embryo.

Throughout this chapter the focus has been on mechanisms of direct determination of embryonic development by the genomic regulatory code. We first considered pan-bilaterian features of embryonic developmental control systems. However, the several modes of pregastrular embryogenesis entail entirely different control strategies by which are formed the initial spatial subdivision of territorial regulatory states and hence cell fates. We now turn to the postgastrular processes of adult body part formation. Here the end products of these processes are entirely different in structure and function, but in contrast to early embryogenesis a basic uniformity can be perceived in the fundamental strategies underlying bilaterian body part formation.

REFERENCES

Akhtar, W., Veenstra, G.J., 2011. TBP-related factors: a paradigm of diversity in transcription initiation. Cell. Biosci. 1, 23.

Bae, S., Reid, C.D., Kessler, D.S., 2011. Siamois and Twin are redundant and essential in formation of the Spemann organizer. Dev. Biol. 352, 367–381.

Barolo, S., Posakony, J.W., 2002. Three habits of highly effective signaling pathways: principles of transcriptional control by developmental cell signaling. Genes. Dev. 16, 1167–1181.

Bauer, D.V., Huang, S., Moody, S.A., 1994. The cleavage stage origin of Spemann's organizer: analysis of the movements of blastomere clones before and during gastrulation in *Xenopus*. Development 120, 1179–1189.

Beh, J., Shi, W., Levine, M., Davidson, B., Christiaen, L., 2007. FoxF is essential for FGF-induced migration of heart progenitor cells in the ascidian *Ciona intestinalis*. Development 134, 3297–3305.

Ben-Tabou de-Leon, S., Su, Y.H., Lin, K.T., Li, E., Davidson, E.H., 2013. Gene regulatory control in the sea urchin aboral ectoderm: spatial initiation, signaling inputs, and cell fate lockdown. Dev. Biol. 374, 245–254.

Blythe, S.A., Cha, S.W., Tadjuidje, E., Heasman, J., Klein, P.S., 2010. Beta-catenin primes organizer gene expression by recruiting a histone H3 arginine 8 methyltransferase, Prmt2. Dev. Cell 19, 220–231.

Bourlat, S.J., Nielsen, C., Economou, A.D., Telford, M.J., 2008. Testing the new animal phylogeny: a phylum level molecular analysis of the animal kingdom. Mol. Phylogenet. Evol. 49, 23–31.

Bowerman, B., Draper, B.W., Mello, C.C., Priess, J.R., 1993. The maternal gene skn-1 encodes a protein that is distributed unequally in early *C. elegans* embryos. Cell 74, 443–452.

Brachet, J., 1933. Recherche sur la synthese de l'acid thymonucleique pendant le development de l'oef d'oursin. Arch. Biol. 44, 519.

Brachet, J., Ficq, A., Tencer, R., 1963. Amino acid incorporation into proteins of nucleate and anucleate fragments of sea urchin eggs: effects of parthenogenetic activation. Exp. Cell Res. 32, 168–170.

Brannon, M., Gomperts, M., Sumoy, L., Moon, R.T., Kimelman, D., 1997. A beta-catenin/XTcf-3 complex binds to the siamois promoter to regulate dorsal axis specification in *Xenopus*. Genes. Dev. 11, 2359–2370.

Broitman-Maduro, G., Owraghi, M., Hung, W.W., Kuntz, S., Sternberg, P.W., Maduro, M.F., 2009. The NK-2 class homeodomain factor CEH-51 and the T-box factor TBX-35 have overlapping function in *C. elegans* mesoderm development. Development 136, 2735–2746.

Calestani, C., Rogers, D.J., 2010. *Cis*-regulatory analysis of the sea urchin pigment cell gene polyketide synthase. Dev. biol. 340, 249–255.

Chan, T.M., Longabaugh, W., Bolouri, H., Chen, H.L., Tseng, W.F., Chao, C.H., Jang, T.H., Lin, Y.I., Hung, S.C., Wang, H.D., Yuh, C.H., 2009. Developmental gene regulatory networks in the zebrafish embryo. Biochim. Biophys. Acta 1789, 279–298.

Chazaud, C., Yamanaka, Y., Pawson, T., Rossant, J., 2006. Early lineage segregation between epiblast and primitive endoderm in mouse blastocysts through the Grb2-MAPK pathway. Dev. Cell 10, 615–624.

Chen, H., Xu, Z., Mei, C., Yu, D., Small, S., 2012. A system of repressor gradients spatially organizes the boundaries of Bicoid-dependent target genes. Cell 149, 618–629.

Cheung, L.S., Schüpbach, T., Shvartsman, S.Y., 2011. Pattern formation by receptor tyrosine kinases: analysis of the Gurken gradient in *Drosophila* oogenesis. Curr. Opin. Genet. Dev. 21, 719–725.

Christiaen, L., Davidson, B., Kawashima, T., Powell, W., Nolla, H., Vranizan, K., Levine, M., 2008. The transcription/migration interface in heart precursors of *Ciona intestinalis*. Science 320, 1349–1352.

Christiaen, L., Stolfi, A., Davidson, B., Levine, M., 2009. Spatio-temporal intersection of Lhx3 and Tbx6 defines the cardiac field through synergistic activation of Mesp. Dev. Biol. 328, 552–560.

Christiaen, L., Stolfi, A., Levine, M., 2010. BMP signaling coordinates gene expression and cell migration during precardiac mesoderm development. Dev. Biol. 340, 179–187.

Chuang, C.K., Wikramanayake, A.H., Mao, C.A., Li, X., Klein, W.H., 1996. Transient appearance of *Strongylocentrotus purpuratus* Otx in micromere nuclei: cytoplasmic retention of SpOtx possibly mediated through an alpha-actinin interaction. Dev. Genet. 19, 231–237.

Coffman, J.A., Coluccio, A., Planchart, A., Robertson, A.J., 2009. Oral-aboral axis specification in the sea urchin embryo III. Role of mitochondrial redox signaling via H2O2. Dev. biol. 330, 123–130.

Coffman, J.A., McCarthy, J.J., Dickey-Sims, C., Robertson, A.J., 2004. Oral-aboral axis specification in the sea urchin embryo II. Mitochondrial distribution and redox state contribute to establishing polarity in *Strongylocentrotus purpuratus*. Dev. biol. 273, 160–171.

Collart, C., Owens, N.D., Bhaw-Rosun, L., Cooper, B., De Domenico, E., Patrushev, I., Sesay, A.K., Smith, J.N., Smith, J.C., Gilchrist, M.J., 2014. High-resolution analysis of gene activity during the *Xenopus* mid-blastula transition. Development 141, 1927–1939.

Crease, D.J., Dyson, S., Gurdon, J.B., 1998. Cooperation between the activin and Wnt pathways in the spatial control of organizer gene expression. Proc. Natl. Acad. Sci. U.S.A. 95, 4398–4403.

Cui, M., Li, E., Siriwion, N., Davidson, E.H., Peter, I.S., 2014. Specific Functions of the Wnt Signaling System in Global Gene Regulatory Networks of Sea Urchin Embryos.

Damle, S., Davidson, E.H., 2011. Precise *cis*-regulatory control of spatial and temporal expression of the alx-1 gene in the skeletogenic lineage of *S. purpuratus*. Dev. Biol. 357, 505–517.

Davidson, E.H., 1968. Gene Activity in Early Development, first ed. Academic Press, New York.

Davidson, E.H., 1986. Gene Activity in Early Development, third ed. Academic Press/Elsevier, San Diego, CA.

Davidson, E.H., 1990. How embryos work: a comparative view of diverse modes of cell fate specification. Development 108, 365–389.

Davidson, E.H., 2006. The Regulatory Genome. Gene Regulatory Networks in Development and Evolution. Academic Press/Elsevier, San Diego, CA.

Davidson, E.H., 2010. Emerging properties of animal gene regulatory networks. Nature 468, 911–920.

Davidson, E.H., Britten, R.J., 1971. Note on the control of gene expression during development. J. Theor. Biol. 32, 123–130.

Davidson, E.H., Peterson, K.J., Cameron, R.A., 1995. Origin of bilaterian body plans: evolution of developmental regulatory mechanisms. Science 270, 1319–1325.

Denny, P.C., Tyler, A., 1964. Activation of protein biosynthesis in non-nucleate fragments of sea urchin eggs. Biochem. Biophys. Res. Commun. 14, 245–249.

Dunican, D.S., Ruzov, A., Hackett, J.A., Meehan, R.R., 2008. xDnmt1 regulates transcriptional silencing in pre-MBT *Xenopus* embryos independently of its catalytic function. Development 135, 1295–1302.

Fan, M.J., Grüning, W., Walz, G., Sokol, S.Y., 1998. Wnt signaling and transcriptional control of siamois in *Xenopus* embryos. Proc. Natl. Acad. Sci. U.S.A. 95, 5626–5631.

Farley, B.M., Pagano, J.M., Ryder, S.P., 2008. RNA target specificity of the embryonic cell fate determinant POS-1. RNA 14, 2685–2697.

Farley, B.M., Ryder, S.P., 2008. Regulation of maternal mRNAs in early development. Crit. Rev. Biochem Mol. Biol. 43, 135–162.

Fujioka, M., Gebelein, B., Cofer, Z.C., Mann, R.S., Jaynes, J.B., 2012. Engrailed cooperates directly with Extradenticle and Homothorax on a distinct class of homeodomain binding sites to repress sloppy paired. Dev. Biol. 366, 382–392.

Fukushige, T., Hawkins, M.G., McGhee, J.D., 1998. The GATA-factor elt-2 is essential for formation of the *Caenorhabditis elegans* intestine. Dev. Biol. 198, 286–302.

Goldstein, B., 1995. An analysis of the response to gut induction in the *C. elegans* embryo. Development 121, 1227–1236.

Good, K., Ciosk, R., Nance, J., Neves, A., Hill, R.J., Priess, J.R., 2004. The T-box transcription factors TBX-37 and TBX-38 link GLP-1/Notch signaling to mesoderm induction in *C. elegans* embryos. Development 131, 1967–1978.

Grobben, K., 1908. Die Systematische Einteilen de Tierreiches. Verh. Kais.-Koniglichen zool.-bot. Gesellshaft Vien 58, 491–511.

Gross, P.R., Cousineau, G.H., 1964. Macromolecule synthesis and the influence of actinomycin on early development. Exp. Cell Res. 33, 368–395.

Henry, J.J., Perry, K.J., Fukui, L., Alvi, N., 2010. Differential localization of mRNAs during early development in the mollusc, *Crepidula fornicata*. Integr. Comp. Biol. 50, 720–733.

Huang, N.N., Mootz, D.E., Walhout, A.J., Vidal, M., Hunter, C.P., 2002. MEX-3 interacting proteins link cell polarity to asymmetric gene expression in *Caenorhabditis elegans*. Development 129, 747–759.

Huynh, J.R., St Johnston, D., 2004. The origin of asymmetry: early polarisation of the *Drosophila* germline cyst and oocyte. Curr. Biol. 14, R438–R449.

Imai, K.S., Stolfi, A., Levine, M., Satou, Y., 2009. Gene regulatory networks underlying the compartmentalization of the *Ciona* central nervous system. Development 136, 285–293.

Iwafuchi-Doi, M., Yoshida, Y., Onichtchouk, D., Leichsenring, M., Driever, W., Takemoto, T., Uchikawa, M., Kamachi, Y., Kondoh, H., 2011. The Pou5f1/Pou3f-dependent but SoxB-independent regulation of conserved enhancer N2 initiates Sox2 expression during epiblast to neural plate stages in vertebrates. Dev. Biol. 352, 354–366.

de Joussineau, C., Bataillé, L., Jagla, T., Jagla, K., 2012. Diversification of muscle types in *Drosophila*: upstream and downstream of identity genes. Curr. Top. Dev. Biol. 98, 277–301.

Jaeger, J., 2011. The gap gene network. Cell. Mol. Life Sci. 68, 243–274.

Jaeger, J., Blagov, M., Kosman, D., Kozlov, K.N., Manu, Myasnikova, E., Surkova, S., Vanario-Alonso, C.E., Samsonova, M., Sharp, D.H., Reinitz, J., 2004. Dynamical analysis of regulatory interactions in the gap gene system of *Drosophila melanogaster*. Genetics 167, 1721–1737.

Jaeger, J., Manu, Reinitz, J., 2013. *Drosophila* blastoderm patterning. Curr. Opin. Genet. Dev.

Jiménez, G., Shvartsman, S.Y., Paroush, Z., 2012. The Capicua repressor – a general sensor of RTK signaling in development and disease. J. Cell Sci. 125, 1383–1391.

King, M.L., Messitt, T.J., Mowry, K.L., 2005. Putting RNAs in the right place at the right time: RNA localization in the frog oocyte. Biol. Cell 97, 19–33.

Kingsley, E.P., Chan, X.Y., Duan, Y., Lambert, J.D., 2007. Widespread RNA segregation in a spiralian embryo. Evol. Dev. 9, 527–539.

Koide, T., Hayata, T., Cho, K.W., 2005. *Xenopus* as a model system to study transcriptional regulatory networks. Proc. Natl. Acad. Sci. U.S.A. 102, 4943–4948.

Kotkamp, K., Mössner, R., Allen, A., Onichtchouk, D., Driever, W., 2014. A Pou5f1/Oct4 dependent Klf2a, Klf2b, and Klf17 regulatory sub-network contributes to EVL and ectoderm development during zebrafish embryogenesis. Dev. Biol. 385, 433–447.

Kugler, J.E., Gazdoiu, S., Oda-Ishii, I., Passamaneck, Y.J., Erives, A.J., Di Gregorio, A., 2010. Temporal regulation of the muscle gene cascade by Macho1 and Tbx6 transcription factors in *Ciona intestinalis*. J. Cell Sci. 123, 2453–2463.

Langdon, Y.G., Mullins, M.C., 2011. Maternal and zygotic control of zebrafish dorsoventral axial patterning. Annu. Rev. Genet. 45, 357–377.

Laurent, M.N., Blitz, I.L., Hashimoto, C., Rothbächer, U., Cho, K.W., 1997. The *Xenopus* homeobox gene twin mediates Wnt induction of goosecoid in establishment of Spemann's organizer. Development 124, 4905–4916.

Leichsenring, M., Maes, J., Mössner, R., Driever, W., Onichtchouk, D., 2013. Pou5f1 transcription factor controls zygotic gene activation in vertebrates. Science 341, 1005–1009.

Lemaire, P., 2009. Unfolding a chordate developmental program, one cell at a time: invariant cell lineages, short-range inductions and evolutionary plasticity in ascidians. Dev. Biol. 332, 48–60.

Li, E., Cui, M., Peter, I.S., Davidson, E.H., 2014. Encoding regulatory state boundaries in the pregastrular oral ectoderm of the sea urchin embryo. Proc. Natl. Acad. Sci. U.S.A. 111(10): E906–E913.

Li, E., Materna, S.C., Davidson, E.H., 2012. Direct and indirect control of oral ectoderm regulatory gene expression by Nodal signaling in the sea urchin embryo. Dev. Biol. 369, 377–385.

Linnaeus, C., 1758. Systema Naturae, tenth ed.

Logan, C.Y., Miller, J.R., Ferkowicz, M.J., McClay, D.R., 1999. Nuclear beta-catenin is required to specify vegetal cell fates in the sea urchin embryo. Development 126, 345–357.

Loose, M., Patient, R., 2004. A genetic regulatory network for *Xenopus* mesendoderm formation. Dev. Biol. 271, 467–478.

Lu, F.I., Thisse, C., Thisse, B., 2011. Identification and mechanism of regulation of the zebrafish dorsal determinant. Proc. Natl. Acad. Sci. U.S.A. 108, 15876–15880.

López-Schier, H., St Johnston, D., 2001. Delta signaling from the germ line controls the proliferation and differentiation of the somatic follicle cells during *Drosophila* oogenesis. Genes. Dev. 15, 1393–1405.

Maduro, M.F., 2010. Cell fate specification in the *C. elegans* embryo. Dev. Dyn. 239, 1315–1329.

Maduro, M.F., Rothman, J.H., 2002. Making worm guts: the gene regulatory network of the *Caenorhabditis elegans* endoderm. Dev. Biol. 246, 68–85.

Materna, S.C., Ransick, A., Li, E., Davidson, E.H., 2013. Diversification of oral and aboral mesodermal regulatory states in pregastrular sea urchin embryos. Dev. Biol. 375, 92–104.

McGhee, J.D., Fukushige, T., Krause, M.W., Minnema, S.E., Goszczynski, B., Gaudet, J., Kohara, Y., Bossinger, O., Zhao, Y., Khattra, J., Hirst, M., Jones, S.J., Marra, M.A., Ruzanov, P., Warner, A., Zapf, R., Moerman, D.G., Kalb, J.M., 2009. ELT-2 is the predominant transcription factor controlling differentiation and function of the *C. elegans* intestine, from embryo to adult. Dev. Biol. 327, 551–565.

Merritt, C., Rasoloson, D., Ko, D., Seydoux, G., 2008. 3′ UTRs are the primary regulators of gene expression in the *C. elegans* germline. Curr. Biol. 18, 1476–1482.

Monroy, A., Tyler, A., 1963. Formation of active ribosomal aggregates (polysomes) upon fertilization and development of sea urchin eggs. Arch. Biochem. Biophys. 103, 431–435.

Murray, J.I., Boyle, T.J., Preston, E., Vafeados, D., Mericle, B., Weisdepp, P., Zhao, Z., Bao, Z., Boeck, M., Waterston, R.H., 2012. Multidimensional regulation of gene expression in the *C. elegans* embryo. Genome Res. 22, 1282–1294.

Nam, J., Su, Y.H., Lee, P.Y., Robertson, A.J., Coffman, J.A., Davidson, E.H., 2007. *Cis*-regulatory control of the nodal gene, initiator of the sea urchin oral ectoderm gene network. Dev. Biol. 306, 860–869.

Newport, J., Kirschner, M., 1982. A major developmental transition in early *Xenopus* embryos: II. Control of the onset of transcription. Cell 30, 687–696.

Nishida, H., 2002. Patterning the marginal zone of early ascidian embryos: localized maternal mRNA and inductive interactions. Bioessays 24, 613–624.

Nishita, M., Hashimoto, M.K., Ogata, S., Laurent, M.N., Ueno, N., Shibuya, H., Cho, K.W., 2000. Interaction between Wnt and TGF-beta signalling pathways during formation of Spemann's organizer. Nature 403, 781–785.

Ochoa-Espinosa, A., Yucel, G., Kaplan, L., Pare, A., Pura, N., Oberstein, A., Papatsenko, D., Small, S., 2005. The role of binding site cluster strength in Bicoid-dependent patterning in *Drosophila*. Proc. Natl. Acad. Sci. U.S.A. 102, 4960–4965.

Oliveri, P., Tu, Q., Davidson, E.H., 2008. Global regulatory logic for specification of an embryonic cell lineage. Proc. Natl. Acad. Sci. U.S.A. 105, 5955–5962.

Onichtchouk, D., Geier, F., Polok, B., Messerschmidt, D.M., Mössner, R., Wendik, B., Song, S., Taylor, V., Timmer, J., Driever, W., 2010. Zebrafish Pou5f1-dependent transcriptional networks in temporal control of early development. Mol. Syst. Biol. 6, 354.

Owraghi, M., Broitman-Maduro, G., Luu, T., Roberson, H., Maduro, M.F., 2010. Roles of the Wnt effector POP-1/TCF in the *C. elegans* endomesoderm specification gene network. Dev. Biol. 340, 209–221.

Paix, A., Yamada, L., Dru, P., Lecordier, H., Pruliere, G., Chenevert, J., Satoh, N., Sardet, C., 2009. Cortical anchorages and cell type segregations of maternal postplasmic/PEM RNAs in ascidians. Dev. Biol. 336, 96–111.

Perkins, T.J., Jaeger, J., Reinitz, J., Glass, L., 2006. Reverse engineering the gap gene network of *Drosophila melanogaster*. PLoS Comput. Biol. 2, e51.

Peter, I.S., Davidson, E.H., 2009. Modularity and design principles in the sea urchin embryo gene regulatory network. FEBS Lett. 583, 3948–3958.

Peter, I.S., Davidson, E.H., 2010. The endoderm gene regulatory network in sea urchin embryos up to mid-blastula stage. Dev. Biol. 340, 188–199.

Peter, I.S., Davidson, E.H., 2011. A gene regulatory network controlling the embryonic specification of endoderm. Nature 474, 635–639.

Peter, I.S., Davidson, E.H., 2013. Transcriptional network logic: the systems biology of development. In: Walhout, A.J.M., Vidal, M., Dekker, J. (Eds.), Handbook of Systems Biology. Concepts and Insights. Academic Press/Elsevier, pp. 211–228.

Peter, I.S., Faure, E., Davidson, E.H., 2012. Feature Article: predictive computation of genomic logic processing functions in embryonic development. Proc. Natl. Acad. Sci. U.S.A. 109, 16434–16442.

Philippe, H., Brinkmann, H., Copley, R.R., Moroz, L.L., Nakano, H., Poustka, A.J., Wallberg, A., Peterson, K.J., Telford, M.J., 2011. Acoelomorph flatworms are deuterostomes related to Xenoturbella. Nature 470, 255–258.

Rabinowitz, J.S., Lambert, J.D., 2010. Spiralian quartet developmental potential is regulated by specific localization elements that mediate asymmetric RNA segregation. Development 137, 4039–4049.

Raff, R.A., 1987. Constraint, flexibility, and phylogenetic history in the evolution of direct development in sea urchins. Dev. Biol. 119, 6–19.

Rafiq, K., Cheers, M.S., Ettensohn, C.A., 2012. The genomic regulatory control of skeletal morphogenesis in the sea urchin. Development 139, 579–590.

Rafiq, K., Shashikant, T., McManus, C.J., Ettensohn, C.A., 2014. Genome-wide analysis of the skeletogenic gene regulatory network of sea urchins. Development 141, 2542.

Range, R., Lapraz, F., Quirin, M., Marro, S., Besnardeau, L., Lepage, T., 2007. *Cis*-regulatory analysis of nodal and maternal control of dorsal-ventral axis formation by Univin, a TGF-beta related to Vg1. Development 134, 3649–3664.

Ransick, A., Davidson, E.H., 2006. *Cis*-regulatory processing of Notch signaling input to the sea urchin glial cells missing gene during mesoderm specification. Dev. Biol. 297, 587–602.

Reid, C.D., Zhang, Y., Sheets, M.D., Kessler, D.S., 2012. Transcriptional integration of Wnt and Nodal pathways in establishment of the Spemann organizer. Dev. Biol. 368, 231–241.

Revilla-i-Domingo, R., Oliveri, P., Davidson, E.H., 2007. A missing link in the sea urchin embryo gene regulatory network: hesC and the double-negative specification of micromeres. Proc. Natl. Acad. Sci. U.S.A. 104, 12383–12388.

Rossant, J., Tam, P.P., 2009. Blastocyst lineage formation, early embryonic asymmetries and axis patterning in the mouse. Development 136, 701–713.

Rota-Stabelli, O., Campbell, L., Brinkmann, H., Edgecombe, G.D., Longhorn, S.J., Peterson, K.J., Pisani, D., Philippe, H., Telford, M.J., 2011. A congruent solution to arthropod phylogeny: phylogenomics, microRNAs and morphology support monophyletic Mandibulata. Proc. Biol. Sci. 278, 298–306.

Roth, S., 2003. The origin of dorsoventral polarity in *Drosophila*. Philos. Trans. R. Soc. Lond B Biol. Sci. 358, 1317–1329 discussion 1329.

Roth, S., Lynch, J.A., 2009. Symmetry breaking during *Drosophila* oogenesis. Cold Spring Harb. Perspect. Biol. 1, a001891.

Rothbächer, U., Bertrand, V., Lamy, C., Lemaire, P., 2007. A combinatorial code of maternal GATA, Ets and beta-catenin-TCF transcription factors specifies and patterns the early ascidian ectoderm. Development 134, 4023–4032.

Sander, V., Reversade, B., De Robertis, E.M., 2007. The opposing homeobox genes Goosecoid and Vent1/2 self-regulate *Xenopus* patterning. EMBO J. 26, 2955–2965.

Sardet, C., Paix, A., Prodon, F., Dru, P., Chenevert, J., 2007. From oocyte to 16-cell stage: cytoplasmic and cortical reorganizations that pattern the ascidian embryo. Dev. Dyn. 236, 1716–1731.

Satou, Y., Satoh, N., Imai, K.S., 2009. Gene regulatory networks in the early ascidian embryo. Biochim. Biophys. Acta 1789, 268–273.

Schroeder, M.D., Pearce, M., Fak, J., Fan, H., Unnerstall, U., Emberly, E., Rajewsky, N., Siggia, E.D., Gaul, U., 2004. Transcriptional control in the segmentation gene network of *Drosophila*. PLoS Biol. 2, E271.

Segal, E., Raveh-Sadka, T., Schroeder, M., Unnerstall, U., Gaul, U., 2008. Predicting expression patterns from regulatory sequence in *Drosophila* segmentation. Nature 451, 535–540.

Shoguchi, E., Hamada, M., Fujie, M., Satoh, N., 2011. Direct examination of chromosomal clustering of organ-specific genes in the chordate *Ciona intestinalis*. Genesis 49, 662–672.

Smith, J., Davidson, E.H., 2009. Regulative recovery in the sea urchin embryo and the stabilizing role of fail-safe gene network wiring. Proc. Natl. Acad. Sci. U.S.A. 106, 18291–18296.

Smith, J., Kraemer, E., Liu, H., Theodoris, C., Davidson, E., 2008. A spatially dynamic cohort of regulatory genes in the endomesodermal gene network of the sea urchin embryo. Dev. Biol. 313, 863–875.

Stern, C., 2004. Gastrulation. Cold Spring Harbor Laboratory Press.

Sudou, N., Yamamoto, S., Ogino, H., Taira, M., 2012. Dynamic in vivo binding of transcription factors to *cis*-regulatory modules of cer and gsc in the stepwise formation of the Spemann-Mangold organizer. Development 139, 1651–1661.

Sulston, J.E., Schierenberg, E., White, J.G., Thomson, J.N., 1983. The embryonic cell lineage of the nematode *Caenorhabditis elegans*. Dev. Biol. 100, 64–119.

Swiers, G., Chen, Y.H., Johnson, A.D., Loose, M., 2010. A conserved mechanism for vertebrate mesoderm specification in urodele amphibians and mammals. Dev. Biol. 343, 138–152.

Sánchez, L., Thieffry, D., 2001. A logical analysis of the *Drosophila* gap-gene system. J. Theor. Biol. 211, 115–141.

Telford, M.J., Copley, R.R., 2011. Improving animal phylogenies with genomic data. Trends Genet. 27, 186–195.

Tseng, W.F., Jang, T.H., Huang, C.B., Yuh, C.H., 2011. An evolutionarily conserved kernel of gata5, gata6, otx2 and prdm1a operates in the formation of endoderm in zebrafish. Dev. Biol. 357, 541–557.

Weaver, C., Kimelman, D., 2004. Move it or lose it: axis specification in *Xenopus*. Development 131, 3491–3499.

Weitzel, H.E., Illies, M.R., Byrum, C.A., Xu, R., Wikramanayake, A.H., Ettensohn, C.A., 2004. Differential stability of beta-catenin along the animal-vegetal axis of the sea urchin embryo mediated by dishevelled. Development 131, 2947–2956.

Wray, G.A., Bely, A.E., 1994. The evolution of echinoderm development is driven by several distinct factors. Dev. Suppl. 97–106.

Yang, J., Tan, C., Darken, R.S., Wilson, P.A., Klein, P.S., 2002. Beta-catenin/Tcf-regulated transcription prior to the midblastula transition. Development 129, 5743–5752.

Zhang, J., King, M.L., 1996. *Xenopus* VegT RNA is localized to the vegetal cortex during oogenesis and encodes a novel T-box transcription factor involved in mesodermal patterning. Development 122, 4119–4129.

Chapter 4

Genomic Control Processes in Adult Body Part Formation

Genomic Control Process
http://dx.doi.org/10.1016/B978-0-12-404729-7.00004-6

Body parts are complex machines each composed of multiple subparts, which execute diverse functions. Their development is fundamentally unlike the construction of the complex machines that humans build, in that in our manufacturing processes the subcomponents are first built separately and are then assembled in a spatial arrangement that is specified a priori in the plan or blueprint. This includes in final detail the relative positions of all the subcomponents in respect to the spatial coordinates of the whole machine. In contrast, bilaterian body parts form from one or two progenitor cell populations which bear no functional resemblance to the characteristics of any of the subcomponents that descend from them, and all their spatial organization has to be generated de novo, beginning with the basic coordinate system of the body part. Indeed there is even a prior regulatory requirement, which is to place the location where the body part will form into the spatial coordinate system of the whole animal.

1.   Common Principles of Body Part Formation

The first step in body part formation is to define a progenitor field, a contiguous domain of dividing cells from the descendants of which a particular body part will form. The cells of the initial progenitor field express distinct regulatory genes which are not expressed by cells across the boundaries which separate the progenitor field from the adjacent tissues. In bilaterians the progenitor field often includes cell populations of different embryological origin, for example, the epidermal and mesodermal mesenchymal cells of the vertebrate limb bud. Ultimately, the body part will consist of cells descending from those of the progenitor field, plus in some cases migratory or invasive immigrant cells of distant origin. For example, the skeletal muscles of the vertebrate limb derive from immigrant muscle cells of somitic origin and the nerves and endothelial blood vessels also grow in from outside. Yet the spatial patterning system determining the developmental coordinates of the limb originates in the limb bud, just as the spatial patterning system determining the conformation of the jaw originates in the anterior pharyngeal arches even though much of the ultimate structure of the jaw is built by immigrant skeletogenic neural crest cells (discussed explicitly in Chapter 7). A general review of progenitor fields in adult body part formation can be found elsewhere (Davidson, 2001). In this chapter we take up the regulatory functions leading to formation of many different kinds of body part, in *Drosophila* and in vertebrates. Without exception, each of these processes begins with establishment of a progenitor field expressing a dedicated regulatory state.

1.1   Regulatory logic in progenitor field specification

Classical embryology experimentally determined the cells from which body parts originate in various embryos, and these embryonic cells were thus defined by the fates of their descendants. Since the location of the progenitor field in the embryo was known for each body part at each stage, this led to investigation of how it is positioned. It was quickly realized from cut and paste embryology that for many body parts the position of the progenitor field depends on signaling. Only relatively recently was it recognized that the role of signaling is to specify causally the expression of unique and developmentally novel regulatory states. We thus escape the phenomenological dead end in which body parts were defined simply as outcomes of signaling events. The regulatory logic is general and abstract: indeed, signals do contribute to progenitor field specification, but their specific function in each developmental context is determined by their effect on the expression of given regulatory genes in given locations. The specific initial regulatory states causally define the progenitor fields for given body parts. The genomic code for positioning the progenitor field, and thus the body part, is therefore resident in the *cis*-regulatory sequences of two classes of genes. The genes encoding the signals respond by their *cis*-regulatory target sites to regulatory inputs located at specific coordinate positions on the axes of the embryo. The genes encoding the transcription factors generating the dedicated regulatory states of each progenitor field are activated through their *cis*-regulatory target sites in response to signaling inputs and/or localized prior regulatory states (including *hox* genes for example; see below and Chapter 7). We shall see examples of both in what follows.

Downstream of the initially expressed components of the progenitor field regulatory state more complex gene regulatory networks (GRNs) are established. But it is interesting to note that typically there is only a very small number of regulatory genes initially activated to set up given progenitor fields. The transcription factors encoded by these genes of course cannot per se build a body part. They merely trigger activation of other immediately downstream genes initiating the functional GRN that will control the next stage of body part formation. The overly simplistic concept of "master regulatory genes" makes it necessary to emphasize that the information for establishing a localized progenitor field regulatory state is resident in networks of genes and in their widely distributed genomic control sequences, rather than in any one gene or transcription factor or signaling factor.

1.2   Internal regulatory coordinates

Once the progenitor field regulatory state is initiated at the correct locality within the coordinates of the embryo, the most important general control function that ensues in the process of body part formation is establishing its own internal coordinates (we return to this hierarchical process in an evolutionary context in Chapter 7; see Fig. 7.1). Ultimately, development of the body part will require the construction of a complex internal landscape of spatial regulatory states. The geometric organization of the regulatory states within the forming body part foreshadows the geometric regionalization of functionally diverse subparts of the body part, such as for example the chambers and outflow tracts of the heart, or the proximal-distal and anterior-posterior components of the limb, and so forth. This regulatory landscape will be used by downstream genes to activate effector gene cassettes in different spatial domains of the body part.

The initial spatial organization of regulatory states subdividing the progenitor field depends on inputs provided by regulatory genes previously expressed with respect to the axes of the embryo. To provide a heuristic example, the progenitor fields for the *Drosophila* wing, the wing imaginal discs, use the previously specified metameric stripes of regulatory gene expression to define the anterior/posterior boundaries within the discs. Cells on the posterior side of the metameric boundaries in the embryo express the *engrailed* gene. This regulatory asymmetry is used in the imaginal disc primordium to define a posterior section expressing *engrailed*. In the process of generating the differential regulatory landscapes within the progenitor field, and thus defining its future subparts as domains of diverse regulatory state, regional transcriptional activation of regulatory genes within the progenitor field is invariably followed by expression of signaling systems. Thus to continue our example, the *engrailed* expressing posterior cells of the wing imaginal disc activate transcription of the *hedgehog* signaling gene. By such means the domains present at each point in time affect the regulatory states in adjacent domains, a progressive, reiterative process which ends only when all the subparts of the body part are blocked out as regulatory states in space. The process of progenitor field specification is a spectacular example of an abstract control mechanism: the system goes through a multistep set of signaling and regulatory transactions the sole object of which is to construct a spatial mosaic of regulatory states, patterns of combinatorial regulatory gene expression.

1.3   Effector gene cassettes in body part formation

Net growth is almost always a major attribute of body part formation. Progenitor fields, since they are embedded in embryonic spatial coordinates, are small structures, consisting at first of a modest, finite number of cells. For instance, there are only about 30 cells in each of the first stage imaginal discs of *Drosophila*. Growth within the subdomains of the body part has to be controlled, and in many mammalian body parts for instance, some subdomains undergo more replications than do others. This raises two regulatory issues, how regional growth is controlled by the GRNs that operate in each locality; and how regulatory states are maintained across cell division such that the daughter cells are the same as their parents. The first of these issues is how the cell cycle is controlled locally in space. There are a number of "checkpoints" at which the biochemical activities of various enzymes, many of them kinases, either stimulate or block progression through the cell cycle. Ultimately, cell cycle activity is determined by the presence or absence of some dedicated transcription factors. Many examples are now known, some illustrated below, in which one or another of the regulatory genes expressed in a given developmental GRN has the function of directly determining the expression of these positively or negatively acting cell cycle checkpoint entities. Thus cell cycle control is among the outputs of GRNs.

How do daughter cells retain the transcriptional program of their immediate parents, until a new signal or a repressor enters the system and causes changes in the program? The answer to this question has several parts. The primary mechanism is that the transcription factors present in the parental cell are normally divided between the daughter cells, and thus the same regulatory state obtains in these as in the parental cell with the result that their own transcriptional program will be the same as that of the parental cell. From the dynamics discussion in Chapter 2 it will be clear that a temporary twofold drop in factor concentration per cell will generally be immaterial and indeed until late in development most regulatory genes are haplosufficient; one copy suffices to permit normal development (including of body parts). In addition there are circuitry features that contribute to maintenance of transcriptional programs across cell division, such as the prevalence of feedback loops in developmental GRNs (see discussion of these subcircuits in Chapter 6). In daughter cells feedback loops will rapidly reestablish the same concentration of the factors that mutually enforce each other's transcription as in the parental cell.

The form of the body part depends on the morphogenetic functions called into play during its growth and regulatory subdivision, particularly toward the end of the process. Its functional attributes are generated by the differentiation gene batteries deployed within each subdomain. As discussed in Chapter 3, GRNs control cassettes of morphogenesis genes by regulating the expression of several key effectors in each pathway. The functionally crucial control of differentiation gene sets requires that late in adult body part formation, each of the terminal subdomains must activate the drivers of the differentiation gene batteries it will express. These drivers then come to dominate the final regulatory states found in different regions of the body part, and their continued transcription, frequently also locked down by feedback loops, reflects the terminal GRNs that will continue to operate indefinitely.

2.   Limbs in Amniotes

The paired appendages of vertebrates provide particularly clear examples of many of the general aspects of adult body part formation, beginning with the specification of their progenitor fields, the limb buds. Here we focus on the regulatory transactions determining appendage formation at successive stages of development, from spatial specification of the limb bud to axial patterning of the limb. By this route we can perceive how major aspects of this spatial process of development are directly encoded in the genome. Though appendage development has also been studied extensively in fish and frogs, the following is confined to amniotes (chick and mouse); in Chapter 7 we take up the evolutionary process that led from fins and fin buds to limbs and limb buds.

2.1   Genomic code for limb bud progenitor field specification in amniotes

The initial problem is how the positions of the forelimb and hindlimb buds are established with respect to the axial coordinates of the amniote torso. We would ideally like to understand exactly how these positions are encoded in regulatory A's, C's, G's, and T's. The route to answering this question is to identify the initial, most upstream regulatory genes the activation of which specifically marks the limb buds, and to determine the *cis*-regulatory inputs that account for the spatial expression of these genes. In gnathostome embryogenesis the limb buds arise from lateral plate mesoderm, which early in development is divided into pharyngeal, cardiac, and posterior domains marked by distinct regulatory states (ph, anterior lateral plate mesoderm, and posterior lateral plate mesoderm in Fig. 4.1(A); Tanaka, 2013). The forelimb bud arises at the anterior end of the posterior lateral plate mesoderm from which the heart field is excluded by retinoic acid (RA) signaling; in the experimental absence of this signal the heart field spreads posteriorly and the forelimb bud does not form. However, RA does not cause the forelimb bud to form, it merely indirectly prevents its exclusion by the heart GRN (Tanaka, 2013). The initial regulatory functions that define the limb progenitor fields are expression of *tbx5* in the prospective forelimb bud and of *pitx1* and *tbx4* in the hindlimb bud (Logan et al., 1998). Their expression can be seen in Fig. 4.1(B) in very early limb buds and after outgrowth has started, and this expression is required for the initiation of limb development. In both forelimb and hindlimb buds the initial direct function of the *tbx* genes is installation of an essential Fgf10-Fgf8 feedback loop between mesenchymal mesoderm and the overlying ectoderm of the buds, which is essential for limb outgrowth. The limbs, though serially homologous, have some structural differences (Fig. 4.1(C)), and their patterning depends on complex developmental processes by which regional spatial regulatory states are installed. Illustrative evidence exists to demonstrate how the steps in limb development are encoded in the genome.

Initial specification of the forelimb bud in amniotes

It has long been suspected, based on much indirect evidence, that the location and morphogenetic processes of limb bud development depend on axial inputs from *hox* genes, which are expressed differentially in the lateral plate mesoderm depending on A/P position. A famous series of experiments in the 1990s (reviewed by Davidson (2001), Duboc and Logan (2011b)), showed that the lateral plate mesoderm of the chick embryo between the forelimb (wing) and hindlimb bud can give rise to an ectopic limb if a source of Fgf is implanted into the flank, and the character of the limb depends on its axial position; if more anterior it expresses *tbx5* and develops the structures of a wing, and if more posterior it expresses *tbx4* and becomes an ectopic leg, so the axial position is determinant. The direct connection between the position of the forelimb progenitor field and the A/P axis of the mouse embryo has been experimentally demonstrated, providing mechanistic evidence for a determinant role of *hox* genes in positioning the *tbx5* expression domain, i.e., the forelimb bud proper. Thus, a *cis*-regulatory study of an intronic enhancer in the mouse *tbx5* gene demonstrates that its expression in the forelimb bud is driven by *hox* genes of paralog groups 4 and 5 (Fig. 4.1(D); Minguillon et al., 2012). These Hox factors could also activate the *tbx5* reporter ectopically in electroporation experiments in the chick embryo. *Tbx5* expression in the forelimb bud and ultimately the location of the whole appendage results from a direct readout of local *hox* gene activation at this level of the anterior/posterior axis. The significance of the GRN linkage between *hox4/5* genes and *tbx5* is thus to correctly position the forelimb appendage within the amniote body plan.

Initial specification of the hindlimb bud

Many observations on knockout mice as well as ectopic overexpression in chicken embryos (Logan and Tabin, 1999) suggested that *pitx1* is upstream of *tbx4* expression in the hindlimb bud. Direct *cis*-regulatory investigation (Menke et al., 2008) reveals two conserved enhancers in the *tbx4* gene which in expression constructs produce hindlimb bud expression, one upstream of the transcription start site that is conserved only within mammals, and one >50 kb downstream of the last *tbx4* exon. This control module is evolutionarily older, as it is highly conserved from sharks to humans. A site-specific mutation experiment identified a crucial input at a conserved *trans*-mammalian *pitx1* site that is responsible for a significant fraction of the upstream enhancer activity, and chromatin immunoprecipitation (ChIP) data show that both enhancers

Figure 4.1

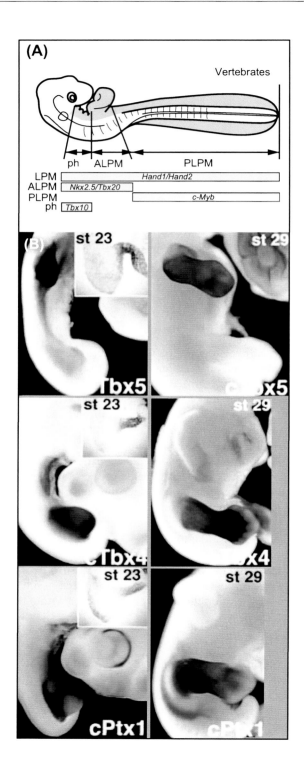

Figure 4.1

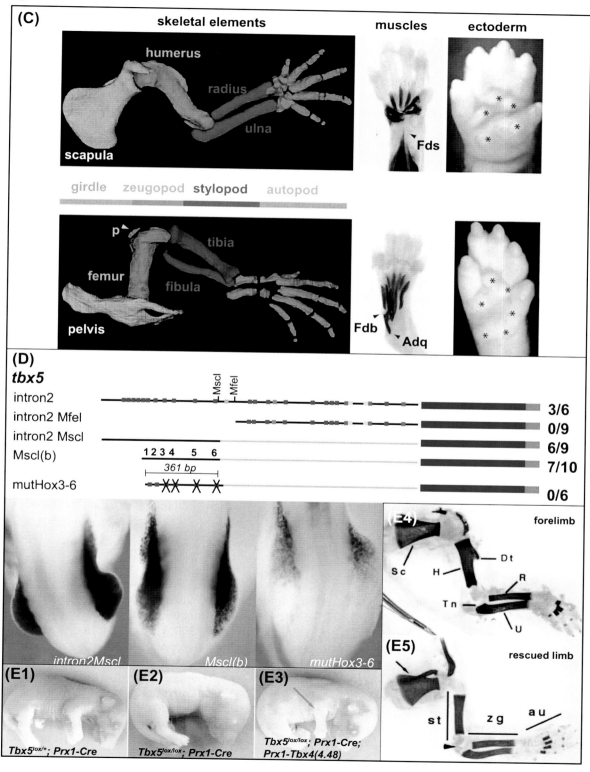

Figure 4.1 Initial specification of amniote limb buds. (A), Regions of the lateral plate mesoderm (LPM, pink). ALPM and PLPM, anterior and posterior LPM; ph, pharyngeal region. The forelimb bud arises within the PLPM *(from Tanaka (2013))*. (B), Expression of initial transcriptional limb bud regulators, the *tbx5* gene in forelimb (wing) bud, and the *tbx4* and *pitx1* genes in hindlimb buds of chick embryos, visualized by in situ hybridization at stages 23 and 29 as indicated. The insets in the stage 23 panels show cross-sections of the respective buds, distal down and dorsal side to left *(from Logan et al. (1998))*. (C), Axial structure of forelimb (top) and hindlimb (bottom) in mouse. The color coding in middle indicates major skeletal elements; bones of hand and foot in orange (autopod). The limbs are homologous but are distinguished by differences such as the patella (p), muscles (Fds, flexor digitorum digitalis; Fdb, flexor digitorum brevis; Adq, abductor quinti) and differences in footpads (asterisks) *(from Duboc and Logan (2011b))*. (D), *cis*-regulatory analysis of *tbx5* enhancer conferring forelimb expression under direct control of *hox* gene products expressed at this axial position. The enhancer is located in intron 2. Constructs containing indicated regions express in transgenic mice at the frequencies indicated at right. Blue squares, putative Hox protein-binding sites; green squares, Hox/Meis sites; red squares, Pbx/Meis sites; yellow squares, Hox/Pbx sites. As shown in the photographic images below, the Hox sites alone confer strong forelimb expression most of which is abolished by mutation of these sites (X) *(from Minguillon et al. (2012))*. (E), Demonstration that the *tbx4* gene, normally expressed in hindlimb, is capable of complete rescue of forelimb development in conditional deletion of *tbx5*. (E1), mouse pup displaying normal phenotype, *tbx5* heterozygote. (E2), Homozygous *tbx5* conditional deletion, lacking forelimbs. (E3), Same homozygous deletion, forelimb development rescued by transgene expressing *tbx4*. (E4,E5), Comparison of normal and rescued forelimb skeletal structure; though shape is not quite normal, articulation is forelimb-like: Sc, scapula; H, humerus; Dt, deltoid tubercle; Tn, trochlear notch; R, radius; U ulna; St, stylopod; Zg, zeugopod; Au, autopod *(from Minguillon et al. (2005))*.

bind Pitx1 factor (Infante et al., 2013). The two enhancers do not have identical spatial activity; both are required for complete hindlimb bud expression; and deletion from a bacterial artificial chromosome (BAC) carrying the whole locus shows they are both necessary for expression of *tbx4* in the early hindlimb bud. In addition to *pitx1*, a second hindlimb bud-specific regulatory gene, *islet1*, provides a required input to *tbx4* expression in the hindlimb bud (Kawakami et al., 2011). These two essential regulatory inputs at least partially account for initiation of hindlimb bud *tbx4* expression, and therefore contribute directly to the specification of the hindlimb progenitor field. Whether the posterior *hox* genes of paralog groups 10 and 11, which are expressed in lateral plate mesoderm in the hindlimb bud region, are directly responsible for activation of either or both of these two known inputs, or also directly control *tbx4*, remains a matter for speculation (Minguillon et al., 2012).

Initial functions of *tbx4* and *tbx5*

Though the genomic wiring utilized to activate *tbx5* in the forelimb bud and *tbx4* in the hindlimb bud is demonstrated to be different, these genes execute interchangeable functions since they can replace one another in rescue experiments. Figure 4.1(E1–E3) (Minguillon et al., 2005) shows that *tbx5* is absolutely required for forelimb formation, but forelimb formation in conditional *tbx5* deletions can be perfectly rescued by ectopic activation of *tbx4* in the forelimb bud. The form of the rescued limb skeleton is decisively that of the forelimb (Fig. 4.1(E4,E5)), so these genes do not respectively determine forelimb and hindlimb skeletal morphologies (cf. Fig. 4.1(C)). Similarly, in a *pitx1*$^{-/-}$ mutant hindlimb, the resulting partial loss of *tbx4* expression results in defective limb outgrowth and this can be rescued by ectopic hindlimb expression of *tbx5* (Duboc and Logan, 2011a). Thus the significance of the limb bud-specific expression of *tbx4* and *tbx5* is only that their different *cis*-regulatory apparatus ensures the expression in both pairs of limb buds of these essentially interchangeable Tbx factors. The downstream function of *tbx4* and *tbx5* is to drive expression of the *fgf10* gene in the limb bud mesenchyme. This is most clearly seen in the forelimb bud, and in the hindlimb bud there may be an additional Tcf1/β-catenin mediated and/or Islet1 input required (Kawakami et al., 2011). In the forelimb bud *cis*-regulatory evidence is available for direct Tbx5 input into the *fgf10* gene (Agarwal et al., 2003). Mice lacking a *tbx5* gene fail to express *fgf10* in the forelimb bud, and a conserved region of the *fgf10* gene extending 7 kb upstream of the start site suffices to promote construct expression. This region contains Tbx sites required for high-level expression of constructs in Tbx transactivation assays, as well as synergistically active Tcf1 target sites. The *tbx5* gene likely drives *fgf10* expression in a feed-forward circuit, as *tbx5* is also required for *tcf1* gene expression in the forelimb bud (Agarwal et al., 2003).

After *fgf10* expression has been initiated, the *tbx4* and *5* genes are no longer needed to maintain it, since *fgf10* is wired into a positive feedback circuit which goes as follows. The Fgf10 signal is received in the overlying limb bud ectoderm (the AER, apical ectodermal ridge), where its ultimate effect is to cause expression of the *fgf8* gene, by two separate routes. Fgf10 activates *sp8* (*buttonhead*) and paralog *sp9* which in turn activate the *fgf8* gene. Fgf10 also activates the *wnt3* gene, and Wnt/β-catenin signaling in turn positively regulates both *fgf8* and *sp8* in a feed-forward circuit (Kawakami et al., 2004). *Pbx1,2* expression is also required for AER expression of the *fgf8* gene (Capellini et al., 2011). In turn, Fgf8 signals back to the underlying mesodermal tissue and activates the expression of *fgf10*. Thus, when the Fgf8 ligand is received in the mesenchyme, its signal transduction system apparently activates Lim homeodomain transcription factors produced by the *lhx2* and *lhx9* genes and the *ldb* gene as well, which encodes a Lim cofactor (Tzchori et al., 2009). These genes now control mesenchymal *fgf10* expression in a positive inter-tissue transcriptional feedback circuit.

In summary, though there are many more players, often of so far ill-defined roles, we can dissect out a more or less unbroken chain of genomic causality extending from the placement and installation of transcription of the early *tbx* regulatory genes of forelimb and hindlimb buds to a stable signaling circuit characteristic of vertebrate limb buds. The *cis*-regulatory apparatus of the pioneer *tbx* genes reads the local regulatory states at the body axis positions of the future limb buds, and once expressed, *cis*-regulatory data and much other evidence indicate these genes both contribute crucially to transcriptional activation of the *fgf10* signaling gene in the limb bud mesenchyme. The indirect consequence in the AER ectoderm is transcription of the *fgf8* gene there. In turn this ligand alters the regulatory state in the mesenchyme receiving the Fgf8 signal, activating Lim transcription factors which now mediate *fgf10* transcription. Thus, as so often found in other contexts, the initial inputs provided by the expression of *tbx* genes in the limb bud are no longer needed because they have been supplanted downstream by inputs from a stable feedback. The Fgf10 generated by the continuing *fgf10* ↔ *fgf8* feedback transcriptional system is subsequently required to drive mesodermal limb outgrowth in the proximal/distal axis during limb development. Its significance is that it ensures the expression of the *fgf8* gene in the AER together with other *fgf* genes (*fgf4*, *fgf9* and *fgf17*). However, combinatorial conditional knockouts of these genes show that *fgf8* alone suffices to rescue normal limb outgrowth (Mariani et al., 2008).

2.2   *Cis*-regulatory organization of the anterior/posterior specification system in the amniote limb bud

As summarized at an abstract level in the first section of this chapter, the crucial processes that follow establishment of the progenitor field for adult body parts are its spatial subdivision into the regulatory state domains that foreshadow its various internal parts and structures. The programs that accomplish this always deploy multiple signaling interactions, together with (signal mediated) growth/cell division control. Amniote limb development again provides an object example, although it is yet impossible to construct complete or completely coherent GRNs underlying the progression of spatial regulatory states and morphogenetic episodes ending in the construction of the stylopod (upper limb), zeugopod (lower limb), and autopod (hand or foot with its digits; Fig. 4.1(C)). Nonetheless much has been learned in recent years in regard to the specific genomic code underlying key aspects of the signal-driven spatial subdivision process which have been incorporated into a GRN model (Rabinowitz and Vokes, 2012). We focus here on how the internal regulatory coordinates that control anterior/posterior and proximal/distal specification are organized.

Signaling, growth, and the internal regulatory coordinates of the limb

The initial signaling interrelationship to be set up is the limb mesenchyme-AER Fgf8-Fgf10 feedback circuit, which initiates proximal-distal polarity and outgrowth. Epidermal Wnt signaling to the underlying mesenchyme takes place as well. Genetic studies utilizing inactivation of the *bmp4* gene at precise times show that for the AER to form and Fgf8 to be produced in this epidermal region there is an early requirement (e9.0 in mouse forelimb, just as limb bud outgrowth begins) for mesenchymal expression of this *bmp* gene (Bénazet et al., 2009). Next, at the posterior margin of the limb bud a mesenchymal center of *sonic hedgehog* (*shh*) transcription is set up, the mechanistic basis of which we see below. The existence of this signaling center, called the Zone of Polarizing Activity (ZPA) was recognized on the basis of tissue transplantation experiments in chick wings almost two-thirds of a century ago (Saunders, 1948). The overall significance of *shh* transcription and secretion of the diffusible ligand is twofold: it contributes crucially to anterior/posterior patterning across the limb (in us, the thumb is the anterior of the autopod and the radius is the anterior long bone of the zeugopod); and it sets up an equally crucial signaling interaction with the AER required for growth and patterning events in proximal/distal patterning. These functions, their spatial relationships to one another and to the morphogenesis of limb structures in both axes of the mouse limb bud, are summarized in the models of Fig. 4.2(A,B) (Zeller et al., 2009). Figure 4.2(A1) shows the limb bud at an early outgrowth stage in a lateral view and in cross-section (for the mouse forelimb about e9.25). The structure is yet sufficiently small so that all the mesenchyme is exposed to ectodermal Wnt signaling. Genes that will provide transcriptional inputs to the *shh* gene, viz. posterior *hox* genes and the *hand2* gene, are active posteriorly, and the Gli3 transcription factor, the Shh response factor, is present throughout. The portrayal of the early specification stage in Fig. 4.2(A2), about embryonic day (e) 9.5 on, shows the distinction between the proximally specified mesenchyme and the more distal growing mesenchyme within range of AER Fgf signaling. Expression of the *shh* gene is occurring in the ZPA, resulting in a posterior to anterior gradient of Shh signaling ligand (Fig. 4.2(A2)); the positions of the future anterior/posterior bony structures are indicated in the cross-section. There is now an anterior Shh-free domain where the default function of the Gli3 factor is target gene repression and here the first digit (thumb) and radius will form. In the hindlimb, formation of digit one and anterior long bones is driven by circuitry requiring Irx3 and Irx5 expression, and this Irx-dependent function is prevented by Shh signaling, so it can only occur in the most anterior parts of the limb bud (Li et al., 2014a). Just posterior of this is a domain with low Shh signaling where the second and third digits will form. In the posterior high Shh domain the ulna and digits four and five will form. Further regulatory spatial subdivision of the limb bud occurs with proliferation, growth, and differentiation from e9.75 on, as shown in Fig. 4.2(A3). Here the distal mesenchyme within range of Fgf and Wnt signaling from the AER remains proliferative and undifferentiated, while more proximally under a BMP regimen *sox9* is activated, collagen synthesis and chondrogenesis begin the process of building skeletal elements as shown, in proximal to distal sequence. Figure 4.2(A4) shows terminal morphogenesis in the autopod (e12 on).

The early specification stage is critically dependent on interlocking signal loops, the dynamics of which affect both growth and spatial patterning. The architecture of these signaling loops is abstractly summarized in Fig. 4.2(B1), and their deployment in space and time is shown in Fig. 4.2(B2–B4) (Bénazet and Zeller, 2009; Zeller et al., 2009). During limb bud initiation Bmp4 signaling in the mesenchyme results in activation of a gene encoding an extracellular antagonist of Bmp, Gremlin, which rapidly down regulates

Figure 4.2

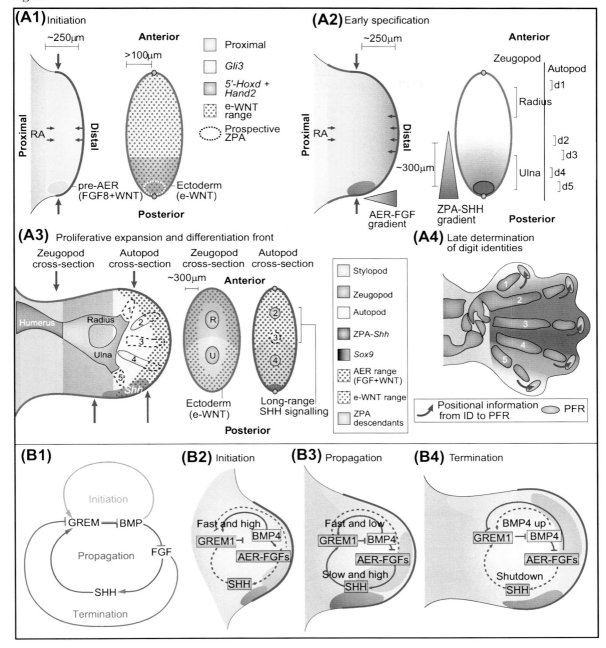

Figure 4.2 Signaling and gene interactions in anterior/posterior and proximal/distal subdivisions of the mouse embryo forelimb bud. (A), Model for forelimb bud development *(from Zeller et al. (2009))*. (A1), Fgf8 and Wnt signaling from the AER is received by the limb bud mesenchyme, and Shh signaling is initiated in the future ZPA. A diagrammatic cross-section (right) at the position marked by the vertical arrows in the side view on the left shows that the bud is still shallow enough that the entire mesenchyme is exposed to the AER Wnt signal (e-Wnt), and the Gli3 transcription factor is also present throughout the mesenchyme.

Figure 4.2

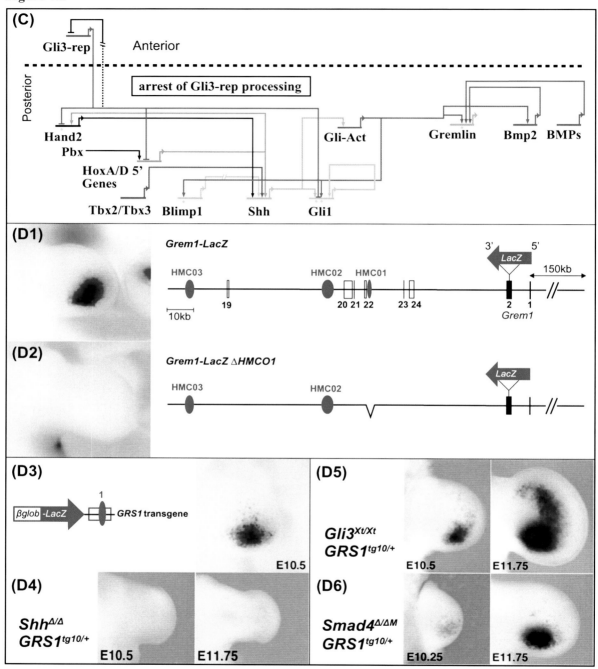

5′ *hox* genes and the *hand2* gene become active in the posterior end of the bud (after e9), where they drive initial expression of the *shh* gene. (A2), Growth and establishment of the distal to proximal Fgf8 gradient (indicated by green shading), and of the posterior to anterior Shh gradient emanating from the ZPA (red, from e9.5). At right a diagrammatic cross-section indicates the range of the Shh gradient (red shading) and the positions with respect to this gradient from which radius and ulna, and in the future autopod the five digits (d1–d5), will form. (A3), Proliferative outgrowth of the bud, a response to Wnt and Fgf signaling from the AER and Shh signaling from the ZPA (from e9.75). At the stage pictured (~e11.5) in the proximal/distal dimension only the most distal region remains within range of Fgf8 and Wnt signals, remaining in an undifferentiated state, while proximal to this the *sox9* gene is activated, initiating chondrogenic condensations. The radius and ulna skeletal elements and then the digits are determined in accord with their anterior/posterior positions. As can be seen in the zeugopod and autopod cross-sections each of these skeletal elements forms in a domain unique in regard to developmental ancestry and signal regimen. (A4), Formation of digits (>e12), according to the prior specification states of the progenitor cells as well as Bmp signaling from interdigitary (ID) mesenchyme to the phalange forming regions (PFR, green ovals). (B), Signaling feedback loops during forelimb bud development. (B1), Initiation, propagation, and termination loops are color coded *(from Bénazet and Zeller (2009))*. Grem, Gremlin, an extracellular BMP antagonist; the propagation loop prevents mesenchymal Bmp from down regulating Fgf8 expression in the AER ectoderm, and this in turn is required in the posterior mesenchyme of the ZPA for enhanced *shh* expression, which in turn results in *grem* expression. Eventually high Fgf8 levels stop *grem* expression, allowing BMP to shut this loop down. (B2, B3, B4), Signaling feedback loops in the context of limb bud outgrowth *(from Zeller et al. (2009))*. *shh* expression, red; Grem, purple; Bmp, blue; AER and AER Fgf8, green; signaling loops operating at each stage indicated by solid lines. Initially Bmp4 activates *grem* expression, but is soon limited by Grem. The propagation loop involves an epithelial (AER)/mesenchymal (ZPA) interaction. In addition to the termination phase effect of high Fgf on *grem* expression, growth plays a role as well since the proliferating descendants of the ZPA now occupying the central regions of the bud do not express *grem*. (C), GRN model indicating interactions of regulatory genes and the specific effects of absence of Shh, mediated by the repressive (Gli3$_{rep}$), and where Shh is present the activating form of Gli3 (Gli3$_{act}$) *(slightly modified from Vokes et al. (2008))*. (D), *cis*-regulatory control of the *grem1* gene *(from Zuniga et al. (2012))*. (D1, D2), Demonstration of functional necessity of a conserved *cis*-regulatory module (HMCO1) using transgenic mice carrying LacZ BAC reporters. Expression in hindlimb bud is shown at left; (D1), Control; (D2), Deletion of HMCO1. (D3), A short construct operating with a βglobin promoter and containing HMCO1/GRS1 placed 9 kb downstream (diagram) expresses accurately in posterior forelimb bud mesenchyme. (D4, D5, D6), Requirement of construct shown in (D3) for Shh- and Bmp-mediated inputs. (D4), Absence of expression in homozygous conditional deletion of *shh* gene. (D5), Ectopic anterior expression of transgene in limb buds conditionally lacking Gli3, thus demonstrating the necessity of Gli3$_{rep}$ for repression of *grem* outside the anterior range of Shh signaling. (D6), Demonstration that initiation of expression is indeed due to Bmp4 signaling (cf. B2), since in the conditional absence of the Smad transcription factor which mediates Bmp4 signaling, early expression is extremely weak.

Bmp activity ("initiation" loop in Fig. 4.2(B1,B2)). The significance of this is to permit the activation of the most important of the limb bud growth phase signaling interactions. Lowered Bmp4 permits enhanced Fgf signaling in the AER, which in the recipient mesenchyme drives *shh* transcription, and this in turn enhances *gremlin* transcription, blocking Bmp4 activity (Fgf > Shh > gremlin-|Bmp4-|Fgf; "propagation" loop in Fig. 4.2(B1,B3)). Thus the expression of Shh and Fgf are accelerated by an indirect positive feed-back circuit. This loop is required for quantitative control of Shh signaling, and hence for specification of the posterior digits as well as for distal developmental processes (Bénazet et al., 2009; Bénazet and Zeller, 2009; Duboc and Logan, 2009). Its spatial significance (Fig. 4.2(B3)) is that it produces a Gremlin zone of low Bmp in the posterior limb bud mesenchyme right under the AER enabling this all-important feedback signaling circuit. It is very interesting that the same Lim domain transcription factors and cofactor (Lhx2, Lhx9, and Ldb) that regulate *fgf10* transcription in the limb bud mesenchyme are also required for *gremlin* transcription in response to Shh signaling (Tzchori et al., 2009). The functional integration of growth geometry with the signaling interactions leads to the eventual abrogation of the propagation loop (Bénazet and Zeller, 2009): continued proliferation in the ZPA region results in the accumulation of cells anterior to the ZPA which are refractory to activation of *gremlin* by Shh. Thus as they grow out they separate the source of Shh from the *gremlin* transcribing cells, thereby exposing the *gremlin* gene to the default repressive form of Gli3, breaking the propagation loop and terminating the indirect positive feedback between Fgf and Shh. Furthermore, in these cells high Fgf also results in inhibition of *gremlin* transcription ("termination" loop in Fig. 4.2(B1,B4)).

What causes these signaling events which are so necessary for anterior/posterior limb bud patterning to occur, and how do they lead to limb development? The answers to these fundamental "why" questions lie in the genomic regulatory code for amniote limb development, and are beginning to be revealed at genetic and *cis*-regulatory levels.

The genomic basis of the "propagation loop"

Cis-regulatory evidence obtained with expression constructs of various kinds, including site-specific mutation, chromatin binding, and construct expression in multiple genetic backgrounds, provides many of the desired explanations. Among these are explanations of why the *shh* gene and its spatial and quantitative drivers are expressed posteriorly in the ZPA of forelimb and hindlimb buds and not elsewhere, and mechanisms of Shh control of *gremlin* expression, as well as other features. The genomically encoded circuitry model shown in Fig. 4.2(C) (modified from Vokes et al. (2008)) provides valuable insights into what is at base a paradigmatic example of how the progenitor field for an adult body part is divided into diverse spatial regulatory states.

We begin with spatial inputs in the posterior limb bud upstream of the *shh* gene, which as indicated above is required for installation of posterior-anterior polarity in the developing limb bud. The programming is evident from the inputs in the *cis*-regulatory module which controls Shh expression in the limb bud. Expression of the *shh* gene in the ZPA is mediated by an enhancer 800 kb upstream from the *shh*-coding exons, and when isolated, this enhancer generates reporter expression in the ZPA. It contains binding sites for Hand2, a bHLH transcription factor that is initially broadly expressed in lateral plate mesoderm. Hand2 itself probably dimerizes with Twist when it functions in the limb bud, as *twist* expression is also required for limb development. Also present on the *shh* enhancer is Hoxd13 (Galli et al., 2010) and the Hox coactivators, Pbx. These are necessary for the initial spatial specification of *shh* expression in the posterior limb bud (Capellini et al., 2011).

Upstream of the *shh* enhancer, some of the inputs causing expression of hand2 and the 5′ *hox* genes are known as well. Control of 5′ *hox* gene expression (posterior *hox* genes, including *hoxd13*) in the posterior limb bud is in turn mediated by a series of distant enhancers, as discussed at length in Chapter 7. The *hand2* gene is activated in parallel with the 5′ *hox* genes: in the forelimb bud all four of the *hox9* paralogs must be expressed for *hand2* (and *shh*) transcription to occur (Xu and Wellik, 2011); in the hindlimb bud, the driver of *hand2* expression is the *islet1* gene (Kawakami et al., 2011). Once posterior *shh* expression is activated, a self-reinforcing cross-regulatory network is set up, as indicated in Fig. 4.2(C). The *cis*-regulatory system of *hand2* contains functional target sites for Gli3 through which it is silenced, except where Shh signaling prevents cleavage of Gli3 to the repressive form, i.e., in the posterior limb bud mesoderm. Thus where Shh signaling occurs, *hand2* is expressed and Hand2 in turn provides a positive input into the *shh* gene. Other *cis*-regulatory studies show that a feed-forward is set up in the posterior limb bud: the 5′ *hox* genes drive *hand2* as well as *shh* directly, while as we have seen *hand2* also drives *shh* (Salsi et al., 2008). There are two additional known inputs into the limb bud transcriptional apparatus of *shh*. One is a feedback loop established between the *blimp1* gene and the *shh* gene. Blimp1 is expressed in the limb bud prior to *shh* and provides a positive boost to *shh* expression, while it is itself responsive to Gli3 in its active form (Fig. 4.2(C)). This loop controls the level of *shh* transcription (Vokes et al., 2008). The second additional input originates from the *tbx2* gene, expressed in the most posterior region of the subepidermal mesenchyme because of an unknown signal from the ectodermal border cells posterior to the AER; this input evidently ensures the tightly localized disposition of the *shh* expressing cells of the ZPA (Nissim et al., 2007). Even though it is not yet complete (Fig. 4.2(C) does not explain the return link of the "propagation loop", from the AER Fgf's to the *shh* gene shown in Fig. 4.2(B)), we see how tightly wired is the spatial control of *shh* expression.

Expression of genes of the *hoxA* and *hoxD* clusters is required for the initial activation of *gremlin* (Sheth et al., 2013). According to the succession of signaling loops discussed above, transcription of this key gene becomes controlled by both activator and repressor forms of Gli3 (Gli3$_{act}$ and Gli3$_{rep}$), and also by Bmp4 input. The genomic basis of this logic, and thus of its implications for spatial subdivision of the limb bud, has also been revealed at the *cis*-regulatory level (Zuniga et al., 2012; Li et al., 2014b). Once again the relevant enhancers are very distant from the coding region of the gene. Figure 4.2(D1,D2) show in a BAC deletion experiment the essential requirement for a conserved *cis*-regulatory module ("GSR1") 70 kb downstream from the *gremlin*-coding sequence. Inserted in a short construct, this DNA fragment suffices to generate spatially appropriate expression (Fig. 4.2(D3)). The following experiments (Zuniga et al., 2012) show that expression of this enhancer is dependent on Shh, and its activity is lost in the absence of Shh expression (Fig. 4.2(D4)). In embryos deficient for Gli3, expression of this enhancer spreads to the anterior cells, revealing the normal role of Gli3 repression there (Fig. 4.2(D5)). *Cis*-regulatory experiments on this enhancer also confirm that *gremlin* expression depends on Bmp input, since its expression is diminished in the absence of the BMP signal transducer Smad4 (Fig. 4.2(D6)). Another consequence of Shh signaling and the reduction of Gli3$_{rep}$ in the posterior domains of the limb bud, is that it permits expression of posterior *hoxA* and *hoxD* genes. This expression, aside from stimulating *shh* transcription as above, also provides positive inputs (direct or indirect) into *fgf10* as well as into *gremlin* in addition to any effects mediated by Gli3 (Sheth et al., 2013). Thus the expression of three key propagation loop genes, *shh*, *fgf10*, and *gremlin*, are maintained by *hoxA* and *hoxD* gene expression.

From regulatory state patterns to morphogenesis

Downstream of the relatively intensely studied patterns of regulatory gene and signaling gene expression far less is known, relative to the molecular complexity and the qualitative richness of the ensuing morphogenetic processes suggested in the drawings of Fig. 4.2(A3,A4). Not surprisingly an immense number of effector genes is expressed during limb development, as several global studies have shown. In one such study on mouse limb development (Taher et al., 2011) some 3250 genes (many unknown) were suddenly upregulated at day e10.5–e11.5, many of which appear to be effector genes having to do with skeletogenesis, and other aspects of morphogenesis such as epithelial transformations, proliferation, etc. Similarly, hundreds of genes of many types of function are expressed downstream of Lmx1b, a specifier of dorsal fates in the limb, and the target gene sets differ along the proximal-distal axis (Krawchuk and Kania, 2008). Most indicative is the range of effector gene functions controlled by given regulatory genes. For example *pitx1*, involved as we have seen in the earliest aspects of hindlimb bud specification, continues to operate in hindlimb development throughout. This gene is required for subtle patterning of soft tissues, particularly for the presence of specific small muscles and tendons (Duboc and Logan, 2011a), and it also controls metatarsal morphogenesis by modulating growth rate in the different digits. A global study in which all mouse enhancers known to drive expression in limb buds were assessed for Pitx1 binding in ChIP seq experiments demonstrated that >67% of these enhancer sequences include a Pitx1 peak (Infante et al., 2013). These kinds of observation highlight what is a next frontier in development of body parts, a frontier of which we are yet mainly ignorant: what control theorems will relate expression of given regulatory genes to mobilization of hundreds of diverse effector genes; how are their own subtle and detailed patterns of expression going to be controlled?

We have a few good examples. In careful studies of the effects of Cre-mediated loss of Gli3, specifically during autopod development, a new role for the repressive form of this regulator was discovered (Lopez-Rios et al., 2012). This is the control of the G to S transition in cell cycling. Shh and Fgf stimulate expression of required cell cycle genes such as *cdk6* in the digit progenitors, and Gli3$_{rep}$ restricts this stimulation. Cells in which proliferation is thus negatively controlled instead enter into chondrogenic differentiation. This function has direct effects on digit size and patterning when perturbed. A detailed *cis*-regulatory study identified another target of the Gli factors in autopod development, which encodes a hyaluronic acid synthase (Liu et al., 2013). Delicately localized Shh signal transduction during autopod development evidently mediates expression of this gene, which if incorrect causes severe mis-positioning of the phalange joints. Gli3$_{rep}$ activity is required to prevent ectopic expression, while Gli1$_{act}$ is required for expression. Much earlier knockout phenotypic evidence suggests that the ultimate locus of control of patterning of joints and bones in the autopod depends on 5′ *hox* gene regulation, but the details of this are only beginning to be understood (cf. Chapter 7), and it is clear that additional regulators will be involved in this aspect of their expression, among which are the mammalian orthologs of the *spalt* (*sal*) genes of *Drosophila* (Kawakami et al., 2009). This last work suggests that *hox13* genes repress the *sal* genes, and that Hox13 and Sal factors compete for a *cis*-regulatory target site in a module of the gene encoding the receptor Eph4a. The antagonism between the *hox13* and *sal* genes could provide a regional mechanism affecting local cartilaginous morphogenesis in the autopod. In considering the deployment of effector genes in limb morphogenesis, a recurrent theme is that their expression is controlled by some of the same regulatory factors as utilized for the patterning of the early limb bud, for example Tbx factors, Pitx1, Gli, and Hox factors.

We have here seen parts of the control system for the three successive major regulatory stages in adult body part formation, progenitor field specification, regulatory spatial subdivision, and spatial mobilization of effector gene sets. From the intensive studies on amniote limb formation touched on here have emerged insights into the encoded logic for the specification of the limb bud progenitor fields, for the initial proximal-distal and anterior-posterior regulatory state subdivision, and for morphogenetic functions. This is a profound advance compared to the preceding age of pure signaling phenomenology. It tells us how the basic patterns of gene expression underlying the form and position of the body part are read out from the inherited developmental instructions in the genome.

3.   Fly Legs

The six legs of *Drosophila* develop from three pairs of thoracic imaginal discs which in the late embryo consist of invaginated patches of 20–30 cells each, set aside discretely from the surrounding epidermis. The imaginal discs are the progenitor fields for the appendages in *Drosophila*. During larval life these cells increase in number rapidly, and they come to express a complex set of concentric, two-dimensional regulatory state patterns which foreshadow the eventual subparts of the leg, as shown by fluorescent immunocytology and summarized in the diagram in Fig. 4.3(A). The proximal-distal components of the leg are generated morphologically on eversion of the folded disc epithelium, such that the most distal components arise from what were the cells at the center of the disc prior to eversion (Fig. 4.3(A1)). The process by which the spatial domains of gene expression are set up thus long precede the ultimate morphogenesis of the leg parts. The necessity of imaginal disc expression of the regulatory genes *distalless* (*dll*), *dachshund*, (*dac*), and *homothorax* (*hth*) respectively (Fig. 4.3(A2,A3)), for the subsequent formation of distal, medial and proximal leg structures, has been demonstrated by mutational as well as experimental studies over the last decades. These studies nicely illustrate the causality of the sequence of patterning events. Regulatory state patterning of the imaginal discs during the larval stages is progressive and dynamic. The problem of understanding the causal spatial mechanisms underlying the formation of this adult body part thus boils down to the problem of understanding how the progression of spatial regulatory states in the leg imaginal discs is encoded.

Figure 4.3

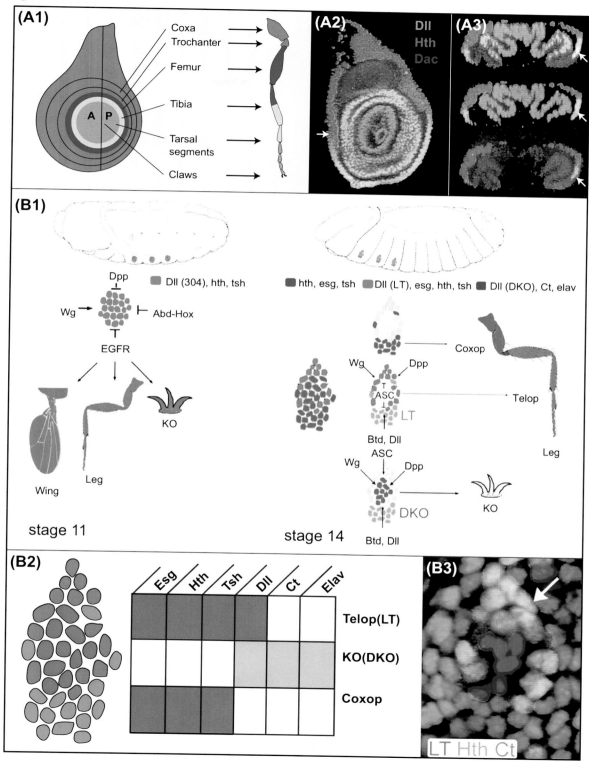

Figure 4.3

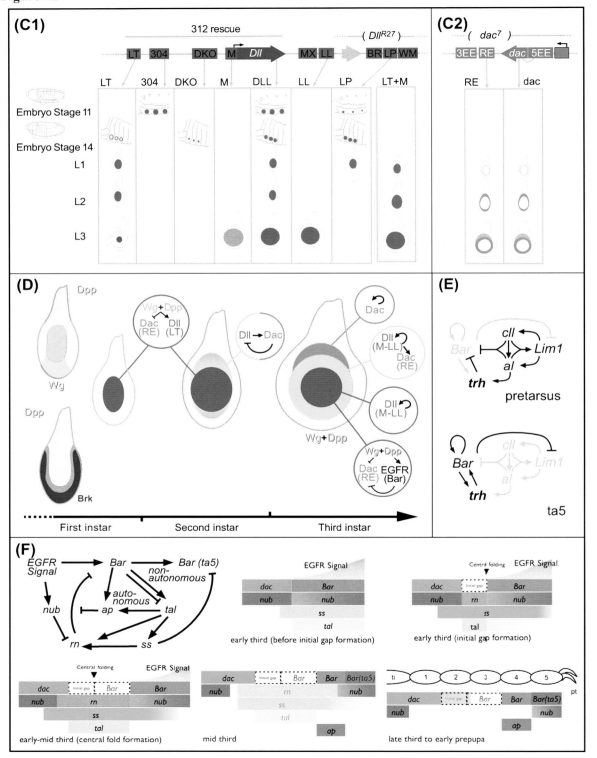

Figure 4.3 Genomically encoded regulatory states of the *Drosophila* leg imaginal disc. (A), Imaginal disc regulatory states and the parts of the leg. (A1), Diagram of third instar disc on left, anterior (A)/posterior (P) axis indicated, and concentric regulatory state domains color coded as in the immunostains of Figures (A2, A3). At right the parts of the leg to which each domain gives rise can be seen: the most distal (pre-tarsus or claws) and the five tarsal segments arise from the center of the disc, followed more proximally by the tibia, femur, and coxa *(from Lecuit and Cohen (1997))*. (A2), Fluorescent immunocytological display showing the central domain where Dll is present (green); a surrounding tarsus of Dll + Dac (light blue); the Dac domain (dark blue); and the Hth domain (red). Overlap of Hth and Dll appears yellow. The spatial distribution of regulatory states exactly indicates the map of the morphological subparts that will derive from this disc. (A3), Cross-sections of the disc (at arrow in (A2)), displaying all three stains at top, Dll and Hth in middle; Dac and Hth at bottom. The disc epithelium is folded to accommodate its two dimensional area *(A2, A3 from Wu and Cohen (1999))*. (B), Fate maps of the leg imaginal disc progenitors *(from McKay et al. (2009))*. (B1), Progressive fate maps. Stage 11 (6 h post fertilization): the pluripotential cells of the early thoracic imaginal disc will give rise to wing, Keilin's Organ (KO), and both coxopodite (coxa and trochanter; Coxop) and telopodite (Telop) or leg proper. These cells all express *dll*, driven by the *dll*304 enhancer, which responds to the indicated inputs (see text). Stage 14 (4 h later): the cells that will give rise to the dorsal appendages, i.e., wing and haltere, migrate dorsally and cease to express *dll*, and the remaining cells of the leg imaginal disc now constitute several regulatory state and fate domains, here color coded. As indicated, descendants of the *hth* expressing cells (green) produce the coxopodite; descendants of the cells expressing *dll* due to activity of the *dll*LT enhancer (red) produce the whole of the telopodite; and descendants of the cells expressing *dll* under control of the *dll*DKO enhancer (blue) and proneural genes such as *cut* (*ct*) produce the sensory Keilin's organs; ASC, *achaete-scute* proneural gene products. (B2), Regulatory states of the cells of the stage 14 leg disc: *esg, escargot; hth, homothorax; tsh, teashirt; elav* (a neuronal marker). Telop (LT), telopodite, *dll*LT enhancer active; KO (DKO), Keilin's Organ *dll*DKO enhancer active; Coxop, coxopodite. (B3), High magnification stage 14 leg disc stained for Hth but not Dll (green), Cut (blue), and *dll*LT activity (red). Some cells expressing *dll*LT also express *hth*, thus appearing yellow (arrow). Cells that express *ct* and *dll*DKO do not express *dll*LT, as this enhancer is repressed by proneural genes. (C), *Cis*-regulatory architecture of leg disc *dll* and *dac* genes; (C1), *dll*. The chart shows the spatial and temporal range of each enhancer; L1, L2, L3, first, second, and third larval instar. In the map at top, Dll represents the *dll* gene itself, and the enhancers are the labeled red boxes upstream and downstream of the gene; see text for their specific developmental roles. During larval life the *dll*LT and the maintenance element (LT + M) account for most *dll* expression. (C2), *dac*. Expression driven by the *dac* Ring Enhancer, RE, is portrayed. (D), Summary of some regulatory gene interactions required to establish median and distal positions along the proximal/distal axis. Signaling inputs and responses, for Wg, transcriptionally mediated by a Tcf factor; and for Dpp, transcriptionally mediated by the Mad factor, are indicated. The *dll* enhancers active at each instar and location are given in parentheses (cf. C1). *(C and D from Estella et al. (2012))*. (E, F), Specification of pre-tarsus and of the five tarsal segments. (E), Network circuitry distinguishing pre-tarsus and the adjacent fifth tarsal segment; *al, aristaless; cll, clawless; trh, trachealess*; see text for discussion *(from Tajiri et al. (2007))*. (F), Dynamic establishment of segment-specific tarsal regulatory states *(from Natori et al. (2012))*. At upper left interactions between genes included in the five spatial diagrams of this Figure are shown: *ap, apterous; tal, tarsaless; ss, spineless; rn, rotund; nub, nubbins*. The spatial diagrams indicate the relative locations on a linear scale of the radial expression domains of these genes in the third instar imaginal disc (central in third instar disc and distal in future leg, at right; peripheral in disc and medial in future leg at left). The future tarsal segments are defined by regulatory state in a developmental sequence as discussed in text, by means of the progressive network of repressive interactions. In the end, each of the five segments and the medial tibia has acquired a unique regulatory state.

3.1   Regulatory state patterning in the *Drosophila* leg imaginal disc

From an external point of view the *dll* gene executes different and distinct functions during leg development, which is to say that the network linkages in which this gene participates change progressively. Furthermore, at each stage, *dll* is expressed in a unique spatial manner, a direct consequence of the array of distinct *cis*-regulatory modules which control *dll* expression, each with its own positive and repressive inputs. As summarized in Fig. 4.3(B1) *dll* is initially expressed in all the cells of each thoracic imaginal disc (stage 11, ~6 h after egg laying). Lineage tracing experiments show that descendants of the *dll* expressing cells of the second thoracic imaginal disc give rise to both the dorsal wing disc and the ventral leg disc. The *cis*-regulatory module driving this initial spatial and temporal phase of *dll* expression, "*dll*304", essentially defines these imaginal discs. This *cis*-regulatory module requires Wg input, is repressed in the abdomen by Ubx and AbdA, is repressed dorsally downstream of Dpp signaling, and is repressed ventrally downstream of EGF signaling (McKay et al., 2009; Estella et al., 2012). But *dll*304 activity shortly fades out. From 10 h on *dll* expression is driven by another *cis*-regulatory module, *dll*LT ("leg trigger"; Estella et al., 2008), which continues to be active during the three successive stages of larval development (instars), and also by an additional module ("DKO") active only in late embryogenesis (Fig. 4.3(B1,C1); Galindo et al., 2011). Henceforth the regulatory state map of the disc is also its fate map. Lineage tracing experiments show that the descendants of *dll*LT expressing cells form all the distal segments of the leg (i.e., distal to the coxa and trochanter, Fig. 4.3(A1)), the "telopodite", red in Fig. 4.3(B1). The cells that do not express *dll* but do express *homothorax* (*hth*) form the proximal leg structures (green in Fig. 4.3(B)). Thus at this early stage, the future proximal and distal cells of the leg are already distinct in regulatory state and the information to distinguish these states resides within *dll*LT. Other cells within the same imaginal disc express *dll* under control of a different *cis*-regulatory module, *dll*DKO. These cells do not contribute to the leg but form larval sensory organs (Keilin's Organs, KO, blue in Fig. 4.3(B)), and they express neurogenic genes such as *cut*. In Fig. 4.3(B2,B3) we can see that by late embryogenesis the cells that will give rise to the coxopodite, the telopodite, and Keilin's Organs are differentially localized within the imaginal disc and that each type expresses a distinct regulatory state.

Both Tcf downstream of Wg signaling, and Mad downstream of Dpp signaling, have been demonstrated to provide direct inputs into *dll*LT (Estella et al., 2008). Expression of *dll* under control of this module occurs in the center of the disc where Wg and Dpp domains meet and both are present at high levels (but the Dpp signal is required only to counteract the repressive effects of Brinker, for which there are also sites in *dll*LT). However, dependence on Wg and Dpp inputs is only transient, and after first instar (L1, Fig. 4.3(C1)) an auto-regulatory module *dll*M, together with *dll*LT maintain *dll* expression as the disc grows (Estella et al., 2012). Other modules such as *dll*LL (late leg), which also responds to Dll as an auto-regulatory driver, function even later (Galindo et al., 2011). Another input into *dll*LT is the Zn finger factor Sp1, which is required for leg development, but suppresses wing developmental fates. The whole panoply of leg disc *dll* *cis*-regulatory modules and their spatial expression functions is summarized in Fig. 4.3(C1) (Estella et al., 2012).

The early third instar leg disc in Fig. 4.3(A2,A3) displays four domains of gene expression, marked by *hth*, *dac*, *dac* + *dll*, and *dll* expression; in the future telopodite the femur arises from the *dac* domain; the future tibia from the *dac* + *dll* domain; the five future tarsus segments from the *dll* only domain (Fig. 4.2(A1)). In other words, the medial proximal-distal progenitors express *dac*, the far distal progenitors express *dll*. At this point Wg + Dpp are confined to a few cells in the center of the imaginal disc (where *dll*LT runs), while most of the much larger *dll* expression domain (Fig. 4.3(C1)) is independent of these signals and depends instead on auto-activation mediated by *dll*M and *dll*LL. *Cis*-regulatory analysis of *dac* expression in the leg has revealed an enhancer, *dac*RE, (Ring Enhancer; Fig. 4.3(C2)) the functions of which explain *dac* medial patterning (Giorgianni and Mann, 2011). The successive, spatial responses of the *dac*RE *cis*-regulatory module are summarized in Fig. 4.3(D) (Estella et al., 2012). The *dac*RE module does not depend on Wg or Dpp signaling, and instead *dac*RE is activated directly by Dll. According to lineage tracing, all cells in the growing disc which express *dac* including those which later do not overlap the *dll* expression domain derive from progenitors which did express *dll*. Earlier, the Wg and Dpp inputs that activated *dll* repressed *dac*, and its transcriptional expression begins only as growth produces peripheral cell populations that escape this repression. After this, the signal-independent, auto-regulatory *dll* transcriptional apparatus (*dll*M and *dll*LL) generates the positive Dll input into *dac*RE, and as these cells grow out of the *dll* domain, *dac* too runs on auto-regulatory input, producing the *dac*-only as well as the *dac* + *dll* domains. The *dac* gene cannot be transcribed in the central *dll*-only region because *dac*RE is now repressed by Bar factors (Fig. 4.2(D)), which as we see below, are expressed in the future tarsal region. Thus is produced the annular medial *dac* expression pattern that we see in Fig. 4.3(A) and (C2).

It was for a long time assumed that positional proximal/distal information in the leg disc is provided by imaginal disc Wg and Dpp morphogen gradients, but this is clearly not the case. Seen through the lens of *cis*-regulatory results, the functions of Wg and Dpp are specifically to start *dll* expression off, and early on, to repress *dac*. The definitive proximal-distal pattern generated downstream is made by auto- and cross-regulatory gene interactions combined with cell growth. As the leg grows out, many additional genes are activated in its individual segments and structures such as joints, and regional *hox* gene inputs further define these segments. For example, the *ap2* gene is required both for tarsal joint formation and upper leg outgrowth, and its regional expression is governed by a complex series of segment-specific *cis*-regulatory modules (Ahn et al., 2011); this means that segment-specific GRNs come to generate different regulatory landscapes in each portion of the leg. But though many players are known, the local GRNs remain to be defined.

3.2 Further proximal-distal patterning in the leg disc

Although there is only limited *cis*-regulatory evidence (e.g., Kozu et al., 2006), we have at least a partial image of the gene interactions by which are programmed the much finer spatial regulatory states instituted by late third instar in the leg disc. This evidence includes clonal gain and loss of function analyses, temporally high-resolution spatial expression data, and genetic observations on effects of multiple mutations. The regulatory state patterns of the late third instar disc define the future five distinct tarsal segments, and the pre-tarsus (terminal claw domain), as well as other aspects of developmental regionalization in the leg. As above, the most distal future structures derive from the central regions of the disc, and thus the most central of these regions gives rise to the future claws. Patterning of the distal tarsal region depends on EGFR signaling (Estella et al., 2012), in the center of the *dll* expression domain of the disc (Fig. 4.3(D)), under control of Dll, and directly or indirectly, Wg, and Dpp signaling. In response to EGFR signals, several homeodomain regulatory genes are activated in the center of the disc, including *clawless* (*cll*) and *aristaless* (*al*), which form a protein complex (Kojima et al., 2005) that has two functional consequences. This complex represses *Bar* (*BarH1* and *BarH2*) in the pre-tarsus, in concert with the product of the bHLH gene *trachealess* (*trh*), in a direct interaction with a *cis*-regulatory module of the *Bar* gene (Tajiri et al., 2007); and it engages in a positive feedback loop with the *lim1* gene, locking in the *cll/al/lim1* regulatory state of the pre-tarsus (Fig. 4.3(E), top). In the adjacent region, the future fifth tarsal segment (ta5), *cll* and *al* are not expressed, and hence *Bar* will be allowed to become fully active driven by auto-activation plus a positive input from *trh*, additionally ensuring the exclusion of the pre-tarsal state by Bar repression of *lim1* (Tajiri et al., 2007).

An overview of the progressive, differential regulatory specification of the five tarsal domains is reproduced in Fig. 4.3(F) (Natori et al., 2012). Essentially this spatial regulatory apparatus operates as a transcriptional repression cascade. At early third instar Bar repression delimits the distal extent of *dac* expression, which as we have seen is a direct *cis*-regulatory interaction in *dac*RE (Giorgianni and Mann, 2011), the *Bar* genes having been activated by EGFR signaling. The region of continued *dac* expression ultimately defines the first tarsal segment, ta1. The *nubbin* gene is also activated by EGFR signaling in the tarsal region, the relevant function of which is to exclude the expression of a gene, *rotund* (*rn*), encoding another repressor. But when *nubbin* expression retracts from the boundary region separating the *dac* and *Bar* expression domain, possibly as an indirect result of cellular growth removing this region from the range of EGF signaling (Natori et al., 2012), the *rn* gene can be expressed, and its effect is to repress *Bar* and clear its expression from the future ta2,3 region. Another key spatial patterning function of the *Bar* gene is to activate the *tarsal-less* (*tal*) gene nonautonomously, not only where *Bar* is expressed, but in adjacent cells as well, in domains extending anteriorly up to the *dac* expression border (Pueyo and Couso, 2008). The *tal* gene encodes short peptides which nonautonomously but locally activate preexistent transcription factor(s), thereby ultimately causing local transcription of three different additional regulatory genes, first *spineless* (ss), then *rn* and *apterous* (*ap*; see Fig 4.3(F)). Ss functions as a transient activator of the repressive *rn* gene. Two feed-forward subcircuits ensure *rn* and *ap* expression, one downstream of *Bar* regulating *ap* expression (*Bar* > *tal* > *ap*; *Bar* > *ap*), and one downstream of *tal* regulating *rn* expression (*tal* > *ss* > *rn*; *tal* > *rn*). Distally the activation of *ap* blocks *rn* from repressing *Bar*; the *Bar*/*ap* domain becomes ta4. *Tal* expression is then extinguished, and the *tal*/*ss*/*rn* loop consequently decays. Weak *dac* expression now is allowed to spread distally; this region becomes ta2, and the adjacent region lacking Bar expression becomes ta3 (Fig. 4.3(F)). Thus a regulatory state code generated by successive mutual transcriptional repressions defines the spatial domain of each tarsal segment.

The joints in the articulated leg between segments are built at the sites of Delta/Notch signaling. At these sites the signaling and receiving cells are sharply delimited, and *tal* plays an important role in this spatial regulatory function as well. Delta and Serrate are expressed in the distal regions of the regulatory state rings in the late third instar disc, and N signaling occurs in the adjacent cells. A downstream consequence is activation of *tal*, possibly as a direct Su(H) target in the Delta/Notch signal receiving cells. In the joint forming regions of the leg proper, the Tal peptides activate their response transcription factor, Shavenbaby (Svb), by posttranscriptional modification (Pueyo and Couso, 2011). The consequence is to shut down Delta expression in these cells, causing a sharp distinction between Delta expressing and adjacent nonexpressing rings of cells without which joints fail to form. During pupariation a second negative feedback circuit initiated in the N signaling cells maintains the permanent interface between N$^+$/Dl$^-$ and Dl$^+$ cells (Greenberg and Hatini, 2009).

To summarize, the processes by which the many spatial state domains form along the proximal/distal axis of the *Drosophila* leg before pupariation are understood mechanistically to an unusual extent, and these processes causally foreshadow the morphogenesis of this complex structure during pupariation. Successive proximal/distal morphological domains are defined completely de novo within the imaginal disc, since the differential regulatory axis of the leg is orthogonal to the body axis, and therefore cannot use body axis regulatory states for spatial subdivision. The patterning processes take place in two dimensions in the single cell thick (though folded) epithelium of the disc, and the patterns that form are largely nested rings of regulatory gene expression. Though knowledge of the network linkages and of the repertoire of participants is yet incomplete, we can see the outlines in some specific detail: Locally transcribed signals initiate regional regulatory gene transcription; regional GRNs are activated by cross-regulation; and as one of their main functions, these GRNs proceed to generate exclusion of one another's regulatory states by transcriptional repression.

4. Establishment of Spatial Regulatory States in Early Development of Fly and Mammalian Brains

The major regions of the brain, such as forebrain and hindbrain, are blocked out in abutting regulatory state domains early in development. Where we have any mechanistic information, this patterning process occurs in postgastrular to early mid-embryogenesis, or early in larval life in indirectly developing animals. The genomically encoded programs for the subsequent processes of brain development are obviously going to be extensive. For example, a recent study identified >4000 putative enhancers of the telencephalon alone in the day e11.5 mouse, of which at least a third are very likely to be active (Visel et al., 2013). Here we focus on those insights that have been gained into the encoded embryonic circuitry by which boundaries separating the diverse regulatory state territories of the early embryonic neuroepithelium are set up. These territories give rise respectively to given parts of the brain. Most of the following concerns *Drosophila* and mice, though useful descriptions of regulatory gene expression in early brain development are becoming available from many other sources as well.

4.1 Commonalities in regulatory gene expression patterns in bilaterian brains

Both chordate and arthropod brains are conventionally considered to display a tripartite organization, consisting of forebrain (telencephalon/diencephalon in mammals, or protocerebrum in *Drosophila*), midbrain (mesencephalon, or in *Drosophila* deutocerebrum), and hindbrain (rhombencephalon, or in *Drosophila* tritocerebrum). This has led to a common though not universally accepted assumption, which is that a tripartite brain is a pleisiomorphic, pan-bilaterian character shared with the last common bilaterian ancestor. Though brains in different animal clades differ greatly in morphology, extensive similarities in the region-specific expression patterns of a canonical set of regulatory genes have been described during early brain development, across the bilaterian world. However, our interest here is not primarily evolutionary, but rather on the patterning mechanisms per se, in that comparative evidence for extreme conservation of given specific regulatory linkages adds a further dimension to the functional significance of these mechanisms.

Tripartite regulatory states in early brain development

Though, as we shall see, many other regulatory genes are involved, the basic idea of the tripartite brain "ground plan" is as follows (Lichtneckert and Reichert, 2005). From flies to mice, orthologous *otd/otx* and *ems/emx* genes are thought to define the anterior regions that will form the forebrains, while a series of colinear *hox* gene expression boundaries define sequential positions in the hindbrain. A posterior brain anlagen gene, *unplugged/gbx*, is expressed in hindbrain and anteriorly until its domain of expression meets that of *otd/otx*; and where these two domains briefly overlap and then abut is considered the location where the definitive future midbrain will form from an intermediate zone of the neuroepithelium, in which region a third specific cohort of regulatory genes is expressed. In addition the most anterior region of the forebrain is initially defined by *optix/six3* expression (Steinmetz et al., 2010). These and other spatial expression relationships of similar general import are illustrated for various bilaterians in Fig. 4.4.

There are remarkable similarities in regulatory gene expression domains during brain development in an ecdysozoan arthropod, a lophotrochozoan annelid, and a deuterostome hemichordate (Fig. 4.4(A)). In each case the most anterior regulator to be expressed is *six3*, followed by an *otx* ortholog, a *gbx* ortholog. and a succession of *hox* or closely related genes. It is interesting that every gene in these diagrams encodes a homeodomain regulator. Remarkably, as we discuss in detail in Chapter 7, though the hemichordate *Saccoglossus* has no centralized brain, similar anterior/posterior patterning appears in its developing subepidermal nervous system, covering the whole larval body. Homology in gene expression pattern across the Bilateria is shown in Fig. 4.4(B,C) for the midbrain/hindbrain junction, marked by abutting *otx* and *gbx* domains, where many additional genes are expressed in common as well. The key boundary enforcing mutual repression between *otx* and *gbx* is demonstrated by mutation for both mouse and *Drosophila* (Fig. 4.4(D)). In the hindbrain of both *Drosophila* and mouse an equally remarkable similarity exists: in both, the hindbrain is obviously segmented, and the regulatory states of the successive segments are marked by boundaries of nested *hox* gene expression as summarized in Fig. 4.4(E). *Trans*-bilaterian homology in the relative spatial expression patterns of another set of regulatory genes is also observed in the mediolateral direction, as

Figure 4.4

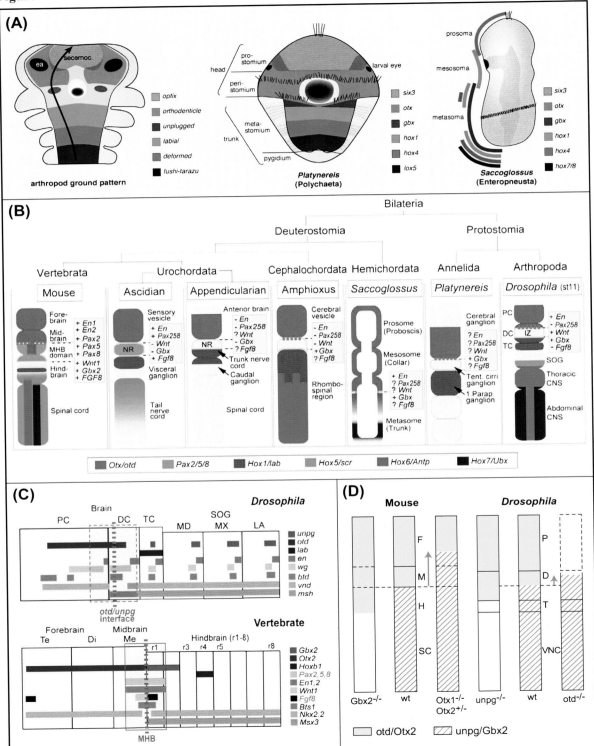

Figure 4.4

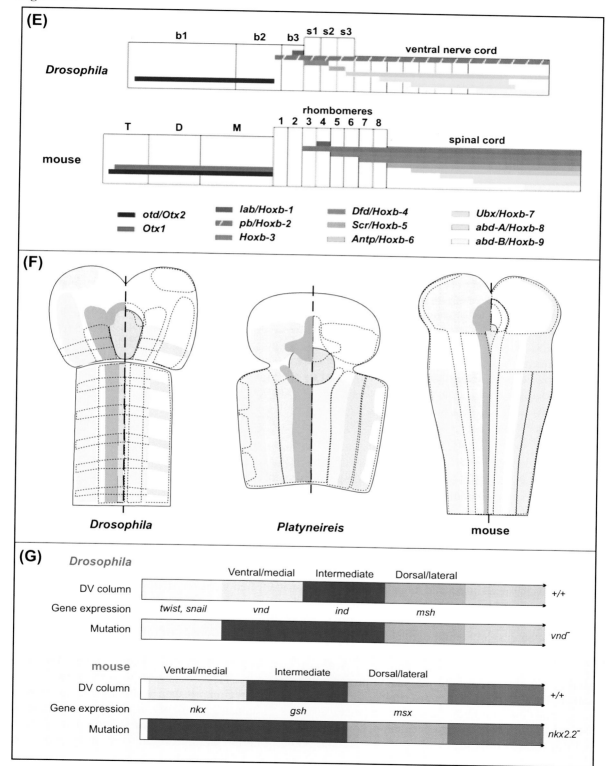

Figure 4.4 Regulatory state patterns in early development of the bilaterian brain. (A), Selected anterior/posterior expression domains in developing brains of a generalized arthropod, left; of the *Platynereis* (polychaete annelid) larva, center; and of the *Saccoglossus* (hemichordate) larva, right. In *Saccoglossus* the nervous system expressing these genes is a subepidermal net rather than a centralized brain. Expression domains are color coded to indicate orthologous gene pairs, *Drosophila* nomenclature first: *optix/six3*; *otd/otx*; *unplugged/gbx*; *labial/hox1*; *dfd/hox4*; *ftz/lox5/hox7* or *8 (from Steinmetz et al. (2010))*. (B), Comparison of regulatory states in midbrain/hindbrain junction domain (yellow boxes) in seven different bilaterians, as indicated. Gene expression patterns are color coded and orthologous gene pairs are shown in the key, mouse nomenclature first *(from Urbach (2007))*. (C), More detailed analyses of differences and similarities in regulatory states of *Drosophila* and mouse midbrain/hindbrain region, and hindbrain; orthologous genes are color coded. PC, protocerebrum; DC, deutocerebrum; TC, tritocerebrum; MD, MX, and LA are mandibular, maxillary, and labial neuromeres of the subesophageal ganglion, SOG; Te, telencephalon; Di, diencephalon; Me, mesencephalon; r denotes rhombomeres of hindbrain *(from Urbach (2007))*. (D), Evidence for reciprocal repression of *gbx2* and *otx* in mouse and of their orthologs *unpg* and *otd* in *Drosophila* across the midbrain/hindbrain boundary, from mutation results; in each case mutation of *otx* orthologs causes the anterior boundary of *gbx/unpg* expression to move anteriorly (arrows) while the reverse occurs with mutation of *gbx/unpg*. F, M, H, SC, forebrain, midbrain, hindbrain, spinal cord; P, D, T, VNC, protocerebrum, deutocerebrum, tritocerebrum, ventral nerve cord *(from Lichtneckert and Reichert (2005))*. (E), Conserved expression of *hox* genes in *Drosophila* and mouse hindbrains, where the "hindbrain" of *Drosophila* can be taken as the tritocerebrum (b3) plus the subesophageal neuromeres (s1, s2, s3); see also (C). Orthologous genes are color coded below *(from Hirth and Reichert (1999))*. (F), Mediolateral columnar regulatory states in developing brains of *Drosophila*, left, ventral view; the polychaete annelid *Platynereis*, center, ventral view; mouse, right, dorsal view, neural tube unfolded. Patterns are of course bilaterally symmetrical but expression of different genes is illustrated in right and left sides of each figure. Gene expression is color coded: *nk2.2/2.1*, orange; *pax6*, light violet; *gooseberry/pax3/7*, green; *msh/msx*, blue; *nk6*, yellow *(from Arendt et al. (2008))*. (G), Evidence for repression of *ind/gsh* by *vnd/nkx* in both *Drosophila* and mouse; in both when *vnd/nkx* is mutated the expression of *ind/gsh* spreads laterally in developing CNS *(from Cornell and Ohlen (2000))*.

illustrated in Fig. 4.4(F,G). The evidence in Fig. 4.4 is mainly descriptive but has high information content due to the geometrical homologies of orthologous gene expression domains. These data generate an indirect argument that the spatial patterning circuitry in early brain development derives from a program that was encoded in the genome of the bilaterian ancestor.

Similar program signatures downstream of early patterning circuitry

Similarities between insect and mouse brains have been identified in the distribution of neuronal cell types, the morphological organization of the brain, and the developmental expression of orthologous regulatory genes in specific brain subregions and structures. These further suggest homologous descent from a common encoded developmental regulatory program. Computer aligned maps of regulatory gene expression have been generated for the developing mushroom body of the larval brain of the annelid *Platynereis* (Tomer et al., 2010). The mushroom body is a prominent bilateral structure containing densely packed neurons found in the anterior regions of protostome brains, which is associated with processing olfactory sensory inputs, and olfactory discrimination and learning. The regulatory state profile of annelid (and also insect) mushroom bodies during development displays an amazing similarity to that of the developing pallium of the vertebrate telencephalon, some portions of which have been noted to bear both structural and functional similarities with insect mushroom bodies. Pallium and mushroom body progenitors arise from similarly positioned *emx* and *pax6* expressing subregions in both mouse and *Platynereis*; in both, they combinatorially express *dach, tll, ngn, ash2, bf1*, etc.; and they both give rise to glutaminergic neuronal cell types. More generally, multiple homologous neuronal cell types, expressing the same signature transcription factors, arise from the equivalent mediolateral gene expression domains (Fig. 4.4(F)) in early developing *Platynereis* and mouse brains (Denes et al., 2007; Arendt et al., 2008). It is also the case that in both flies and mice the *ems/emx* genes are essential for development of peripheral and central olfactory neurons, while the *otx/otd* genes are essential for both peripheral and central visual system neurons (Sen et al., 2013). Furthermore, in fundamental organization, ganglional network connections, and neuronal functions, a multitude of similarities between the basal ganglia of the mammalian forebrain and the central complex of the arthropod brain have been identified (Strausfeld and Hirth, 2013). It follows that at least some elements of the encoded genomic programs utilized in the development of the brain descend from a common origin.

4.2 Regulatory state boundary circuitry in the developing *Drosophila* brain

The regulatory state pattern defining the initial regional specification of the *Drosophila* brain consists essentially of a somewhat warped two-dimensional grid, each component of which is required for subsequent development of the appropriate types of neuroblasts, and for downstream developmental organization. The major anterior to posterior components (neuromeres), the protocerebral, deuterocerebral, and tritocerebral brain neuromeres, are followed by mandibular and maxillary neuromeres, and then the ventral nerve cord of the trunk (cf. Fig. 4.4(E)). The *Drosophila* brain is bilateral and forms from about 105 procephalic neuroblasts. In their fates, foreshadowed by regulatory state patterns, the brain regions are divided on each side into ventral (medial) and dorsal proterocerebrum; ventral, intermediate, and dorsal deuterocerebrum; and ventral, intermediate, and dorsal tritocerebrum. These spatial regulatory state subdivisions form progressively, as indicated in Fig. 4.5(A). Here the expression of *vnd, ind*, and *msh*, the cardinal "columnar" genes of the medial, intermediate, and distal ventral nerve cord (Sprecher and Hirth, 2006; Sprecher et al., 2006; Seibert and Urbach, 2010), is shown in the neuroblasts of the early developing brain. Figure 4.5(A1), right, shows a ventral view of the bilateral central nervous system, flattened and opened out, anterior to the top (for comparison at an approximately similar stage, see Fig. 4.4(F)). The developmentally essential boundaries of all these regions form as shown in Fig. 4.5(A2–A4) beginning at stage 5 after cellularization (3h) and continuing to early stage 11 (about 6h).

A canonical form of GRN subcircuit is required for some of these boundaries to form and to be maintained in the developing *Drosophila* brain. These subcircuits contain mutually repressive pairs of regulatory genes such that a key repressor transcribed in subdomain A prevents expression there of a repressor expressed in subdomain B, while in subdomain B this same repressor blocks transcription of the subdomain A repressor (see Chapter 6). The corollaries are that the regulatory system of each of these two genes has target sites for the other gene's product, and that the boundaries depend on the mutual repression, so that if expression of either of these repressors is experimentally prevented, the alternative regulatory state domain spreads across the normal boundary into the opposing terrain. This is not to be confused with a "bi-stable switch" since there is no bi-stability, and the system is, like the rest of early development, entirely determinate: in normal development the A repressor is always exclusively expressed in subdomain A and the B repressor is always exclusively expressed in subdomain B. Figure 4.5(B–E) show four examples, of many, in which the boundaries between regulatory state domains are violated by the across-boundary regulatory state in genetic perturbations where one of the genes of a mutually repressing pair is mutated or ectopically expressed (Seibert et al., 2009; Seibert and Urbach, 2010). In Fig. 4.5(B) is shown the antagonism between the head gap gene *ems* and the columnar gene *vnd*, which is especially interesting because it illustrates the functional intersection between the anterior/posterior and dorsal/ventral patterning coordinate system. At

Figure 4.5

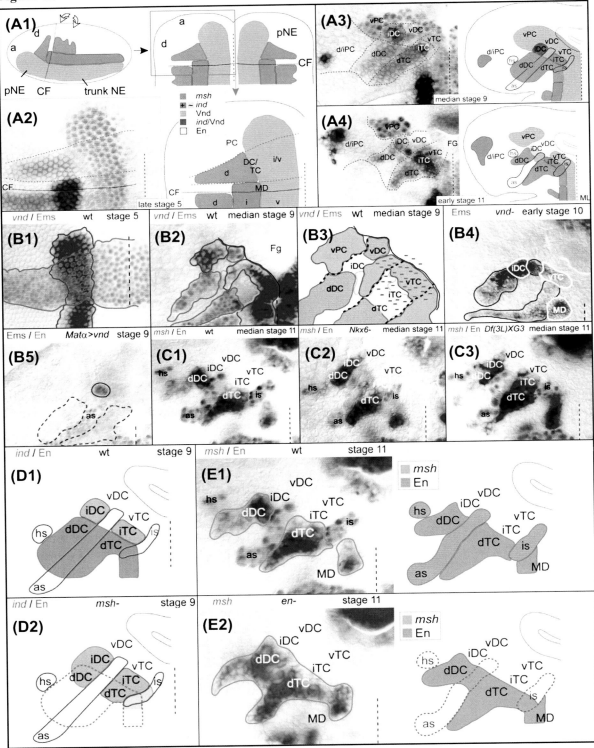

Figure 4.5

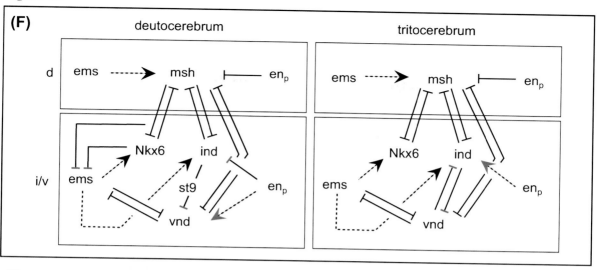

Figure 4.5 Reciprocal transcriptional repression across regulatory state boundaries in the embryonic _Drosophila_ brain. (A), Subdivision of regulatory state territories, coded by spatial expression of various homeobox genes _(from Seibert and Urbach (2010))_. (A1), Diagrammatic side view (left) and flattened ventral view of early developing brain, anterior up (right), stage 5. Black dotted line represents midline; this orientation and symbolism pertains in remainder of Figure: pNE, procephalic ectoderm; CF, cephalic furrow; NE, neuroectoderm, a, anterior; d, dorsal. The red box marks the region from which the brain will develop. (A2), _msh_ and _ind_ in situ hybridization, Vnd protein immunocytological display (left), color code and diagrammatic summary of spatial expression observed (right), stage 5. DC, deutocerebrum; TC, tritocerebrum; PC, protocerebrum; i, intermediate; v, ventral, other abbreviations as in (A1). (A3), same displays, midstage 9; _engrailed (en)_ stripes (outlined in black), are labelled respectively, head spot (hs) at posterior boundary of protocerebrum, antennal stripe (as) at posterior boundary of deutocerebrum, and intercalary stripe (is) at posterior boundary of tritocerebrum. (A4), Same displays, with definitive territories at early stage 11; FG, foregut, represented by lobes at upper right of diagrams here and in (A3). (B), Mutual repression between _ems_ and _vnd_, in situ hybridization for _vnd_ mRNA and immunocytology for Ems protein _(from Seibert et al. (2009))_. (B1), Simultaneous expression of both genes at stage 5 in neuroectoderm of future ventral/intermediate DC and TC, when _ems_ provides positive inputs into several brain regulatory genes (see text). (B2), Strictly exclusive expression domains have emerged by midstage 9. (B3), Brain regions expressing _vnd_ and _ems_ are labeled in diagram, which shows that expression of _ems_ has been quenched in the ventral TC (brown bars), while that of _vnd_ has been quenched in the intermediate TC (blue bars). (B4), In _vnd⁻_ embryo _ems_ expression spreads to the dorsal DC and TC; (B5), when _vnd_ is overexpressed in the neuroectoderm using a Gal4/UAS system _ems_ expression almost disappears. (C), Territorial repression of _msh_ by _nkx6_ and _ind_, orientation of diagrams and abbreviations as in (B) _(from Seibert and Urbach (2010))_. (C1), Wild type; (C2), in _nkx6⁻_ embryo, _msh_ is expressed ectopically in intermediate DC and in part of intermediate TC; (C3), embryo bearing a deletion that includes both _ind_ and _nkx6_ genes in which _msh_ is expressed ectopically throughout the intermediate TC. (D), Repression of _ind_ (blue) by _msh_ (purple) in dorsal DC and TC, En expression stripes shown in white _(from Seibert and Urbach (2010))_. (D1), wild type; (D2), _msh⁻_ embryo. (E), Repression of _msh_ by _en_ in DC and TC shown by in situ hybridization (left) and diagrammatically (right) _(from Seibert and Urbach (2010))_. (E1), Wild type; (E2), _en⁻_. (F), Networks of repressive gene interactions in DC and TC _(from Seibert and Urbach (2010))_. Each of the repressive interactions demonstrated experimentally in (B)–(E) is included in this network. Red arrows (activation) and bars (repression) indicate interactions that are specific to DC or TC. The only positive inputs resulting in gene activation are indicated by dashed arrows. Boxes on top indicate dorsal DC and TC, those on bottom ventral and intermediate (i/v) DC and TC. The dotted line from _en_ to _vnd_ indicates maintenance.

stage 5 they are coexpressed while at stage 9 they have exclusive domains of expression corresponding to the ventral versus dorsal domains of all three brain neuromeres (Fig. 4.5(B1–B3)). In *vnd⁻* embryos expression of *ems* spreads to the dorsal tritocerebral domain, where vnd is usually expressed (Fig. 4.5(B4), compare to Fig. 4.5(B3)), and in *vnd* overexpression, *ems* expression in the brain almost disappears (Fig. 4.5(B5)). Figure 4.5(C1,2) show that *nkx6* precisely excludes *msh* expression from the intermediate deutocerebral domain, and from the anterior intermediate tritocerebral domain, while Fig. 4.5(C3) shows that both *nkx6* and *ind* repress *msh* in the posterior tritocerebral domain. Similarly *msh* prevents expression of *ind* precisely in the dorsal domains of both these neuromeres (Fig. 4.5(D)), and *en* expression excludes *msh* expression, its stripes of expression forming the anterior and posterior boundaries of both deutocerebrum and tritocerebrum (Fig. 4.5(E)). The overall network of mutual repressions across the boundaries between dorsal versus intermediate/ventral domains of both deutocerebrum and tritocerebrum is summarized in Fig. 4.5(F) (Seibert and Urbach, 2010). With the exception of the positive inputs from *ems* to *nkx6*, *ind*, and *msh*, and the differing positive inputs from *en* in the two neuromeres, only repressive interactions are shown here. The circuitry is different in deutocerebrum from that in the tritocerebrum. Curiously, again every gene in these diagrams encodes a homeodomain transcription factor.

4.3   Rhombomere specification in the vertebrate hindbrain

During its early development the vertebrate hindbrain is transiently segmented into seven rhombomeres, which later give rise to specific neuronal cell types and ganglionic architectures, as well as to specific subsets of cranial neural crest. These developmental products are generated according to anterior/posterior position in the linear rhombomere sequence, and each rhombomere expresses a specific regulatory state (Schneider-Maunoury et al., 1997; Pasqualetti et al., 2007; Chambers et al., 2009; Tumpel et al., 2009). Much study has led to the conclusion that the nested set of *hox* gene expressions established dynamically in the rhombomeres over the time of their formation, and then maintained by cross-regulatory interactions, is required for their differential specification. Each rhombomere (r) expresses given *hox* genes, with the exception of r1 which expresses none: r2 expresses *hoxa2*; r3 expresses *hoxa2* and *b2*; r4 expresses *hoxa1* (only briefly), *hoxb1*, *a2*, and *b2*; r5 and r6 express *hoxa2*, *b2*, *a3*, *b3*, and *d3*; and r7 expresses all of these plus *hoxa4*, *b4*, and *d4*, as shown in Fig. 4.6(A). In addition r3 and r5 express the *krox20* regulatory gene, and r5 and r6 express the *kreisler* (*mafB*) regulatory gene. The anterior boundaries of the *hox* gene expression domains correspond with the morphological rhombomere boundaries: once again the process of subdivision of the progenitor field resolves into the establishment of boundaries separating adjacent, different, regulatory states. However, these expression domains are installed asynchronously and progressively, and are of various intensities. Inter-rhombomere signaling provides specific inputs into the GRNs operating in each, and signals from adjacent tissues also are instrumental in the initial assignment of *hox* gene expression patterns (Tumpel et al., 2009).

Multiple regulatory linkages contribute to the spatial specification of rhombomere regulatory states. Very generally the process can be summarized as one in which multiple early inputs initiate *hox* gene expression in the neuroepithelium of the future hindbrain, followed by activation of rhombomere-specifc GRNs of which a prominent component is *hox* gene interactions. While these GRNs are incompletely known, the genomic *cis*-regulatory basis of the spatial expression of the various *hox* genes is unusually well understood.

Some initial anterior/posterior spatial specification mechanisms in hindbrain rhombomeres

Anterior/posterior patterning in the hindbrain depends on initial signals from the adjacent paraxial mesoderm. In amniote embryos anterior *hox* gene expression domains are established in the paraxial

Figure 4.6

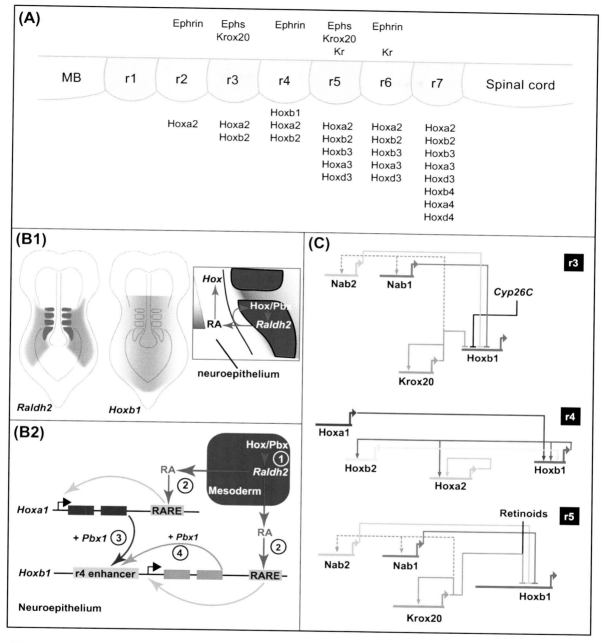

Figure 4.6 *Cis*-regulatory analysis of developing hindbrain GRNs. (A), The hindbrain "*hox* code" *(from Tumpel et al. (2009))*. *Hox* gene expression is listed below the illustration of the seven rhombomeres; note the alternating expression of Ephrin and Ephs and expression of Krox20 in r3/ r5 and Kreisler (Kr) in r5/r6. (B), Control of RA synthesis by Hox factors acting on the *cis*-regulatory system of the *raldh2* gene *(from Vitobello et al. (2011))*. (B1), Distribution of early *hoxb1* and *raldh2* expression in lateral plate mesoderm and somites (blue shading). Boxed inset shows generation of RA gradient (red triangle) due to diffusion from these mesodermal tissues to the neuroectoderm, in

Figure 4.6

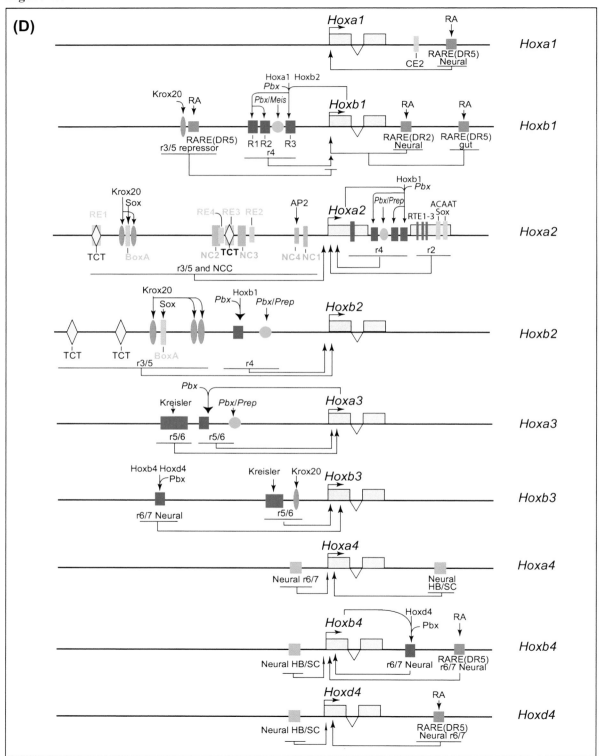

consequence of the positive *hox* input into the *raldh2* RA synthesis gene. Since RA causes *hox* gene expression (see text and B2) a positive feedback is implied in the mesoderm. (B2), *Cis*-regulatory activation of *hoxa1* and *hoxb1* by RA, via their RAREs (retinoic acid response elements). In addition, Hoxa1 (plus its Pbx cofactor), activates the *hoxb1* gene through its r4 control element, generating a feed-forward relationship. (C), GRNs controlling *hoxb1* expression in r4 after e8.75, and ensuring the silence of this gene in r3 and r5 *(from Tumpel et al. (2009))*. (D), *Cis*-regulatory modules of anterior *hox* genes including those responsible for rhombomere expression. The inputs where known are indicated above the lines representing the regulatory DNA and the spatial locations of the expression these modules produce are indicated below *(from Tumpel et al. (2009))*.

mesoderm as the precursors of these cells immigrate through the primitive streak, and prior to any *hox* gene expression in the neuroepithelium of the future hindbrain. In this mesoderm, which directly flanks the neuroepithelium, Hoxa1 together with its cofactors Pbx1/2 and Meis2, directly activate a known *cis*-regulatory control element of the *raldh2* gene, encoding the enzyme that generates RA from the retinal-dehyde precursor (Vitobello et al., 2011). RA is presented in a dynamic graded pattern in the mesoderm, whence it diffuses into the neuroectoderm and there leads to activation of the *hoxa1* and *hoxb1* genes via their *cis*-regulatory RA response elements (RARE). As illustrated in the diagrams of Fig. 4.6(B), the *cis*-regulatory architecture features a feed-forward subcircuit such that RA activates both *hoxa1* and *hoxb1*, while *hoxa1* also provides an activating input to *hoxb1* (Tumpel et al., 2009). At this early stage, *hox* gene expression in the future hindbrain, is initiated exclusively in the precursors of r7-r4, reaching only up to the anterior border of r4 and initially defining the r3/r4 boundary. The availability of RA extends thus far, and the sharpness of the boundary is enforced by expression in pre-r3 of P450 enzymes (e.g., Cyp26 enzymes) that degrade RA (Glover et al., 2006). An additional feature (Fig. 4.6(B1)) may be the feedback in the paraxial and then somitic mesoderm from RA to *hoxa1* thus stimulating further RA synthesis.

A key early player in the development of sharply bounded rhombomeres 3 and 4 is *krox20*, which encodes a Zn Finger transcription factor that is activated in two prominent stripes corresponding to what will become r3 and r5. Its expression in these domains is mediated by separate, distant *cis*-regulatory modules (Chomette et al., 2006). By e8 in the mouse, activation of *krox20* in pre-r3 and r5 is required for the maintenance of the r3/r4 and r4/r5 boundaries by controlling *hoxb1* expression (Wassef et al., 2008; Tumpel et al., 2009). The r3 and r4 regulatory states remain exclusive, as *krox20* expression precludes *hoxb1* expression (via its activation of corepressors Nab1, Nab2) in both r3 and r5 (see network in Fig. 4.6(C)). Conversely, in r4, *hoxb1* expression is vigorously maintained by a positive feedback circuit between *hoxb2* and *hoxb1*, as well as by auto-activation, and it represses *krox20* (Barrow et al., 2000). Thus once again we see that a mutual exclusion circuit is called into play together with auto- and cross-regulation to generate regulatory state boundaries.

R2 is distinguished by expression of *hoxa2*, the only *hox* gene expressed there (Fig. 4.6(A)). Curiously, the enhancer driving this expression is located in a nonconserved coding region of the protein, and is apparently a vertebrate invention (Tümpel et al., 2008), since the active sequences are absent from amphioxus *hox2*. In transgenic mice amphioxus *hox2* fails to express in r2. The drivers of this enhancer include Sox factors expressed in this region. No *hox* genes are expressed anterior to r2, and r1 is specified by signaling from the midbrain/hindbrain boundary (Moreno-Bravo et al., 2013), which is discussed separately below.

Expression of additional regulatory genes distinguishes r5–r7, and here the prominent early driver is a MafB transcription factor, the product of the *kreisler (kr)* gene, activated in the future r5 and then r6 regions. Upstream of *kr* are Fgf signaling from r4, and the Vhnf1 homeodomain regulator, which directly activates an early *kr* enhancer (Kim et al., 2005). MafB, a linchpin of r5 and r6 specification, in turn provides an essential input into the r5 enhancer of the *krox* gene, and it also activates the *hoxa3* and *hoxb3* genes (Fig. 4.6(A)).

The elegance of the genetic regulatory program for rhombomere specification is explicitly seen in the *hox* gene *cis*-regulatory maps of Fig. 4.6(D) (Tumpel et al., 2009), which include the modules controlling hindbrain expression of 9 out of the 10 *hox* genes expressed there (all except *hoxd3*). As can be seen at a glance *hox* gene interactions (including those of Meis and Pbx cofactors) account for a major fraction of the spatially significant inputs, with additional anterior/posterior spatial information transduced through RARE target sites, Krox20 target sites and MafB (Kr) target sites. These interactions are in some cases positive and others negative, as we have seen. Figure 4.6(D) shows explicitly the genomic regulatory information underlying the rhombomere "*hox* code". In this diagram, individual *cis*-regulatory modules are shown with the particular regulatory inputs which determine their specific spatial activity in rhombomeres or other locations. For example, the *hoxb1 cis*-regulatory system accounts for the functions shown in the network model in Fig. 4.6(C), namely repression of this gene in r3 and r5 and its activity in r4. Expression of *hoxb1* is essential for maintenance of the identity of r4, and a number of direct *hoxb1* targets have been identified, including both transcriptional regulatory genes and signaling genes such as *eph2a*.

4.4   The midbrain/hindbrain boundary and other encoded regionalization functions

Control of boundary formation

The second stage of the process of body part development, following primary establishment of the progenitor field, is setting regulatory state/fate boundaries within this field, and there is no better illustration of such a mechanism than formation of the vertebrate midbrain/hindbrain boundary (customarily, "MHB"). This is at the exact anterior terminus of rhombomere 1. Rhombomeres 1–3 contribute to the cerebellum, while the region of the neural tube anterior to the MHB has a totally nonoverlapping fate and becomes the midbrain or mesencephalon. It was established over a decade ago in the mouse that the MHB forms where the posterior limit of *otx2* expression directly abuts the anterior limit of *gbx2* expression (e8, 6 somite stage), and that indeed these two regulatory genes establish each other's expression boundaries by mutual repression (Fig. 4.4(C,D); Wurst and Bally-Cuif, 2001). Two signals are expressed in cells adjacent to the MHB. At the anterior end of the *gbx2* domain the *fgf8* gene is expressed, and in the immediately abutting *otx2* domain *wnt1* is expressed (Fig. 4.7(A1)). As can be seen in Fig. 4.7(A2), in an *otx2* mutant the boundary of *gbx2* expression moves far anteriorly, as do the adjacent *fgf8* and *wnt1* expression domains, with the consequence that a giant cerebellum forms while mesencephalon is lost. In a *gbx2* mutant, the posterior border of *otx2* expression moves posteriorly to the anterior boundary of rhombomere 4 (Fig. 4.7(A3)). This causes the loss of r1-3 and of the cerebellum, while the midbrain instead extends posteriorly into what normally should be the r1-3 region of the hindbrain. However, *gbx2* expression does not per se suffice to induce the specification of the r1-3 region of the hindbrain, to which its expression is confined after e8.5. Thus, ectopic expression of *gbx2* in r4, forced by placing its expression under control of a *hoxb1* rhombomeric enhancer, fails to convert r4 to r1-3 development (Li et al., 2005). As we see in the following, additional regulatory genes are required for the formation of the MHB and for development anterior and posterior of it. The experiment highlights the specific patterning function of the *gbx2* gene, which as Fig. 4.7(A) shows, is required for positioning of the developmental process that follows under the control of other genomic regulatory apparatus. The major roles of *gbx2* are to block expression of *otx2*, by a mechanism we come to below, and to provide an input necessary for expression of *fgf8* at the boundary, but *gbx2* expression also has multiple downstream consequences. An experiment was carried out to visualize descendants of all cells that transcribe *gbx2* in the neuroepithelium (this begins at e7.75) by detection of an engineered β-galactosidase marker (Sunmonu et al., 2011). This experiment revealed that *gbx2* is required to prevent these cells from migrating across the MHB and invading the mesencephalon domain, which these same descendants do if *gbx2* has been conditionally deleted from them. The abnormal mixing

Figure 4.7

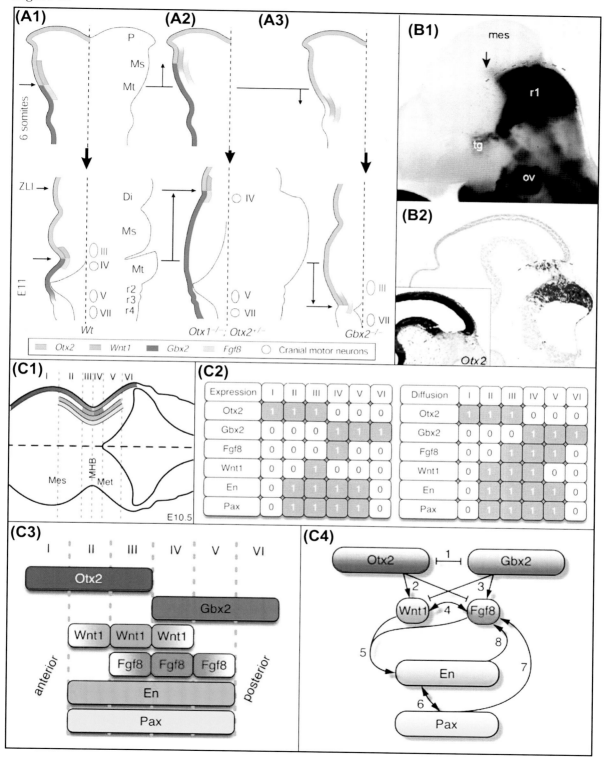

Figure 4.7

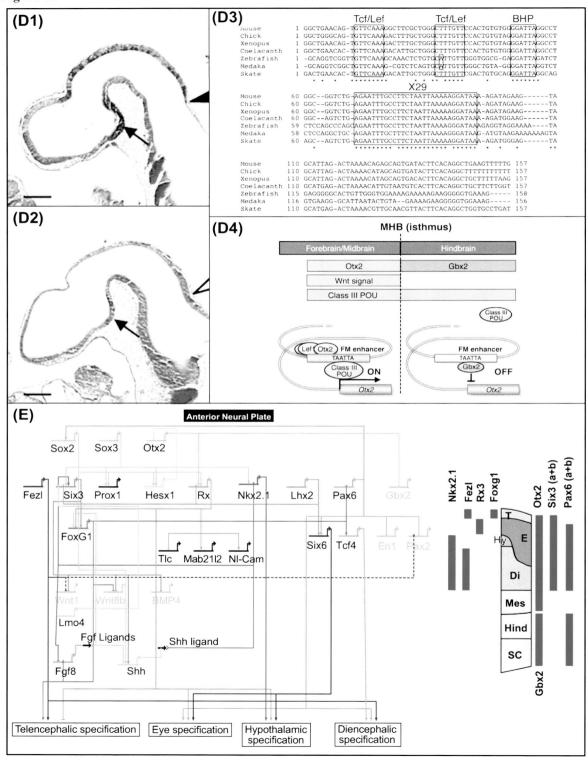

Figure 4.7 Spatial exclusion of regulatory state domains in mouse brain development. (A), Genetic evidence for formation of the midbrain/hindbrain boundary (MHB) by interaction between *otx2* and *gbx2 (from Wurst and Bally-Cuif (2001))*. Expression of *otx2*, *gbx2*, *fgf8*, and *wnt1* genes are indicated diagrammatically; top row, e8.5, six somites; bottom row, e11. (A1), Wild type. (A2), *otx* deficiencies (*otx1* homozygous deletion, *otx2* heterozygous deletion; these mice display loss of mesencephalon). (A3), *gbx2* homozygous deletion. Note loss and misplacement of cranial nerves in A2, A3 (open ovals, roman numerals). (B), Restriction of mixing across MHB by descendants of cells that had earlier expressed *gbx2 (from Sunmonu et al. (2011))*. Mice engineered to express β-galactosidase once they had expressed *gbx2*, conditionally on application of tamoxifen at e8.5, were stained with X-gal at e10.5, in order to locate the descendants of *gbx2* positive cells with respect to the MHB (mes, mesencephalon; tg, trigeminal ganglion; ov, otic vesicle; r1, rhombomere 1). (B1), Whole mount; (B2) section. Inset shows exactly complementary pattern of *otx2* transcripts visualized by in situ hybridization. (C), Gene expression domains around MHB and authenticated interactions *(from Wittmann et al. (2009))*. (C1), Diagrammatic summary of gene expressions around MHB, in six arbitrary zones, I–VI, three anterior, and three posterior of the MHB. Expression domains are color coded as in (C3); Mes, mesencephalon; Met, metencephalon (hindbrain). (C2), Boolean chart of expression domains as in (C1). (C3), Localization of protein products of the same genes taking into account ligand diffusion (note posterior expansion of Wnt1 domain and anterior expansion of Fgf8 domain across the MHB). (C4), Gene interactions from literature sources (Wittmann et al. (2009)): (1) mutual transcriptional repression between *gbx2* and *otx2*; (2) activation of *wnt1* and repression of *fgf8* by *otx2*; (3) activation of *fgf8* and repression of *wnt1* by *gbx2*; (4) mutual maintenance of expression of *fgf8* by Wnt1 signaling and of *wnt1* by Fgf8 signaling; (5) synergistic activation of *en* genes by Fgf8 and Wnt1 signaling; (6) positive feedback between *en* and *pax* genes; (7) activation of *fgf8* gene by *pax* genes; (8) positive feedback of *en* on *fgf8* gene. (D), MHB *cis*-regulatory *otx2* switch *(from Inoue et al. (2012))*. (D1), Expression of "FM" *otx2 cis*-regulatory construct driving lacZ in transgenic mouse neuroepithelium, 18 somite stage; arrow head indicates MHB (arrow denotes a region of the ventral diencephalon where the enhancer is also active); scale bar, 200 μ. (D2), Loss of expression on mutation of conserved 29 bp enhancer sequence (X29). (D3), Sequence of FM enhancer, conserved regions boxed. Vertebrate species are listed at left. (D4), Model of transcription switch mediated by FM enhancer; see text for discussion. The switch ensures *otx2* silence in r1–3 where Gbx2 is present in nuclei, and *otx2* expression in mesencephalon, where inputs from brain POU factors, Wnt signaling, and auto-stimulatory Otx2 are available. (E), BioTapestry representation of partial GRN controlling initial forebrain regionalization, late gastrula stage *(from Beccari et al. (2013))*. Diagram at right indicates regional expression of indicated genes, anterior up: T, telencephalon; E, eye domain; Hy, hypothalamus domain; Mes, mesencephalon; Met, metencephalon; SC, spinal cord.

of *gbx2*⁻ cells is evidently due to *gbx2* control of *fgf8* expression, which in turn likely affects cadherin expression and cell adhesion behavior, resulting in sorting at the MHB. The normal effectiveness of the MHB compartment restriction is illustrated dramatically in Fig. 4.7(B1,2).

The Boolean spatial expression domains of some key genes that generate opposing regulatory states across the MHB are shown in Fig. 4.7(C1,2) (Wittmann et al., 2009). The ligands Fgf8 and Wnt1 are expressed posteriorly and anteriorly of this boundary respectively, and their products will diffuse across it as indicated spatially in Fig. 4.7(C3). The circuitry among these genes is indicated in Fig. 4.7(C4), as established in various prior studies (Wittmann et al., 2009). Here we see that as above, *otx2* and *gbx2* directly repress one another. Each of these regulators respectively activates the signaling gene expressed on its side of the MHB, while repressing expression of the reciprocal gene. Furthermore, these signaling genes maintain each other's expression by feedback relations between one another and also with *engrailed* (*en1* and *en2*). *Pax* (*pax2* and *pax5*) and *en* genes activate each other by positive feedback circuitry and provide a positive regulatory input into *fgf8* expression. The MHB circuitry is thus recursively wired, as Fig. 4.7(C4) includes one negative feedback and three positive feedbacks among just these few genes. A Boolean model based on these gene interactions successfully encompasses the results of various loss of function experiments, so they are likely necessary, and at least with respect to these genes, sufficient (Wittmann et al., 2009).

The mechanistic key to the spatial positioning of the MHB is the top interaction in Fig. 4.7(C4), i.e., the transcriptional antagonism between *otx2* and *gbx2* expression. An enhancer of *otx2* which displays extremely conserved sequence features has recently been found which explains how this antagonism is programmed in the genomic *cis*-regulatory sequence (Inoue et al., 2012). This enhancer recreates *otx2* spatial expression in the anterior neuroepithelium of transgenic mice (Fig. 4.7(D1)), a function which requires the presence of a 29 bp *cis*-regulatory sequence element retained almost identically in ortholo-gous enhancers from skate to mouse (Fig. 4.7(D2,3)). The highly conserved sequence reacts with POU III domain factors (probably Brn2 and Oct 6) in the future mesencephalon, where it also responds to its own gene product, Otx2, as well as to Tcf/Lef evidently provided at the MHB by Wnt1 signaling. But as shown diagrammatically in Fig. 4.7(D4), the same enhancer binds Gbx2 where it is present in the immediately abutting posterior side of the MHB, competitively displacing the POU III factors and dominantly shutting down *otx2* transcription probably through its interaction with the obligate corepressor Groucho (Heimbu-cher et al., 2007). The MHB follows.

Genomic regulatory nodes for forebrain regionalization

In Fig. 4.7(E) is a BioTapestry GRN model summarizing the initial regulatory basis for several of the major regionalizations of the vertebrate forebrain during early development (Beccari et al., 2013). The regions considered are the telencephalon, the future optic domain, the hypothalamus, and the diencephalon. This is to be regarded as an initial framework for a future developmental GRN model, which is certain to involve a great many interactions. Nonetheless Fig. 4.7(E) contains further examples of boundary formation by repressive regulatory gene interactions. Thus, *otx2* is repressed in the anterior most forebrain by *hesx1*; and *otx2* drives *rx* expression in the optic domain, where Rx excludes the expression of the telencephalic gene *foxg1*. Gene by gene, detailed *cis*-regulatory analyses explain the spatial patterns of regulatory gene expression that underlie regionalization of the brain, and for example recent studies provide direct evidence of the DNA sequence code that controls expression of *six3*, one of the most important of forebrain specification genes (Chao et al., 2010; Lee et al., 2013).

Though this discussion is necessarily far from providing comprehensive mechanisms for brain development in vertebrates, our intent is to show that explanatory islands of knowledge are emerging, which demonstrate how regional specification in early brain development is encoded in genomic regulatory sequence. Regional GRNs that operate early in brain development form exclusive boundaries, and lead to the establishment of the regulatory state domains that functionally define the major divisions of the vertebrate brain. A common leitmotif is the progressive subdivision of bounded regulatory state domains by transcriptional repression. This is in general just as in the development of all other body parts.

5. The Vertebrate Heart

5.1 Spatial specification in early heart development

Allocation of progenitor cell populations

The mammalian heart develops from three main sources, the first and second heart field (FHF and SHF), and immigrant neural crest, with additional minor contributions from mesenchymal cells at the posterior end of the SHF (Black, 2007; Vincent and Buckingham, 2010). The initially forming heart tube, which begins pumping blood and nutrients early in development, is generated mainly from the FHF, while the SHF is the source of a continuing flow of differentiating myocardial cells. Ultimately the FHF descendants contribute mainly the left ventricle, the SHF the right ventricle and the interventricular septum, the outflow tracts and the major aortas, while descendants of both constitute the right and left atria (Fig. 4.8(A)). The neural crest immigrates via the anterior pharyngeal arches and contributes the smooth muscle of the great arteries, and the endocardial cushions from which develop the heart valves. Neural crest also produces signals which affect SHF developmental processes during heart morphogenesis. Maps of the successive stages of heart field disposition, the folding of the heart tube that brings the originally posterior region of the SHF to the anterior (outflow) end of the forming heart, and a diagrammatic summary of the contributions of these fields to the different structures of the developing heart are shown in Fig. 4.8(A) (Vincent and Buckingham, 2010). While the course of events displayed here refers specifically to mouse (and quite

Figure 4.8

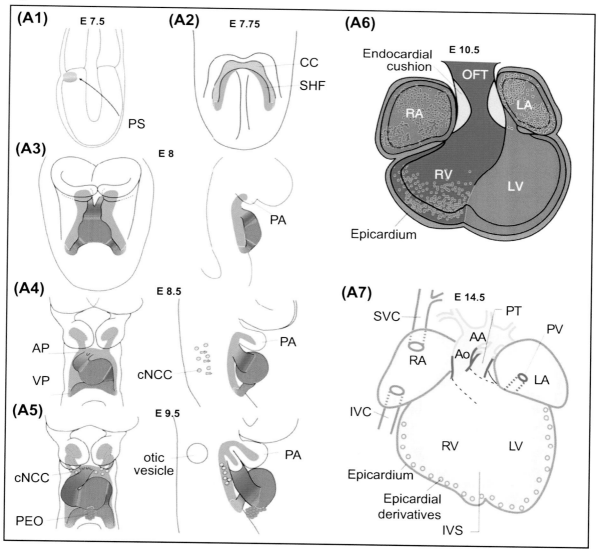

Figure 4.8

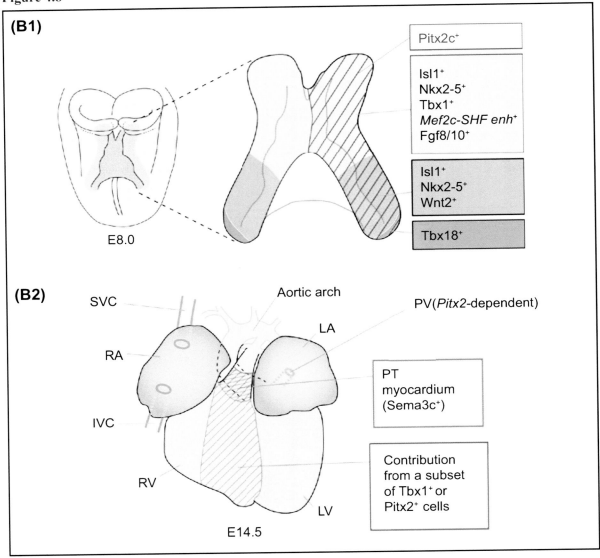

Figure 4.8 Early heart development and core circuitry determining specification of mammalian heart fields. (A), First and second heart fields (FHF and SHF) and their developmental derivatives *(from Vincent and Buckingham (2010))*. (A1), Migration of mesodermal precursors to anterior dorsal region of mouse embryo. (A2), Cardiac crescent (CC, orange), and medially positioned SHF. (A3–A5), FHF, red, and SHF, green, during successive stages of heart tube looping, shown in ventral (left) and lateral (right views): PA pharyngeal arch; AP and VP, arterial and venous poles; cNCC, yellow, cardial neural crest cells; PEO, proepicardial organ. (A6), Fully looped heart tube and cardiac compartments, color coded to indicate contributions from FHF (red), SHF (green), and NCC (yellow): RA and LA, right and left atria; RV and LV, right and left ventricles; OFT, outflow tract. (A7), Mature heart: SVC and IVC, superior and inferior caval vein; PV, pulmonary vein; AA, aortic arch; Ao, aorta; IVS, interventricular septum. (B), Regulatory state domains of

Figure 4.8

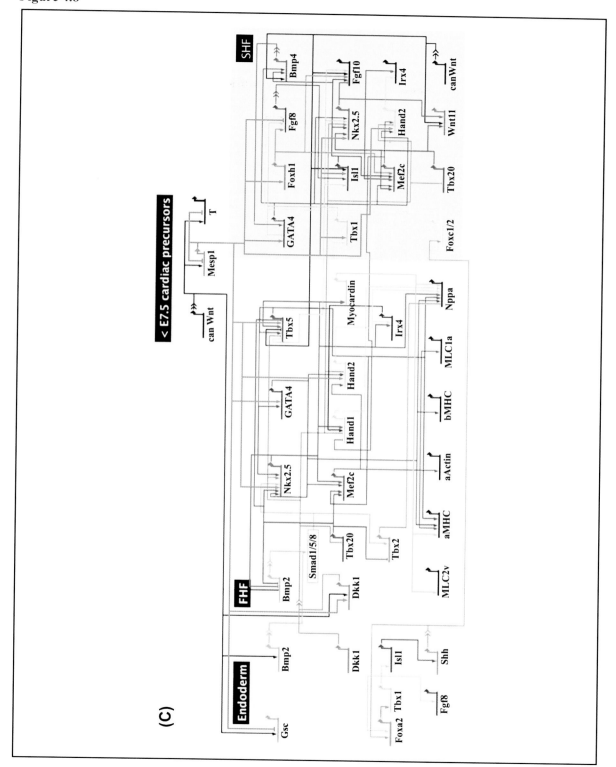

(C)

Figure 4.8

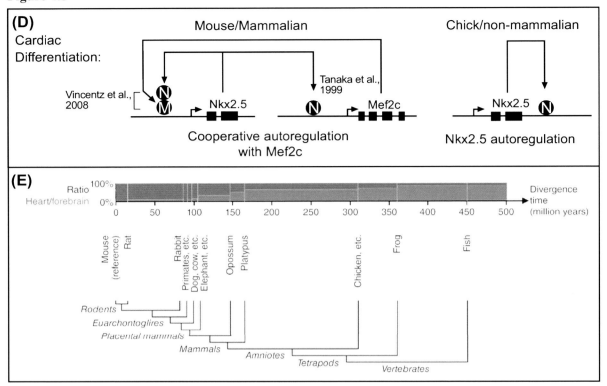

the SHF before looping and their later derivatives *(from Vincent and Buckingham (2010))*. (B1), E8.0 regulatory states: anterior, light green, expresses genes indicated in light green box; *pitx2c*, expressed on left side, indicated by diagonal hatched bars; posterior, dark green, expresses genes indicated in green box; far posterior and lateral, *tbx18* domains. (B2), SHF contributions to the four-chambered E14.5 heart, abbreviations as in (A). Atrial myocardium contains descendants of formerly posterior SHF cells that expressed *isl1* and *nkx2.5*. Myocardium of caval veins derives from the *tbx18* SHF cells. Pulmonary vein myocardium derives from posterior SHF cells also expressing *pitx2c*. The right ventricle and aortic arch derive from formerly anterior SHF cells. The central hatched region expressing *pitx2/tbx1* will form the interventricular septum and the myocardium at the base of the pulmonary trunk (PT; tan), while the PT expresses Semaphorin3c (Sema3c), which is required for neural crest immigration into the outflow tract and future valves. (C), BioTapestry models for FHF and SHF GRNs as indicated *(from Herrmann et al. (2012))*. Signals from the endoderm are indicated at left. Symbolism is as in other BioTapestry presentations reproduced in this book (see Fig. 6.1 for detailed lexicography). (D), Evolutionary change in SHF feedback regulation of *nkx2.5 (from Clark et al. (2013))*. The relevant portions of the SHF *nkx2.5* maintenance enhancers are shown diagrammatically for mouse and chick: M, Mef2c factor bound to its cognate target site in the mouse *nkx2.5* gene; N, Nkx2.5 factor, left to right, bound to the Mef2c factor protein on the mouse *nkx2.5* gene, to its cognate DNA target site in the mouse *mef2c* gene or to its cognate target site in the chick *nkx2.5* gene. (E), Relative conservation of mouse forebrain and atrial heart enhancers isolated by P300 ChIP-seq *(from Blow et al. (2010))*. The colored bars indicate the maximum fraction of forebrain enhancers (blue) and cardiac enhancers (red) which are conserved and can be aligned at each evolutionary distance from mouse shown on the abscissa in real time and phylogenetically in the tree below.

similarly to other amniotes such as the chick). The zebrafish heart has only a single ventricle, and the initial heart tube derives from an initial wave of progenitor cells. But an SHF exists in zebrafish as well, marked by early expression of many of the same regulatory genes that we discuss below. Genetic lineage tracing shows that a second wave of medially located mesodermal heart progenitors, equivalent to the SHF, generates the arterial pole of the developing zebrafish heart (Lazic and Scott, 2011; Guner-Ataman et al., 2013).

Regional regulatory states of the SHF and vectorial signal inputs

The diverse developmental fates of different regions of the heart presuppose the establishment of diverse regulatory state domains along both anterior/posterior and mediolateral axes. A diagrammatic portrayal of these domains in the e8.0 mouse heart is shown in Fig. 4.8(B1), and the contributions to the structures of the developed heart are shown color coded in Fig. 4.8(B2) (Vincent and Buckingham, 2010). To understand this diagram it is necessary to remember that because of the looping of the developing heart tube the posterior domains of Fig. 4.8(B1) become the antrerior structures of Fig. 4.8(B2). Thus, from posterior to anterior in Fig. 4.8(B1) the domain expressing *tbx18* gives rise to the great inflow veins of the right atrium; the atrial myocardium includes desendants of cells expressing *nkx2.5* and *islet1*; the anterior domain of the SHF expressing *tbx1*, *fgf8/10*, and *islet1* gives rise to the right ventricle and the outflow tract (see legend for further details). In the mediolateral direction, a *pitx2* (+*tbx1*) domain on the left side gives rise to the interventricular septum and the pulmonary vein, while *tbx1* positive cells contribute the pulmonary trunk myocardium of the right ventricle. These progenitor fields are positioned by signaling, from both within and without the SHF (for comprehensive review, and primary references see Vincent and Buckingham (2010)). In the anterior SHF RA signaling affects *tbx1* expression, and is in turn controlled by nested anterior to posterior expression of *hoxb1*, *hoxb1* + *hoxa1*, and *hoxb1* + *hoxa1* + *hoxb3*, which are ultimately reflected in atrial and outflow tract fates (Bertrand et al., 2011). The expression of *pitx2c* on the left side of the SHF (Fig. 4.8(B1)) is activated by Nodal signaling from the lateral plate mesoderm, which installs the specification of left versus right atrial identity. Shh signaling is required to maintain left atrial *pitx2c* expression and prevent bilateral *pitx2c* atrial expression. Bmp signaling plays a major role in the activation of heart regulatory genes. Sources of the Bmp signals vary: early in FHF specification, the initial mesodermal cardiac progenitors are exposed to Bmp2 from the immediately adjacent endoderm, and to Bmp4/7 from the overlying ectoderm at about the stage indicated in Fig. 4.8(A2) (Schultheiss et al., 1997). Then cells of the FHF express the *bmp2* gene at regulated levels, and cells of the SHF express the *bmp4* gene, both of which produce direct intradomanic inputs into key regulatory genes of these fields, as we see explicitly below. Another aspect of the role of external signals early in heart development is that Bmp signals emanating from the lateral plate mesoderm affect the laterally positioned FHF cells, promoting initial cardial specification, while the more medially located SHF receives Wnt signals from the medial neural tube which transiently promote proliferation and delay differentiation. In later SHF development, in phases beyond the scope of this discussion, Wnt, Bmp, Shh, and Fgf's all have regional inputs into particular domains and affect differentiation and morphogenesis, particularly in the complex structures of the outflow tracts, and a whole additional set of external signaling inputs, including Fgfs and Notch, affects neural crest derivative specification in the arterial pole region.

5.2 GRNs for first and second heart field regulatory states

The GRNs

A model for the genomically encoded network of gene interactions causing the appearance of FHF and SHF regulatory states in the mouse embryo is shown in Fig. 4.8(C) as a BioTapestry compilation, derived from literature sources (Vincent and Buckingham, 2010; Herrmann et al., 2012). Almost all linkages discussed in the following can be seen in this GRN model. The initial signal inputs from the underlying endoderm are shown at the left, namely Bmp2 which as noted above provides an initial input into the core regulatory genes of the FHF, and Dkk1, an antagonist of Wnt signal reception. An intrinsic initial regulatory input expressed in the mesoderm cells which will constitute both the FHF and the SHF is provided by the *mesp1* gene. This gene is expressed transiently in the migrating mesoderm cells which will give rise to the cardiac lineages. In these cells Mesp1 directly activates key FHF and SHF genes, though elsewhere it also promotes early hematopoietic and muscle differentiation (Bondue et al., 2008; Chan et al., 2013). The *bmp2* and *dkk1* signaling genes are activated in the FHF by Mesp1. An initial tier of regulatory genes activated by Mesp1 in the FHF include *gata4*, *tbx5*, and *nkx2.5*, which also receives an input from Bmp2 signaling, and these core genes establish an encoded feedback lockdown circuit: *nkx2.5* activates *tbx5* which feeds back on it, as well as on *gata4*; and in addition *gata4* feeds back on *nkx2.5*. The genes of this circuit provide inputs to a new tier of cardiac regulatory genes, including the *hand1/2* genes, *mef2c*, *tbx20*, *myocardin*, and *irx4*, which also feeds back on *hand1*. At the bottom of the FHF GRN model are shown differentiation genes that are driven by subsets of the latter tier of regulatory genes as well as by genes higher up in the FHF hierarchy. An interesting aspect of FHF regulatory development is that it is not bilaterally symmetric: differentiation functions are inhibited on the right side of the cardiac crescent. The mechanism unexpectedly has to do with expression of *hoxa10* on the right side, which apparently inhibits multiple Nkx2.5 transcriptional functions (Behrens et al., 2013).

In the SHF domain Mesp1 targets in parallel a number of key regulatory genes, including *foxh1*, *tbx1*, *islet1*, *nkx2.5*, *hand2*, and *gata4*. Then, in a feedback circuit which functions analogously to that set up in the FHF, *gata4* stabilizes its own expression by activating Bmp4 signaling within the SHF, which in turn generates an activating input into it. Inputs from *gata4* also go to the *nkx2.5* and *hand2* genes, in feed-forward constellations. Another linkage specific to the SHF is the input to *mef2c* from *foxh1*. As in the FHF, *nkx2.5* activates *mef2c*. *Islet1* also feeds forward to *nkx2.5* and *mef2c*. Thus in both FHF and SHF GRNs, following the initial inputs, heavily recursive feedback, feed-forward and other cross-wiring is installed, ensuring the regulatory states produced. As we discuss later (Chapter 6) a Boolean model built on the basis of the GRN models in Fig. 4.8(C) and tested against the results of genetic perturbations, demonstrate the explanatory sufficiency of these GRN models (Herrmann et al., 2012).

Figure 4.8(C) represents the control system for only the beginning of heart development. As differentiation proceeds many cell types appear, each with its own downstream GRN. For example, differentiation of the specialized cells that determine cardiac rhythm, the "conduction system" of the right atrium, is programmed by specific regulatory circuitry, different from that of atrial or ventricular myocytes (Munshi, 2012). In these circuits many of the same regulatory genes just considered operate downstream effector functions, including *tbx5*, *tbx18*, *nkx2.5*, and *gata4*. Differentiation gene networks of great complexity control all aspects of structure and function as the heart matures, and as discussed in Chapter 1, heart miRNAs are frequently involved in both later developmental and physiological functions (Liu and Olson, 2010; Schlesinger et al., 2011).

GRN kernel for heart field specification

A remarkable apparent orthology can be seen between the mouse heart GRNs and the GRN for early *Drosophila* heart development, in the repertoire of regulatory genes and in the retention of some of the regulatory linkages. This has contributed to the concept of the GRN kernel (Davidson and Erwin, 2006), defined as a core developmental regulatory circuit dedicated to early stages of specification of given body parts, and displaying unusual evolutionary conservation, as discussed in Chapter 7. In the case of the heart kernel, an amazing fraction of the same auto- and cross-regulatory linkages among the core regulatory genes in mammalian and fly heart GRNs, argue for a common pan-bilaterian origin of the heart kernel (for specific circuitry, Davidson (2006)). For example, the core circuitry in both GRNs includes the *nkx2.5* (*Drosophila tinman*), *islet* (*Drosophila tup*), *mef2*, *tbx2/3/5* (*Drosophila doc1/2/3*), *tbx20* (*Drosophila* mid/H15), bmp2/4 (*Drosophila dpp*), hand1/2 (*Drosophila hand*), gata4/6 (*Drosophila pnr*), and *wnt* (*Drosophila wg*) genes. In both, *nkx2.5* is activated by a *gata* gene input; in both *nkx2.5* is linked into feedback circuits with *tbx* genes; in both *mef2* and *hand* are together activated by *nkx2.5* and *gata*; etc. (for the *Drosophila* GRN see Olson (2006), Bryantsev and Cripps (2009), Reim and Frasch (2010)). In addition to conservation due to the vital embryonic role of the evolutionarily ancient, early acting program for heart development, recent studies in both systems have revealed a mechanistic explanation for the extreme conservation of the cardiac regulatory states. Genome-wide ChIP experiments have uncovered combinatorial binding of many kernel factors to *cis*-regulatory elements of genes active during heart development. For example, in a mouse atrial cell line various combinations of Nkx2.5, Mef2, Tbx5, Hand2, Gata4, Tbx20, Srf (a target of Myocardin) were bound together in the enhancers of many hundreds of cardiac effector genes (He et al., 2011). A similar result was obtained for the *Drosophila* heart kernel regulators Doc (mammalian Tbx2/3/5), Pnr (Gata4/6), Tin (Nkx2.5), dTcf (effector of Wg signaling), and Mad (effector of Dpp signaling) (Junion et al., 2012). Here, putative enhancers that were bound by all five factors when tested were preferentially expressed in cardiac lineages (see Chapter 5, Fig. 5.1(B)).

Evolutionary augmentation of heart GRNs in the vertebrates

Evidence from *cis*-regulatory studies of kernel cardiac specification genes indicates that different modules direct expression of given genes in different regions of the heart. For example, the pivotal *nkx2.5* gene employs multiple enhancers which are each specifically active in regional subdivisions of the FHF and SHF, such as the right ventricle lateral wall, the early anterior cardiac crescent and the outflow tracts, and other domains (Schwartz and Olson, 1999; Olson, 2006). Expression of *nkx2.5* occurs throughout the developing heart, but its expression in each domain of the heart is mediated by region-specific enhancers responding to the particular specification GRNs operating in that domain. Thus during the evolutionary processes leading to the amniotes, as the heart became structurally more complex, additional *cis*-regulatory modules must have appeared in the control systems of these genes. Changes in specific linkages have happened even within the amniotes. Thus in chick and other birds, as well as in *Xenopus*, and in *Drosophila*, *nkx2.5* auto-regulates by direct *cis*-regulatory binding of the transcription factor it encodes. In contrast, in mouse (and other placental mammals but not monotremes) direct auto-regulation is replaced with feedback cross-regulation by Mef2c (Clark et al., 2013). The *mef2c* gene is a target of Nkx2.5 in mouse and in the chick. But in mouse the relevant *cis*-regulatory maintenance module of *nkx2.5*, which is active in the SHF, includes a Mef2c target site, which is bound by a Mef2c:Nkx2.5 protein:protein complex (Fig. 4.8(D); not shown in Fig. 4.8(C)), and this site is confined to placental mammals. Further afield within amniotes, additional regulatory changes have occurred. Phylogenetic observation of *tbx5* expression in the developing heart of a turtle provides a specific case of the relation between regulatory change affecting this core FHF gene, and developmental form of the heart (Koshiba-Takeuchi et al., 2009). The turtle lacks an interventricular septum, and *tbx5* is expressed almost uniformly in its developing heart, in contrast to mammals where ventricular expression occurs only in the structure to which the FHF gives rise, the left ventricle. If uniform *tbx5* expression is forced in the mouse heart, an atavistic one-chambered ventricle phenotype results. Since the advent of vertebrates, hearts have evolved morphologically from one-chambered to four-chambered structures, with complex intake and outtake morphology. The implication is that much of the regulatory apparatus used to build the mammalian four-chambered heart might be relatively recent in evolutionary terms. Genome-wide confirmation of the recent evolution of heart enhancers emerges from a comparison of sequence conservation of hundreds of enhancers recovered by P300 ChIP from heart and from forebrain (Blow et al., 2010). As we see in Fig. 4.8(E) the large majority of forebrain enhancers display alignable sequence from fish to mouse, while the majority of heart enhancers can be aligned only within the placental mammals. Nonetheless, since these relatively nonconserved enhancers were in fact usually active in heart when tested, they provide clear evidence that the placental mammals share a large array of heart enhancers specific to them.

6.   Spatial Regulatory State Subdivision in and Around the *Drosophila* Ocellus

Ocelli are simple light–dark sensitive eyes with low spatial resolution, occurring throughout pan-arthropods, which are located on top of the head and which may contribute to control of locomotion. The ocellus complex of *Drosophila* consists of three simple eyes arranged at the apices of a forward pointing equilateral triangle, one anterior and two bilaterally positioned behind it, on the dorsal surface of the head (Fig. 4.9(A1)). These eyes form from the fused anterior dorsal domains of the eye-antennal imaginal discs, in a field initially defined by expression of a *pax6* gene (*twin of eyeless*) and an *otx* gene (*orthodenticle*). The *hh* gene is expressed in the center of the future ocellar triangle in what will become the interocellar domain, and is required for formation of the ocellar complex. Thus when Hh signaling is severely reduced, the ocelli (and the accompanying bristles) fail to form (Fig. 4.9(A2)). The retinal determination genes *eyes absent* (*eya*) and *sine oculis* (*so*) operate together to specify the ocellar fields per se (for comparison of *Drosophila* ocellar and eye specification GRNs, see Chapter 7). Expression of these genes is accompanied by repression of an ocellar specification antagonist, *homothorax* (*hth*) which is active in the periocellar region. A recent study has uncovered the encoded network circuitry by which the spatial interocellar, ocellar, and periocellar regulatory state domains are established in response to Hh emanating from the central interocellar region (Aguilar-Hidalgo et al., 2013).

The logic behind this tri-domanic transcriptional response to high, intermediate, and insignificant Hh levels is shown in the GRN model of Fig. 4.9(B) (Aguilar-Hidalgo et al., 2013). The encoded circuit connections are portrayed in Fig. 4.9(B1), and the linkages active and inactive in the three regions are seen in Fig. 4.9(B2). In the interocellar region (right) where the *hh* gene is expressed, and where high Hh signal is available, the *en* gene is turned on in response to Ci activator (Ci$_{act}$). However, as the wiring indicates, En then functions to extinguish Hh signal transduction, by preventing expression of the genes encoding the Hh receptor (*ptc*) and the Hh signal transducer (*ci*). In order to ensure that *en* expression continues in the interocellar domain even after its activating Ci driver has disappeared, the *en* gene once turned on activates itself, and this auto-activation is enhanced by Delta-Notch signaling, present throughout the region. Thus in the interocellar domain, *en* is on, the *hh* gene is on, but Hh signal transduction is off and therefore Ci$_{act}$ is no longer available. Thus in the interocellar domain *eya* is off, because *eya* expression requires Ci$_{act}$, and *so* expression is off because it requires *eya* expression (Fig. 4.9(B1,B2)). In the ocellar domain *en* is not activated at the available Hh concentrations (nor of course in the peripheral periocellar domain), but intermediate levels of Ci$_{act}$ are available from Hh signal transduction, which here is not repressed because *en* is not active. Therefore in the ocellar domain, *eya* is expressed and activates *so*, which feeds back to ensure *eya* activity. In addition, the Eya/So complex excludes expression of the antagonistic *hth* gene: here is where the ocelli are allowed to form. In the periocellar domain, lack of Hh signal precludes *eya* and *so* expression and instead *hth* expression is permitted, driven by Wg signaling. Hth functions specifically to clamp down on *eya* expression, in both the periocellar and interocellar domains. This elegant device ensures that the alternative regulatory states are excluded in each domain by active repression. The key is the special feature that high Hh signal is used to activate *en* expression so as to preclude Hh signal transduction in that region, while in the next domain the same signal activates the retinal determination system. More generally, it is interesting that two of the three external signals in play, Wg and Delta/Notch, which have no decisive spatial input, are used indirectly by activating default repressors, and the spatial control input comes from the process by which two levels of Hh are used to mediate repression versus activation. Ultimately the circuit logic results in three states: the interocellar state, where Hh signaling turns expression of *en* on, and En turns Hh signaling off, with the consequence that the retinal specification genes are off; the ocellar state, where Hh signaling promotes expression of the retinal specification genes *eya* and *so*, which in turn exclude *hth* expression; and the periocellar state, where there is no Hh signaling and Hth represses the retinal specification genes. Figure 4.9 illustrates a canonical case of Boolean ON/OFF switches producing adjacent spatial regulatory states.

Figure 4.9

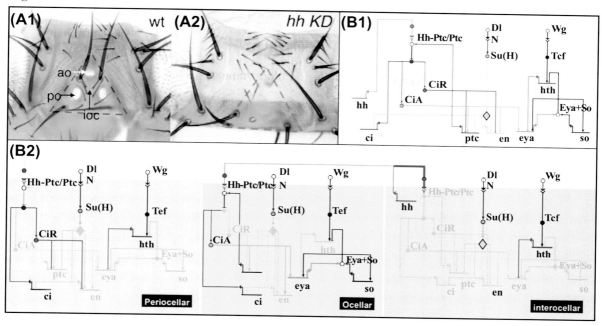

Figure 4.9 Model for the spatial specification GRN for _Drosophila_ ocellar complex *(from Aguilar-Hidalgo et al. (2013)).* (A), Ocellar complex. (A1), Dorsal view of normal ocellar complex (including interocellar bristles), ocelli appearing light in color: ao, anterior ocellus; po, posterior ocellus; ioc, interocellar cuticle; dashed triangle indicates the dimensions of the complex. (A2), Effect of severe knockdown of Hh signaling by overexpression in ocellar field of mutated *ptc* gene: the domain of the ocellar complex (red dashed lines) is empty of the normal features. (B), GRN circuitry. (B1), "View from the genome", showing all gene interactions or functional linkages programmed in the *cis*-regulatory sequence, BioTapestry presentation. (B2), GRN linkages functional in periocellar, ocellar, and interocellar domains, BioTapestry "view from the nuclei" presentation. Genes are indicated by abbreviations in text; CiR, Cubitus interruptus repressor form; CiA, Cubitus interruptus activator form. The diamond linking Delta/Notch pathway to *en* gene indicates the positive stimulation by this pathway of the auto-activation of *en* by its own gene product. For explanation of mode of function of this network, see text.

7. The Vertebrate Gut

The gut may easily serve as a paradigm for organogenesis, one which encompasses many of the features we have discussed in this chapter. Here the specification of progenitor fields and their progressive subdivision into discrete regulatory state domains is particularly accessible because of the almost linear organization of the gut tube. In amniote vertebrates, the endodermal organ system encompasses not only the gut tube but also a number of organs which develop from the foregut endoderm, for example liver and pancreas. During gut development, the progenitor fields for these different organs are specified by the expression of particular regulatory states long before these organs become morphologically apparent. These discrete organs evolved within the higher vertebrates, but the extensive patterning of the gut in all Bilateria indicates a similar functional diversification. Almost all Bilateria have through guts and thus the genome of the last common Bilaterian ancestor must have encoded a program for the development of the gut. Only a few aspects of the program for gut development are understood mechanistically in any bilaterian, but some regional processes have indeed been heavily studied. An example is the development of the pancreas, as we review below.

7.1   Nodal signaling and the early endoderm specification GRN

The earliest expression of endodermal regulatory genes depends in all vertebrates so far studied on Nodal signaling. Induction of endoderm specification thus has to take place in proximity to a source of Nodal signaling ligand, but where Nodal signaling occurs differs among vertebrate embryos. In zebrafish, Nodal signal derives at first from the yolk, and provides an initial input into both mesodermal and endodermal fate specification. In mice, Nodal is expressed in the node and cells passing through the primitive streak are specified as mesodermal and endodermal progenitor cells. A similar set of regulatory genes is expressed downstream of Nodal signaling in endomesodermal precursor cells of embryos of jawed vertebrates, including *brachyury* in the mesoderm and *mix*, *foxa*, *sox*, and *gata* genes in the endoderm. Some of the regulatory interactions among these genes differ among species and obviously the regulation of these genes relies on additional inputs (Zorn and Wells, 2007). But despite differences in the early endoderm GRNs among vertebrates, initial localization of endoderm specification always depends on Nodal signaling. This is a vertebrate invention, since use of Nodal in early endomesoderm specification is not found even in invertebrate chordates nor in any other invertebrate embryo (Cattell et al., 2012). Instead, localized maternal β-catenin is widely used in nonvertebrate bilaterians as the initial spatial input in specification of embryonic endoderm (Hudson et al., 2013).

Mesoderm specification in axolotl embryos is initiated with the activation of *brachyury* (*bra*) expression, controlled by the circuit shown in Fig. 4.10(A) (Swiers et al., 2010). This circuitry is significantly simpler than the equivalent *Xenopus* circuit (Loose and Patient, 2004; Koide et al., 2005) because of the multiple copies of the key *nodal* and *mix* genes in *Xenopus*, in contrast to the single copy axolotl orthologs. Since urodeles (newts, salamanders) are closer to the basal tetrapod ancestor than are anurans (frogs, toads), the axolotl developmental program is likely to be less derived. As can be seen in Fig. 4.10(A) the *cis*-regulatory system operating the embryonic expression of *bra* combines the input from vegetal Nodal signaling, itself downstream of VegT (cf. Chapter 3), with an input from the *mix* gene, also downstream of both *nodal* and VegT. The Nodal (Smad) and Mix inputs are processed by AND logic and *bra* is thus at the combinatorial terminus of a feed-forward subcircuit. To this are added additional inputs from *gsc* and VegT. In classic fashion, once activated, *bra* ensures its own continued expression in mesodermal cells by two positive feedback interactions (Fig. 4.10(A)). Bra controls its own expression by auto-activation, and also by activating an *fgf* gene, the signal from which in turn stimulates *bra* transcription. Bra then turns off its initial driver, *mix,* in what are now going to be mesoderm founder cells. Since *mix* is also upstream of *gata* and *sox17* genes, which as we see below are essential endoderm specification genes, this subcircuit constitutes a mesoderm/endoderm decision point. Observations on the consequences of loss of *bra* expression in vertebrate embryos show that it affects multiple developmental aspects of downstream mesodermal function, including posterior somite formation, notochord formation, gastrular cell migration, left/right asymmetry, and tail extension. ChIP measurements on zebrafish and mouse embryos reveal a large array of mesodermal Bra targets associated with these functions (Fig. 4.10(B); Morley et al., 2009; Lolas et al., 2014).

In zebrafish activation of endoderm specification genes occurs in cells immediately adjacent to the yolk syncytial layer (YSL), which is the source of maternally encoded Nodal signaling. In response to Nodal signaling, both endodermal and mesodermal regulatory genes are initially expressed in prospective endodermal cells. Figure 4.10(C) shows three stages of endoderm specification process (Chan et al., 2009b). At 5 h, Nodal signaling has activated the *mix* genes *og9x* and *bon, gata5,* and *sox32* as well as the Nodal signaling genes *ndr1* and *ndr2* (Fig. 4.10(C1)). Smad2/3, Smad4, and their cofactor Foxh1 collaborate to transduce this signal at the target gene *cis*-regulatory level (Pereira et al., 2012). An additional input is provided by the maternally encoded Tbox factor Eomesodermin. A cross-regulatory linkage from *gata5* to *sox32* is also set up. By the end of gastrulation 5 h later (Fig. 4.10(C2)), multiple additional genes that will define the presumptive endoderm regulatory state have been activated. These include *sox17*, *foxa2*, and *gata6*, which are expressed downstream of Gata5, Sox32, and continued Nodal signaling (Chan et al., 2009a,b). As in the axolotl, zebrafish *brachyury* (*ntl*) is activated by a *mix* gene (*og9x*), which also turns on another mesodermal gene, *gsc*. Feedback and cross-regulatory linkages have appeared: *sox32* auto-regulates; it is now engaged in a feedback relation with *gata5*; *og9x* and *bon* both *trans*-activate *sox32*. In addition, *gata6* has been activated by *gata5* and latch-like, feeds back on *sox32*. Sox32 plays the important role of extinguishing expression of the mesodermal genes *gsc*, *bra/ntl*, *tbx6* as well as *foxh1* in endodermal precursor cells (Fig. 4.10(C3)). It is interesting that expression of the initial tier of endodermal regulators, *sox32*, *og9x*, and *bon* is subsequently extinguished, perhaps due to interference of Lefty with Nodal signaling and the loss of *foxh1* expression.

In mice, few linkages are known, but it is clear that many of the same regulatory genes participate in the initial process of endoderm specification as in other vertebrates. Again, Nodal expression is upstream of endodermal gene activation and a *mix* gene (*mixl1*) and *foxa2* are among the earliest required regulatory genes expressed in endodermal precursors. Lack of *mixl1* in mice leads to an expansion of *brachyury* expressing presumptive mesodermal cells, and to fewer *sox17* expressing presumptive endodermal cells. Mixl1 therefore seems to affect the cell fate decision between endoderm and mesoderm (Hart et al., 2002; Tam et al., 2007). Vice versa, Brachyury interferes with the transcriptional activation function of Mixl1 by protein–protein interaction (Pereira et al., 2011). The import of this interaction is similar to the repression

Figure 4.10

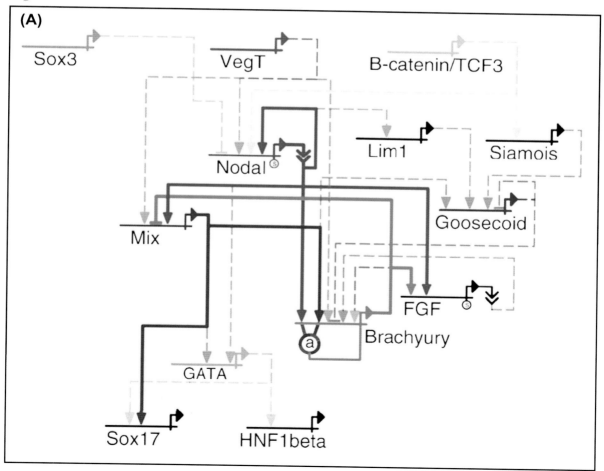

Figure 4.10

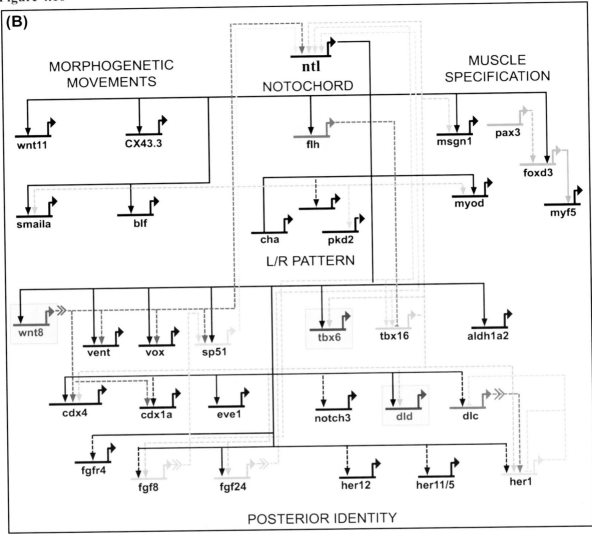

Figure 4.10 GRNs for initial mesoderm and endoderm specification in vertebrate embryos. (A), Early endoderm and mesoderm specification in the axolotol *(from Swiers et al. (2010))*. The network shown indicates regulatory transactions that lock on expression of the key mesoderm identity gene *brachyury* (*bra*), and that exclude the alternative endoderm fate. Solid lines indicate linkages confirmed at the *cis*-regulatory level; dashed lines indicate results from perturbation experiments. (B), Regulatory and signaling gene targets of *bra* (here *ntl*) gene in zebrafish embryos *(from Morley et al. (2009))*. Solid lines indicate that both ChIP-binding data and genetic inference supports link; dashed lines indicate that one or the other type of evidence supports link. (C), GRNs for early endoderm specification in zebrafish *(from Chan et al. (2009b))*. YSL, yolk syncytial layer; Ndr1, Ndr2, nodal signaling ligands. Note endoderm genes, blue background; mesoderm genes, orange background; genes expressed in both, orange with blue borders, e.g., *gata6*. Colored linkages indicate transcriptional activity, grayed out gene names and linkages indicate inactivity at the given stage. Data were compiled from literature. (C1), Early mesendoderm specification at 5h; (C2), Gastrula stage zygotic mesendoderm GRN when both endoderm and mesoderm genes are being expressed downstream of Nodal; (C3), Repression of early endoderm regulatory genes and extinction of mesoderm regulatory gene expression in endoderm cells.

Figure 4.10

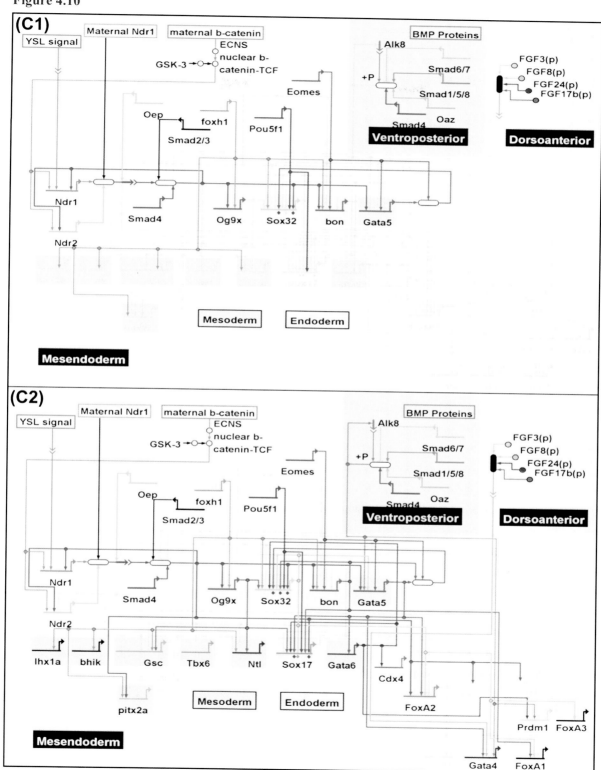

Figure 4.10

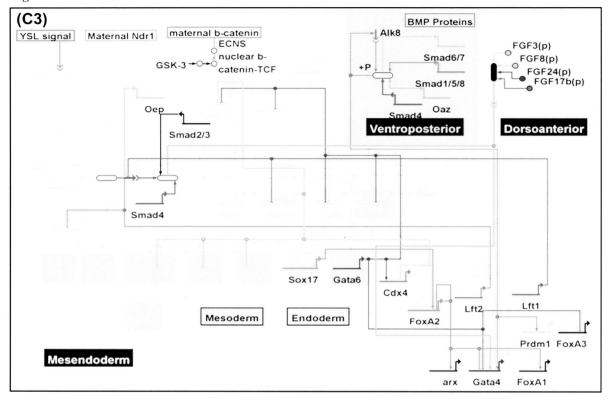

of *mix* by Brachyury in axolotl (Fig. 4.10(A)). In contrast to other bilaterians, where Gata factors are universally found to be important for early endoderm specification (cf. Chapter 3), in mice Gata factors of the 4,5,6 type appear only to have later functions in the specification of particular endoderm derivatives such as the liver (Woodland and Zorn, 2008). By e7.0 *sox17* is expressed in all definitive endoderm cells, but very quickly anterior and posterior endodermal regulatory states become distinct. Thus by e8.0 *sox17* accumulates in posterior endoderm cells while *foxa2* is required specifically for the development of the future foregut cells (Kanai-Azuma et al., 2002; McKnight et al., 2010).

7.2   Regulatory regionalization of the mammalian gut

Evidence for the rapid diversification of endodermal fates along the anterior-posterior and dorsal-ventral axis derives from cell lineage analyses (Lewis and Tam, 2006). Differential specification of endodermal cells according to their future position within the gut tube is already initiated during gastrulation with the separation of anterior and posterior endoderm, and proceeds rapidly with the specification of additional subdomains while endodermal cells are still arranged within a two-dimensional sheet of cells. Labeling of individual cells at e7.5–8.0 and following their migration until e8.5 has demonstrated that very shortly after gastrulation there are at least 13 different cell populations that localize to discrete bilaterally arranged spatial domains of the forming gut (Franklin et al., 2008).

Prominent derivatives of the mammalian gut are illustrated in Fig. 4.11(A). Several organs of distinct function are formed within the foregut endoderm, including the thymus, liver, pancreas, lung, and thyroid. These all arise as localized buds at positions defined by the expression of particular regulatory states.

Figure 4.11

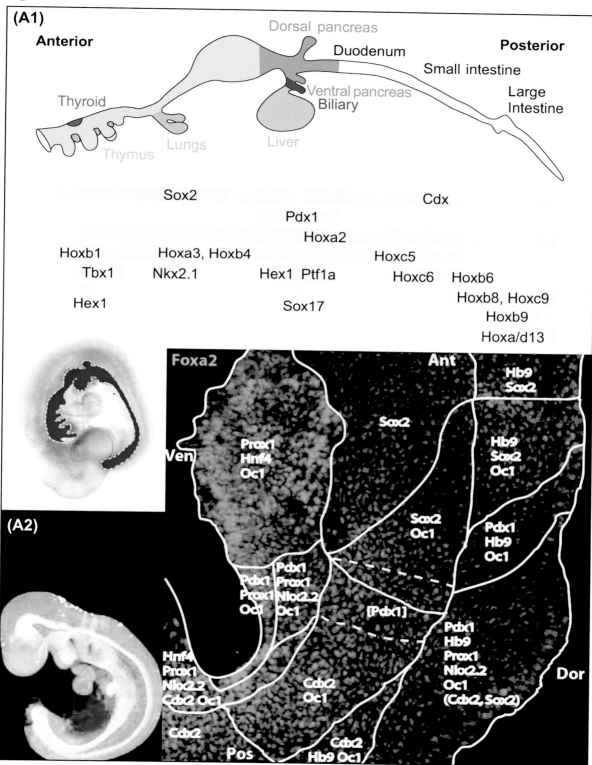

Figure 4.11

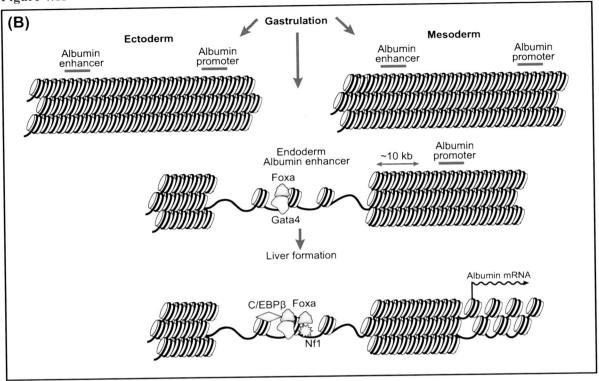

Figure 4.11 Mammalian gut regulatory states and pancreatic developmental GRNs. (A), Regulatory state diversification along the murine gut. (A1), Regional organogenesis *(from Zorn and Wells (2009))*. Below left is a false colored image of an e9.5 mouse embryo illustrating the whole of the gut tube. The diagram shows budding organs derived from the foregut at e10.5, colored, and gut regions forming posteriorly in outline. Below are approximate limits of expression domains of some regulatory genes important for gut regionalization, as can be inferred by the correspondence between the boundaries of expression and the locations of the foregut organ primordia. (A2), Regulatory states in the pancreas/liver region of the murine foregut, e9.5 *(from Sherwood et al. (2009))*. High magnification image of the posterior foregut region of the gut stained for *foxa* expression (green); the region shown is indicated in the inset. Axes are indicated: Ven, ventral; Dor, dorsal; Ant, anterior; Pos, posterior. White lines separate fields of cells expressing the indicated regulatory gene sets (some but not all cells in the region included in dashed white lines express *pdx1* at a low level). Dorsal and ventral pancreas buds derive from the *pdx1*[+]/*nkx2.2*[+] regions, and the liver primordium from the ventral/anterior *hnf4*[+]/*prox1*[+] region. *Cdx2* and *sox2* are shown in parentheses dorsally because they are expressed in this region at e8.75, but not e9.5. (B), "Pioneer" function of FoxA2, illustrated in an enhancer of the *albumin* effector gene *(from Zaret (2002))*. Three states of the *albumin* gene are indicated: prior to gut development and in other tissues the gene is transcriptionally silent and the enhancer is covered with tightly packed nucleosomes. The pioneer FoxA factor (together with Gata4) functions during liver development, invading the chromatin structure of the enhancer to bind its target site. This is insufficient to promote transcription but it causes opening of the chromatin structure, here portrayed as the appearance of nucleosome-free DNA. This structural change potentiates the binding of liver-specific factors that do initiate transcription, here C/Ebpβ and Nf1. The general role of FoxA in foregut genes is in this way to induce competence for sequence-specific interaction with regional- and organ bud-specific regulatory gene products. (C), GRN models for pancreatic specification and cell differentiation, compiled from literature sources. Functionally significant features

Figure 4.11

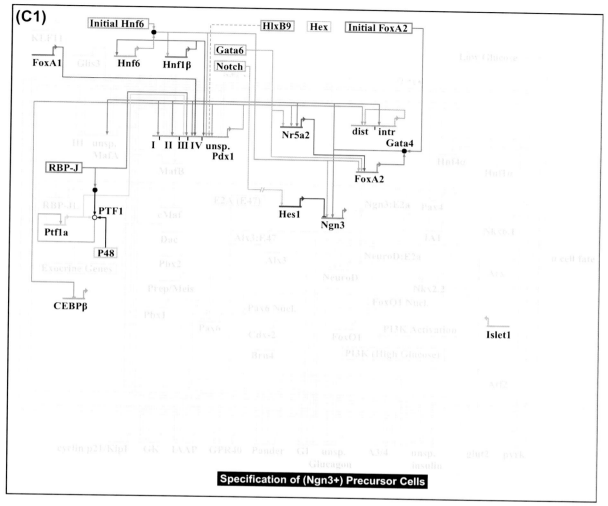

Specification of (Ngn3+) Precursor Cells

of the GRN circuitry are discussed in text. (C1), Early specification events (e9–e13), including the causal inputs that result in activation of the key regulatory genes *foxa2, pdx1, gata4, nr5a2*, and *ngn3*. (C2), Progression toward endocrine regulatory states that will produce α and β pancreatic cell types from *ngn3* expressing cells; and generation of distinct exocrine specification state (e13–e15). In the endocrine GRN *nkx2.2, neuroD, pax4, nkx6.1, arx*, and *islet1* are activated. (C3), GRN models for differentiation of insulin synthesizing β-cells and of glucagon synthesizing α-cells (e15 to post partum). Sources: For references to linkages that had been established earlier, see caption to predecessor of this GRN (Davidson, 2006, p. 179). More recent sources of data supporting the linkages shown in this GRN model, both positive and negative, are as follows: PTF1 to *ptf1a*, PTF1 to exocrine genes, PTF1 to RBP-JL (Beres et al., 2006); (*ptf1a*, RBP-JL) to *pdx1* (Miyatsuka et al., 2007; Wiebe et al., 2007); (*mafa, mafb*) to *pdx1* (Vanhoose et al., 2008); (*foxa1, foxa2*) to *pdx1* (Gao et al., 2008); (*foxa, nkx2.2, pdx1*) to *mafa* (Raum et al., 2006); *glis3* to *mafa*, *glis3* to *insulin*, *glis3* to *pdx1*, *glis3* to *ngn3* (Fernandez-Zapico et al., 2009; Kang et al., 2009); (*islet1, hnf4a*) to *insulin*, (Eeckhoute et al., 2006; Zhang et al., 2009); (*meis, pbx1, pbx2*) to *pax6* (Zhang et al., 2006; Delporte et al., 2008); *nkx6.1* to *hnf1α* (Donelan et al., 2010); (*nkx2.2, ngn3*) to *neuroD* (Anderson et al., 2009); (*pdx1, gata4*) to *gata4* (Rojas et al., 2009); (*nkx6.1, pdx1, pax6*) to *mafa* (Raum et al., 2010); *foxa* to *gata4* (Rojas et al., 2010); *pax6* to *mafb*, *pax6* to *c-maf*, *pax6* to *neuroD1*, and

Figure 4.11

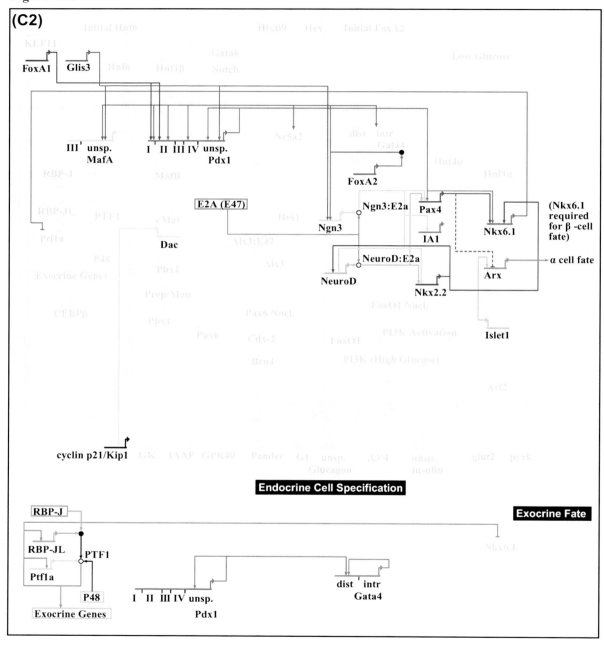

(*mafb, neuroD1, foxa1, foxa2*) to *glucagon* (Gosmain et al., 2010); *islet1* to *arx* (Liu et al., 2011); *dac* to *cyc21* (Kalousova et al., 2010); *hnf1a* to *mafa* (Hunter et al., 2011); *pdx1* to *pax6* (Delporte et al., 2008); *ngn3* to *ia1* (Mellitzer et al., 2006); *hnf4a* to hnf1α (Lu et al., 2008); *pax6* to *glucagon* (Grapp et al., 2009); *pbx1* to *glucagon* (Liu et al., 2006); (*pax6, cmaf*) to *glucagon* (Gosmain et al., 2007); *nk6.1* to *glucagon* (Gauthier et al., 2007); *brn4* to *glucagon*, (*pdx1, mafa, neuroD*) to *pander* (Burkhardt et al., 2008); *alx3* to *insulin* (Mirasierra and Vallejo, 2006; Mirasierra et al., 2011); (*pdx1, neuroD*) to *gpr40* (Bartoov-Shifman et al., 2007); *pax6* to *insulin* (Wolf et al., 2010); *atf2* to *insulin* (Han et al., 2011).

Figure 4.11

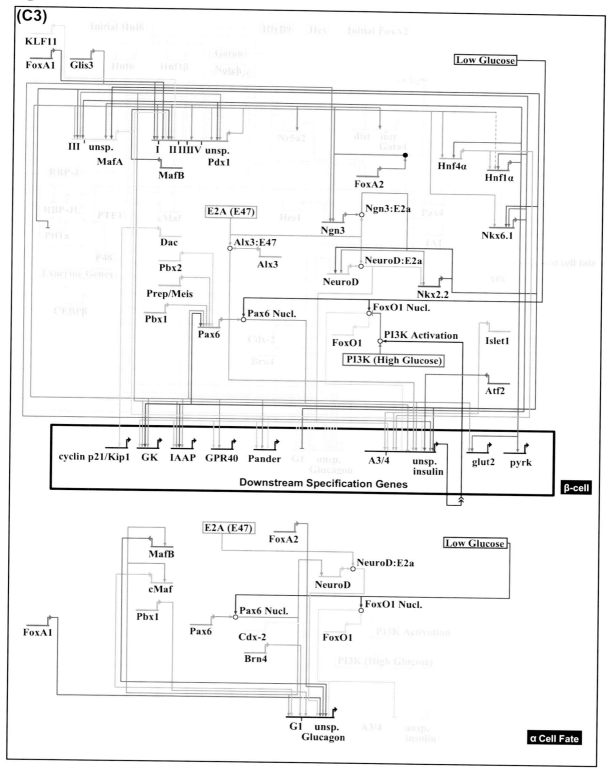

A map of regulatory gene expression in the foregut endoderm at e9.5 vividly illustrates the complexity of the regionalization process (Fig. 4.11(A2); Sherwood et al., 2009). Cells which derive from paired lateral domains of the foregut give rise to liver and ventral pancreatic cell fates (Deutsch et al., 2001). Early ventral foregut endodermal cells express *hhex* and *foxa2* (e8.5). The decision between pancreatic and liver cell fates is affected by signals expressed in adjacent mesodermal cells. These cells produce Bmps, which induce expression of specific genes in liver as well as ventral pancreas precursors. Fgf signals, to which the more anterior portion of this foregut region is preferentially exposed, activate liver-specific genes and suppress the specification of ventral pancreas fates in these cells (Deutsch et al., 2001). Liver precursor cells are distinct from pancreatic precursor cells by e8.5, when the endodermal expression of *hhex* is restricted to the liver primordium and the expression of *pdx1* defines the pancreatic primordium on the ventral side. An additional separation of cell fates follows soon after that, as cells of the ventral pancreatic primordium segregate further into two different regulatory state domains. At e8.5 the regulatory genes *pdx1* and *sox17* are coexpressed in the ventral foregut endoderm adjacent to the *hhex* expressing liver primordium, but by e9.5 these two regulatory genes are expressed in separate cellular domains (Spence et al., 2009). By e10.5, foregut expression of *sox17* is restricted to cells giving rise to the biliary tree and gall bladder and *pdx1* is expressed specifically in ventral and dorsal pancreatic cells. The separation of *pdx1* and *sox17* expression is of functional importance for the development of the pancreas and gall bladder. In the experimental absence of *sox17*, *pdx1* expression expands spatially within the ventral side of the gut and no biliary primordium is formed. On the other hand, ectopic expression of *sox17* in *pdx1* expressing cells interferes with pancreatic development and suppresses the expression of *nkx2.2*, which as we see below is required for development of the pancreas (Spence et al., 2009). Other boundaries within the gut tube are similarly marked by exclusive domains of regulatory gene expression, for example *sox2* is expressed in the foregut endoderm down to the boundary with posterior endoderm throughout which *cdx* is expressed.

The *foxa2* gene has an especially interesting role in endoderm specification. As we have seen, it is among the earliest expressed endoderm regulatory genes and continues to be expressed throughout the foregut region shown in Fig. 4.11(A2). Thus *foxa2* does not participate directly in the regionalization of the foregut. But, FoxA2 has been shown to function as a pioneer factor, that is, it will bind to its specific target sites even in the presence of nucleosomes, requiring no prior transcription factor–DNA interaction (see Chapter 1). Instead, FoxA2 facilitates interaction with regulatory genomic sequence of the factors that determine regional specificity, once it has bound and altered the nucleosomal conformation at enhancer sequences (Fig. 4.11(B); Zaret, 2002; Zaret and Carroll, 2011).

7.3   GRN for pancreatic specification and cell type differentiation

The process of specification in the pancreatic buds and the subsequent development of endocrine cell types are relatively well studied at the regulatory level and can be taken as an example for later organogenesis in the gut. In the following, we consider three phases of activity of the GRN controlling pancreatic development, from the initial specification of pancreatic precursor cells to the terminal differentiation of insulin secreting β-cells. Among the early regulatory genes expressed in the foregut endoderm and activating pancreatic specification are *hnf6*, *foxa2*, and *gata6* (Fig. 4.11(C1); references for all the regulatory linkages discussed in the following are to be found in the Figure caption). All three of these genes participate in setting up positive feedback circuits operating at upper levels in the pancreas GRN hierarchy. Hnf6 activates the expression of *hnf1β*, which in turn feeds back on the *hnf6* gene. The other feedback circuit is initiated by Gata6, which activates both *nr5a2* and *foxa2*, two genes which then drive one another's expression. *Hnf6* provides an additional input into *foxa2*, linking the two feedback circuits. Both of these feedback circuits activate the expression of *pdx1* and *ngn3*, by the direct inputs from *hnf6* and *foxa2*. A further feedback circuit is provided by *pdx1*, which drives both itself and *nr5a2*. Again, FoxA2 provides an activating input into almost all genes in this network, perhaps here too executing the role of a pioneer factor, as illustrated in Fig. 4.11(B). It is clear that the logic of this GRN shown in Fig. 4.11(C1) is designed to ensure the expression of the early pancreatic drivers Ngn3 and Pdx1.

The multipotent pancreatic precursor cells defined by the network modeled in Fig. 4.11(C1) (e9–e13) give rise to exocrine and multiple endocrine cell types as well as to duct cells. The expression of *pdx1* and *ngn3* leads to the activation of a unique set of further regulatory genes. The circuitry by which this is achieved is illustrated in Fig. 4.11(C2) (e13-e15). Thus, *ngn3* encodes a bHLH factor which dimerizes with the ubiquitously present E2A and this complex activates two new regulatory genes, *neuroD* and *nkx2.2*. These proceed to cross-regulate, generating another positive feedback loop. Ngn3:E2A also activates *pax4*, which in turn drives the expression of *nkx6.1*, while both *pax4* and *nkx6.1* are direct targets of Pdx1. Thus *nkx6.1* expression is the output of a feed-forward circuit (*pdx1* > *nkx6.1*; *pdx1* > *pax4* > *nkx6.1*). Besides this, *nkx6.1* also receives a positive input from the *ngn3* target gene *nkx2.2*. This heavily recursive wiring ensures the expression of a new regulatory state which now defines endocrine cell fate. Indeed, *nkx6.1* directly excludes exocrine cell fates by repressing *ptf1a*, a key driver of exocrine genes. Reciprocally, the Ptf1a factor (in complex with another bHLH protein) excludes expression of *nkx6.1* in presumptive exocrine cells. This mutual repression circuit thus functions to distinguish endocrine and exocrine cell fates at the regulatory level. Another output of the GRN model in Fig. 4.11(C2) occurring downstream of *pax4* expression, is the repression of *arx*, which encodes an essential driver of α endocrine cell fate.

After e15, downstream differentiation genes of the β-cell type begin to be expressed as additional regulatory genes and new regulatory circuitry comes into play (Fig. 4.11(C3)). The terminal GRN needs to accomplish three objectives: it has to install the apparatus for expression of downstream effector genes; it has to ensure the permanence of the regulatory state driving effector gene expression; and it has to install the physiological circuitry that enables the change of effector gene expression according to ambient circumstances. Most famously, this last includes variation of *insulin* expression level in response to glucose concentration in the blood. The maintenance of the regulatory state expressed in terminally differentiated pancreatic β-cells is mediated by positive feedback circuitry. The *neuroD-nkx2.2* feedback circuit downstream of *ngn3,* which was initiated at the previous stage, remains active as does the linkage from *nkx2.2* into *nkx6.1*. Downstream of *nkx6.1* expression, a new positive feedback circuit is initiated: *nkx6.1* activates *hnf1α*, setting up a feedback linkage between *hnf4α* and *hnf1α*. The expression of *hnf1α* also activates an additional important feedback circuit, the outcome of which is to ensure the continued expression of pdx1. Thus *hnf1α* provides a positive input into *pdx1* and also into *mafa*, whereupon these two genes also form a positive feedback loop. Almost all regulatory genes in this network are known to provide direct inputs into the downstream effector genes such as *insulin, IAAP,* and *glucokinase* (Fig. 4.11(C3)). For example, *cis*-regulatory inputs into the best known effector gene in these cells, *insulin*, include Nkx2.2, MafA, Pdx1, Hnf4α and NeuroD:E2A. Additional regulatory genes function as drivers of downstream genes including *insulin*, such as *alx3, atf2,* and *islet1*, although their interactions with the rest of the GRN remain to be determined. Thus major components of the terminal β-cell GRN act both to ensure the permanence of the regulatory state through the linked positive feedback circuitry in which they participate, and to drive the expression of the downstream effector genes by direct *cis*-regulatory interactions.

The metabolic response functions so important to β-cell transcriptional performance require additional GRN circuitry which is activated at this stage. Insulin mediates numerous responses to glucose throughout the body and expression of the *insulin* gene is required to be active only under high glucose conditions, while under low glucose conditions, transcription of the gene must be downregulated. This is accomplished by at least two routes. In low glucose conditions, FoxO1 is translocated to the nucleus where it represses the *insulin* gene at the *cis*-regulatory level (Meur et al., 2011). In high glucose, a cellular response results in the activation of PI3 kinase, which phosphorylates FoxO1, resulting in its exclusion from the nucleus. This allows the high level of expression of the *insulin* gene driven by the positive inputs of the β-cell GRN mentioned above. In addition, nuclearization of Pax6 is also affected by high glucose levels, such that it is present in the nucleus only at low glucose concentrations (Wolf et al., 2010). Pax6 directly represses *insulin* transcription by competing with Pdx1 for binding to the same *cis*-regulatory sequences. Furthermore, Pdx1, a direct driver of insulin and an important regulator of the whole GRN, is degraded more rapidly under conditions of low glucose and accumulates under condition of high glucose. The same glucose-response system controlling nuclear levels of Pax6 and FoxO1 is also utilized in α-cells to drive the expression of *glucagon*. Only here, instead of preventing gene expression under low glucose conditions, it activates *glucagon* gene expression, thus generating opposing transcriptional responses in α- and β-cells under the same physiological conditions (Fig. 4.11(C3)). Thus in summary, we can see in this unusually well-known case how the regulatory linkages of Fig. 4.11(C) encode pancreatic specification, cell type diversification, and ultimately the differentiation and metabolic response of β-cells.

* * *

In this chapter we see that GRNs determining the earlier phases of adult body part formation universally display an overriding, primary function. This is to accomplish the particular spatial patterning required to set up Boolean regulatory states in the diverse future subparts of the organ or structure. Signaling is always involved in the initial steps of each phase of the patterning process, but the diffusible signals so commonly utilized in vertebrate development do not per se generate Boolean regulatory state patterns: GRN circuitry, invariably including repressive transcriptional interactions, provide the determinative mechanisms by which the form and organization of each body part is specified, according to its genomically encoded program of gene interactions. For the first time in the history of biology it is becoming possible to see exactly how these spatial patterning processes are encoded in the genome, i.e., to understand them in terms of GRN structure. In this chapter we have seen many diverse examples that show how this is achieved. Adult body part formation is the process by which the phylogenetically definitive morphological characters of each animal clade are generated, and we return to this subject in a specifically evolutionary context in Chapter 7 of this book. We now address the control machinery underlying cell type specification, a very different process than that required for spatial patterning of the parts and subparts of the organism.

REFERENCES

Agarwal, P., Wylie, J.N., Galceran, J., Arkhitko, O., Li, C., Deng, C., Grosschedl, R., Bruneau, B.G., 2003. Tbx5 is essential for forelimb bud initiation following patterning of the limb field in the mouse embryo. Development 130, 623–633.

Aguilar-Hidalgo, D., Domínguez-Cejudo, M.A., Amore, G., Brockmann, A., Lemos, M.C., Córdoba, A., Casares, F., 2013. A Hh-driven gene network controls specification, pattern and size of the *Drosophila* simple eyes. Development 140, 82–92.

Ahn, Y., Zou, J., Mitchell, P.J., 2011. Segment-specific regulation of the *Drosophila AP-2* gene during leg and antennal development. Dev. Biol. 355, 336–348.

Anderson, K.R., Torres, C.A., Solomon, K., Becker, T.C., Newgard, C.B., Wright, C.V., Hagman, J., Sussel, L., 2009. Cooperative transcriptional regulation of the essential pancreatic islet gene *NeuroD1* (beta2) by Nkx2.2 and neurogenin 3. J. Biol. Chem. 284, 31236–31248.

Arendt, D., Denes, A.S., Jékely, G., Tessmar-Raible, K., 2008. The evolution of nervous system centralization. Philos. Trans. R. Soc. London B Biol. Sci. 363, 1523–1528.

Barrow, J.R., Stadler, H.S., Capecchi, M.R., 2000. Roles of *Hoxa1* and *Hoxa2* in patterning the early hindbrain of the mouse. Development 127, 933–944.

Bartoov-Shifman, R., Ridner, G., Bahar, K., Rubins, N., Walker, M.D., 2007. Regulation of the gene encoding GPR40, a fatty acid receptor expressed selectively in pancreatic beta cells. J. Biol. Chem. 282, 23561–23571.

Beccari, L., Marco-Ferreres, R., Bovolenta, P., 2013. The logic of gene regulatory networks in early vertebrate forebrain patterning. Mech. Dev. 130, 95–111.

Behrens, A.N., Iacovino, M., Lohr, J.L., Ren, Y., Zierold, C., Harvey, R.P., Kyba, M., Garry, D.J., Martin, C.M., 2013. Nkx2-5 mediates differential cardiac differentiation through interaction with hoxa10. Stem Cells Dev. 22, 2211–2220.

Beres, T.M., Masui, T., Swift, G.H., Shi, L., Henke, R.M., MacDonald, R.J., 2006. PTF1 is an organ-specific and notch-independent basic helix-loop-helix complex containing the mammalian suppressor of hairless (RBP-J) or its paralogue, RBP-L. Mol. Cell. Biol. 26, 117–130.

Bertrand, N., Roux, M., Ryckebüsch, L., Niederreither, K., Dollé, P., Moon, A., Capecchi, M., Zaffran, S., 2011. *Hox* genes define distinct progenitor sub-domains within the second heart field. Dev. Biol. 353, 266–274.

Black, B.L., 2007. Transcriptional pathways in second heart field development. Semin. Cell. Dev. Biol. 18, 67–76.

Blow, M.J., McCulley, D.J., Li, Z., Zhang, T., Akiyama, J.A., Holt, A., Plajzer-Frick, I., Shoukry, M., Wright, C., Chen, F., Afzal, V., Bristow, J., Ren, B., Black, B.L., Rubin, E.M., Visel, A., Pennacchio, L.A., 2010. ChIP-Seq identification of weakly conserved heart enhancers. Nat. Genet. 42, 806–810.

Bondue, A., Lapouge, G., Paulissen, C., Semeraro, C., Iacovino, M., Kyba, M., Blanpain, C., 2008. Mesp1 acts as a master regulator of multipotent cardiovascular progenitor specification. Cell Stem Cell 3, 69–84.

Bryantsev, A.L., Cripps, R.M., 2009. Cardiac gene regulatory networks in *Drosophila*. Biochim. Biophys. Acta 1789, 343–353.

Burkhardt, B.R., Cook, J.R., Young, R.A., Wolf, B.A., 2008. PDX-1 interaction and regulation of the pancreatic derived factor (PANDER, FAM3B) promoter. Biochim. Biophys. Acta 1779, 645–651.

Bénazet, J.D., Bischofberger, M., Tiecke, E., Gonçalves, A., Martin, J.F., Zuniga, A., Naef, F., Zeller, R., 2009. A self-regulatory system of interlinked signaling feedback loops controls mouse limb patterning. Science 323, 1050–1053.

Bénazet, J.D., Zeller, R., 2009. Vertebrate limb development: moving from classical morphogen gradients to an integrated 4-dimensional patterning system. Cold Spring Harb. Perspect. Biol. 1, a001339.

Capellini, T.D., Zappavigna, V., Selleri, L., 2011. Pbx homeodomain proteins: talented regulators of limb patterning and outgrowth. Dev. Dyn. 240, 1063–1086.

Cattell, M.V., Garnett, A.T., Klymkowsky, M.W., Medeiros, D.M., 2012. A maternally established SoxB1/SoxF axis is a conserved feature of chordate germ layer patterning. Evol. Dev. 14, 104–115.

Chambers, D., Wilson, L.J., Alfonsi, F., Hunter, E., Saxena, U., Blanc, E., Lumsden, A., 2009. Rhombomere-specific analysis reveals the repertoire of genetic cues expressed across the developing hindbrain. Neural Dev. 4, 6.

Chan, S.S., Shi, X., Toyama, A., Arpke, R.W., Dandapat, A., Iacovino, M., Kang, J., Le, G., Hagen, H.R., Garry, D.J., Kyba, M., 2013. Mesp1 patterns mesoderm into cardiac, hematopoietic, or skeletal myogenic progenitors in a context-dependent manner. Cell Stem Cell 12, 587–601.

Chan, T.M., Chao, C.H., Wang, H.D., Yu, Y.J., Yuh, C.H., 2009a. Functional analysis of the evolutionarily conserved *cis*-regulatory elements on the *sox17* gene in zebrafish. Dev. Biol. 326, 456–470.

Chan, T.M., Longabaugh, W., Bolouri, H., Chen, H.L., Tseng, W.F., Chao, C.H., Jang, T.H., Lin, Y.I., Hung, S.C., Wang, H.D., Yuh, C.H., 2009b. Developmental gene regulatory networks in the zebrafish embryo. Biochim. Biophys. Acta 1789, 279–298.

Chao, C.H., Wang, H.D., Yuh, C.H., 2010. Complexity of *cis*-regulatory organization of six3a during forebrain and eye development in zebrafish. BMC Dev. Biol. 10, 35.

Chomette, D., Frain, M., Cereghini, S., Charnay, P., Ghislain, J., 2006. Krox20 hindbrain *cis*-regulatory landscape: interplay between multiple long-range initiation and autoregulatory elements. Development 133, 1253–1262.

Clark, C.D., Zhang, B., Lee, B., Evans, S.I., Lassar, A.B., Lee, K.H., 2013. Evolutionary conservation of Nkx2.5 autoregulation in the second heart field. Dev. Biol. 374, 198–209.

Cornell, R.A., Ohlen, T.V., 2000. Vnd/nkx, ind/gsh, and msh/msx: conserved regulators of dorsoventral neural patterning? Curr. Opin. Neurobiol. 10, 63–71.

Davidson, E.H., 2001. Genomic Regulatory Systems: Development and Evolution. Acdemic Press, San Diego.

Davidson, E.H., 2006. The Regulatory Genome. Gene Regulatory Networks in Development and Evolution. Academic Press/Elsevier, San Diego, CA.

Davidson, E.H., Erwin, D.H., 2006. Gene regulatory networks and the evolution of animal body plans. Science 311, 796–800.

Delporte, F.M., Pasque, V., Devos, N., Manfroid, I., Voz, M.L., Motte, P., Biemar, F., Martial, J.A., Peers, B., 2008. Expression of zebrafish pax6b in pancreas is regulated by two enhancers containing highly conserved cis-elements bound by PDX1, PBX and PREP factors. BMC Dev. Biol. 8, 53.

Denes, A.S., Jekely, G., Steinmetz, P.R., Raible, F., Snyman, H., Prud'homme, B., Ferrier, D.E., Balavoine, G., Arendt, D., 2007. Molecular architecture of annelid nerve cord supports common origin of nervous system centralization in bilateria. Cell 129, 277–288.

Deutsch, G., Jung, J., Zheng, M., Lóra, J., Zaret, K.S., 2001. A bipotential precursor population for pancreas and liver within the embryonic endoderm. Development 128, 871–881.

Donelan, W., Koya, V., Li, S.W., Yang, L.J., 2010. Distinct regulation of hepatic nuclear factor 1alpha by NKX6.1 in pancreatic beta cells. J. Biol. Chem. 285, 12181–12189.

Duboc, V., Logan, M.P., 2009. Building limb morphology through integration of signalling modules. Curr. Opin. Genet. Dev. 19, 497–503.

Duboc, V., Logan, M.P., 2011a. Pitx1 is necessary for normal initiation of hindlimb outgrowth through regulation of Tbx4 expression and shapes hindlimb morphologies via targeted growth control. Development 138, 5301–5309.

Duboc, V., Logan, M.P., 2011b. Regulation of limb bud initiation and limb-type morphology. Dev. Dyn. 240, 1017–1027.

Eeckhoute, J., Briche, I., Kurowska, M., Formstecher, P., Laine, B., 2006. Hepatocyte nuclear factor 4 alpha ligand binding and F domains mediate interaction and transcriptional synergy with the pancreatic islet LIM HD transcription factor Isl1. J. Mol. Biol. 364, 567–581.

Estella, C., McKay, D.J., Mann, R.S., 2008. Molecular integration of wingless, decapentaplegic, and autoregulatory inputs into Distalless during *Drosophila* leg development. Dev. Cell 14, 86–96.

Estella, C., Voutev, R., Mann, R.S., 2012. A dynamic network of morphogens and transcription factors patterns the fly leg. Curr. Top. Dev. Biol. 98, 173–198.

Fernandez-Zapico, M.E., van Velkinburgh, J.C., Gutiérrez-Aguilar, R., Neve, B., Froguel, P., Urrutia, R., Stein, R., 2009. *MODY7* gene, KLF11, is a novel p300-dependent regulator of Pdx-1 (MODY4) transcription in pancreatic islet beta cells. J. Biol. Chem. 284, 36482–36490.

Franklin, V., Khoo, P.L., Bildsoe, H., Wong, N., Lewis, S., Tam, P.P., 2008. Regionalisation of the endoderm progenitors and morphogenesis of the gut portals of the mouse embryo. Mech. Dev. 125, 587–600.

Galindo, M.I., Fernández-Garza, D., Phillips, R., Couso, J.P., 2011. Control of Distal-less expression in the *Drosophila* appendages by functional 3' enhancers. Dev. Biol. 353, 396–410.

Galli, A., Robay, D., Osterwalder, M., Bao, X., Bénazet, J.D., Tariq, M., Paro, R., Mackem, S., Zeller, R., 2010. Distinct roles of Hand2 in initiating polarity and posterior *Shh* expression during the onset of mouse limb bud development. PLoS Genet. 6, e1000901.

Gao, N., LeLay, J., Vatamaniuk, M.Z., Rieck, S., Friedman, J.R., Kaestner, K.H., 2008. Dynamic regulation of Pdx1 enhancers by Foxa1 and Foxa2 is essential for pancreas development. Genes Dev. 22, 3435–3448.

Gauthier, B.R., Gosmain, Y., Mamin, A., Philippe, J., 2007. The beta-cell specific transcription factor Nkx6.1 inhibits *glucagon* gene transcription by interfering with Pax6. Biochem. J. 403, 593–601.

Giorgianni, M.W., Mann, R.S., 2011. Establishment of medial fates along the proximodistal axis of the *Drosophila* leg through direct activation of dachshund by Distalless. Dev. Cell 20, 455–468.

Glover, J.C., Renaud, J.S., Rijli, F.M., 2006. Retinoic acid and hindbrain patterning. J. Neurobiol. 66, 705–725.

Gosmain, Y., Avril, I., Mamin, A., Philippe, J., 2007. Pax-6 and c-Maf functionally interact with the alpha-cell-specific DNA element G1 in vivo to promote *glucagon* gene expression. J. Biol. Chem. 282, 35024–35034.

Gosmain, Y., Marthinet, E., Cheyssac, C., Guérardel, A., Mamin, A., Katz, L.S., Bouzakri, K., Philippe, J., 2010. Pax6 controls the expression of critical genes involved in pancreatic {alpha} cell differentiation and function. J. Biol. Chem. 285, 33381–33393.

Grapp, M., Teichler, S., Kitz, J., Dibaj, P., Dickel, C., Knepel, W., Krätzner, R., 2009. The homeodomain of PAX6 is essential for PAX6-dependent activation of the rat *glucagon* gene promoter: evidence for a PH0-like binding that induces an active conformation. Biochim. Biophys. Acta 1789, 403–412.

Greenberg, L., Hatini, V., 2009. Essential roles for lines in mediating leg and antennal proximodistal patterning and generating a stable notch signaling interface at segment borders. Dev. Biol. 330, 93–104.

Guner-Ataman, B., Paffett-Lugassy, N., Adams, M.S., Nevis, K.R., Jahangiri, L., Obregon, P., Kikuchi, K., Poss, K.D., Burns, C.E., Burns, C.G., 2013. Zebrafish second heart field development relies on progenitor specification in anterior lateral plate mesoderm and nkx2.5 function. Development 140, 1353–1363.

Han, S.I., Yasuda, K., Kataoka, K., 2011. ATF2 interacts with beta-cell-enriched transcription factors, MafA, Pdx1, and beta2, and activates *insulin* gene transcription. J. Biol. Chem. 286, 10449–10456.

Hart, A.H., Hartley, L., Sourris, K., Stadler, E.S., Li, R., Stanley, E.G., Tam, P.P., Elefanty, A.G., Robb, L., 2002. Mixl1 is required for axial mesendoderm morphogenesis and patterning in the murine embryo. Development 129, 3597–3608.

He, A., Kong, S.W., Ma, Q., Pu, W.T., 2011. Co-occupancy by multiple cardiac transcription factors identifies transcriptional enhancers active in heart. Proc. Natl. Acad. Sci. U.S.A. 108, 5632–5637.

Heimbucher, T., Murko, C., Bajoghli, B., Aghaallaei, N., Huber, A., Stebegg, R., Eberhard, D., Fink, M., Simeone, A., Czerny, T., 2007. Gbx2 and Otx2 interact with the WD40 domain of Groucho/Tle corepressors. Mol. Cell. Biol. 27, 340–351.

Herrmann, F., Groß, A., Zhou, D., Kestler, H.A., Kühl, M., 2012. A boolean model of the cardiac gene regulatory network determining first and second heart field identity. PLoS One 7, e46798.

Hirth, F., Reichert, H., 1999. Conserved genetic programs in insect and mammalian brain development. Bioessays 21, 677–684.

Hudson, C., Kawai, N., Negishi, T., Yasuo, H., 2013. β-Catenin-driven binary fate specification segregates germ layers in ascidian embryos. Curr. Biol. 23, 491–495.

Hunter, C.S., Maestro, M.A., Raum, J.C., Guo, M., Thompson, F.H., Ferrer, J., Stein, R., 2011. Hnf1α (MODY3) regulates β-cell-enriched MafA transcription factor expression. Mol. Endocrinol. 25, 339–347.

Infante, C.R., Park, S., Mihala, A.G., Kingsley, D.M., Menke, D.B., 2013. Pitx1 broadly associates with limb enhancers and is enriched on hindlimb *cis*-regulatory elements. Dev. Biol. 374, 234–244.

Inoue, F., Kurokawa, D., Takahashi, M., Aizawa, S., 2012. Gbx2 directly restricts Otx2 expression to forebrain and midbrain, competing with class III POU factors. Mol. Cell. Biol. 32, 2618–2627.

Junion, G., Spivakov, M., Girardot, C., Braun, M., Gustafson, E.H., Birney, E., Furlong, E.E., 2012. A transcription factor collective defines cardiac cell fate and reflects lineage history. Cell 148, 473–486.

Kalousova, A., Mavropoulos, A., Adams, B.A., Nekrep, N., Li, Z., Krauss, S., Stainier, D.Y., German, M.S., 2010. Dachshund homologues play a conserved role in islet cell development. Dev. Biol. 348, 143–152.

Kanai-Azuma, M., Kanai, Y., Gad, J.M., Tajima, Y., Taya, C., Kurohmaru, M., Sanai, Y., Yonekawa, H., Yazaki, K., Tam, P.P., Hayashi, Y., 2002. Depletion of definitive gut endoderm in Sox17-null mutant mice. Development 129, 2367–2379.

Kang, H.S., Kim, Y.S., ZeRuth, G., Beak, J.Y., Gerrish, K., Kilic, G., Sosa-Pineda, B., Jensen, J., Pierreux, C.E., Lemaigre, F.P., Foley, J., Jetten, A.M., 2009. Transcription factor Glis3, a novel critical player in the regulation of pancreatic beta-cell development and *insulin* gene expression. Mol. Cell. Biol. 29, 6366–6379.

Kawakami, Y., Esteban, C.R., Matsui, T., Rodríguez-León, J., Kato, S., Izpisúa Belmonte, J.C., 2004. Sp8 and Sp9, two closely related buttonhead-like transcription factors, regulate *Fgf8* expression and limb outgrowth in vertebrate embryos. Development 131, 4763–4774.

Kawakami, Y., Marti, M., Kawakami, H., Itou, J., Quach, T., Johnson, A., Sahara, S., O'Leary, D.D., Nakagawa, Y., Lewandoski, M., Pfaff, S., Evans, S.M., Izpisua Belmonte, J.C., 2011. Islet1-mediated activation of the β-catenin pathway is necessary for hindlimb initiation in mice. Development 138, 4465–4473.

Kawakami, Y., Uchiyama, Y., Rodriguez Esteban, C., Inenaga, T., Koyano-Nakagawa, N., Kawakami, H., Marti, M., Kmita, M., Monaghan-Nichols, P., Nishinakamura, R., Izpisua Belmonte, J.C., 2009. *Sall* genes regulate region-specific morphogenesis in the mouse limb by modulating Hox activities. Development 136, 585–594.

Kim, F.A., Sing, l, A., Kaneko, T., Bieman, M., Stallwood, N., Sadl, V.S., Cordes, S.P., 2005. The vHNF1 homeodomain protein establishes early rhombomere identity by direct regulation of *Kreisler* expression. Mech. Dev. 122, 1300–1309.

Koide, T., Hayata, T., Cho, K.W., 2005. Xenopus as a model system to study transcriptional regulatory networks. Proc. Natl. Acad. Sci. U.S.A. 102, 4943–4948.

Kojima, T., Tsuji, T., Saigo, K., 2005. A concerted action of a paired-type homeobox gene, *aristaless*, and a homolog of *Hox11/tlx* homeobox gene, clawless, is essential for the distal tip development of the *Drosophila* leg. Dev. Biol. 279, 434–445.

Koshiba-Takeuchi, K., Mori, A.D., Kaynak, B.L., Cebra-Thomas, J., Sukonnik, T., Georges, R.O., Latham, S., Beck, L., Henkelman, R.M., Black, B.L., Olson, E.N., Wade, J., Takeuchi, J.K., Nemer, M., Gilbert, S.F., Bruneau, B.G., 2009. Reptilian heart development and the molecular basis of cardiac chamber evolution. Nature 461, 95–98.

Kozu, S., Tajiri, R., Tsuji, T., Michiue, T., Saigo, K., Kojima, T., 2006. Temporal regulation of late expression of bar homeobox genes during *Drosophila* leg development by spineless, a homolog of the mammalian dioxin receptor. Dev. Biol. 294, 497–508.

Krawchuk, D., Kania, A., 2008. Identification of genes controlled by LMX1B in the developing mouse limb bud. Dev. Dyn. 237, 1183–1192.

Lazic, S., Scott, I.C., 2011. Mef2cb regulates late myocardial cell addition from a second heart field-like population of progenitors in zebrafish. Dev. Biol. 354, 123–133.

Lecuit, T., Cohen, S.M., 1997. Proximal-distal axis formation in the *Drosophila* leg. Nature 388, 139–145.

Lee, B., Song, H., Rizzoti, K., Son, Y., Yoon, J., Baek, K., Jeong, Y., 2013. Genomic code for Sox2 binding uncovers its regulatory role in Six3 activation in the forebrain. Dev. Biol. 381, 491–501.

Lewis, S.L., Tam, P.P., 2006. Definitive endoderm of the mouse embryo: formation, cell fates, and morphogenetic function. Dev. Dyn. 235, 2315–2329.

Li, D., Sakuma, R., Vakili, N.A., Mo, R., Puviindran, V., Deimling, S., Zhang, X., Hopyan, S., Hui, C.C., 2014a. Formation of proximal and anterior limb skeleton requires early function of Irx3 and Irx5 and is negatively regulated by Shh signaling. Dev. Cell 29, 233–240.

Li, J.Y., Lao, Z., Joyner, A.L., 2005. New regulatory interactions and cellular responses in the isthmic organizer region revealed by altering *Gbx2* expression. Development 132, 1971–1981.

Li, Q., Lewandowski, J.P., Powell, M.B., Norrie, J.L., Cho, S.H., Vokes, S.A., 2014b. A Gli silencer is required for robust repression of *gremlin* in the vertebrate limb bud. Development 141, 1906–1914.

Lichtneckert, R., Reichert, H., 2005. Insights into the urbilaterian brain: conserved genetic patterning mechanisms in insect and vertebrate brain development. Hered. (Edinb) 94, 465–477.

Liu, J., Hunter, C.S., Du, A., Ediger, B., Walp, E., Murray, J., Stein, R., May, C.L., 2011. *Islet-1* regulates Arx transcription during pancreatic islet alpha-cell development. J. Biol. Chem. 286, 15352–15360.

Liu, J., Li, Q., Kuehn, M.R., Litingtung, Y., Vokes, S.A., Chiang, C., 2013. Sonic hedgehog signaling directly targets hyaluronic acid synthase 2, an essential regulator of phalangeal joint patterning. Dev. Biol. 375, 160–171.

Liu, N., Olson, E.N., 2010. MicroRNA regulatory networks in cardiovascular development. Dev. Cell 18, 510–525.

Liu, T., Branch, D.R., Jin, T., 2006. Pbx1 is a co-factor for Cdx-2 in regulating *proglucagon* gene expression in pancreatic A cells. Mol. Cell. Endocrinol. 249, 140–149.

Logan, M., Simon, H.G., Tabin, C., 1998. Differential regulation of T-box and homeobox transcription factors suggests roles in controlling chick limb-type identity. Development 125, 2825–2835.

Logan, M., Tabin, C.J., 1999. Role of Pitx1 upstream of Tbx4 in specification of hindlimb identity. Science 283, 1736–1739.

Lolas, M., Valenzuela, P.D., Tjian, R., Liu, Z., 2014. Charting Brachyury-mediated developmental pathways during early mouse embryogenesis. Proc. Natl. Acad. Sci. U.S.A. 111, 4478–4483.

Loose, M., Patient, R., 2004. A genetic regulatory network for Xenopus mesendoderm formation. Dev. Biol. 271, 467–478.

Lopez-Rios, J., Speziale, D., Robay, D., Scotti, M., Osterwalder, M., Nusspaumer, G., Galli, A., Holländer, G.A., Kmita, M., Zeller, R., 2012. GLI3 constrains digit number by controlling both progenitor proliferation and BMP-dependent exit to chondrogenesis. Dev. Cell 22, 837–848.

Lu, P., Rha, G.B., Melikishvili, M., Wu, G., Adkins, B.C., Fried, M.G., Chi, Y.I., 2008. Structural basis of natural promoter recognition by a unique nuclear receptor, HNF4alpha. Diabetes gene product. J. Biol. Chem. 283, 33685–33697.

Mariani, F.V., Ahn, C.P., Martin, G.R., 2008. Genetic evidence that FGFs have an instructive role in limb proximal-distal patterning. Nature 453, 401–405.

McKay, D.J., Estella, C., Mann, R.S., 2009. The origins of the *Drosophila* leg revealed by the *cis*-regulatory architecture of the *Distalless* gene. Development 136, 61–71.

McKnight, K.D., Hou, J., Hoodless, P.A., 2010. Foxh1 and Foxa2 are not required for formation of the midgut and hindgut definitive endoderm. Dev. Biol. 337, 471–481.

Mellitzer, G., Bonné, S., Luco, R.F., Van De Casteele, M., Lenne-Samuel, N., Collombat, P., Mansouri, A., Lee, J., Lan, M., Pipeleers, D., Nielsen, F.C., Ferrer, J., Gradwohl, G., Heimberg, H., 2006. IA1 is NGN3-dependent and essential for differentiation of the endocrine pancreas. EMBO J. 25, 1344–1352.

Menke, D.B., Guenther, C., Kingsley, D.M., 2008. Dual hindlimb control elements in the *Tbx4* gene and region-specific control of bone size in vertebrate limbs. Development 135, 2543–2553.

Meur, G., Qian, Q., da Silva Xavier, G., Pullen, T.J., Tsuboi, T., McKinnon, C., Fletcher, L., Tavaré, J.M., Hughes, S., Johnson, P., Rutter, G.A., 2011. Nucleo-cytosolic shuttling of FoxO1 directly regulates mouse *Ins2* but not *Ins1* gene expression in pancreatic beta cells (MIN6). J. Biol. Chem. 286, 13647–13656.

Minguillon, C., Del Buono, J., Logan, M.P., 2005. Tbx5 and Tbx4 are not sufficient to determine limb-specific morphologies but have common roles in initiating limb outgrowth. Dev. Cell 8, 75–84.

Minguillon, C., Nishimoto, S., Wood, S., Vendrell, E., Gibson-Brown, J.J., Logan, M.P., 2012. *Hox* genes regulate the onset of *Tbx5* expression in the forelimb. Development 139, 3180–3188.

Mirasierra, M., Fernández-Pérez, A., Díaz-Prieto, N., Vallejo, M., 2011. Alx3-deficient mice exhibit decreased insulin in beta cells, altered glucose homeostasis and increased apoptosis in pancreatic islets. Diabetologia 54, 403–414.

Mirasierra, M., Vallejo, M., 2006. The homeoprotein Alx3 expressed in pancreatic beta-cells regulates *insulin* gene transcription by interacting with the basic helix-loop-helix protein E47. Mol. Endocrinol. 20, 2876–2889.

Miyatsuka, T., Matsuoka, T.A., Shiraiwa, T., Yamamoto, T., Kojima, I., Kaneto, H., 2007. Ptf1a and RBP-J cooperate in activating *Pdx1* gene expression through binding to Area III. Biochem. Biophys. Res. Commun. 362, 905–909.

Moreno-Bravo, J.A., Perez-Balaguer, A., Martinez-Lopez, J.E., Aroca, P., Puelles, L., Martinez, S., Puelles, E., 2013. Role of Shh in the development of molecularly characterized tegmental nuclei in mouse rhombomere 1. Brain Struct. Funct.

Morley, R.H., Lachani, K., Keefe, D., Gilchrist, M.J., Flicek, P., Smith, J.C., Wardle, F.C., 2009. A gene regulatory network directed by zebrafish no tail accounts for its roles in mesoderm formation. Proc. Natl. Acad. Sci. U.S.A. 106, 3829–3834.

Munshi, N.V., 2012. Gene regulatory networks in cardiac conduction system development. Circ. Res. 110, 1525–1537.

Natori, K., Tajiri, R., Furukawa, S., Kojima, T., 2012. Progressive tarsal patterning in the *Drosophila* by temporally dynamic regulation of transcription factor genes. Dev. Biol. 361, 450–462.

Nissim, S., Allard, P., Bandyopadhyay, A., Harfe, B.D., Tabin, C.J., 2007. Characterization of a novel ectodermal signaling center regulating Tbx2 and Shh in the vertebrate limb. Dev. Biol. 304, 9–21.

Olson, E.N., 2006. Gene regulatory networks in the evolution and development of the heart. Science 313, 1922–1927.

Pasqualetti, M., Díaz, C., Renaud, J.S., Rijli, F.M., Glover, J.C., 2007. Fate-mapping the mammalian hindbrain: segmental origins of vestibular projection neurons assessed using rhombomere-specific Hoxa2 enhancer elements in the mouse embryo. J. Neurosci. 27, 9670–9681.

Pereira, L.A., Wong, M.S., Lim, S.M., Sides, A., Stanley, E.G., Elefanty, A.G., 2011. Brachyury and related Tbx proteins interact with the Mixl1 homeodomain protein and negatively regulate Mixl1 transcriptional activity. PLoS One 6, e28394.

Pereira, L.A., Wong, M.S., Mei Lim, S., Stanley, E.G., Elefanty, A.G., 2012. The Mix family of homeobox genes–key regulators of mesendoderm formation during vertebrate development. Dev. Biol. 367, 163–177.

Pueyo, J.I., Couso, J.P., 2008. The 11-aminoacid long Tarsal-less peptides trigger a cell signal in *Drosophila* leg development. Dev. Biol. 324, 192–201.

Pueyo, J.I., Couso, J.P., 2011. Tarsal-less peptides control notch signalling through the shavenbaby transcription factor. Dev. Biol. 355, 183–193.

Rabinowitz, A.H., Vokes, S.A., 2012. Integration of the transcriptional networks regulating limb morphogenesis. Dev. Biol. 368, 165–180.

Raum, J.C., Gerrish, K., Artner, I., Henderson, E., Guo, M., Sussel, L., Schisler, J.C., Newgard, C.B., Stein, R., 2006. FoxA2, Nkx2.2, and PDX-1 regulate islet beta-cell-specific mafA expression through conserved sequences located between base pairs -8118 and -7750 upstream from the transcription start site. Mol. Cell. Biol. 26, 5735–5743.

Raum, J.C., Hunter, C.S., Artner, I., Henderson, E., Guo, M., Elghazi, L., Sosa-Pineda, B., Ogihara, T., Mirmira, R.G., Sussel, L., Stein, R., 2010. Islet beta-cell-specific MafA transcription requires the 5'-flanking conserved region 3 control domain. Mol. Cell. Biol. 30, 4234–4244.

Reim, I., Frasch, M., 2010. Genetic and genomic dissection of cardiogenesis in the *Drosophila* model. Pediatr. Cardiol. 31, 325–334.

Rojas, A., Schachterle, W., Xu, S.M., Black, B.L., 2009. An endoderm-specific transcriptional enhancer from the mouse *Gata4* gene requires GATA and homeodomain protein-binding sites for function in vivo. Dev. Dyn. 238, 2588–2598.

Rojas, A., Schachterle, W., Xu, S.M., Martín, F., Black, B.L., 2010. Direct transcriptional regulation of Gata4 during early endoderm specification is controlled by FoxA2 binding to an intronic enhancer. Dev. Biol. 346, 346–355.

Salsi, V., Vigano, M.A., Cocchiarella, F., Mantovani, R., Zappavigna, V., 2008. Hoxd13 binds in vivo and regulates the expression of genes acting in key pathways for early limb and skeletal patterning. Dev. Biol. 317, 497–507.

Saunders, J.W., 1948. The proximo-distal sequence of origin of the parts of the chick wing and the role of the ectoderm. J. Exp. Zool. 108, 363–403.

Schlesinger, J., Schueler, M., Grunert, M., Fischer, J.J., Zhang, Q., Krueger, T., Lange, M., Tönjes, M., Dunkel, I., Sperling, S.R., 2011. The cardiac transcription network modulated by Gata4, Mef2a, Nkx2.5, Srf, histone modifications, and microRNAs. PLoS Genet. 7, e1001313.

Schneider-Maunoury, S., Seitanidou, T., Charnay, P., Lumsden, A., 1997. Segmental and neuronal architecture of the hindbrain of Krox-20 mouse mutants. Development 124, 1215–1226.

Schultheiss, T.M., Burch, J.B., Lassar, A.B., 1997. A role for bone morphogenetic proteins in the induction of cardiac myogenesis. Genes. Dev. 11, 451–462.

Schwartz, R.J., Olson, E.N., 1999. Building the heart piece by piece: modularity of *cis*-elements regulating Nkx2-5 transcription. Development 126, 4187–4192.

Seibert, J., Urbach, R., 2010. Role of en and novel interactions between msh, ind, and vnd in dorsoventral patterning of the *Drosophila* brain and ventral nerve cord. Dev. Biol. 346, 332–345.

Seibert, J., Volland, D., Urbach, R., 2009. Ems and Nkx6 are central regulators in dorsoventral patterning of the *Drosophila* brain. Development 136, 3937–3947.

Sen, S., Reichert, H., Vijayraghavan, K., 2013. Conserved roles of *ems/Emx* and *otd/Otx* genes in olfactory and visual system development in *Drosophila* and mouse. Open Biol. 3, 120177.

Sherwood, R.I., Chen, T.Y., Melton, D.A., 2009. Transcriptional dynamics of endodermal organ formation. Dev. Dyn. 238, 29–42.

Sheth, R., Grégoire, D., Dumouchel, A., Scotti, M., Pham, J.M., Nemec, S., Bastida, M.F., Ros, M.A., Kmita, M., 2013. Decoupling the function of Hox and Shh in developing limb reveals multiple inputs of *Hox* genes on limb growth. Development 140, 2130–2138.

Spence, J.R., Lange, A.W., Lin, S.C., Kaestner, K.H., Lowy, A.M., Kim, I., Whitsett, J.A., Wells, J.M., 2009. Sox17 regulates organ lineage segregation of ventral foregut progenitor cells. Dev. Cell 17, 62–74.

Sprecher, S.G., Hirth, F., 2006. Expression and function of the columnar patterning gene *msh* in late embryonic brain development of *Drosophila*. Dev. Dyn. 235, 2920–2929.

Sprecher, S.G., Urbach, R., Technau, G.M., Rijli, F.M., Reichert, H., Hirth, F., 2006. The columnar gene *vnd* is required for tritocerebral neuromere formation during embryonic brain development of *Drosophila*. Development 133, 4331–4339.

Steinmetz, P.R., Urbach, R., Posnien, N., Eriksson, J., Kostyuchenko, R.P., Brena, C., Guy, K., Akam, M., Bucher, G., Arendt, D., 2010. Six3 demarcates the anterior-most developing brain region in bilaterian animals. Evodevo 1, 14.

Strausfeld, N.J., Hirth, F., 2013. Deep homology of arthropod central complex and vertebrate basal ganglia. Science 340, 157–161.

Sunmonu, N.A., Li, K., Guo, Q., Li, J.Y., 2011. Gbx2 and Fgf8 are sequentially required for formation of the midbrain-hindbrain compartment boundary. Development 138, 725–734.

Swiers, G., Chen, Y.H., Johnson, A.D., Loose, M., 2010. A conserved mechanism for vertebrate mesoderm specification in urodele amphibians and mammals. Dev. Biol. 343, 138–152.

Taher, L., Collette, N.M., Murugesh, D., Maxwell, E., Ovcharenko, I., Loots, G.G., 2011. Global gene expression analysis of murine limb development. PLoS One 6, e28358.

Tajiri, R., Tsuji, T., Ueda, R., Saigo, K., Kojima, T., 2007. Fate determination of *Drosophila* leg distal regions by trachealess and tango through repression and stimulation, respectively, of bar homeobox gene expression in the future pretarsus and tarsus. Dev. Biol. 303, 461–473.

Tam, P.P., Khoo, P.L., Lewis, S.L., Bildsoe, H., Wong, N., Tsang, T.E., Gad, J.M., Robb, L., 2007. Sequential allocation and global pattern of movement of the definitive endoderm in the mouse embryo during gastrulation. Development 134, 251–260.

Tanaka, M., 2013. Molecular and evolutionary basis of limb field specification and limb initiation. Dev. Growth Differ. 55, 149–163.

Tomer, R., Denes, A.S., Tessmar-Raible, K., Arendt, D., 2010. Profiling by image registration reveals common origin of annelid mushroom bodies and vertebrate pallium. Cell 142, 800–809.

Tumpel, S., Wiedemann, L.M., Krumlauf, R., 2009. *Hox* genes and segmentation of the vertebrate hindbrain. Curr. Top. Dev. Biol. 88, 103–137.

Tzchori, I., Day, T.F., Carolan, P.J., Zhao, Y., Wassif, C.A., Li, L., Lewandoski, M., Gorivodsky, M., Love, P.E., Porter, F.D., Westphal, H., Yang, Y., 2009. LIM homeobox transcription factors integrate signaling events that control three-dimensional limb patterning and growth. Development 136, 1375–1385.

Tümpel, S., Cambronero, F., Sims, C., Krumlauf, R., Wiedemann, L.M., 2008. A regulatory module embedded in the coding region of Hoxa2 controls expression in rhombomere 2. Proc. Natl. Acad. Sci. U.S.A. 105, 20077–20082.

Urbach, R., 2007. A procephalic territory in *Drosophila* exhibiting similarities and dissimilarities compared to the vertebrate midbrain/hindbrain boundary region. Neural Dev. 2, 23.

Vanhoose, A.M., Samaras, S., Artner, I., Henderson, E., Hang, Y., Stein, R., 2008. MafA and MafB regulate Pdx1 transcription through the area II control region in pancreatic beta cells. J. Biol. Chem. 283, 22612–22619.

Vincent, S.D., Buckingham, M.E., 2010. How to make a heart: the origin and regulation of cardiac progenitor cells. Curr. Top. Dev. Biol. 90, 1–41.

Visel, A., Taher, L., Girgis, H., May, D., Golonzhka, O., Hoch, R.V., McKinsey, G.L., Pattabiraman, K., Silberberg, S.N., Blow, M.J., Hansen, D.V., Nord, A.S., Akiyama, J.A., Holt, A., Hosseini, R., Phouanenavong, S., Plajzer-Frick, I., Shoukry, M., Afzal, V., Kaplan, T., Kriegstein, A.R., Rubin, E.M., Ovcharenko, I., Pennacchio, L.A., Rubenstein, J.L., 2013. A high-resolution enhancer atlas of the developing telencephalon. Cell 152, 895–908.

Vitobello, A., Ferretti, E., Lampe, X., Vilain, N., Ducret, S., Ori, M., Spetz, J.F., Selleri, L., Rijli, F.M., 2011. Hox and Pbx factors control retinoic acid synthesis during hindbrain segmentation. Dev. Cell 20, 469–482.

Vokes, S.A., Ji, H., Wong, W.H., McMahon, A.P., 2008. A genome-scale analysis of the *cis*-regulatory circuitry underlying sonic hedgehog-mediated patterning of the mammalian limb. Genes Dev. 22, 2651–2663.

Wassef, M.A., Chomette, D., Pouilhe, M., Stedman, A., Havis, E., Desmarquet-Trin Dinh, C., Schneider-Maunoury, S., Gilardi-Hebenstreit, P., Charnay, P., Ghislain, J., 2008. Rostral hindbrain patterning involves the direct activation of a Krox20 transcriptional enhancer by Hox/Pbx and Meis factors. Development 135, 3369–3378.

Wiebe, P.O., Kormish, J.D., Roper, V.T., Fujitani, Y., Alston, N.I., Zaret, K.S., Wright, C.V., Stein, R.W., Gannon, M., 2007. Ptf1a binds to and activates area III, a highly conserved region of the Pdx1 promoter that mediates early pancreas-wide *Pdx1* expression. Mol. Cell. Biol. 27, 4093–4104.

Wittmann, D.M., Blochl, F., Trumbach, D., Wurst, W., Prakash, N., Theis, F.J., 2009. Spatial analysis of expression patterns predicts genetic interactions at the mid-hindbrain boundary. PLoS Comput. Biol. 5, e1000569.

Wolf, G., Hessabi, B., Karkour, A., Henrion, U., Dahlhaus, M., Ostmann, A., Giese, B., Fraunholz, M., Grabarczyk, P., Jack, R., Walther, R., 2010. The activation of the rat insulin gene II by BETA2 and PDX-1 in rat insulinoma cells is repressed by Pax6. Mol. Endocrinol. 24, 2331–2342.

Woodland, H.R., Zorn, A.M., 2008. The core endodermal gene network of vertebrates: combining developmental precision with evolutionary flexibility. Bioessays 30, 757–765.

Wu, J., Cohen, S.M., 1999. Proximodistal axis formation in the *Drosophila* leg: subdivision into proximal and distal domains by Homothorax and Distal-less. Development 126, 109–117.

Wurst, W., Bally-Cuif, L., 2001. Neural plate patterning: upstream and downstream of the isthmic organizer. Nat. Rev. Neurosci. 2, 99–108.

Xu, B., Wellik, D.M., 2011. Axial Hox9 activity establishes the posterior field in the developing forelimb. Proc. Natl. Acad. Sci. U.S.A. 108, 4888–4891.

Zaret, K.S., 2002. Regulatory phases of early liver development: paradigms of organogenesis. Nat. Rev. Genet. 3, 499–512.

Zaret, K.S., Carroll, J.S., 2011. Pioneer transcription factors: establishing competence for gene expression. Genes Dev. 25, 2227–2241.

Zeller, R., López-Ríos, J., Zuniga, A., 2009. Vertebrate limb bud development: moving towards integrative analysis of organogenesis. Nat. Rev. Genet. 10, 845–858.

Zhang, H., Wang, W.P., Guo, T., Yang, J.C., Chen, P., Ma, K.T., Guan, Y.F., Zhou, C.Y., 2009. The LIM-homeodomain protein ISL1 activates insulin gene promoter directly through synergy with BETA2. J. Mol. Biol. 392, 566–577.

Zhang, X., Rowan, S., Yue, Y., Heaney, S., Pan, Y., Brendolan, A., Selleri, L., Maas, R.L., 2006. Pax6 is regulated by Meis and Pbx homeoproteins during pancreatic development. Dev. Biol. 300, 748–757.

Zorn, A.M., Wells, J.M., 2007. Molecular basis of vertebrate endoderm development. Int. Rev. Cytol. 259, 49–111.

Zorn, A.M., Wells, J.M., 2009. Vertebrate endoderm development and organ formation. Annu. Rev. Cell Dev. Biol. 25, 221–251.

Zuniga, A., Laurent, F., Lopez-Rios, J., Klasen, C., Matt, N., Zeller, R., 2012. Conserved *cis*-regulatory regions in a large genomic landscape control SHH and BMP-regulated Gremlin1 expression in mouse limb buds. BMC Dev. Biol. 12, 23.

Chapter 5

Genomic Strategies for Terminal Cell Fate Specification

1.   Circumstances of Terminal Cell Fate Specification

Specification of cell types poses problems different from those of body part patterning. Here our focus is on what is required at the gene regulatory level to establish the specific functions of a cell type. To begin where this process terminates, differentiated function is mediated by the expression of particular cohorts of downstream effector genes. This in turn requires installation of a terminal regulatory state, which consists of the transcription factors that serve as drivers of effector gene expression. Because the terminal regulatory state defines the differentiated cell type, this state cannot be operative in precursor cells which also have the capacity to give rise to other cell types. The differentiated regulatory state must therefore be established during the transition from the last multipotent precursor to the differentiated cell type. Mechanistically the problem is therefore to understand the gene regulatory network (GRN) circuitry that allows such precursor cells to activate alternative cell fate specification programs, depending on developmental context. Terminal cell fate specification is mediated by both, the internal regulatory state of the multipotent precursor cell, and by the particular exogenous signals to which it is exposed. Considered in terms of spatial regulatory state patterns, a fundamental difference between the process of body part formation

Genomic Control Process
http://dx.doi.org/10.1016/B978-0-12-404729-7.00005-8

and the process of cell type specification, is that progenitor field regulatory states are always by definition transient, whereas regulatory states in differentiated cells end up being stable, sometimes for life. Thus, cell type specification is not body part formation, even though body parts all consist of multiple cell types.

Within the hierarchy of a developmental GRN, the specification programs discussed in this chapter occur at or near the periphery of the network. At the periphery, the control system operates in one direction, from the drivers to the effector genes. The set of effector genes activated in given cells may include both differentiation gene batteries and cell biology cassettes that execute morphogenetic processes. Differentiated cells depend on the expression of particular sets of effector genes that encode proteins required for specialized functions, for example, photoreceptor pigments, globins, or muscle contractile proteins. Other classes of effector genes encode cell biology functions which are deployed in many cell types and at many stages throughout embryogenesis, for example, those conferring motility or epithelial–mesenchymal transitions.

Much was learned in previous decades from studies on differentiation processes in tissue culture, even though in vitro contexts are often bizarre and unnatural, such as when permanent lines of cancerously transformed cells are made to "differentiate" in plastic bottles by application of "cocktails" of growth factors. In vitro, differentiation appears as a functional property of individual cells, and in some systems it occurs stochastically, such that in the same medium individual cells embark upon diverse differentiation pathways. But usually there is not much that is stochastic about the deployment of the right cell specification state in the right spatial domain of a developing animal body part, although we shall encounter some famous exceptions where stochastic processes do affect the exact positions of alternative cell types even while their differentiation pathways and proportions are genetically ordained. In general, the genomically hardwired apparatus utilized in normal developmental cell fate specification is visible only if studied in context. In the embryo given spatial domains provide particular signaling contexts, and these direct invariant cell fate choice in specified precursor cells. Thus within a given signaling context a multipotential precursor will always give rise to the same terminally differentiated cell type, and only when exposed to an alternative context will it give rise to an alternative cell type. For this reason, when confronted with artificial contexts (cell culture or ectopic transplantation), the same precursors often display multiple alternative differentiation potentialities depending on the signals they are exposed to. Because of the experimentally demonstrable flexibility of such precursors, their diverse descendants have often been treated as the output of "bi-stable switches". But in normal development embryonic specification systems are rarely actually bi-stable: the same cell types always appear in the same locations because the programs that direct their fate divergences are there to ensure that each of the alternative fates are in fact installed in the right places. Whatever stochastic or fate-flipping behavior can be produced by experimental manipulations may tell us something about the molecular levers that need be thrown to elicit the generation of one cell type-specific regulatory state or the other. In life, however, it is the spatial or positional signaling cues which are imposed by the surrounding contextual tissues that are used to throw these levers. The following examples illustrate how multipotential precursor cells operate regulatory programs that are capable of giving rise irreversibly to alternative pathways of differentiation, and these programs are used in a developmentally determinate way to produce the same spatial array of differentiated cell types in each individual of a species. Our focus is on the network circuitry that mediates these fate choice systems, and thereby on the encoded genomic programs directing these aspects of development.

In what follows we deal with aspects of cell fate specification that pertain directly to normal developmental processes. This chapter is divided into several separate discussions. As the bedrock and background upon which the remainder rests, we begin where we left off in the brief discussion of differentiation gene batteries in Chapter 3, Section 1.5. From a few well-studied examples emerge the principles of the cis-regulatory wiring which controls the developmentally coordinate expression of the individual effector genes constituting given differentiation gene batteries. We then take up four prominent examples of embryonic cell type specification process, in which transient multipotential populations of specific precursor cells arise in given locations, in the already partially formulated morphology of the developing animal. The descendants of these precursors may differentiate alternatively into various cell types, determined, as just noted, by developmental context.

We have included a brief discussion of adult stem cells. These are here defined in the strict sense, as cells present in adult tissues for the whole of life, that replicate in a suspended state of specified but incomplete differentiation, until their differentiation is required. They then generate differentiating daughter cells and sometimes further copies of the stem cell type. Stem cell systems are also utilized in certain niches in normal embryonic development. Throughout postembryonic life adult stem cells are essential in various animal forms for generation of hematopoietic and other cell types, for normal turnover and for regenerative repair. Adult stem cells, like the differentiated cells of which the mature animal is largely composed, are also the product of embryonic programs of progenitor cell differentiation, except that their development has been arrested prior to the terminal readout of those programs.

Finally the genomic regulatory logic that allows the same effector gene batteries to be deployed in different domains of the developing organism requires mention. These different domains are formulated by diverse GRNs, and yet they may come to express the same effector gene batteries. Common examples are the muscle, neuron, and cartilage cell types that occur throughout the vertebrate body plan. This is a quite general phenomenon, particularly if the discussion is broadened to include morphogenetic and cell cycling functions that are used repeatedly. At the *cis*-regulatory level, the strategic solution to the problem of deploying the same functions in specific response to diverse developmental regulatory states is certainly among the fundamental aspects of animal genome organization, and the solution enables both development and evolution.

The world of gene regulation in adult animal forms also includes unique obverse mechanisms in which cells of identical regulatory state express different genes of given gene families. These examples we note only in passing. The coding sequence elements of these gene families are arranged in large genomic clusters, and specific mechanisms determine which of the genes or combinations of genes will be expressed, varying from cell to cell. The most prominent examples are those afforded by adaptive immune systems, where in each cell the genes encoding the immune receptors are uniquely rearranged. In other cases the mechanism is different, but the result is similar, in that cells expressing the same sets of transcription factors nonetheless express different proteins of given effector gene families. Among such examples are clustered smell receptor genes in mammals individually expressed by stochastic promoter choice (Shykind et al., 2004); and the endless variety of neuronal cadherin isotypes produced by splicing variations, dependent on promoter choice (Tasic et al., 2002). As we are concerned here with cell type specification in development of the body plan, which is not mediated by such mechanisms, this fascinating subject lies outside our scope.

2. Combinatorial *Cis*-Regulatory Definition of Differentiation Gene Batteries

Many available examples indicate a fundamental commonality in the way differentiation gene batteries are wired. Their regulatory structure has been revealed by experimental examination of the *cis*-regulatory systems controlling the expression of multiple effector genes, which respond to the stable regulatory states presented in differentiated cell types. Although the following examples pertain to sets of functionally similar differentiation genes, since this is where extensive *cis*-regulatory analyses are available, we should be aware that differentiated cells utilize a broad range of diversely functioning effector genes, all of which may respond to the same small set of drivers (see, for example, amniote serotonergic differentiation gene batteries; Deneris and Wyler, 2012). The regulatory principles underlying the operation of differentiation gene batteries are simpler than those applying to any other aspect of GRN function. They can be briefly summarized (cf. Chapter 3, Section 1.5): (1) differentiation gene batteries consist of more or less coordinately regulated sets of effector genes that all respond to combinatorial inputs from a small, dedicated set of regulatory gene drivers (undoubtedly plus other general factors); (2) the deployment in time and space of any given differentiation gene battery depends solely on the expression of its set of drivers, and thus on the *cis*-regulatory control of these driver genes; (3) expression of the driver regulatory state is a terminal output of the upstream developmental GRN; (4) drivers of differentiation gene batteries usually act positively on the expression of effector genes. However, exceptions occur when very similarly differentiated cells are required to diversify their terminal gene expression and effector genes associated with alternative cell fates are subject to direct repression. For example, direct effector gene repression occurs in terminal photoreceptor specification in *Drosophila*, as discussed below; and in specification of the ASE left and right smell neurons of Caenorhabditis *elegans* (Hobert, 2014).

Following are three exemplary cases that illuminate the regulatory wiring of differentiation gene batteries.

2.1   Lens crystallin gene battery

Lens crystallins constitute about 90% of the water-soluble protein of the amniote eye lens. In mammals these proteins are encoded by three small families of crystallin genes, α, β, and γ. Birds and reptiles also utilize δ-crystallins which belong to an entirely different protein family. The crystallin genes are expressed at relatively low levels in many other tissues besides lens, where they apparently function in various ways in stress amelioration; thus for example, the α-crystallin family of lens proteins is similar to the small heat shock proteins, and one of these genes responds directly to a xenobiotic response factor (aryl hydrocarbon receptor; Liu and Piatigorsky, 2011). In addition other genes that encode housekeeping enzymes contribute to the lens crystallin assemblage. The mammalian lens itself develops from the optic placode (see below), which invaginates when contacted by bilateral optic vesicle out-pocketings from the diencephalon (e9.5 in the mouse). Crystallin gene expression begins soon after in the lens epithelium, and accelerates as lens fiber cells differentiate. This process culminates with the degradation of all cellular organelles, eventually including the nucleus, producing an optically clear syncytium of crystallin filled lens fibers, overlain with a thin epithelium which controls water and ion balance and also serves continuously as the source of new lens fiber cells. Anterior to the lens is the cornea, which contains different optically clear proteins known as corneal crystallins.

The lens crystallins are regulated coordinately (Templeton et al., 2013), and constitute a bona fide gene battery by the above definition. A prominent member of their regulatory drivers is Pax6, which in addition to its major role in the spatial regulatory specification of the eye early in development directly controls many effector genes in the lens (Xie et al., 2013). The *cis*-regulatory modules (CRMs) required for expression of the *α-crystallin* (mouse αA and αB), *γ-crystallin* (mouse γF), and *β-crystallin* (chicken βB1) genes in the lens have been analyzed; other CRMs direct expression of these same genes in different tissues. Figure 5.1(A1–A4) details the occupancy of these enhancers by the driver transcription factors expressed in differentiating lens placode epithelium and in differentiated lens fibers (Cvekl and Duncan, 2007). The *cis*-regulatory occupancy configurations present in the lens fiber are those operating at the highest rates. For each of these genes, *cis*-regulatory analyses have revealed the combinations of inputs required for expression during lens development, as well as the changes in inputs responsible for the increase of expression levels at terminal differentiation. The common drivers utilized by all these genes are Pax6 and Maf factors. In addition, the *αA-crystallin* enhancer utilizes CREB (Fig. 5.1(A1)), the *αB-crystallin* enhancer utilizes at early states a Retinoic Acid Receptor (RAR) and later a heat shock factor (HSF4; Fig. 5.1(A2)), the *γ-crystallin* enhancer utilizes RAR early and Sox factors, as well as HSF4 (Fig. 5.1(A3)), and the chicken *βB1-crystallin* enhancer in addition utilizes a Prox factor (Fig. 5.1(A4)). Increase in expression levels during differentiation is often due to minor changes in the populations and/or concentrations of bound factors. For instance, *αA*-gene expression is enhanced

Figure 5.1

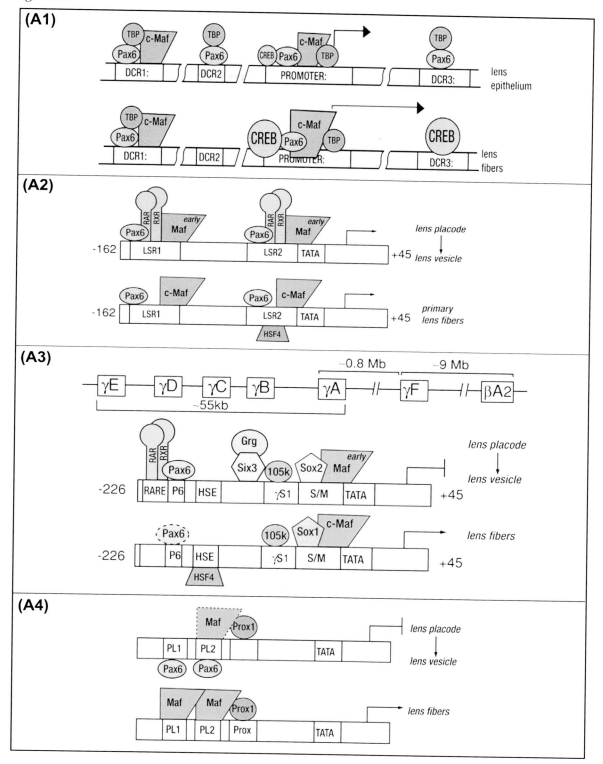

Figure 5.1

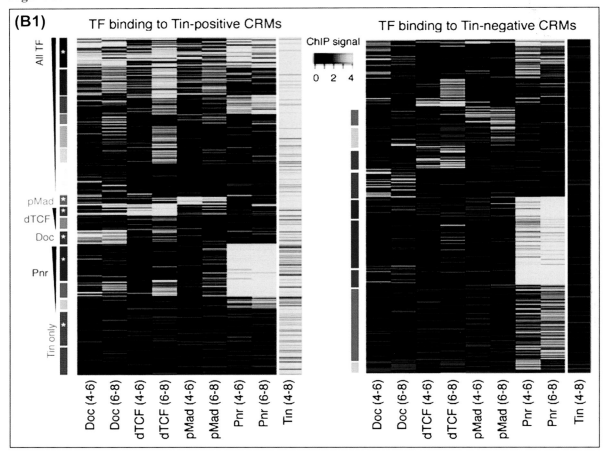

Figure 5.1 *Cis*-**regulatory architecture of differentiation gene batteries.** (A), *cis*-regulatory modules of lens crystallin genes *(from Cvekl and Duncan (2007))*. (A1), *cis*-regulatory control systems of mouse *αA-crystallin* genes active in lens epithelium and in differentiating lens fibers, where the genes run at highest levels; DCR, distal control region. The size of the symbols indicates relative strength (concentrations) of transcriptional inputs. (A2), Mouse *αB-crystallin* gene; RAR, RXR, retinoic acid receptors which provide activating inputs early in development of the lens (lens placode, top). HSF, heat shock factor 4; LSR, lens-specific region. (A3), Mouse *γ-crystallin* genes. Organization of the locus is shown at top. There is an early input from RAR in lens placode as in (A2), while the high expression configuration active in lens fibers, when Pax6 input is no longer important, is shown at bottom; HSE, heat shock element. (A4), Chicken *βB1-crystallin* gene (one of six *βB-crystallin* genes); PL, Maf-responsive element. (B), Cardiac enhancers of *Drosophila* embryos. (B1), Cohorts of transcription factors binding to *Drosophila* cardiac enhancers active in heart development, which according to ChIP assays, contain (left panel) or lack (right panel) Tinman (Nkx2.5) binding *(from Junion et al. (2012))*. The respective right-hand columns verify the selection criteria, Tin⁺ and Tin⁻ enhancers. Strength of signal is indicated by heat map, yellow highest, black, background. Each row indicates an enhancer, and these are clustered according to similarity of site content. The rectangles at left indicate sets of enhancers with similar constellations of binding sites, for the indicated factor the most prominent of which in each class is marked by an asterisk. Columns represent ChIP signals demonstrating binding of Dorsocross (Doc), Tcf (dTCF, activated by Wnt signaling), phosphorylated Mad (pMad) activated by Dpp signaling, Pannier

Figure 5.1

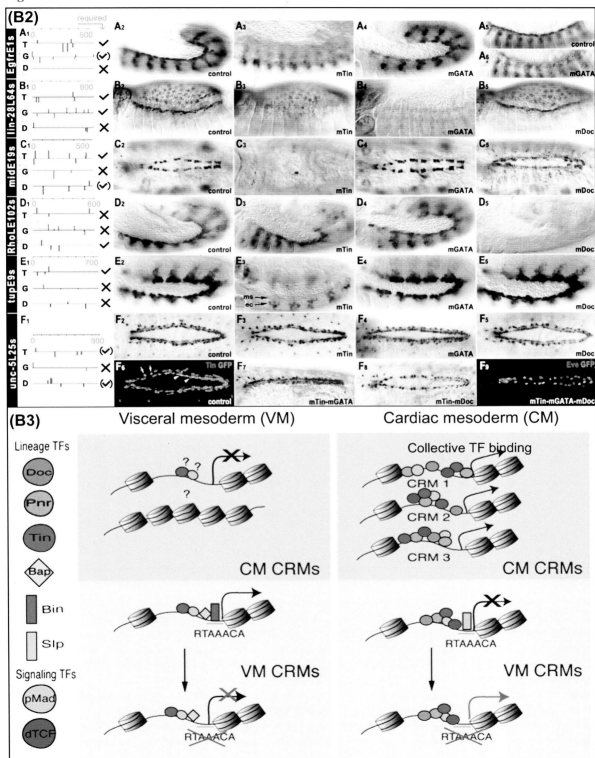

Figure 5.1

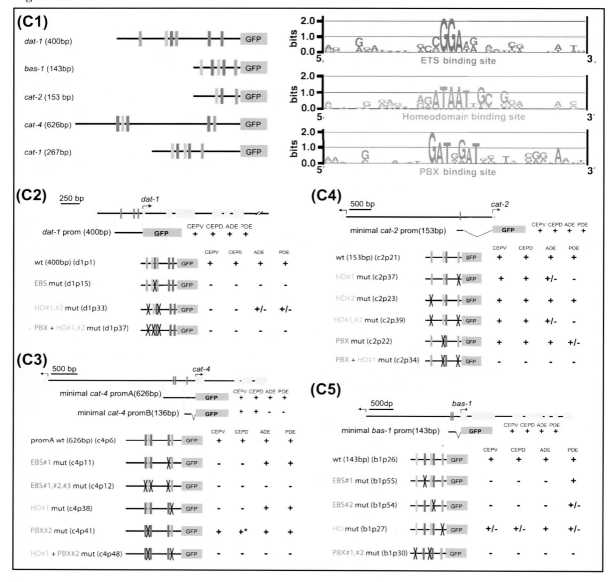

(Pnr, Gata4/6), and Tinman (Tin), at indicated time intervals (h). (B2), Functional test of normal and mutated cardiac enhancers in transgenic *Drosophila* embryos *(from Jin et al. (2013))*. Six enhancers in which ChIP observations indicated binding of Tinman, Pannier, and Dorsocross were incorporated in GFP constructs and tested in vivo in their native form, and after sites for each of these factors were mutated. Results are summarized in (A_1–F_1) at left: Locations of sites in enhancers is indicated by colored vertical bars (red Tinman sites; blue Gata (Panier) sites; violet Dorsocross sites). The heights of the bars represent strengths of binding signal, and symbols to right summarize results of the mutational expression assays illustrated in columns 3–5 (m indicates sites mutated). Here checkmarks indicate that inputs are required for expression; (checkmark) indicates contributory input function; X indicates no evident effect of site mutations. As seen in the control patterns shown in panels of column 2, *egfr* enhancer is expressed in cardiac mesoderm (A_2); *lin28* enhancer in cardioblasts and amnioserosa cells

(B$_2$); *mid* enhancer in cardioblasts (C$_2$); *rho* enhancer in cardiogenic and dorsal somatic mesoderm (D$_2$); *tup* enhancer in dorsal and cardiogenic mesoderm (E$_2$); *unc* enhancer in cardioblasts (F$_2$). Results in transgenic embryos of *cis*-regulatory site mutations are shown in columns 3–5: mTin, Tinman site mutations; mGata, Pannier site mutations; mDoc, Dorsocross site mutations. In F$_6$ embryos were also stained for Tin expression and cardioblasts, where construct is expressed, are shown by green GFP fluorescence (arrows), and pericardial cells by red Tin signal (arrow heads). In F$_9$, embryos were also stained for Eve, which is expressed in cardioblasts and the expression of a triple *cis*-regulatory mutation (mTin/mGata/mDoc) of the *unc*-GFP construct are shown. (B3), The combinatorial cardiac *cis*-regulatory code and role of repression in control of expression in cardiac versus visceral mesoderm *(from Junion et al. (2012))*. The diagrams show the enhancer target sites devoid of nucleosomes (gray cylindrical symbols). Cardiac enhancers bind all five of the factors indicated by colored ovals (Doc, Pnr, Tin, pMad, dTcf), while this whole set is not expressed in visceral mesoderm where these cardiac enhancers are silent. Enhancers active in visceral mesoderm use a different combination of inputs (Tin, pMad, Bagpipe (Bap/NK3), and Biniou (Bin/FoxF)). Mutational *cis*-regulatory experiments show that Biniou is required for their activity in visceral mesoderm. The same site is bound in cardiac mesoderm by Sloppy paired (Slp), which acts as a dominant repressor of these enhancers; these enhancers are expressed in cardiac mesoderm if the Bin/Slp site is mutated. (C), Regulatory organization of the dopaminergic neuron effector gene battery in C. *elegans (from Doitsidou et al. (2013))*. (C1), *cis*-regulatory organization of target sites for the three transcription factors that together constitute the informational code defining this gene battery, shown at left in diagrams of five GFP constructs utilized for site-specific mutational assays in transgenic worms. At right PWMs for the respective color-coded target sites are shown: the Ets factor is Ast1 (mouse Etv1); the homeodomain factor is ceh43 (*C. elegans* Distalless). (C2–C5), Mutational analysis of target sites in four of the five effector genes: *dat-1* (C2); *cat-4* (C3); *cat-2* (C4); *bas-1* (C5). Color code as in (C1); genomic organization is shown in a box at the top of each panel, indicating the location of the enhancer required in each gene for expression in the dopaminergic neurons (CEPV, CEPD, ADE, PDE). Below are expression results for the indicated *cis*-regulatory mutations.

by elevated levels of Creb and c-Maf factors (Yang et al., 2006). In the *αB-* and *γF-crystallin* genes, increase in expression level is accompanied by the appearance of HSF4 on the enhancer. The driver set for a chicken *δ-crystallin* gene is also Pax6 plus Sox2 (Kamachi et al., 2001). As an aside, it is interesting that the minimal chicken *δ-crystallin* enhancer also works in the cone cells of the *Drosophila* eye though here the Pax-binding site is occupied by Pax2 (Blanco et al., 2005). For a corneal crystallin gene, *aldehyde dehydrogenase 3a1*, the drivers are again Pax6, plus Oct1 and Klf4 (Davis et al., 2008; Swamynathan et al., 2008).

The comparisons in Fig. 5.1(A) illustrate two canonical features of the CRMs of differentiation gene batteries. First, they all require inputs from at least two of the three canonical lens crystallin driver regulatory genes, *pax6*, *sox1/2*, and *maf*. Second, no two of the CRMs are exactly alike, and their distinctions account for subtle differences in timing of their expression during lens development (Yang and Cvekl, 2005).

2.2 Cardiac enhancers

Cardiac-specific effector gene enhancers have been studied extensively at the *cis*-regulatory level, and as reviewed earlier (Davidson, 2006) a set of vertebrate examples including genes encoding cardiac Troponin T, Atrial Natriuretic Factor (ANF) and cardiac Adenylate synthase, display the same overall regulatory features as just noted for the crystallin gene battery. That is, the enhancers of these genes utilize subsets of a cardiac-specific driver set, which in this case consists of Nkx2.5, Srf, and its collaborator Myocardin (Wang et al., 2003), Gata, Mef2, and a bHLH heterodimer including a Hand factor. Again, no two of these enhancers are identical in structure, and additional factors besides the canonical members of the cardiac driver set, that are also expressed in developing heart, are bound to some of them. Examples are Pitx2 which interacts with Mef2a in an *ANF* enhancer (Toro et al., 2004); Tbx5 which interacts with Myocardin in the cardiac *myosin heavy chain* gene enhancer (Wang et al., 2011); and heart HMG and other factors that function in the enhancers of several other effector genes (Harlan et al., 2008). Genes encoding heart miRNAs also belong to the heart gene batteries, relying for their transcriptional control on the same drivers, Mef2C and Srf (Liu and Olson, 2010).

Genome-wide studies on heart enhancers have now broadened our perspective on the regulatory definition of cardiac gene batteries, particularly in *Drosophila* (Junion et al., 2012; Jin et al., 2013). As noted in Chapter 4, the heart field is specified by a core heart GRN kernel, essential circuit elements of which are shared across the Bilateria (Davidson and Erwin, 2006). Among these conserved elements are feedback relations that link expression of *tinman/nkx2.5*, *pnr/gata4/6* and *dorsocross* (*doc*)/*tbx4/5/6* to one another. Also conserved are signaling inputs from Smad (i.e., in consequence of Dpp/Bmp signaling) and dTcf/Tcf (in consequence of Wg/Wnt signaling). Once again regulatory genes earlier engaged in an upstream patterning GRN are also heavily involved in effector gene deployment at later stages in the same body part. In ChIP studies analyzing the binding of these drivers to Tinman-bound CRMs in *Drosophila*, a remarkable frequency of co-occupancy by all five of these factors on effector gene heart enhancers was observed, which does not occur in enhancers selected for the absence of Tinman binding (Fig. 5.1(B1)). The correlation was so strong that the binding of all five factors plus Tinman can be taken as definitive for cardiac-specific expression (Junion et al., 2012). Nonetheless, there is no requirement for given order, architecture, or organization of the enhancers animating cardiac effector gene batteries, and each is structured differently. In another ChIP study of *Drosophila* heart enhancers (Jin et al., 2013), 39 out of 51 of the best Tinman-binding genomic regions were shown to be expressed in mesodermal domains where the *tin* gene is developmentally expressed. Of these, 16 were active in cardiac mesoderm during heart development, and activity of the respective constructs required *tin* expression in vivo. Six of these cardiac enhancers were examined in detail. These all included Gata sites for Pannier, Tbox sites for Dorsocross (or Mid/H15/Tbx20), as well as Tinman sites, and the functional requirement of these species of transcription factor-binding sites were all tested by mutation. Results can be seen in Fig. 5.1(B2), the overall conclusion from which is that multiple members of the core cardiac set interact (at multiple sites) within each enhancer, even if the import of some of these interactions is not revealed by the qualitative metric employed. As seen here, for each enhancer certain of the factors are required, other factors have ancillary functions while mutation of the binding sites for some factors did not interfere with spatial expression. However, each of these enhancers has a unique design, but all utilize similar inputs and all produce cardiac-specific expression patterns. The ChIP study of mammalian heart enhancers cited in Chapter 4 similarly revealed co-occupancy by multiple members of the core cardiac driver set, viz. Gata4, Tbx5, Srf, and Mef2, as well as Nkx2.5 (He et al., 2011).

An additional interesting insight into the distinction between Tinman-binding enhancers active in *Drosophila* visceral mesoderm and those active in cardiac tissues is summarized in Fig. 5.1(B3) (Junion et al., 2012). Here we see that the visceral mesoderm enhancers but not the heart enhancers include a target site which in visceral mesoderm binds the positively acting factor Biniou/FoxF, while in cardiac mesoderm it binds Sloppy Paired (also related to Forkhead-type factors). But because Slp acts as a dominant repressor, these enhancers are inactive in cardiac mesoderm even though they bind the same set of positively acting drivers in both visceral and cardiac mesoderm. This neat mechanism distinguishes the regulation of gene batteries expressed in visceral mesoderm from those expressed in cardiac mesoderm, and if the FoxF site is destroyed, the visceral mesoderm genes run in the cardiac mesoderm while remaining silent in the visceral mesoderm.

2.3 Dopaminergic neuron effector gene battery

Completeness is a wonderful advantage in explanatory regulatory biology. Although whole genome analyses such as those just mentioned provide overall views of differentiation gene battery regulation, causal *cis*-regulatory validation must almost always be confined to exemplary subsets of the genes in each battery. But in a recent work, every relevant effector gene in a differentiation gene battery of dopaminergic neurons in *C. elegans* was examined (Doitsidou et al., 2013). This gene battery, which is expressed in all eight dopaminergic neurons in the worm, consists of genes encoding the enzymes required for neurotransmitter synthesis, for vesicular packaging, and for Dopamine reuptake. There are five such genes in this differentiation gene battery, and the *cis*-regulatory requirements for specific expression in all eight of these neurons have been studied for all five genes. Three of the genes code for biosynthesis enzymes (*tyrosine hydroxylase, cat-2*; GTP *cyclohydrolase, cat-4*; *aromatic amino acid decarboxylase, bas-1*), a fourth is required for packaging (*vesicular monoaminergic transporter, cat-1*), and the fifth for Dopamine reuptake (*dopamine transporter, dat-1*). Three of the four classes of dopaminergic neurons are located in the head, where they arise during embryogenesis, while the fourth, in the midregion of the body, is generated postembryonically.

The CRMs controlling each of these genes contain functional target sites for direct interactions with three different factors, and the logic underlying the specific regulation of this gene battery in all four neuron classes is encompassed therein. This is shown diagrammatically in Fig. 5.1(C1). All five genes require direct inputs from an Ets family factor (Ast-1) for function (Flames and Hobert, 2009), but the Ast-1 sites alone do not suffice to cause all of these genes to be expressed in dopaminergic neurons, and the *ast-1* gene is itself expressed in multiple other neurons. Using the GFP constructs diagrammed in the Figure, the exquisite cell type specificity of expression of the five genes is in this work very clearly shown to depend on direct combinatorial inputs from two additional regulators plus Ast-1: one, the product of a *C. elegans distalless* gene (*ceh-43*) and the other, the product of *pbx* genes (in the head *ceh-40* and *ceh-20*, only *ceh-20* in the midbody). This combination of factors is sufficient to cause expression driven by the dopaminergic CRMs even in a heterologous yeast transcription system. The regulatory states expressed combinatorially in all dopaminergic neurons include these three factors, though these factors are also expressed in one nondopaminergic neuron where a repressive interaction may prevent dopaminergic gene expression. Furthermore, exactly as we have seen in the other differentiation gene batteries discussed in this Section, no particular *cis*-regulatory organization of target sites is required for cell type-specific function. Another general feature illustrated in detail here is the individuality of the *cis*-regulatory elements driving the five genes. No two are the same, and the necessity/sufficiency of the individual inputs varies among them. This is shown in the mutational studies summarized in Fig. 5.1(C2–C5). The response of these CRMs to the individual inputs also varies among the diverse dopaminergic neurons, as we can see by glancing at the mutation results in these figures.

Thus we are left with a very simple rule. Differentiation gene batteries expressed coordinately in a developmental (spatial) sense are defined by *cis*-regulatory target sites for combinations of small sets of driver factors. In general there are no other defining features or structural constraints. Many differentiation gene batteries have now been partially defined at the *cis*-regulatory level. However, most current developmental data on control of effector genes refer only to the targets of single individual transcription factors. The objective of such measurements is to identify all effector genes that might respond to that factor in given contexts, genome wide. This approach may identify possible or putative members of differentiation gene batteries. But it does not provide the decisive evidence obtained by *cis*-regulatory analysis of effector gene CRMs that demonstrates the causal dependence of differentiation gene expression on combinatorial inputs from the driver gene set. The informational requirement for unique definition of given differentiation gene batteries, and the developmental requirement for regulatory instructions that will ensure the unique deployment of the battery at the right time and place, converge here on the essential feature of combinatorial control. No differentiation gene battery, and no effector gene, is controlled by a single "master" regulatory factor. Nor could they be, since every regulatory gene in the animal genome is utilized multiply and diversely in development.

As we shall see, deep evolutionary consequences emerge from the logical simplicity of this mechanism of achieving coordinate developmental expression of functionally related but diverse effector genes. In evolution, addition of a new effector gene to an old differentiation gene battery may occur simply by acquisition of *cis*-regulatory response elements to the driver regulatory state for that battery, while novel deployment of the whole differentiation gene battery may occur by changes in GRN architecture such that the driver regulatory state is expressed in a new developmental context. In other words, it is the driver regulatory state that determines the deployment of differentiation gene batteries in both development and evolution.

3. Cell Type Specification in Multipotential Embryonic Precursors

This section is focused on the genomic code by which cell fate choices are made in normal embryonic development, in situations where alternative final differentiated regulatory states derive from a common multipotential precursor regulatory state. In each of the four developmental systems we briefly examine in the following, the mechanism has been at least partly resolved at the GRN level. Thus, we can approach the encoded mechanisms by which arise alternative regulatory states driving terminal differentiation pathways.

3.1   Photoreceptor types in the eyes of *Drosophila*

Control functions in *Drosophila* eye development

Drosophila eyes develop from the posterior portions of the bilateral eye-antennal discs. The initial job of the encoded developmental GRN controlling eye formation is to install the regulatory state that mandates eye fate, and to establish the boundaries of the eye field, i.e., its boundary with the antenna-forming imaginal disc epithelium, and with the surrounding future head capsule. This GRN, known as the "retinal determination" GRN, has been extensively studied (Kumar, 2009, 2010), though its linkages are still not completely known. At present the retinal determination GRN consists of 14 regulatory genes and their interactions, and this GRN regulates the early specification of compound eyes. It has two overall functions: to specify eye fate, and to repress alternative fates, as summarized in Fig. 5.2(A) (Weasner and Kumar, 2013). The retinal determination network is discussed from an evolutionary perspective in Chapter 7 (see Fig. 7.5(D) for retinal determination GRN topology). An encoded regulatory control system operates downstream of the retinal determination GRN. In accordance with the theme of this chapter, the following concerns the mechanisms by which are specified the eight diverse photoreceptor neuron types of each of the ~800 ommatidia constituting the compound eye. In addition to these photoreceptors, each ommatidium contains stereotypically arranged bristle cells, cone cells that secrete lens proteins, and pigment cells.

In developmental sequence, specification of eye fate in the eye imaginal disc by the retinal determination GRN occurs during the first two larval instars. By the end of this period, regulatory factors required for the subsequent events of differentiation are specifically expressed in the epithelial cells of the eye imaginal disc. Dorsal/ventral clonal compartments have been established, with consequences noted below, and proliferation has generated a field of about 20,000 cells per disc, from which will be formed the 800 ommatidia of each compound eye. Cell division is controlled during eye imaginal disc development by regulation of the *string* Cdc25 phosphatase gene. Until the beginning of the third instar the cells constituting the eye imaginal disc epithelium remain undifferentiated. At this time, however, a dramatic process of

Figure 5.2

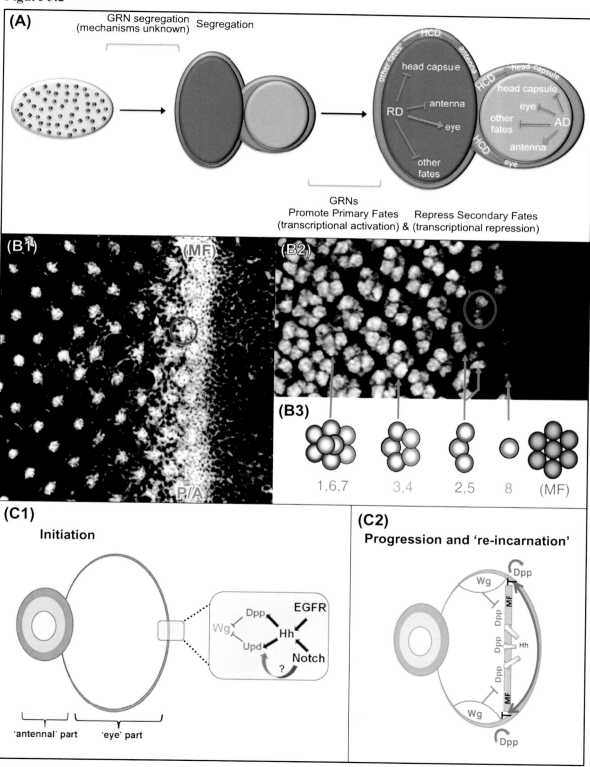

Figure 5.2

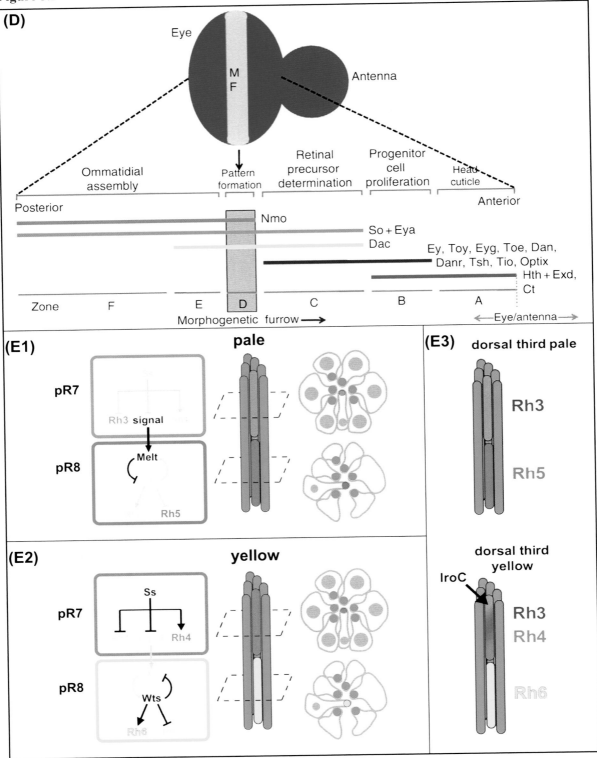

Figure 5.2

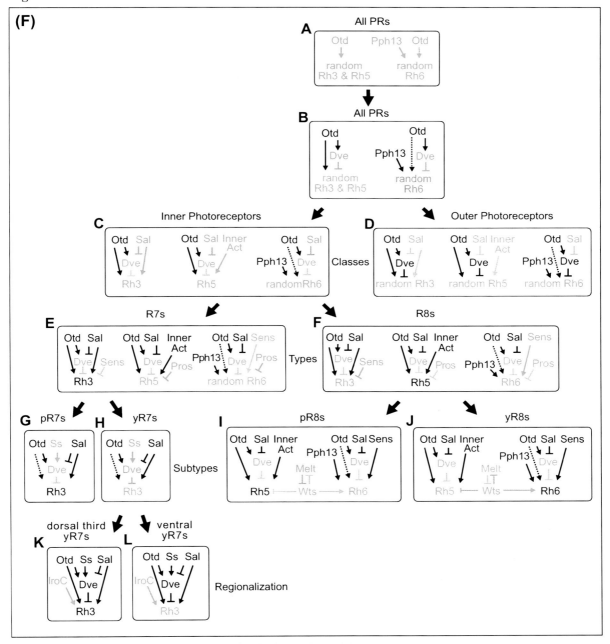

Figure 5.2 Encoding photoreceptor specification in *Drosophila*. (A), Reciprocal exclusion of alternative fates in the eye/antennal imaginal disc *(from Weasner and Kumar (2013))*. Antennal progenitor field, green, anterior; eye progenitor field, red. As indicated, the exclusive GRNs operating in each of these initially multipotential fields promote their respective morphogenetic fates, while also specifically repressing key regulators, expression of which is required for any of the alternative regional fates; RD, retinal determination; AD, antennal determination; HCD, head capsule determination. (B), Morphogenetic furrow (MF) and assembly of photoreceptor types into ommatidia *(from Kumar (2009, 2010))*. (B1), Before and after the

morphogenetic furrow. The furrow progresses from posterior (P) to anterior (A); third instar larval disc, F-actin stain. Anterior, ahead of the furrow no multicellular organization can be discerned. Groups of cells constituting ommatidial primordia coalesce in the furrow (e.g., the primordium in the blue circle), and organize progressively behind the furrow (left as shown). (B2), High-magnification view, neuronal ELAV stain, posterior of the MF. (B3), Sequence of photoreceptor specification. As indicated the furrow (MF) is positioned at the right, where eight unspecified photoreceptor progenitors are shown in a cluster in gray in the schematic color-coded drawing. The blue circle denotes a future ommatidium positioned as that circled in (B1). Photoreceptor specification occurs in the order shown in this diagram, right to left, R8 first (see text). (C), Signaling mechanisms in the initiation and progression of the MF; note that the A/P orientation of this diagram, anterior left, is opposite to that in (A) and (B) *(from Tsachaki and Sprecher (2012))*. (C1), Initiation of the MF at the posterior margin (right in this diagram), where as indicated in inset Egf and Notch signaling initiate *hh* gene expression with the ultimate consequence of Dpp expression and Wg repression (see text); Wg antagonizes ommatidial development. (C2), Furrow progression. Progress of the MF depends on Hh signaling from just specified cells to those immediately anterior (left) of the furrow (blue bar), thereby activating *dpp* expression, which is more distally kept off by Wg signaling (red). "Re-incarnation" of the MF refers to the altered circuitry at the MF margins where the MF must spread dorsally and ventrally as it expands across the convex disc. Here the *dpp* gene activates itself, in turn inhibiting *wg*. (D), Spatial expression of retinal determination genes in the eye disc during MF progression, orientation as in (A) and (B), i.e., anterior right *(from Kumar (2010))*. The eye portion of the disc is divided into six regulatory state zones with respect to the MF (A–F, blue bars), though expression of only some genes of the RD GRN (Fig. 7.5(D)) is indicated. The functional significance of each zone is given above. (E), "Pale", "Yellow", and dorsal type ommatidia expressing diverse photoreceptor rhodopsins *(from Johnston et al. (2011))*. The cylindrical drawings and section diagrams (at levels indicated by the dashed line boxes) display the arrangement of the cytoplasmic projections or rhabdomeres containing rhodopsin (Rh) pigments; R7 photoreceptors are stacked above R8 photoreceptors, surrounded by R1-6 photoreceptors (green). (E1), pale type: R7, blue; R8, red. Rh genes are selected by specific repression (see F). Here Rh3 is allowed to be expressed in R7, and a signal couples this expression with that of Rh5 in the R8 photoreceptor via an antagonist of the Hippo signaling kinase Warts, encoded by the *melted* gene, the result of which is to prevent Warts from effecting the repression of Rh5 (left diagram). (E2), yellow type: R7, purple, expresses Rh4 under control of the regulatory gene *spineless*; R8, yellow, expresses Rh6. (E3), Dorsal third pale and yellow ommatidia: in pale ommatidia R7, blue, expresses Rh3, and R8 expresses Rh5, as in other pale ommatidia; however, in yellow ommatidia of this region R7, blue and purple, expresses both Rh3 and Rh4, while R8 still expresses Rh6. (F), Photoreceptor specification circuitry *(from Johnston et al. (2011))*. The boxes at each tier represent the successive levels of specification indicated by the legends, and the linkages activated anew at each respective level are shown in blue; PR, photoreceptor; yPR, yellow, and pPR, pale ommatidia. Circuitry and genes involved are discussed in text. In the initial tiers (*A,B*) the activating regulatory states appear, and thereafter the *rh* genes are specifically repressed or activated according to the encoded logic shown. Interactions determining expression and repression of rh3, rh5, and rh6 are explicitly as follows: For *rh3*, Otd is an activator (*A*) but Otd also activates *dve* which represses *rh3* unless this interaction is canceled (*B*); in future R7 and R8 *sal* expression blocks *dve* expression allowing *rh3* expression (*C*), while in the outer PR's Dve continues to repress *rh3* (*D*); in R8 the *senseless* gene product represses *rh3* (*F*), so it remains expressed only in R7 (*E*), and in the pale R7 subtype (*G*); but in the yellow R7 subtype expression of *spineless* reactivates its repressor *dve* (*H*), while in dorsal pale R7s only, expression of *iro* drives *rh3* expression despite low level *dve* expression (*K*). For *rh5*, expression is prevented in R7 by expression of the repressor *prospero* (*E*), but not in R8 where an additional activator (Inner Act) is present (*F*); the Melt:Warts system allows expression of *rh5* in pale R8s and prevents it in Yellow R8s, while the reverse is true for *rh6* (I, J). For *rh6*, the key activator is Pph13, but Dve repression is dominant (*A,B*); in the inner but not the outer PRs *sal* expression prevents *dve* expression allowing *rh6* expression (*C,D*); in R7s *prospero* expression prevents *rh6* expression while in R8s *senseless* expression promotes *rh6* expression.

ommatidial organization is initiated, centered on the program of photoreceptor specification and differentiation. Described classically as a "wave" of differentiation, a "morphogenetic furrow" moves across the epithelial field from posterior to anterior. Cells anterior to the morphogenetic furrow remain undifferentiated, whereas in its immediate wake the cells become organized in stereotypic arrays, and thereupon the photoreceptor types differentiate in a sequential order. A close-up view of the morphogenetic furrow is presented in Fig. 5.2(B) (Kumar, 2010).

The morphogenetic furrow is initiated by Spitz/EGFR and Notch signaling at the posterior margin of the eye disc, and in consequence of these signals *hh* is expressed. This is a key step with two indirect consequences (Fig. 5.2(C1,2); Tsachaki and Sprecher, 2012). Hh signaling causes local activation of the *dpp* gene in cells immediately anterior to those expressing *hh*, where Dpp functions together with another Hh target gene, *upd*. This gene encodes a secreted ligand that activates the Jak/Stat pathway. These signal transduction processes result in inhibition of Wg signaling. In second instar discs Wg is expressed throughout the eye disc and it antagonizes the retinal determination GRN, interfering with some of its main outputs, the expression of the *sinoculis* (*so*), *eyes absent* (*eya*), and *dac* genes. Wg has the general role of inhibiting premature differentiation in the eye disc and at the boundaries, where it also promotes head capsule specification. Thus at morphogenetic furrow formation, a primary role of Hh is to allow eye differentiation by blocking Wg signaling via Dpp and Upd, as can be seen in Fig. 5.2(C1) (Tsachaki and Sprecher, 2012). As the morphogenetic furrow advances anteriorly, Hh is expressed in the differentiating cells posterior to it, and Dpp is expressed just anterior to the furrow Fig. 5.2(C2). The retinal determination GRN is linked into morphogenetic furrow function by the gene *optix*, which is driven by the major retinal determination GRN output, the Eya:Sinoculis transcription factor complex. In turn, Optix regulates *dpp* expression (Li et al., 2013). The Dpp ligand diffuses ahead of the furrow, and overcomes a "pre-proneural" state in undifferentiated cells further anterior. This contributes to the progression of the morphogenetic furrow. In these undifferentiated cells, the expression of the proneural gene *atonal*, which is an ultimate target of the retinal determination GRN, is prevented by two repressors, Hairy and Extramacrochaetae. But in response to Dpp signaling the expression of these two factors is repressed and differentiation is allowed to proceed. *Atonal* expression is driven in differentiating retinal cells by identified enhancers that respond to the retinal determination network factors Sinoculis (Six1/2) and Eyeless (Pax6), as well as by other factors such as Dachshund that differ from anterior to posterior in their expression (Zhou et al., 2014).

The spatial regulatory states expressed in different regions during this dynamic spatial process of development are indicated in Fig. 5.2(D) (Kumar, 2010). In this diagram the anterior to posterior positions of the third instar eye imaginal disc are labeled zones A to F. In zone A the eye specification antagonists *hth* and *exd* are expressed and this region becomes head cuticle. The retinal determination GRN might be expressed throughout the remainder of the disc up to the morphogenetic furrow except that in the most anterior region where proliferation continues (zone B) expression of the *hth* repressor gene together with Wg signaling prevents expression of the main outputs of the GRN, *eya* and *so*, and their target gene *dac*. The whole GRN runs in the strip of cells just anterior of the morphogenetic furrow (zone C), where as we see below it activates the expression of regulatory genes that will be required for the subsequent events of photoreceptor differentiation. Posteriorly, *so* and *eya* continue to be expressed in differentiating cells (zones E and F). Genetic evidence indicates that the Eya/So complex is required for some aspects of the photoreceptor specification occurring in this posterior zone, but no other genes of the retinal specification GRN continue to be expressed posterior to the furrow (Pignoni et al., 1997; Kumar, 2010). Thus the retinal determination GRN functions to initiate eye specification, up to the point when ommatidial cells are allowed to differentiate into the various photoreceptor cell types.

Photoreceptors and photoreceptor subtypes

The relative spatial order of the individual photoreceptors indicated in the diagram of Fig. 5.2(B3) is stereotypic, and is preserved throughout the eye, though it is represented in mirror image orientation on either side of the dorsal/ventral boundary which crosses the middle of the eye. The dorsal/ventral boundary is set up earlier, initially through expression of *pannier* in the dorsal compartment, which in turn activates the *iroquois* complex (Singh et al., 2012). In the orthogonal axis, comparing the right versus the left eye, the same ommatidial organization also presents mirror image symmetry. The organization and the uniqueness of each of the eight photoreceptors, suggests that the encoded process by which spatial photoreceptor organization is installed must occur stepwise, that it must involve short range signaling, and that the regulatory state of each of the eight photoreceptors must also be unique during their differentiation. Many molecular aspects of this process are now known (Quan et al., 2012; Tsachaki and Sprecher, 2012). The R8 photoreceptor neuron is specified first, and the remainder of the photoreceptors in the order shown in Fig. 5.2(B3). The outer photoreceptors (R2 and R5, then R3 and R4, then R1 and R6) are specified following an EGF signal (Spitz) emitted by the pioneer photoreceptor, R8. Specification of R8 requires transient expression of *atonal* which in turn activates an R8 regulatory gene, *senseless*, a major role of which is to repress alternative fates. Similarly, each pair of outer photoreceptors expresses specific regulatory genes required to ensure their particular fates: R2 and R5 express *rough*, R3 and R4 express *coup* (*sevenless*), R1 and R6, which together with R7 derive from an additional mitotic division, express *bar*. The last photoreceptor to be specified, in response to regulatory state changes caused by EGF and Notch signaling, is R7, which arises from a group of five cells the others of which become cone cells. An important identity and differentiation factor in R7 is encoded by the *prospero* gene, expression of which is required to prevent R7 from adopting R8 fate. *Cis*-regulatory analysis shows that R7 expression of *prospero* is driven by the product of a runt factor gene, *lozenge*, which is derepressed by EGFR signaling, by the Eya/So complex, and by a factor encoded by a Zn-finger gene *glass*, required in all the photoreceptors (Hayashi et al., 2008). Thus each pair of outer photoreceptors and R7 and R8 have all become uniquely defined by expression of specific regulatory genes.

The function of these diverse regulatory states is to generate the particular light response capabilities of the photoreceptor subtypes. The photoreceptors are connected specifically to the brain according to subtype, which makes logical sense since only a single one of the several *rhodopsin* (*rh*) genes is expressed in each subtype during the final steps of their differential specification, with one exception. The five rhodopsins expressed in the ommatidial photoreceptors differ in the types of light input to which they respond and hence in their roles in the visual system. Encoded network circuitry directly upstream of the *rh* gene *cis*-regulatory systems determines which one is to be expressed in each photoreceptor type, as we see explicitly in the following.

The outer photoreceptors (all but R7 and R8) express the *rh1* gene, which encodes a broad spectrum light response protein, and these photoreceptors function in motion detection. The general control strategy in *rh* gene selection is use of broadly present activators encompassed in network wiring that also causes expression of dominant, photoreceptor-specific repressors. These exclude all but the correct *rh* gene from expression. In the outer photoreceptors at least one driver is the Q50 homeodomain protein Php13 (which binds sites similar to Pax6), and another may be Glass, both of which are expressed generally in photoreceptors (Mishra et al., 2010). The inner photoreceptors, R7 and R8, are used for color vision and there are three main types of R7/R8 pairs, each of which expresses a different combination of *rh* genes. The R7 cell is stacked over the R8 cell in the center of the ommatidium. In the "pale" (P) ommatidial subtype R7 expresses *rh3* which is UV sensitive, and R8 expresses *rh5* which is blue sensitive (Fig. 5.2(E1)); in the "yellow" (Y) subtype R7 expresses *rh4*, also UV sensitive, and R8 expresses *rh6* which is green sensitive (Fig. 5.2(E2)). P and Y ommatidia are randomly distributed according to the stochastic expression of the bHLH regulatory gene *spineless*, but always in a 35:65 ratio (Wernet et al., 2006). In the dorsal third of the eye a variant combination occurs under Iroquois control: the P subtype is present there as well, but in the Y subtype, R7 expresses both *rh3* in addition to *rh4* (Fig. 5.2(E3)), broadening UV wavelength sensitivity in this area.

The encoded regulatory logic which underlies the expression of subtype-specific regulatory states and which controls expression of the various rhodopsin genes is shown in Fig. 5.2(F) (Johnston et al., 2011). This provides a beautiful example of hardwired, cell by cell spatial specification: precise Boolean outputs are generated by the interplay of activators and cell type-specific dominant repressors, which together encode coherent and incoherent feed-forward circuits that terminate in the *cis*-regulatory control systems of the *rh* genes, as detailed in the individual parts of Fig. 5.2(F). At early stages multiple *rh* genes are expressed transiently and randomly in all photoreceptor progenitors under control of their general activators Otd and Php13 (part A, B). But the consequence of *otd* expression is to activate the *dve* (*defective proventriculus*) gene, which encodes a dedicated K50 homeodomain repressor. This repressor blocks transcription of *rh3, rh5,* and *rh6* in the outer photoreceptors, which as we saw above are allowed to express *rh1* (part D). In the inner photoreceptors R7 and R8, the two similar *spalt* genes (collectively *sal*) are expressed and repress *dve*. In consequence, only R7 and R8 cells are permitted to express *rh3, rh5,* or *rh6* (part C). In R7, *rh3* can now be expressed, driven by *otd* and *sal*, but the R7-specific regulator Prospero directly prevents *rh5* or *rh6* from being expressed (part E). On the other hand, in R8s, the regulatory factor Senseless activates the expression of *rh6* and also directly prevents *rh3* from being expressed (part F). The final step accounts for the differences between Y and P ommatidia: Y R7's now express Senseless, which activates *dve* thus repressing *rh3* (part H), while P R7's continue to express *rh3* (part G); P R8's express *rh5* under the influence of a signal (unknown) from P R7 (part I), while Y R8's express *rh6* instead (part J). This switch depends on alternative activation of one of a mutually repressive pair of factors which modulate signaling pathways (Melted and Warts in Fig. 5.2(F)). Melted is dominant over Warts in P R8s while Warts is dominant over Melted in Y R8s. An autonomously expressed regulatory gene (*erect wing*) is directly or indirectly required to promote Melted expression and inhibit Warts expression in P R8s (Hsiao et al., 2013). The result is that P R8's express *rh5* while Y R8's express *rh6* (parts I, J). In the dorsal region *iroquois* complex expression adds *rh3* transcription to that of *rh4* in Y R7's (part K). Note the Boolean consequence of the wiring shown in Fig. 5.2(F): if *sal* is not expressed as in outer photoreceptors, then within the incoherent feed-forward circuits the general activator Otd drives the expression of its target gene *dve* which encodes a repressor of the Otd target genes *rh3, rh5,* and *rh6*. If *sal* is expressed, then double-negative gates are created, and *dve* is repressed, allowing expression of the *rh* genes downstream of coherent feed-forward circuits including *sal* or *senseless* inputs (Johnston et al., 2011). At the very dorsal margin of the eye is a row of ommatidia with special sensitivity to polarized light, and in these ommatidia, unlike in any others, both R7 and R8 express the Rh3 rhodopsin normally expressed only by dorsal R7. The mechanism underlying this depends on activation of *hth* and *exd* expression in R7 and R8 cells of ommatidia adjacent to the dorsal head capsule, the consequence of which is to repress *senseless* in R8, thereby eliminating *rh6* expression and permitting *rh3* expression (Wernet and Desplan, 2014).

Photoreceptor specification in the *Drosophila* compound eye provides concrete evidence of exact logic functions for cell type specification from multipotential precursors. The logic transactions are directly encoded in the genome, and they control the multiple steps of the process. The retinal specification GRN first sets up the bounded regional regulatory state of the eye versus the head capsule and antennal domains, then it generates the outputs necessary for the ensuing differentiation stages, viz. *eya/so, dac,* and *atonal* expression. Subtype-specific regulatory states follow, and in a final, automaton-like manner, elegant and geometrically invariant assembly of ommatidial *rh* gene expression arrays is created, 800 times over in each eye.

3.2   Embryonic erythropoiesis

Formation of red blood cells in the vertebrate embryo occurs in successive steps, by different processes, from diverse populations of multipotential mesodermal precursors. Within jawed vertebrates, major aspects of these processes are conserved at the molecular level, which is to say that at least portions of the underlying genomically encoded regulatory programs are homologous in their descent from the gnathostome common ancestor. The several developmental modes of embryonic erythrocyte generation occur in specific regions of the embryo, where like the processes we dealt with in Chapter 4, they are initiated by exogenous, niche-specific transcriptional regulators. But unlike the patterning of adult body parts, the outcome of this process is the immediate production of a specific differentiating cell type, not a system of spatial regulatory state domains which will only later give rise to multiple differentiating cell types. Furthermore, erythropoietic cells are formed individually and circulate singly rather than being engaged in a spatially coherent parenchyma or epithelium. As single cells, the ultimate maturation of their differentiation phenotype occurs on an individual basis rather than in fields of cells which are locked together by, for example, community effect signaling. Here we are concerned only with developmental blood cell generation in the embryo, rather than with the continuing process of neonatal and adult generation of erythrocytes from hematopoietic stem cells (HSCs), though the following discussion also includes the initial embryonic generation of definitive HSCs (models of blood stem cell GRNs are considered in Chapter 6).

Stages of blood cell specification in fish and mouse embryos

In zebrafish there are three waves of hematopoietic cell generation (Chen and Zon, 2009). A diagrammatic summary of embryonic hematopoiesis in this animal is reproduced in Fig. 5.3(A1). The earliest specification of hematopoietic cells ("primitive" hematopoiesis) is initiated in anterior and posterior lateral mesoderm (yellow in Fig. 5.3(A1)) starting as early as 6 h after fertilization. The anterior lateral mesoderm produces multipotential blood precursors (hemangioblasts) that quickly differentiate into endothelial and myeloid cell types. The posterior lateral mesoderm produces bipotential hemangioblasts that become erythrocytes and endothelial cells (Vogeli et al., 2006; Ciau-Uitz et al., 2010). Thus at very early stages these hemangioblasts produce cells that perform immune functions (myeloid cells), cells that generate blood vessels (endothelial cells) and circulating red blood cells (primitive erythrocytes; 6–24 h, Fig. 5.3(A1)). Expression of the *scl* regulatory gene is required for both types of hemangioblast, while expression of *gata1* uniquely marks cells that will turn into erythrocytes (Fig. 5.3(A1–A3); Ciau-Uitz et al., 2010). The initial waves of about 300 primitive pro-erythroblasts enter circulation at about 1 day and complete their differentiation (see timeline in Fig. 5.3(A1)). This initial phase of embryonic hematopoiesis shuts down at this time but a second transient phase of hematopoiesis becomes active in the tail mesodermal region (posterior blood island, pink in Fig. 5.3(A1)), where *gata1* expression appears, and lasts for another day (Chen and Zon, 2009). The hematopoietic cells produced at this time give rise exclusively to erythropoietic and myeloid derivatives. In contrast to the adult HSCs, these early embryonic hematopoietic precursors do not possess self-renewal capacity, though they are multipotential. The precursors of true HSCs derive from an independent mesodermal source (blue in Fig. 5.3(A1)). Adult hematopoiesis, as in other vertebrates, begins with the delamination of these HSCs from the ventral surface of the dorsal aorta, a region known as the AGM (aorta/gonad/mesonephros). Much evidence indicates that these HSC cells arise in the aorta from precursors of endothelial phenotype, induced by various signals, in particular Bmp emanating from beneath the ventral wall of the aorta (Ciau-Uitz et al., 2010). In zebrafish the initial phase of HSC production continues for about 3 days. From there the hematopoietic precursors migrate to different organs, as indicated in Fig. 5.3(A1) (blue boxes), for their further differentiation. The definitive HSCs are marked by *runx* expression (Fig. 5.3(A4)). Adult-type erythrocytes ("definitive" erythrocytes) descending from these HSCs appear just after 3 days, and gradually replace the circulating primitive erythrocytes. To summarize, in the initial early phase of embryonic hematopoiesis dedicated blast cells are specified from lateral posterior mesoderm, quickly enter the circulation, differentiate into primitive erythrocytes, and provide the embryo with an early oxygen supply as the newly formed heart begins to beat. A brief second wave of erythropoiesis of independent origin follows, and only much later, definitive hematopoiesis occurs after the production of true HSCs in the dorsal aorta. The HSCs seed the entire hematopoietic system and give rise to all adult blood cell types. In *Xenopus* a very similar course of events provides first primitive erythrocytes, and later definitive erythrocytes differentiating from HSCs that also arise in the AGM (Ciau-Uitz et al., 2010). The entirely separate embryological origin of these red cell populations is underlined in *Xenopus* by their distinct cell lineages: most primitive erythrocytes arising in the posterior ventral blood island descend from the 32-cell stage D4 blastomere, while the definitive AGM hematopoiesis occurs in cells descendant from the C3 blastomere (cf. Fig. 3.7(A) for *Xenopus* blastomere map).

Figure 5.3

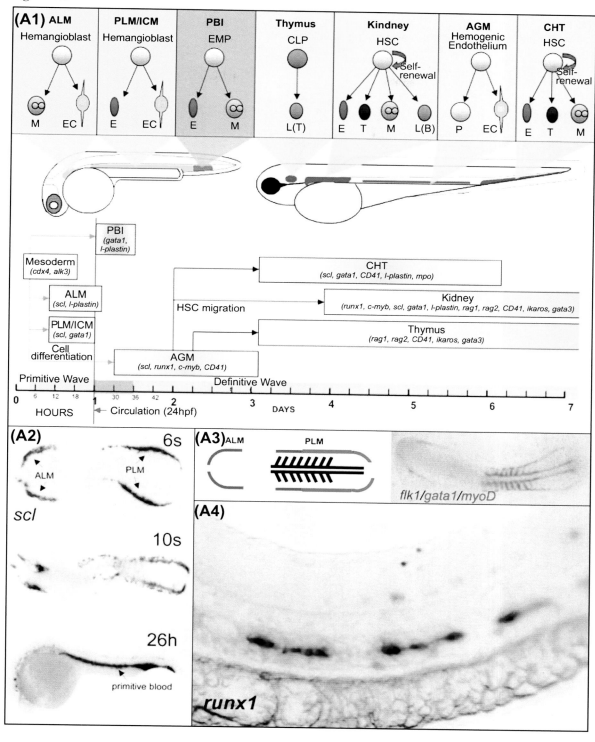

Figure 5.3

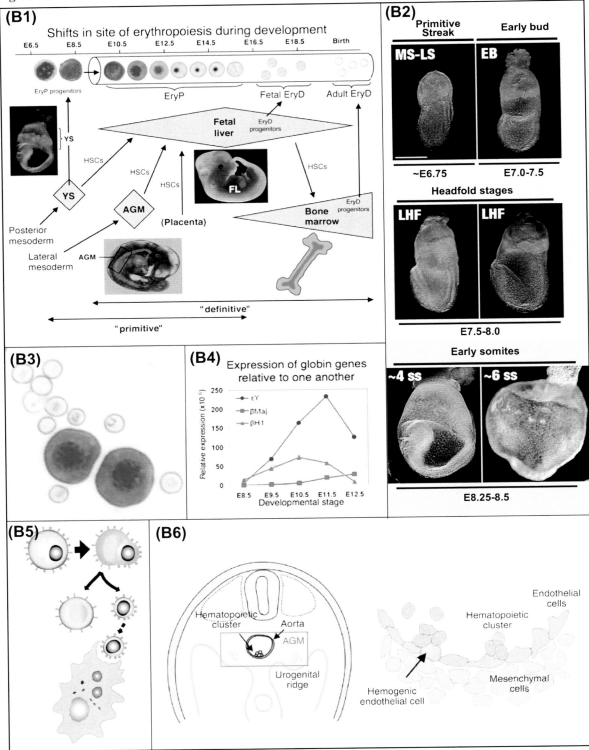

(B1) Shifts in site of erythropoiesis during development

(B2)

(B3)

(B4) Expression of globin genes relative to one another

(B5)

(B6)

Figure 5.3 Embryological origins of blood cell precursors in gnathostomes. (A), Primitive hematopoiesis in zebrafish. (A1), Successive sites of hematopoiesis in zebrafish embryogenesis. Locations and hematopoietic output occur in three waves: primitive hemangioblasts (symbolized by yellow); erythromyeloid precursors (EMP; pink); definitive HSCs (hematopoietic stem cells) arising in the dorsal aortal region (aorta-gonad-mesonephros or AGM, blue). These regions are mapped onto the embryo drawings below; note developmental timescale at bottom. ALM, anterior lateral mesoderm; PLM, posterior lateral mesoderm; ICM "intermediate cell mass", which is the medial repository of PLM products; PBI, posterior blood island; CHT, posterior hematopoietic tissue, which is transiently populated by the multipotent, expanding populations of AGM HSCs. These are the definitive self-renewing stem cells capable of generating all blood cell types and lineages: CLP, lymphoid precursor; E, erythroid; M, myeloid (P, precursor); L, lymphoid; T, T-cell; B, B-cell; EC, endothelial cell. HSCs generated in the AGM seed the kidney and thymus, permanent hematopoietic organs. Within the rectangles indicating the duration of each phase, characteristic regulatory genes and other molecular markers are indicated *(from Chen and Zon (2009))*. (A2), In situ hybridizations identifying *scl* transcripts, which mark hemangioblasts, seen in ALM and PLM (cf. A1); and in primitive blood cells just as circulation begins at 26 h. (A3), Diagram and in situ hybridization displaying location of expression in six somite embryos of the gene encoding the endothelial Vegf receptor *flk1* (blue) in ALM plus PLM, and *gata1* expression in PLM only, which marks erythropoietic fate. (A4), Expression of the HSC gene *runx1* in cells on the ventral wall of the dorsal aorta, 26 somites *(A2–A4 from Ciau-Uitz et al. (2010))*. (B), Embryonic hematopoiesis in the mouse. (B1), Sites of erythropoiesis in the mouse embryo. At top are shown primitive erythroblasts (EryP), the large size of which can be compared to the definitive erythroblasts (EryD) which enter circulation (tube) later; see timescale at top. Primitive erythropoiesis begins in the yolk sac (YS), illustrated by fluorescent reporter expression from an embryonic ε-globin transgene as seen at left. Definitive hematopoietic precursors arise in the lateral mesoderm, enter the AGM region where they generate the main pool of definitive HSCs, some of which may also develop within the placenta. The HSCs migrate to and reproduce and differentiate within the fetal liver (FL), and thence at birth and thereafter in the bone marrow which is seeded from the fetal liver (other hematopoietic organs, spleen, thymus not shown). Definitive erythrocytes of fetal liver origin appear in the circulation only late in gestation and until then the fetus survives on primitive erythrocytes *(from Baron (2013))*. (B2), Proliferation of differentiating primitive erythroblasts in yolk sac, marked by expressing ε-globin transgene as in (B1), MS-LS, midstreak-late streak gastrula stages; E, early; L, late *(from Isern et al. (2011))*. (B3), Nucleated e10.5 primitive erythrocytes mixed with adult erythrocytes for comparison on the same slide. (B4), Globin gene switching in differentiating primitive erythroid cells, as indicated. (B5), Enucleation of primitive erythroid cells, and redistribution of proteins: orange, Ter119, a protein associated with a red cell-specific surface protein glycosylated with sialic acid (Glycophorin A); blue, Integrins. As enucleation proceeds the primitive reticulocyte retains the Ter119, while the membrane covering the nucleus retains Integrins which induce engulfment by macrophages *(B3, 4, 5 from Baron et al. (2012))*. (B6), The AGM and the origins of intra-aortic hematopoietic clusters *(from Dzierzak and Philipsen (2013))*. On left is a cross-section of the dorsal trunk region of an E10.5 embryo; on right a close-up of the ventral domain of the dorsal aorta featuring a bipotential hematopoietic/endothelial cell (blue and brown) and hematopoietic clusters (brown) from which HSCs derive.

In the mouse blood cells similarly originate in successively different places and from different precursors during embryogenesis (Medvinsky et al., 2011; Baron et al., 2012; Baron, 2013; Dzierzak and Philipsen, 2013). A summary is shown in Fig. 5.3(B1). The primitive erythroblasts arise in the yolk sac outside of the embryo proper, in a dense circular band beginning by e7.0. They proliferate actively (Fig. 5.3(B2)), but when the heart begins to beat they enter the circulation and are no longer evident in the yolk sac after e9.0. Amazingly, in disaggregated e8.5 mouse embryos, cells expressing the primitive ε-globin constitute almost half of all cells, having expanded as a population 30,000 fold (Isern et al., 2011). During circulation they express different surface proteins and different globins as they mature, becoming very large compared to definitive erythrocytes and eventually they enucleate, whereupon attendant macrophages engulf the nuclei (Fig. 5.3(B3–B5)). Analogous to the second brief wave of hematopoiesis in zebrafish, the yolk sac of the mouse embryo then transiently produces myeloid and definitive erythroid cells which seed the fetal liver (McGrath et al., 2011; Baron, 2013; Dzierzak and Philipsen, 2013). The major wave of definitive HSC production in the embryo then ensues, of which the most intensely studied anatomical origin is again the AGM (Fig. 5.3(B6); we consider this specification process below). However, there could be additional sites of definitive HSC production, including as Fig. 5.3(B1) suggests, yolk sac and placenta (Medvinsky et al., 2011; Baron et al., 2012; Baron, 2013).

Gene regulatory interactions in primitive erythropoietic specification

The position of the erythrogenic mesodermal zone where the primitive erythrocytes first arise (Fig. 5.3(A1), PLM) in the zebrafish embryo is established with respect to the anterior/posterior body axis downstream of expression of *caudal* genes (*cdx4* and *cdx1a*). These *cdx* genes indirectly and redundantly affect primitive hematopoiesis in the posterior lateral mesoderm in two ways (Davidson and Zon, 2006; de Jong et al., 2010). Loss of *cdx* function (by mutation or morpholino treatment) precludes most primitive erythropoiesis. This effect is due partly to the control of posterior *hox* gene expression by *cdx4*, and thus, for instance, expression of *hoxa9a* rescues erythropoiesis in embryos lacking Cdx4. Much evidence indicates that posterior *hox* gene expression is required for erythropoiesis in zebrafish (and also mouse) as is expression of the *pbx* and *meis* genes, which encode obligatory Hox cofactors (Pillay et al., 2010). *Hox* gene expression is upstream of the canonical erythropoietic regulatory gene *gata1*, and also may be required for initial specification of lateral plate mesoderm. However, *cdx* has another encoded role as well, which is to restrict expression of the *raldh2* gene. As we saw in Chapter 4, *raldh2* encodes the key retinoic acid (RA) synthesis enzyme. A series of experiments is summarized in Fig. 5.4(A) using *cdx4* mutants, RA and an RA signaling inhibitor, in which *gata1* expression was monitored (de Jong et al., 2010). These perturbations show that RA indirectly inhibits expression of *gata1* and blocks erythropoiesis, and that this effect can be rescued by forced expression of *scl*, a direct activator of *gata1* expression. Thus RA synthesis, controlled spatially by repression from *cdx* genes, sets the anterior boundary of the hematopoietic domain of erythropoiesis within the lateral mesoderm.

The specification of the precursors of the primitive erythrocytes (hemangioblasts) requires Bmp signaling, which controls the initial regulatory genes in the specification subcircuit, *gata2* and the Ets family gene *fli1* (Liu et al., 2008; Ciau-Uitz et al., 2010). The circuit (for *Xenopus* and zebrafish) is shown in Fig. 5.4(B). Once activated, *fli1* locks itself on, and activates the essential hemangioblast regulatory genes *scl* and a gene encoding a required Lim domain cofactor of *scl*, *lmo2* (Patterson et al., 2007; Liu et al., 2008). An enhancer of the *scl* gene that expresses specifically in the primitive erythroid lineage in the mouse (Ogilvy et al., 2007) requires Gata + E-box (E2A) inputs for expression, and also includes putative Ets target sites. But this is probably not the only relevant *scl* enhancer, since it can be deleted with only mild effects, while deletion of *scl* per se prevents all primitive erythropoiesis, and is consequently lethal (Ferreira et al., 2013).

Anterior and posterior hemangioblasts, the precursors of myeloid and primitive erythrocytes, respectively, both express Scl and its partner Lmo2. The distinction between the myeloid and erythroid fates is encoded downstream of the Scl/Lmo2 complex. As shown in Fig. 5.4(C1,C2), different GRNs are activated by this complex in the anterior and posterior lateral plate mesoderm regions. Figure 5.4(C1) (Patterson et al., 2007) shows regulatory targets of the zebrafish Scl/Lmo2 complex in the erythroid posterior lateral mesoderm. Figure 5.4(C2) shows the targets of the same regulatory complex in the myeloid anterior mesoderm (cf. Fig. 5.3(A)). The Scl/Lmo2 complex expressed in anterior hemangioblasts directly activates the myeloid driver gene *pu.1*, promoting myeloid differentiation downstream. In the posterior hemangioblasts, this same complex activates *gata1*, *runx1*, and *c-myb* thus promoting erythrocyte

Figure 5.4

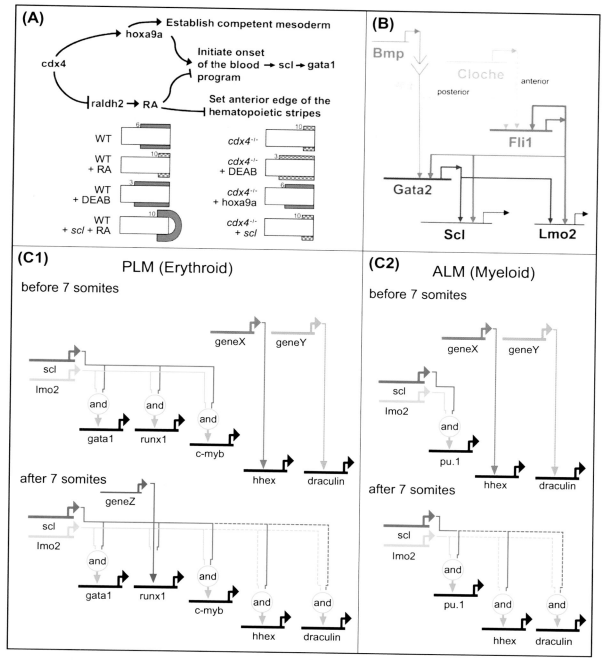

Figure 5.4 GRNs encoding aspects of embryonic hematopoietic development. (A), Interactions of *cdx4*, *raldh2*, and *hoxa9a* genes that are proposed to control anterior boundary of primitive hematopoietic posterior lateral plate mesoderm in zebrafish embryos *(from de Jong et al. (2010))*. The dual functions of *cdx4* are indicated in diagram at top (see text for discussion). Schematics below represent expression of *gata1* in 10 somite embryos in wild type (WT) or homozygous *cdx4* mutant embryos (cdx4⁻/⁻); or if treated with the RA receptor inhibitor DEAB; or if subjected to overexpression of Scl, RA, or Hoxa9a

Figure 5.4

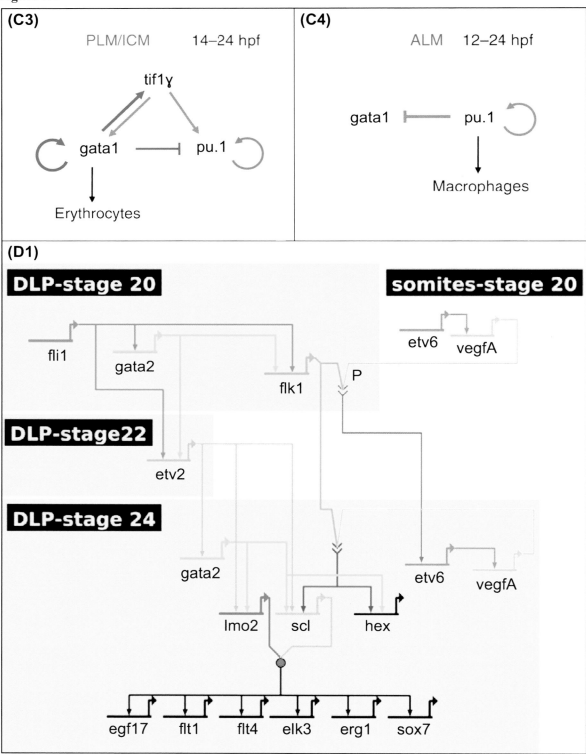

Figure 5.4

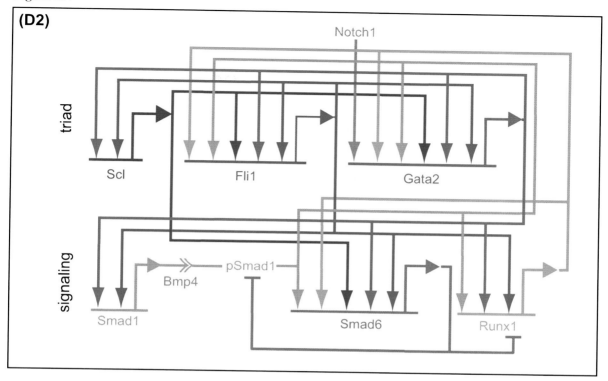

as indicated. Hatched areas represent weak *gata1* expression; numerals represent the somite pair at which the anterior boundary of expression was observed. The top two rows show that *cdx4* expression protects erythropoiesis from antagonistic effects of RA, by inhibition of *raldh2*, and that the anterior hematopoietic boundary is set in this fashion. The bottom two rows show perturbation results consistent with this interpretation. (B), Circuitry for specification of primitive hemangioblasts (from *Xenopus* and zebrafish results), in which the initial key regulatory event required is activation of the *scl* and *lmo2* genes *(from Ciau-Uitz et al. (2010))*. Specification is initiated by the dual signaling pathways indicated (*cloche*, a zebrafish mutation, apparently affects a required signaling event) followed by the encoded double feed-forward cross-regulation that ensures *scl* and *lmo2* expression. (C), GRNs for anterior lateral mesoderm (ALM) and posterior lateral mesoderm (PLM) primitive hematopoietic specification in zebrafish. (C1), Erythroid GRN of PLM before and after 7 somite stage. Dashed lines indicate possibly indirect, solid lines known direct, interactions. Note that *runx* expression is at first directly dependent on *scl/lmo2* expression, then later dependent on another input. The essential erythroid gene *gata1* is robustly driven synergistically by *scl* and *lmo2*, and both *gata1* and *scl* gene expressions are observed to be required for primitive erythropoiesis. (C2), Myeloid GRN of ALP. Here the prime direct target of *scl* and *lmo2* is the *pu.1* gene *(C1 and C2 from Patterson et al. (2007))*. (C3), Dominant *gata1* circuitry in erythroid PLM (and ICM; see Fig. 5.3(A)). Tif1γ is a transcriptional elongation factor-binding protein which positively affects expression of both *pu.1* and *gata1*. Its own transcription is shown in a feedback with *gata1*, i.e., it is proposed to increase *gata1* expression as a function of the intensity of *gata1* expression, while the *gata1* gene also auto-activates. Gata1 directly represses *pu.1* transcription, which thus becomes extinguished in the erythrogenic PLM. (C4), Dominant *pu.1* expression in myeloid ALM: The *pu.1* gene auto-activates and directly represses *gata1* expression, which as we saw in Fig. 5.3(A2) is indeed absent in the ALM; *(C3 and C4 from Monteiro et al. (2011))*. (D), GRNs for development of definitive HSCs. (D1), GRN for initial hemangioblast specification in the dorsal lateral plate (DLP) mesoderm in *Xenopus* embryos *(from Ciau-Uitz et al. (2013))*. As discussed in text specification occurs by both autonomous regulatory gene interactions and signal-mediated inputs into target regulatory genes. The autonomous pathway terminating in the feed-forward activation of *scl* and *lmo2* is shown on the left. An initial target of the early DLP regulatory state is the *flk1* receptor gene which enables these cells to utilize first exogenous (of somitic origin) and then autocrine Vegf signals, eventually contributing directly to *scl* activation. (D2), GRN for specification of HSCs in the AGM based on extensive *cis*-regulatory evidence obtained earlier in mouse cells *(from Narula et al. (2013))*. As discussed in text here and also in Chapter 6, when examined dynamically, this GRN accounts for the developmental transience of *runx* expression by interaction with positive and negative signal transduction components (bottom tier of interactions). It also explains by its wiring configuration the developmental events in the AGM that lead to permanent institution of high-level expression of the canonical HSC "triad" kernel (top tier of interactions).

differentiation. The encoded mechanism accounting for the difference in regulatory activity is shown in Fig. 5.4(C3,C4) (Monteiro et al., 2011): in the erythropoietic domain a stimulatory transcription factor complex, Tif1γ, which promotes polymerase elongation of erythropoietic genes and is required for erythropoiesis, is locked in a positive feedback relation with *gata1* transcription, while *gata1* also positively autoregulates following its activation. These positive inputs enable *gata1* to dominantly prevent *pu.1* expression by repression. In contrast, in the myeloid precursors, only *pu.1* positively autoregulates and is able to dominantly repress *gata1* transcription. In zebrafish embryos treated with *gata1* morpholino, excess myeloid cells are formed from precursors that normally would give rise to erythroid cells (Galloway et al., 2005). Thus while the mutual repression circuitry between *pu.1* and *gata1* contributes to their exclusive expression, these genes are regulated by additional distinct inputs which provide the explanation for their respective spatial expression.

Gata1 is the canonical erythrogenic regulatory driver, and some regulatory linkages downstream of *gata1* expression are known or strongly implied. In mammals and birds the positive functions of Gata1 are directly required for terminal erythropoietic differentiation, and in fact the *gata1* gene was discovered over 20 years ago as the major erythropoietic *cis*-regulatory driver of *globin* genes (Wall et al., 1988; Evans and Felsenfeld, 1989; Tsai et al., 1989). The level of *gata1* expression in primitive erythrogenic cells of mice depends on autocrine Vegf signaling such that overexpression decreases *gata1* expression and erythroid differentiation (Drogat et al., 2010). The Scl/Lmo2 complex also directly activates *runx* (Fig. 5.4(C1)), and a downstream role of *runx* expression in primitive mammalian erythropoiesis was also identified in studies of yolk sac erythrocytes in mice mutant for this gene (Yokomizo et al., 2008). Runx expression is required for yolk sac erythropoiesis as shown by timed conditional deletion (Tober et al., 2013), and quantitative measurements show that in these cells, directly or indirectly, *runx* provides cross-regulatory positive inputs to both *gata1* and *elkf* (*klf1*) (Yokomizo et al., 2008). The latter gene encodes a Zn finger regulator and in mice its expression, together with that of its cousin *klf2*, directly regulates the order of globin gene transcription in primitive erythropoiesis (Fig. 5.3(B4)), binding to CACCC target sites in globin gene *cis*-regulatory sequences. The target genes of Klf include a large number of cell surface, cytoskeletal and other erythrocyte effector genes, plus a regulatory gene, *c-myc*, encoding a potent additional transcriptional regulator of further cohorts of downstream erythrocyte effector genes (Siatecka and Bieker, 2011; Baron et al., 2012; Pang et al., 2012).

HSC specification in the embryonic AGM

Following early demonstrations of the functional pluripotentiality of HSCs of embryonic AGM origin (for example, Delassus et al., 1999), evidence of many kinds has been obtained from different vertebrate species that all adult HSCs derive from embryonic AGM hemangioblasts (Dzierzak and Philipsen, 2013). In the mouse, for example, an inducible Cre-recombinase expressed under the control of the *scl* regulatory system was used to genetically label these cells, and it was shown that HSCs arising during the initial wave of embryonic AGM hematopoiesis (e10.5–e11.5) are ancestral to adult postembryonic HSCs, and that no further de novo specification of HSCs occurs after e14.5 by when AGM HSC generation has ceased (Göthert et al., 2005). In *Xenopus*, hemangioblasts derive originally from precursors in the dorsal lateral plate (DLP) mesoderm lying bilaterally beneath the somites, from which they receive essential Vegf signals required for their specification in the DLP (Leung et al., 2013). Vegf isoforms are also secreted from the hypochord (a transient sub-notochordal rod, of endodermal derivation) which promotes migration to the midline of hemangioblast precursors, and as we see below, autocrine Vegf signaling is also involved in HSC specification from hemangioblasts. The DLP hemangioblasts give rise both to the vascular endothelial cells and the HSCs, which as we see in Fig. 5.3(B6), arise from individual endothelial cells of the inner ventral wall of the dorsal aorta. On adopting HSC fate, these cells lose their epithelial junctions, round up, and form hematopoietic clusters within the aorta.

Figure 5.4(D1) shows a well-supported GRN model of the process by which the early steps of hemangioblast specification within the DLP mesoderm are encoded in *Xenopus* (Ciau-Uitz et al., 2013). Two pathways converge on the *scl* gene: one dependent on exogenous and autocrine Vegf signaling, and one an autonomous, progressive circuit of regulatory gene interactions. Both are triggered by the highest gene in the hierarchical network, *fli1*, an *ets* family gene activated within the DLP mesoderm. Immediately downstream, its initial target in the autonomous pathway is the *gata2* gene. Two sequential feed-forward subcircuits follow: the target of the first is *etv2*, a second *ets* family gene (*fli1* > *gata2*; *gata2* > *etv2*; and *fli1* > *etv2*). The targets of the second are *lmo2* and *scl* (*etv2* > *gata2*; *gata2* > [*lmo2* + *scl*]; *etv2* > [*lmo2* + *scl*]). The initial target of Fli1 in the Vegf pathway, which potentiates the activity of this branch of the network, is the *flk1* gene, encoding the receptor for the VegfA ligand. Another Ets family factor, Etv6, expression of which is required in the somites for hemangioblast development, drives transcription of the *vegfA* gene in somitic cells. The resulting signal is received in the DLP mesoderm cells via their Flk1 receptor, causing *etv6* expression. In consequence, the *etv6* to *vegfA* linkage is established within the DLP cells resulting in autocrine signaling via the same receptor. This signaling generates an additional required input to the *scl* gene. The Lmo2/Scl complex now serves as an activator for the definitive hemangioblast regulatory state shown at the terminus of Fig. 5.4(D1).

Specification and emergence of definitive HSCs in the dorsal aorta in the mouse is driven by a recursive kernel of the three regulatory genes we have already met, *scl, fli1,* and *gata2*, in which the relevant *cis*-regulatory control modules of each are directly bound and activated by the factors produced by the other two to form a complete web of feedback interrelations (Göttgens et al., 2004; Pimanda et al., 2007; Wilson et al., 2011; Spensberger et al., 2012; Narula et al., 2013). This subcircuit (known as "the triad") can be seen in the top portion of Fig 5.4(D2) (Narula et al., 2013): thus *scl* has direct inputs to *fli1* and *gata2*; *gata2* has direct inputs to *fli1* and *scl*; *fli1* has direct inputs to *scl* and *gata2*. In addition *fli1* and *gata2* positively autoregulate. Once operating at high levels this subcircuit runs irreversibly, defining the permanent HSC regulatory state. In developing hemangioblasts, two signaling inputs are required to achieve permanent high-level triad expression. Notch signaling provides a direct input into the *gata2* gene and occurs in the dorsal aorta downstream of Vegf signaling (Leung et al., 2013). Bmp4 signaling is additionally necessary to switch the hemogenic endothelial cells into the high-activity HSC state (i.e., at the ambient levels of Notch signaling). Phosphorylated Smad1 generated by Bmp4 signaling steps up *runx* gene expression, which in turn provides direct positive inputs to *fli1* and *gata2*. These signaling inputs are required only for the transient developmental phase leading to full triad activation. HSC specification is thereafter independent of signaling inputs, and *runx1* expression is no longer required, as shown by a timed Cre-mediated deletion (Chen et al., 2009). The effects of the wiring shown in the lower portion of Fig. 5.4(D2) emerge from a dynamic analysis of this network (Narula et al., 2013), to which, per se, we return in Chapter 6. As we can see qualitatively from the topology of the network, outputs from multiple triad genes also activate *smad6* transcription as well as *runx1* transcription. The key to the subcircuit dynamics is that Smad6 is a negative regulator: thus while a Smad1/Runx complex positively regulates the triad genes, Smad6/Runx and Smad6/Smad1 complexes are targeted for degradation (Knezevic et al., 2011). Thus, as the triad picks up steam, Smad6 expression is accentuated, resulting in the eventual disappearance of both pSmad1 and Runx1. In other words, if we consider the Smad1/Smad6/Runx subcircuit as the control module, this functions initially to turn on the triad subcircuit at high levels in response to Bmp4 signaling, and as its activation functions become superfluous it turns itself off.

Another function of early *runx1* expression is its necessary role in promoting the transition from endothelial to hematopoietic cell type, in which progenitor cells depart from the endothelial wall and assume a rounded up form (Fig. 5.3(B6)). This function turns out to require expression of two Zn finger repressors encoded by *gfi1* and *gfi1b* genes, which are direct transcriptional targets of *runx1* (Wilson et al., 2011; Lancrin et al., 2012). In the absence of *runx1* expression, the cells remain endothelial.

In summary, embryonic specification of primitive erythropoietic cells and of AGM HSCs from their respective mesodermal precursors, can now be considered in the precise terms of encoded processes driven by GRNs, even if further genes and further linkages will be added as knowledge accumulates. The generation of primitive erythrocytes produces a transient, though essential cell type, while the HSCs represent a long-lasting state. The structure of the networks controlling these specification processes is remarkable for the density of their robust interconnections, their feedbacks, feed forwards, and positive and negative cross-regulations.

3.3   Somitic cell fate specification

In jawed vertebrates all skeletal muscles in the trunk and appendages originate during embryogenesis from the somites. These are transient segmented paraxial mesodermal structures, bilaterally arranged on either side of the neural tube. The muscle progenitors arise from the outer layers (dermomyotome) of the somites (see Fig. 5.5(A1)). Muscles of the back (epaxial muscles) derive from the dorsal/medial edge of the somites, adjacent to the neural tube, and differentiate more or less in situ. But quite remarkably, the body wall musculature and all the appendicular musculature (adaxial muscles), that is, by far most of the skeletal musculature, arises from myoblasts that migrate a long distance within the developing body plan from their origins in the distal hypaxial portions of the somites. The ventral/medial portions of the somites (sclerotome) become mesenchymal and later give rise to the cartilaginous and bony structures of the vertebra and associated ribs. Generation of the migratory trunk skeletal musculature is a multistep process, in which waves of myoblasts arise from originally multipotential precursors. The successive waves and the difference in fates between epaxial and adaxial somitic myogenesis are determined by different genetic regulatory programs. Jawless vertebrates such as lampreys lack paired appendages and also migratory precursors of adaxial skeletal muscle. Their muscular body walls nonetheless also derive from somites, which display patterns of gene expression that are in some ways unmistakably similar to those of gnathostome somites (Kusakabe et al., 2011). Thus per se the generation of body muscle from dorsally arising somites is a vertebrate character, while the myogenic epaxial/adaxial specializations and other aspects of genetic programming with which we are now to be concerned are gnathostome characters.

Diverse myogenic and other cell types arising from amniote somites

While there are of course fundamental similarities, somitic diversification processes in *Xenopus* (Della Gaspera et al., 2012) and zebrafish (Hammond et al., 2007; Bryson-Richardson and Currie, 2008; Buckingham and Vincent, 2009) differ in detail from those of amniotes (chick and mouse), and the following is confined to the latter. In amniotes there are three successive phases of myogenesis: embryonic, fetal, and adult (for reviews, Biressi et al. (2007a), Buckingham and Vincent (2009), Buckingham and Rigby

(2014)). The initial myoblasts to differentiate arise from the dorsal/medial (epaxial) somite and these are the first to populate the underlying myotome where they form an anterior/posterior scaffold (Fig. 5.5(A2)). The remaining progenitors of all three phases of myogenesis derive from the dermomyotome, a multipotential external epithelial layer of cells which eventually give rise to endothelial, smooth muscle, dermis and brown fat, as well as skeletal muscle progenitors. The relative spatial disposition of dermomyotome, myotome and sclerotome in the e10.5 mouse embryo is shown in Fig. 5.5(A1) (Buckingham and Vincent, 2009). As shown here, the embryonic myotome is seeded with cells immigrant from the hypaxial lip of the dermomyotome. In the myotome, some cells complete their differentiation and fuse to form myotubes (Fig. 5.5(A2)), while others migrate to give rise to embryonic trunk muscle and embryonic appendage muscles (first phase, see Fig. 5.5(A3); Biressi et al., 2007a; Buckingham and Rigby, 2014). All these cells express *pax3* but not *pax7* (Fig. 5.5(A4); Buckingham and Vincent, 2009; Hutcheson et al., 2009). In the second phase, the central dermomyotome loses its epithelial form (Fig. 5.5(A2); Bryson-Richardson and Currie, 2008) and individual cells, expressing both *pax3* and *pax7*, become the highly proliferative precursors of the fetal skeletal muscle of both trunk and limbs (see Fig. 5.5(A1–A4)). Migratory *pax3-* and *pax7-* expressing cells support adult muscle development and produce the satellite cells required for later muscle growth and adult muscle regeneration after injury (third phase, see Fig. 5.5(A4); Relaix et al., 2004, 2005; Buckingham, 2006; Buckingham and Vincent, 2009).

As can also be seen in Fig. 5.5(A3), the migratory muscle progenitor cells emanating from the myotome are exposed to different signals from adjacent tissues. The epaxial muscles exclusively depend for specification on signals from the adjacent axial structures, namely Shh from the notochord and floor plate of the neural tube and Wnt1 from the neural tube. The hypaxial emmigrants are instead exposed to Fgf and Bmp4. In addition the somites at different anterior/posterior levels are intrinsically different in respect to *hox* gene expression as we see below. The complex head musculature, that is, jaw, extraocular, and facial muscles, have other origins (Buckingham and Vincent, 2009; Tzahor and Evans, 2011; Buckingham and Rigby, 2014). They derive from the mesodermal components of the anterior branchial arches and from prechordal plate mesoderm. Head muscle progenitors differ from the somitic trunk and appendage muscle progenitors in having different founding regulatory states, and different external patterning influences, such as the differentially developing neural crest (Rinon et al., 2007).

Mechanistic links in embryonic muscle specification

It is already clear that even within the somite, the regulatory networks causing myogenesis differ regionally. Thus the epaxial and hypaxial specification circuitry is not the same, though both end with expression of similar downstream drivers of myogenesis and of large sets of muscle cell effector genes (Bryson-Richardson and Currie, 2008). Later muscle subtype diversification is a matter that lies beyond the concerns of the present Section. As noted above the epaxial dermomyotome requires Shh and Wnt1 signaling, and these provide inputs into the *myf5* gene, a lynchpin of dermomyotome muscle specification. This gene has an immensely complex *cis*-regulatory array of regulatory modules stretched out over 140 kb, as we see below (Hadchouel et al., 2003; Bajard et al., 2006; Daubas and Buckingham, 2013; Relaix et al., 2013), including enhancers active in different regions of the somite at different times. Figure 5.5(B1) shows an epaxial subcircuit (Sato et al., 2010), the function of which is to promote expression of the muscle differentiation driver gene *myogenin* via *myf5*, in the epaxial portion of the somite giving rise to the initial myotomal muscle cells (see above). This subcircuit is supported by detailed *cis*-regulatory studies on all the regulatory genes it includes. We see here the primary importance of the gene which lies at the top of the hierarchy, *pax3*. In this domain *pax3* affects myogenic specification via activation of the regulatory gene *dmrt2*, which in turn provides an essential direct input into the regulatory cascade at a particular CRM of the *myf5* gene. Interference with the *dmrt2* subcircuit ultimately affects epaxial back muscles and other myotome products. Figure 5.5(B2) (Relaix et al., 2013) summarizes the different circuitry downstream of *pax3* which operates in muscle progenitors at the distal end of the dermomyotome which are destined to

Figure 5.5

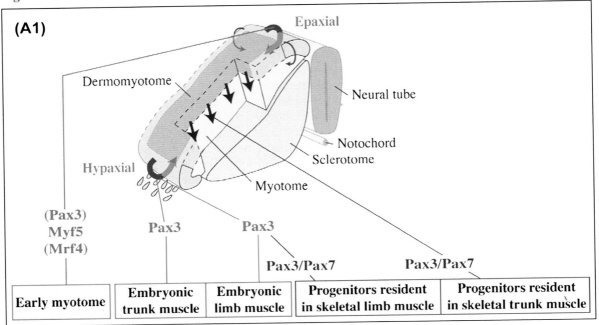

Figure 5.5 Regulatory states and transactions in generation of somitic skeletal muscle precursors and other derivatives. (A1), Cutaway diagram illustrating the general organization of the somite and the origins of its developmental skeletal muscle derivatives in mouse *(from Buckingham and Vincent (2009))*. All somitic progenitors of skeletal muscle have expressed *pax3*. As indicated in red, the early myotome, and early embryonic muscles, derive from cells that express only *pax3*, while the definitive migratory progenitors of limb and trunk skeletal muscles arising in the central dermomyotome (dark green region) express both *pax3* and *pax* 7 (blue arrows), and some *pax3* skeletal muscle precursors activate *pax7* as they enter the limbs (indicated in blue). (A2), Phases of somitic myogenesis in the chick *(from Bryson-Richardson and Currie (2008))*. Top, the earliest "pioneer" myofibrils (blue rods) form from myoblasts arising from somitic tissue (red-brown) ventral to the future dermomyotome (tan). Cells from the four epithelial somite lips migrate inward to form the primary myotome (brown rods), in the first wave of myogenesis, center diagram. Below, the dermomyotome epithelium transforms into mesenchymal myoblasts which produce the second wave of myogenesis and generate replicative migratory myoblasts (tan rods and individual cells). (A3), Distinct fates of epaxial and hypaxial myoblasts in mouse *(from Biressi et al. (2007a))*. The dermomyotome (DM) cells producing the myotome initially express *pax3* and *pax7* but not yet the muscle regulatory factors (MRFs). The myotome consists of these cells, MRF expressing myoblasts (red) derived from the overlying somitic tissue as above, and myotomal muscle cells (green). Epaxial somitic derivatives dorsal and medial to the dashed line produce the epaxial back muscles, under the influence of Shh and Wnt signals (green arrows). Hypaxial dermomyotome myoblasts and *pax3* progenitors, not expressing MRFS until they reach the limb domain, migrate ventrally and laterally to give rise to intracostal and body wall trunk skeletal muscles and limb skeletal muscles. These muscle progenitor cells are exposed to signals indicated in red (red arrow), emanating from the dorsal ectoderm (DE); SF, scatter factor. Notochord (NC) and neural tube (NT) are indicated. (A4), Cell types derived during fetal myogenesis from migrating, pluripotential *pax3* and *pax7* expressing hypaxial progenitors *(from Hutcheson et al. (2009))*. Cell fate is affected differentially by β-catenin (green). (B), Regulatory circuitry in trunk somites. (B1), Inputs into the epaxial myotomal enhancer of the *myf5* gene and their immediately surrounding regulatory network *(from Sato et al. (2010))*. Inputs

Figure 5.5

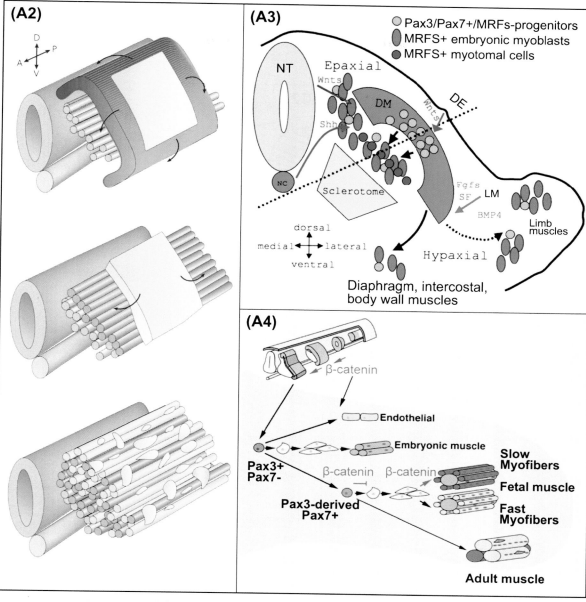

shown in boldface have been established as direct in *cis*-regulatory experiments; thus Myf5 directly activates the *myogenin* muscle regulatory gene, and the *myf5* gene is a direct target of Dmrt2 activation, which is itself directly dependent on Pax3. Downstream targets, the *laminin* and *a6β1integrin* genes, are required for construction of the basal lamina of the myotome. (B2), Circuitry immediately upstream of *MyoD* activation in hypaxial myogenic precursors expressing *pax3 (from Relaix et al. (2013))*. A core feedback subcircuit between *pax3* and *six1/2/4* drives this regulatory process. (C), Muscle vs. vascular (endothelial and smooth muscle) cell fate choice in somitic cells, mediated by mutual repression between *foxc* and *pax3* genes *(from Lagha et al. (2009))*. (C1), Outline of exclusion circuitry. (C2), Early somite in which ventrally located somitic cells expressing *foxc2* contribute to endothelial wall of aorta. Key (colored boxes) refers to both C2 and C3. (C3), Later somite, showing central dermomyotomal region

Figure 5.5

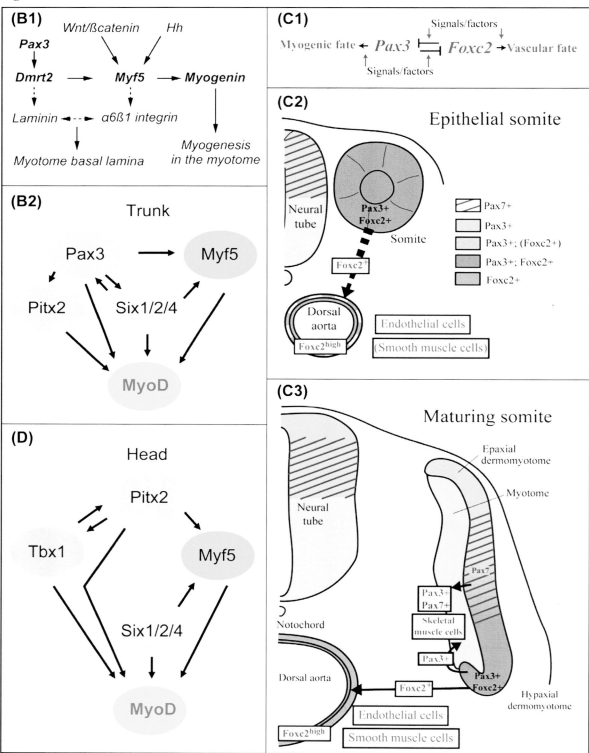

generating *pax3/pax7* expressing myotomal muscle cells (black arrow). Hypaxial cells predominantly expressing *foxc* mainly generate aortal smooth muscle, while cells predominantly expressing *pax3* contribute to myotomal skeletal muscle pathway (black arrow). (D), Additional encoded regulatory circuitry upstream of MyoD activation in the head *(from Relaix et al. (2013))*. Here *pitx2* rather than *pax3* is at the top of the hierarchy, but the circuit architecture is fundamentally similar to that in (B2) (see text).

give rise to limb skeletal muscle. Here *six1/4* (and *eya*) genes are essential, particularly in the hypaxial domain and in the myogenic cells emigrating to the limbs therefrom (Giordani et al., 2007; Daubas and Buckingham, 2013). The interesting structure of this circuit, again heavily bolstered by direct *cis*-regulatory observations on transgenic mouse embryos, places the *six1/2/4* genes in a feedback lock with *pax3*, the result of which is to set up feed-forward positive *cis*-regulation of *myf5* (*pax3* > *myf5*; *pax3* > *six* genes; *six* genes > *myf5*), and thence a direct input of *myf5* into *MyoD*, as well as a second feed-forward activation of *MyoD* (*pax3* > *MyoD*; *pax3* > *six* genes; *six* genes > *MyoD*). Still a third feed-forward subcircuit directly activates *MyoD* via another *pax3* target gene, *pitx2* (L'honoré et al., 2010). The final objective is obviously to get *MyoD* running without failure or reversibility.

The *pax3* gene is multifunctional in somite myogenic specification (Buckingham and Rigby, 2014). Not only does *pax3* participate in determining activation of the myogenic bHLH regulators *myf5*, *mrf4* (Bryson-Richardson and Currie, 2008), *MyoD*, and *myogenin*, but it also directly targets downstream effector genes expressed during skeletal muscle specification such as the transmembrane protein-coding gene *itm2a* (Lagha et al., 2013). In addition, *pax3* acts directly as a fate exclusion device, transcriptionally repressing an alternative fate that dermomyotome cells can adopt, which is to become specified as vascular smooth muscle rather than skeletal muscle. Genetic and experimental perturbation experiments reveal a reciprocal repression between *pax3* and *foxc2*, a driver of endothelial fate (Lagha et al., 2009); the ratio between the activities of these genes determines the choice between skeletal muscle and vascular cell fate (Fig. 5.5(C1)). Early in somite development, multipotent progenitor cells express both *pax3* and *foxc2* at low levels; some of these cells turn off *pax3*, exclusively express *foxc2* and migrate to the aorta where they contribute to the vascular wall (Fig. 5.5(C2)). Later, as the central dermomyotome contributes to the myotome *pax3/pax7*-expressing cells which will give rise to skeletal muscle, hypaxial cells expressing *foxc2* that have turned off *pax3* generate aortal smooth muscle, while reciprocally other hypaxial cells that retain *pax3* expression and have turned off *foxc2* become additional myotomal embryonic skeletal muscle (Fig. 5.5(C3)).

In the head, where (except for extraocular muscles) the craniofacial musculature derives from the head mesoderm, *pax3* plays no role in specification of myogenic precursors. Instead, *pitx2*, under other regulatory control, is at the top of the hierarchy (Sambasivan et al., 2009, 2011). The cranial mesoderm migrates to the arches from the cranial paraxial mesoderm, and expresses the *tbx1* gene (as does also the developing heart). In the craniofacial muscle precursors deriving from the first pharyngeal arch, *pitx2* and *tbx1* operate in a positive feedback relation at the top of the hierarchy. Again, *MyoD* is the end target of the specification network by virtue of direct inputs from the *myf5* and *six1/2/4* genes, as shown in Fig. 5.5(D) (Relaix et al., 2013). Genetic evidence indicates that additional regulatory factors operate downstream of Pitx2 and Tbx1 and contribute to the expression of *MyoD* and *myf5* in craniofacial myogenesis, including Tcf21 and Msc (Moncaut et al., 2012). Comparing the networks operating specification of head versus trunk muscle precursors, we see that the wiring is different in terms of gene identity (Fig. 5.5(B2,D)). Yet it is not so different in terms of circuit structure: in both, two regulatory genes locked into a positive feedback loop are used to drive *myf5* and *MyoD* directly.

A common theme in cell type specification wiring is the engagement of given genes not only in the control circuits that establish the definitive regulatory state of the cell type, but also in driver control circuits that cause effector gene expression in that cell type (Biressi et al., 2007b). A specific characterized example is the role of *pitx2* in the migrating myotomal muscle precursors that give rise to limb skeletal muscle. Downstream gene functions affected by *pitx2* include migration, cell morphology, and cell adhesion, and these functions are mediated by both positive and negative (direct and indirect) inputs (Campbell et al., 2012; Eng et al., 2012).

Gene interactions in the sclerotome

The sclerotome (Fig. 5.5(A1)) gives rise to multiple structures, each requiring generation of certain cell types. Depending on anterior/posterior position in the overall body axis, the identity of the vertebral bony structures (vertebrae and ribs) deriving ultimately from the sclerotome is controlled downstream of *hox* gene inputs, the subject of a detailed discussion in Chapter 7. But each sclerotome must also produce the tendons of the epaxial muscles, cartilage, vertebral masses, and intervertebral discs; these require coordinated differentiation of anterior and posterior domains within each somite as we discuss below. While comprehensive circuitry leading to any of the cell type specification outcomes is yet to be formulated, we can begin to get a glimpse of what is an overarching feature of this process of fate diversification: programmed, regulatory exclusion of alternative cell fates.

The initial fate choice is between the myogenic dermomyotome/myotome and the sclerotome. This depends on axial Shh signaling, which impinges most strongly on the ventral/medial portions of the somite, i.e., the future sclerotome. Experimental manipulations demonstrate that where the levels of Shh are highest in somitic tissue, in the sclerotome, *nkx3.2* is activated, and that where these levels are lowest, in the dermomyotome region, the pro-myogenic *pax3* gene is activated. As indicated by genetic perturbations, this initial bias is converted into a Boolean exclusion by the mutual repression of *pax3* by *nkx3.2*, and of *nkx3.2* by *pax3* (Cairns et al., 2008). It had long been thought that the source of the essential initial Shh signal was the notochord, but a recent experiment remarkably shows that in a mouse mutant in which the notochord degenerates right after floor plate specification, the sclerotome and vertebral development proceed normally over most of the body axis (Ando et al., 2011). Thus Shh signaling from the floor plate (the ventral most region of the neural tube) alone suffices.

A second exclusion sets up the distinction between the anterior and posterior domains of the sclerotome formed in each somite. The vertebral bodies from which the thoracic spine forms are generated by the fusion of the posterior half of a given sclerotome with the anterior half of the next sclerotome. The cartilaginous discs separating the vertebrae are also formed from the posterior domains of the somites. Thus the posterior portion of each sclerotomal body gives rise to the anterior portion of a vertebra plus the intervertebral disc. In cervical somites the posterior sclerotome produces the intervertebral discs and also both anterior- and posterior-most portions of the vertebral bodies (Takahashi et al., 2013). Posterior sclerotome domains are marked by, and their identity depends upon, expression of *unc4.1*. The key sclerotome gene *meox1*, which executes several roles in sclerotome regionalization apparently provides a direct input into *unc4.1* (Skuntz et al., 2009). A further regulatory relationship (direct or indirect) is that *unc4.1* represses the anterior sclerotome regulatory gene *tbx18*, which in turn represses *unc4.1*. However that asymmetry is initiated, since *unc4.1* is preferentially expressed in the posterior portion of the sclerotome, this circuitry ensures that *meox1* and *unc4.1* will be stably expressed posteriorly, while the reverse is true in the anterior sclerotome where *tbx18* is strongly expressed.

Genetic and other studies have shown that *bapx1* is upstream of the cartilage differentiation driver *sox9* in the major mesenchymal portion of the sclerotome, and there *meox1* (also possibly *meox2*) provides a direct *cis*-regulatory input to *bapx1* (Rodrigo et al., 2004). Additional necessary inputs to sclerotomal *bapx1* are provided by *pax1* and *pax9*. These genes are expressed because *unc4.1* may be a driver of *pax9* and Shh, a driver of *pax1* (Takahashi et al., 2007). But the mesenchymal sclerotome also generates a separate domain located immediately beneath the myotome which is the source of the tendons that ultimately link the back muscles between the vertebrae to the lateral vertebral projections (Brent et al., 2003).

Myotomal expression of Fgf4/6 indirectly causes expression of the regulatory gene *scleraxis* in tendon precursor cells (Brent and Tabin, 2004). *Scleraxis* is a dominant tendon specification gene; forced expression of *scleraxis* in mesodermal mesenchyme cells suffices to convert them to tendon progenitors (Alberton et al., 2012). Once again, furthermore, a negative fate exclusion function is encoded: *sox9* expression in the precartilage sclerotomal mesenchyme leads to *sox5/6* expression, but this leads in turn to exclusion of *scleraxis* expression in these cells, confining the tendon-forming cells to a narrow strip abutting the myotome (Brent et al., 2005).

These last sections touch on mechanisms that not so long ago were utterly out of reach. We have dealt here with the genetic machines that underlie specification of multiple cell types, as they emerge from multipotential precursors within the complex developmental events of mid-late vertebrate embryogenesis. There are several take-home lessons, of different conceptual focal lengths. The most general is that despite incomplete knowledge in many areas, it is obvious that the causal explanation for cell type specification lies in the gene interactions encoded in the genomic control sequences of regulatory genes. A second generality that begins to emerge is that we see how signals (the same ones over and over) are used just to kick off a spatial asymmetry in regulatory gene expression, and how this is immediately converted by dominant exclusionary repression and heavily wired cross-regulatory interactions into Boolean cell type progenitor states.

3.4 Cell type specification in multipotential placode and neural crest precursors

Origins: placode versus neural crest

Neural crest and placodes are vertebrate apomorphies, that is, they are characters of all vertebrate classes but of no other chordates (nor other deuterostomes). In jawed vertebrates, neural crest gives rise to many cell types, including neurons and glia of sensory, visceral, and autonomic nervous systems, adrenal secretory cells, melanocytes, smooth muscle cells, portions of the heart outflow apparatus, cartilage, and bone-forming cells that build the structures of the face. Placodes give rise to several discrete anatomical structures, most of which have sensory functions. They include the adenohypophysis (anterior pituitary), the bilateral olfactory, otic, and lens placodes, and the trigeminal and epibranchial ganglia (the latter respectively generate neuronal complexes that will bring sensory information from the face, the viscera, and the oral region). The neural crest contributes all of the glia of these ganglia as well as sensory neurons of the trigeminal ganglion. Thus both neural crest and placodes give rise to cranial sensory neurons. Both derive from a multipotent band of neuroectodermal cells that at gastrula stage abuts the future neural plate, and is itself flanked by nonneuronal epidermis, as in the chick embryo diagrams in Fig. 5.6(A1) (Grocott et al., 2012). Separate regulatory states segregate within this band that produce respectively the neural crest and the placodes. The responsible regulatory interactions can be seen at the bottom of the developmental BioTapestry GRN model in Fig. 5.6(A1). Here we see the preplacodal region in lavender and the neural crest precursors in green. In the late gastrula, precursors of neural crest express a GRN that includes *pax3*, *pax7*, and *msx1* among many other interacting regulatory genes. The preplacodal GRN includes *six1*, *eya*, and *six4*. It is at this point unclear whether these two fates derive from common progenitor cells, or whether their separate progenitors are intermingled, as the gross spatial expression of genes later specific to either neural crest or placodal precursors largely overlaps (Grocott et al., 2012). The segregation process involves complementary effects of three signals, as indicated in the BioTapestry GRN model of Fig. 5.6(A1). The mesoderm underlying the more lateral regions produces Fgfs as well as antagonists to the Bmp and Wnt signals deriving from flanking epidermis and other nearby tissues. According to this network model, in placodal precursors, Fgf signaling promïtes *six1* and *eya1/2* gene expression, while the interference with Wnt signaling allows the expression of

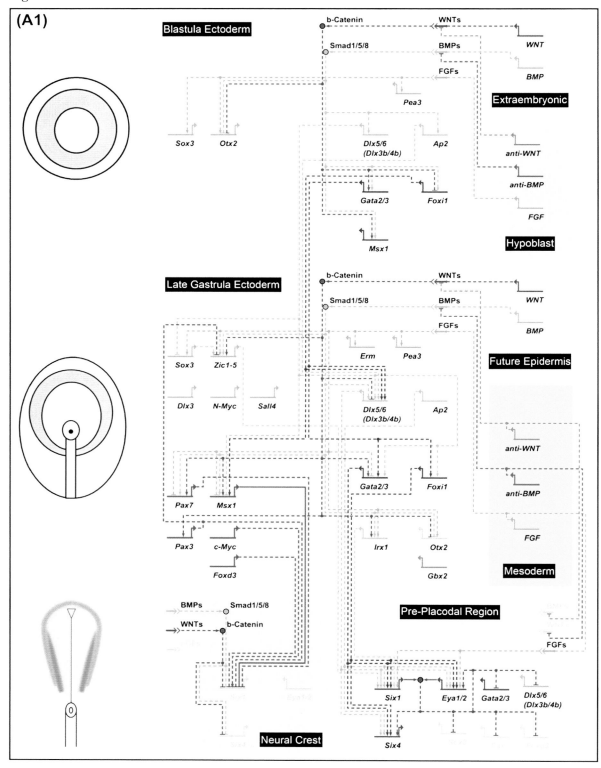

Figure 5.6

Figure 5.6

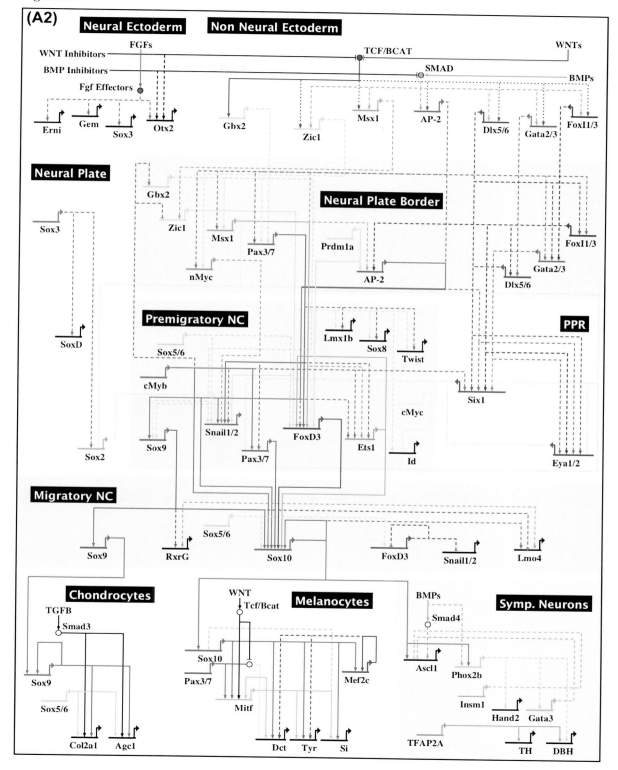

Figure 5.6

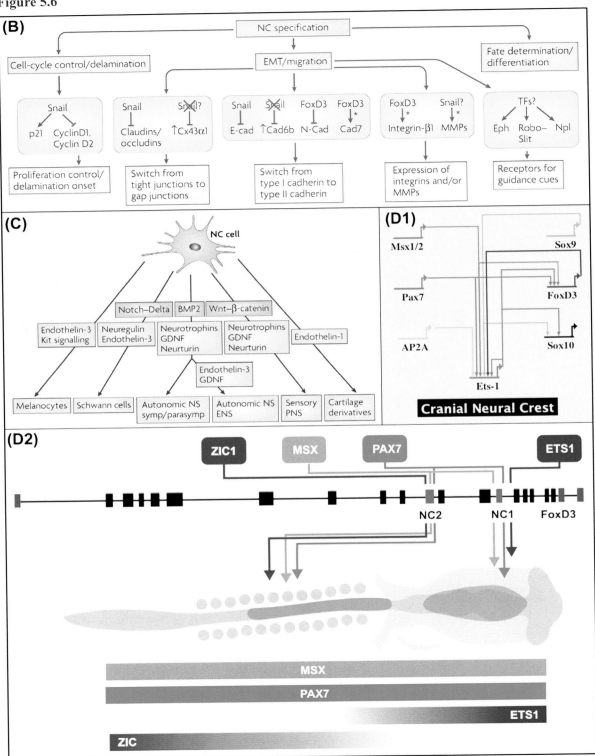

Figure 5.6

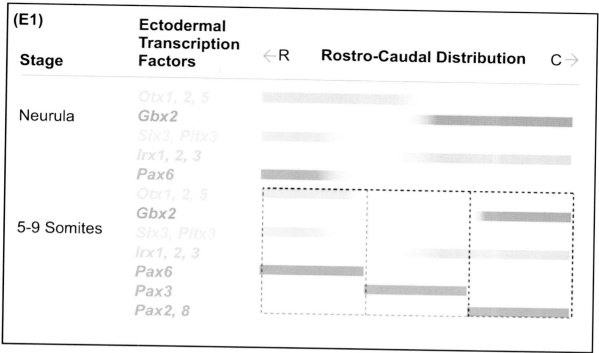

Figure 5.6 Gene regulatory pathways in neural crest and placode formation. (A1), GRN model for separation of neural crest and preplacodal territories in the chick embryo *(from Grocott et al. (2012))*. Embryonic stages at each GRN level are indicated in drawings at left; bottom, eventual neural crest territory green, and preplacodal region lavender. Signaling inputs and tissue sources at each stage are indicated at right. The middle panel shows gene interactions in border region separating neural plate and nonneural ectoderm containing both placodal and neural crest precursors. Dashed lines denote evidence based on perturbations by knockdown or overexpression; data compiled from literature. (A2), BioTapestry GRN model for developing neural crest (kindly provided by Marcos Simões-Costa). Interactions in both trunk and cranial neural crest are incorporated (Betancur et al., 2010a; Betancur et al., 2014; and references in text). The network includes linkages from various vertebrates though most data are from chick. Developmental stage progresses from top to bottom and spatial domains are indicated in labels: NC, neural crest; PPR, preplacodal region; Symp, sympathetic noradrenergic neurons. Terminal differentiation GRN models are shown at the bottom of the model, downstream of the definitive neural crest *sox9* and *sox10* genes. GRN model for specification of chondrocytes, in which *sox9* directly activates collagen (*col2a1*) and proteoglycan (*agc1*) genes; and GRN model for specification of melanocytes, of which the key differentiation effector gene driver is *mitf*, are shown at the bottom of the figure *(from Betancur et al. (2010a))* At right is GRN model for specification of noradrenergic neurons in which *sox10* induces expression of *ascl1* and *phox2b*, the latter of which is a required driver of the effector genes of this neuronal cell type *(from Hendershot et al. (2008))*. (B), Cell biology functions downstream of neural crest specification *(from Sauka-Spengler and Bronner-Fraser (2008))*. The changes in cell behavior required for neural crest delamination and subsequent directed migration are listed in gray boxes at the bottom of diagram. Above these are shown specific effector genes that must be either repressed or activated for these respective changes to take place; MMP, matrix metalloprotease; cad, cadherin; red X indicates result of developmental downregulation of *snail*. In each example shown, except for the control of Eph and

Figure 5.6

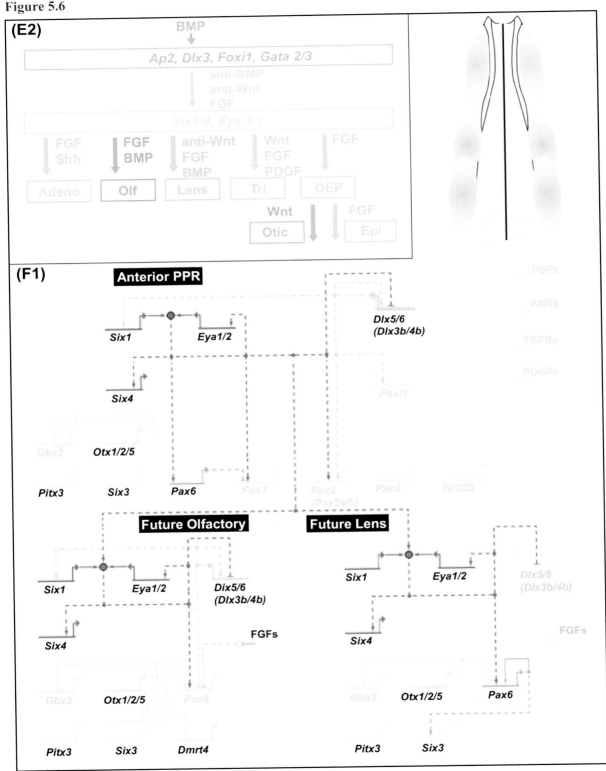

Figure 5.6

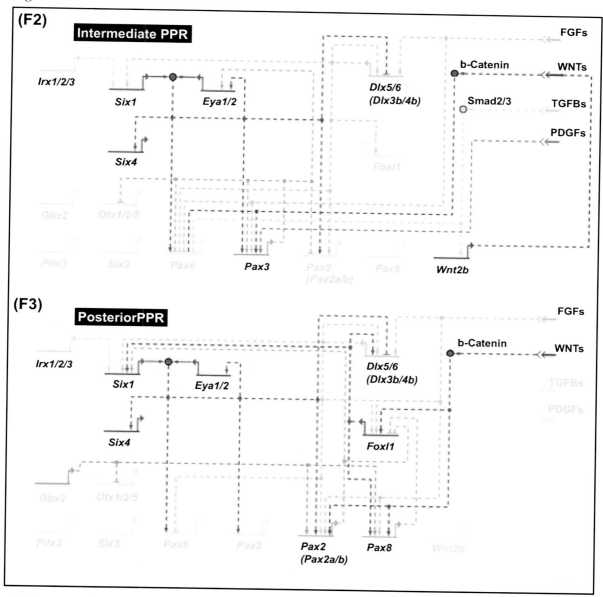

Npl expression where the drivers are unknown, the "neural crest specifier genes" *snail* and *foxd3* provide positive or negative inputs to these effector genes, as indicated; that is, neural crest regulatory specification directly causes the installation of canonical neural crest cell behavior. (C), Signals and neural crest products *(from Sauka-Spengler and Bronner-Fraser (2008))*. Morphological components of the gnathostome body that derive from neural crest are indicated at bottom (green boxes). As neural crest cells migrate they encounter many diverse, locally presented signals (blue boxes) that directly impinge on their ultimate fates, following prior exposure to the more commonly presented Wnt, Bmp, and Delta/Notch signaling. (D), Inputs required for expression of key neural crest regulators. (D1), Encoded circuitry upstream of genes essential for neural crest cell biology, *viz. foxd3* and *ets1* and for differentiation functions, here *sox10 (from Barembaum and Bronner (2013))*. *Cis*-regulatory analyses

Figure 5.6

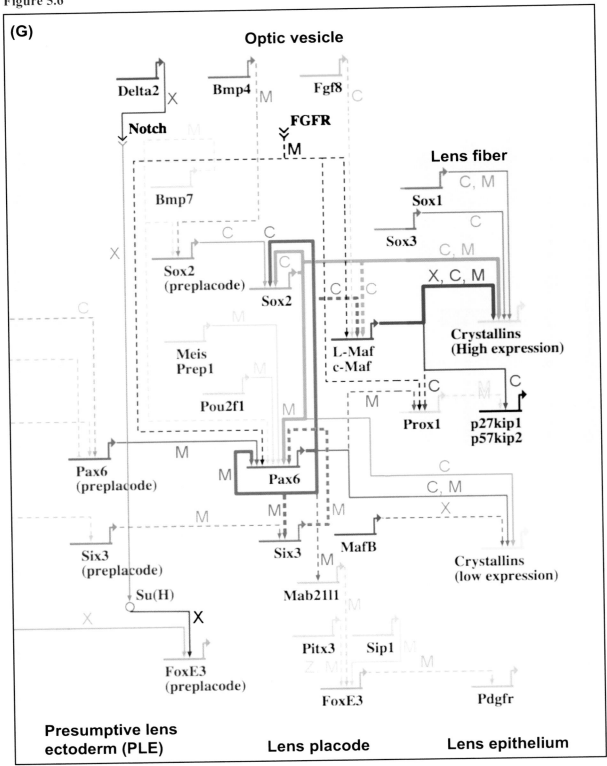

(G)

demonstrate that these are direct inputs. Potential inputs into *msx1/2* and *pax7* have been inferred from the results of perturbation experiments (Betancur et al., 2010a). (D2), Differential activation of the *foxd3* neural crest specifier gene in cranial and trunk neural crest, mediated respectively by *cis*-regulatory modules NC1 and NC2 *(from Simões-Costa et al. (2012))*. Both modules require inputs from the canonical neural crest regulators *pax7* and *msx1/2*. In the cranial neural crest the *ets1* gene is expressed due to the circuitry shown in (D1), and Ets1 is a required input for module NC1. In the trunk neural crest the *zic1* gene is expressed and provides a necessary input to module NC2. (E), Anterior/posterior specification of placodal regions in chick *(from Grocott et al. (2012))*. (E1), Selected components of regulatory state in anterior to posterior domains of preplacodal region; dashed line boxes indicate output of anterior, intermediate, and posterior GRNs shown in (F), by which stage regions are sharply distinct. (E2), Generation of the placodes and diverse signal inputs to each: Adeno, adenohypophysis; Olf, olfactory; Tri, trigeminal; OEP, otic-epibranchial precursor; Epi, epibranchial. (F), Anterior, Intermediate, and Posterior GRN models, and GRN models for lens placode and olfactory placode descendant from anterior domain *(from Grocott et al. (2012))*. At right top is a color-coded diagram indicating morphological locations of the anterior, intermediate, and posterior placodal regions. (F1), GRN model for anterior preplacodal region, above; GRN model for future olfactory and future lens placodes, which descend from the anterior region, below. (F2), GRN model for intermediate preplacodal region. (F3), GRN model for posterior preplacodal region. Dashed line symbolism as in (A1); much direct evidence beyond perturbation data is available for *six1/eya* (solid lines). (G), Encoded regulatory relationships in lens development *(from Ogino et al. (2012))*. Optic vesicle GRN model showing progression from pre-lens placode to lens placode to lens epithelium. Results are collated from literature from mouse (M), chick (C), zebrafish (Z), *Xenopus* (X). Bold solid lines highlight central *pax6/sox2* circuitry; solid lines indicate direct *cis*-regulatory evidence; dashed lines indicate linkages based on perturbation results only.

dlx5/6, another driver of *six1*. In neural crest precursors, on the other hand, the expression of *pax7* and *msx1* is activated by Bmp and Wnt signaling. *Pax3* is an additional target of Wnt signaling. These signals are broadly distributed and variable between species, but contribute to an asymmetry that biases the more lateral regions of the neural plate border toward the placodal fate and more medial regions toward the neural crest fate. Whatever the details of these mechanisms, and of the cellular segregation process, once established, a main function of the respective GRNs is mutual exclusion of neural crest and placodal fates by reciprocal regulatory gene repression: thus as can be seen at the bottom of Fig. 5.6(A1), Six1 represses the neural crest specification genes *pax7*, *foxd3*, and *sox3*, and conversely Pax7, Pax3, and Msx1 repress the *six1* gene. The key genes of each GRN are initially activated by factors expressed in the generalized ectoderm of the gastrula.

In zebrafish as well, the cells in the region separating neural and nonneural ectoderm give rise to both neural crest and preplacodal ectoderm, depending in some way on differential exposure to Wnt and Bmp. In zebrafish, a different mechanism of preplacodal specification results in early installation of a cross-regulatory GRN subcircuit laced with feedbacks and autoregulatory linkages. These form an autonomously maintained network of interactions between *ap2* genes, *foxi1*, and *gata3*, now operating independently of BMP signaling (these genes are also expressed as part of the preplacodal amniote GRN, Fig. 5.6(A1)). Once installed, this circuit locks down preplacodal ectoderm fate by activating a *dlx3* gene (Bhat et al., 2013).

The stepwise control mechanisms of neural crest specification and diversification

The neural crest, like the somitic skeletal muscle progenitors that we consider above, is a transient, migratory, embryonic cell population, that ultimately produces or contributes to many fixed aspects of the body plan. From the point of view of the encoded regulatory apparatus, specification of neural crest progenitor cells and their subsequent diversification and differentiation into various cell types is the output of a GRN (Sauka-Spengler and Bronner-Fraser, 2008; Betancur et al., 2010a; Bronner and LeDouarin, 2012; Simoes-Costa and Bronner, 2014). At least the initial GRN activities must to some extent be distinct in cranial, vagal, and trunk regions, since each ultimately produces or does not produce particular cell types. *Hox* genes have long been known to be expressed differentially along the anterior/posterior axis in migrating neural crest, and there is some evidence that they affect neural crest specification and differentiation (e.g., Gouti et al., 2011). While individual neural crest cells can give rise to multiple cell types, there are some limitations, such as the restriction to cranial neural crest of the potential to form bone and cartilage. The initial GRNs establish "neural plate border" regulatory states, in response to local signals, which in functional terms means expression of a set of genes encoding transcription factors required for the specification of neural crest fate per se. This regulatory state results not only in exclusion of other progenitor cell fates, such as placode or neural tube proper, but also in the activation of genes required to set up the definitive neural crest GRN. The latter, the "premigratory neural crest GRN" has two overall functions: to cause expression of the downstream effector genes required for transition to mesenchymal cell type and migratory behavior; and to activate, and by cross-regulation to maintain the expression of certain canonical neural crest regulatory genes. In local regions where distinct signals become available, these regulatory genes participate in diverse differentiation GRNs, of which they constitute the top of the hierarchy.

Major components of the neural plate border regulatory state that are required for definitive neural crest specification functions include the products of the genes *ap2*, *msx1/2*, *zic1*, and *pax3/7*, which respond indirectly to canonical Wnt signaling (Fig. 5.6(A2); Betancur et al., 2010a; Simoes-Costa and Bronner, 2014). Epistasis and overexpression experiments in *Xenopus* show that *ap2* expression is at the top of the hierarchy generated by this initial set of genes (de Crozé et al., 2011). Following activation of *ap2* these genes become engaged in a cross-regulatory network, ensuring their continued expression. Figure 5.6(A2) shows two positive feedback interactions in the neural plate border network, one between *ap2* and *msx1*, and the other between *ap2* and *foxi1/3*. This will insure the expression of a particularly important set of neural crest specification genes, namely *msx1* and *pax3/7*, and *msx1* in turn activates *zic1*. In the ensuing premigratory neural crest regulatory state, many essential neural crest regulators can be seen to be targets of Zic1, Msx1, and Pax3/7 as well as of Ap2, including *sox9*, *snail1/2*, *foxd3*, *lmx1b*, *sox8*, *twist*, and *ets1* (Bae et al., 2014; Plouhinec et al., 2014; Simoes-Costa and Bronner, 2014). Another study shows that uniquely among these genes, expression of *pax3* plus *zic1* alone suffices to re-specify uncommitted animal caps or even ventral endoderm of amphibian embryos to neural crest fate, demonstrating an essential function of the initial GRN circuitry (Milet et al., 2013).

Cis-regulatory and other evidence causally link several genes of the definitive neural crest GRN model to particular cell biology effector genes. These genes are required to mediate the transition to mesenchymal state and subsequent migratory behavior. Among the drivers of these morphogenetic effector genes is *snail*, which in pre-migratory neural crest appears to be activated downstream of some of the same above genes, viz., *ap2*, *msx1*, and *pax3/7* (Fig. 5.6(A2); Betancur et al., 2010a; Simoes-Costa and Bronner, 2014). *Snail* expression directly controls genes affecting cell adhesion, cell cycle control, delamination, etc., as summarized in Fig. 5.6(B) (Taneyhill et al., 2007; Sauka-Spengler and Bronner-Fraser, 2008). Regulatory inputs into cell biology functions are provided by other genes as well, such as *foxd3* and *zic* genes (Teslaa et al., 2013). From an informational point of view, however, the unique importance of the neural crest specification GRN is activation and cross-regulatory control of a small set of genes that will trigger diverse differentiation pathways, as we see below. These differential outcomes are controlled in part by an unusually complex set of local signaling environments (Fig. 5.6(C)), but depend specifically on the encoded wiring that determines how these signals will be interpreted. The individual cell fate GRNs operate as a set of AND logic switches in which canonical neural crest regulators operate together with locally activated signal response transcription factors to regulate the expression of differentiation drivers and to steer the neural crest progenitors to exclusive different fates. Some key regulatory genes of the neural crest specification GRNs that perform such functions in the ultimate differentiation GRNs are *sox10* and *sox9*, as we see below.

Current *cis*-regulatory evidence on the wiring of the neural crest specification GRN models is illuminating. In the cranial neural crest, the enhancer known to be responsible for *sox10* expression is driven by c-Myb, Ets1, and Sox9 (Betancur et al., 2010b), all of which are linked into the neural crest specification GRN (Fig. 5.6(A2); Betancur et al., 2010a). As Fig. 5.6(D1) shows (Barembaum and Bronner, 2013), the *ets1* input into *sox10* is itself locked on by a feedback loop with another gene of the same specification GRN, *foxd3*, while *ets1* also auto-activates. The key *ets1* gene also receives inputs from additional neural crest regulators, i.e., from *ap2*, *msx1/2*, *sox9*, and *pax7*. Furthermore, *foxd3* and *sox10* in common require inputs (direct or indirect) from another more upstream regulator, Elk3 (Rogers et al., 2013). It is interesting that the same *sox10* CRM functions in the otic placode where it utilizes a different set of paralogous regulators, viz. Sox8 instead of Sox9, and Pea3 instead of Ets1 (Betancur et al., 2011). An aspect additionally illuminated by this *sox10* enhancer is that it operates only in the cranial neural crest, while a different enhancer mediates *sox10* expression in the vagal neural crest, implying diversity in the neural crest regulatory states along the axis. In fact, the cranial neural crest driver of *sox10*, Ets1, is also responsible for the exclusively cranial operation of an enhancer of the *foxd3* gene. Here again a different enhancer causes *foxd3* to be expressed in the trunk neural crest (Simões-Costa et al., 2012). In Fig. 5.6(D2) we see that activity of the cranial enhancer of *foxd3* requires the AND input from anteriorly expressed *ets1* plus *msx* and *pax7* genes, while vagal and trunk *foxd3* enhancer requires the latter two inputs, plus that from a posteriorly expressed *zic* gene. In sum, the neural crest specification GRNs execute multiple roles: they operate suites of effector genes controlling migratory behavior; they mandate the distinct linkages that define cranial versus more posterior neural crest; and they ensure by cross-regulatory wiring the expression of genes utilized in the succeeding diversification of cell fates in their descendants. The GRN encodes the differential transcriptional responses to the regionally expressed signals, which as we see in Fig. 5.6(C) affect the various downstream cell types to which the cranial neural crest gives rise.

The manner in which the neural crest executive genes *foxd3*, *sox10*, *sox9,* and *pax3/7* are utilized, together with other regulatory genes, in cell type specification from neural crest progenitors is illustrated at the bottom of Fig. 5.6(A2). Here we see that melanocyte function depends on *mitf* expression, which directly controls production of the enzyme that synthesizes melanin (Dopachrome tautomerase; Dct). Transcription of *mitf* is positively regulated by Sox10 and Pax3/7 (reviewed by Betancur et al., 2010a; Greenhill et al., 2011). A chondrocyte specification circuit utilizes Sox9 to provide direct inputs into two collagen genes. A circuit in which Sox10 initiates a network controlling specification of noradrenergic sympathetic ganglion neurons is also shown in Fig. 5.6(A2) (Hendershot et al., 2008; Rohrer, 2011). Typically, this circuit is laced with feedbacks that result in expression of regulators upstream of noradrenergic neurotransmitter expression and neurogenesis, of which the key components are *hand2*, *phox2b*, and *gata3*. We see in each of these examples that regulatory components operating upstream in the specification GRN have direct roles in terminal differentiation as well. Other established interrelationships reveal additional functions: for example, *foxd3* acts to exclude melanocyte fate in neuronal pathways by repression of *mitf* (Pavan and Raible, 2012; Nitzan et al., 2013). Additional downstream circuitry for differentiation of other neural crest derivatives has been elucidated but lies beyond the scope of this chapter.

Placodal diversification

Six types of sensory placodes are differentially specified in amniote embryos with respect to anterior/posterior position. The diversification of the bilateral preplacodal region occurs stepwise, starting with its separation into three anterior-posterior regulatory state domains. The anterior (rostral) placodal domain will give rise to adenohypophysis, lens, and olfactory placodes; the intermediate region will give rise to the trigeminal placode; and the posterior (caudal) region will produce the otic and epibranchial placodes (Grocott et al., 2012; Saint-Jeannet and Moody, 2014). Elements of the same anterior/posterior neuroectoderm patterning system as we encountered in Chapter 4 in respect to early brain specification are utilized for the initial specification of placodal precursors; thus anterior *otx* and posterior *gbx2* expression, including their mutual transcriptional antagonism, are important for this process just as for neural plate regionalization. In the chicken, by 5–9 somite stage, three placodal regions are expressing distinct regulatory states (Fig. 5.6(E1)), although earlier than this the boundaries have not yet been established. By the time the transient anterior, intermediate, and posterior placodal GRNs are installed, they differ by expression of key regulatory genes: *six3*, *pitx3*, *pax6,* and *otx* genes are expressed anteriorly; *pax3* and *irx* genes are expressed in the intermediate domain; and *gbx2*, *irx* genes, *pax2*, and *pax8* are expressed in the posterior domain. It is clear that as in neural crest specification, a battery of regionally diverse signals impinges on the diversification process (Fig. 5.6(E2)). There is little direct gene interaction evidence as yet available, and the proposed GRNs in Fig. 5.6(F) indicate only some linkages that may be responsible for these territorial specifications. The various placodes will end up having extremely diverse functions, but at this stage their different identities are manifest only in their regionally differential regulatory gene expression.

In all the three placodal precursor regions indicated in Fig. 5.6(E1,E2), i.e., anterior, intermediate, and posterior, the expression of *six1* and *eya* is required (Grocott et al., 2012). As we recall, this regulatory complex is definitive for the placode precursor regulatory state and exclusive of neural crest specification (Fig. 5.6(A1)). In the anterior domain (Fig. 5.6(F1)) Six1:Eya activates *pax6*, which represses *pax3*; in the intermediate domain (Fig. 5.6(F2)) *pax3* activation is mediated by Six1:Eya and several signals, and Pax3 proceeds to exclude expression of *pax6* and *pax2*; and in the posterior domain (Fig. 5.6(F3)) Six1:Eya and Gbx2 turn on *pax2* and *pax8*, and these *pax* genes directly and indirectly feed back on *six1*. Together these network interactions specifically account for the anterior versus intermediate versus posterior *pax* gene expression patterns. So once again, even though these networks are incomplete, we see the now familiar design strategy in which feedbacks and reciprocal repressions are used to convert early axial (medial/lateral or anterior/posterior) transcriptional and signaling biases into precise Boolean regulatory state patterns. The same proceeds with the separation of olfactory and lens placode fates within the anterior domain: at the anterior end Dlx5/6 and Fgf appear to repress *pax6*, while in the future lens placode *pax6* and *six3* lock each other on in a mutual feedback, an interaction which occurs on a known *pax6* enhancer, while *dlx5/6* is repressed (bottom panels of Fig. 5.6(F1)).

Finally, in the lens placode per se, we can now trace the network wiring all the way to differentiation of the lens fibers. Figure 5.6(G) provides a composite from results obtained in mouse, chick, *Xenopus*, and zebrafish (Ogino et al., 2012). Species differences in earlier stage wiring are not inconsiderable; there is more than one way to arrive at the final Boolean placode regulatory states, and for example, *six1*, *six4*, *eya1,* and *eya2* are all dispensable for lens placode formation in mouse while essential in other vertebrates (reviewed in Ogino et al., 2012). The terminal differentiation of the lens placode into lens epithelium results in lens fibers which transcribe crystallin genes at a high rate, as discussed in detail above. In the presumptive lens ectoderm, Bmp4 signaling from the optic vesicle activates *sox2* expression. Pax6 drives the expression of *maf* genes, and the drivers of these crystallin genes, *pax6*, *sox2,* and *maf* genes are now all active. These regulatory genes engage in extensive feedback cross-regulation during the final stages of lens placode differentiation (highlighted in Fig. 5.6(G)), including *pax6* auto-regulation, a feedback between *pax6* and *sox2,* and a third feedback between *six3* and *pax6*. Though unavoidably sketchy, we glimpse here the genomic logic underlying a process that extends from the early stages of postgastrular preplacode specification to terminal differentiation of one of the several placodes.

4.   A Comment on Stem Cells in Postembryonic Life

The maintenance of some if not all tissues throughout adult life of many animal forms depends on the continuous differentiation of specific cell types from adult stem cells. These cells are specified to become tissue-specific progenitor cells which are not themselves terminally differentiated, but retain the potential to differentiate to one or several terminal fates. Stem cells are also different from all other cells of the tissue in that for the entire adult life span of the animal they retain the capacity to proliferate and give rise to additional stem cells. We have throughout this Chapter encountered already highly specified embryonic progenitor cells which proliferate in a state of incomplete specialization and then differentiate. In many ways, adult stem cell appear to execute similar functions as do embryonic progenitor cells, in that few steps separate them from terminal differentiation. However, the need of adult stem cells to remain undifferentiated and proliferative for very long periods requires them to be specified by particular genomically encoded regulatory programs, different from those utilized to form the embryonic progenitor cells of the equivalent cell types. Therefore, the routes by which particular terminally differentiated cell types are derived from their embryonic progenitor cells, or from adult stem cells in later life, cannot be assumed to be identical.

Adult tissue-specific stem cells indeed differ in multiple ways from embryonic progenitor cells. Embryonic progenitor cells display temporally and developmentally limited phases of intense replicative activity prior to their terminal differentiation, rather than retaining replicative ability for the whole of the animal's life span. Examples are myoblasts of somitic origin during their migratory phase, or migratory neural crest. When embryonic progenitor cells replicate they do not recreate homeostatic ratios of progenitor to differentiated cell type as do adult stem cells. Embryonic progenitor cells, as the examples we have traversed show, are always to some extent multipotent, but never totipotent, nor are they unipotent, as are some true adult tissue-specific stem cells. Some embryonic progenitor cells, particularly those of mesodermal origin, and neural crest, migrate from distant origination points into the locations where they will form components of given body parts, rather than being stored within the adult tissues where they will be required in the future. Embryonic cells must traverse a sequence of multiple, progressively diversifying regulatory states generated by a hierarchical sequence of encoded GRNs, before they arrive at terminal differentiation processes. They give rise to the cell types which constitute the developing body parts and also to the tissue-specific stem cells that are poised very near the termini of their ancestral developmental pathways, just before terminal differentiation. Adult tissue-specific stem cells must deploy special mechanisms to preserve their quiescence (e.g., Montarras et al., 2013). Embryos are not built from stem cells, the particular function of which is to renew differentiated cell types of adult tissues.

When called upon by the demands of tissue renewal or regeneration adult stem cells differentiate to replace the specialized cell types of that tissue. Depending on the tissue, this function may call for occasional replicative activity as in mammary gland stem cells during pregnancy (van Keymeulen et al., 2011), or for continuous replication as in intestinal crypt stem cells (Barker et al., 2012), or for replication in response to external incident e.g., injury, as in skeletal muscle satellite cells (Le Grand and Rudnicki, 2007). Stem cell replicative behavior thus varies enormously (He et al., 2009; Simons and Clevers, 2011; Fuchs and Chen, 2013). So also does mode of replication: the essential requirement for regeneration of tissue-specific stem cells which retain future replicative potential even in circumstances of stem cell differentiation can be met in two ways. The classical mode of asymmetric division in which one daughter cell differentiates and the other becomes a new stem cell, accomplishes this on an autonomous, per stem cell basis. But many types of stem cell replicate continuously and symmetrically without differentiating, while progeny that are pushed or crawl out of their niche into a new microenvironment differentiate, as in *Drosophila* and mammalian testicular stem cells (Spradling et al., 2011). These symmetrical stem cell division systems are ultimately balanced in respect to stem cell replication and differentiation rates, so that perpetual maintenance of the stem cell cohort is achieved on a population rather than a per cell level. Either mechanism accomplishes the definitive function of ensuring continuity and availability of the tissue-specific stem cell complement. Adult tissue-specific stem cells are authenticated in many vertebrate tissues and in other bilaterians as well. Thus, in addition to the prominent examples just mentioned, stem cells are found in mammalian brain (telencephalon) and more extensively in the central nervous systems of amphibians and teleosts (Grandel and Brand, 2013); in mammalian hair follicles (Hsu et al., 2011), sweat glands (Lu et al., 2012), epidermis and stomach (Barker et al., 2010), lung (Zacharek et al., 2011), and many other tissues (Snippert and Clevers, 2011). In *Drosophila*, true stem cells have been extensively described not only in testis (Spradling et al., 2011), but also in ovary (Xie, 2013) and adult midgut (Takashima et al., 2013). Stem cells are a widespread and possibly pan-bilaterian means by which tissues are renewed in the normal course of events throughout adult life. Though tissue-specific stem cells are indeed utilized in regenerative processes, many mechanisms of regeneration of whole body part following injury depend on other strategies than stem cell mobilization in situ, such as dramatic dedifferentiation of previously committed cells at the site of injury, and inward migration of cells of diverse developmental potential from external sites.

Most tissue-specific stem cells possess a very limited degree of multipotency. Some tissue-specific, self-renewing stem cells are in fact unipotent. A good example is provided by mammary gland stem cells. Although mammary stem cells can differentiate into multiple mammary cell types in artificial transplantation assays in vivo, in normal life the major mammary cell types, luminal cells, and myoepithelial cells, are each individually renewed during phases of mammary tissue expansion by dedicated unipotent stem cells (van Keymeulen et al., 2011). Similarly, lineage tracing of specific, genetically marked cells in adult sweat glands shows that these organs regenerate from unipotent stem cells (Lu et al., 2012). In other well-studied cases such as the intestinal crypt, the stem cells are narrowly multipotent, and their descendants form the functional epithelium of the crypt wall which consists of five cell types (Barker et al., 2010).

Pluripotent hematopoietic stem cells (HSCs) give rise to an astonishing array of erythroid, myeloid, and lymphoid cell types, that is, all blood cell types of the adult (for reviews, Graf and Enver (2009), Pimanda and Göttgens (2010), Mercer et al. (2011), Schütte et al. (2012)). These stem cells are specified as such during embryogenesis but they continue to produce additional stem cells as well as the diverse blood cell types during adult life. HSCs are unique in that they are the exclusive source of what is essentially a delocalized organ system distributed to most parts of the body. The process of their differentiation is hierarchical, stepwise, and enormously complex in that it gives rise to many differentiated cell types, often by multiple pathways. In this way HSCs continuously generate a whole adult tissue rather than serving as a source of particular cells in tissue repair or cell type renewal. Their differentiation pathway choices are controlled by genomic circuitries which respond to a variety of external cytokines and other signals that are presented according to physiological circumstance and location within the body. Thus the process of hematopoietic cell type specification is facultative and conditional.

In considering tissue-specific stem cell function in general, prominent mechanistic challenges include: the control levers utilized in late embryogenesis to prevent their differentiation and specify them to stem cell status; the means by which in each case their delayed differentiation is triggered when needed during adult life; how the equilibrium between renewal and differentiation of the stem cell population is achieved; and how the choice between alternative cell types is programmed in the case where the stem cells are multipotential. These are all important aspects particular to stem cell molecular biology and these issues are philosophically illuminating in respect to the biological meaning of postembryonic adult life, renewal, and regeneration.

Another type of stem cell has drawn much interest in recent years, both for medical and biological reasons. These are experimentally generated stem cells such as iPS cells and ES cell lines. Novel experimental approaches have been developed which artificially generate pluripotential cells that do not occur in the natural developmental process. Thus iPS cells (induced pluripotent stem cells) can be formed from diverse already differentiated cell types by forced expression of certain sets of ectopically expressed regulators (Takahashi and Yamanaka, 2006). The so-called ES cells ("embryonic stem cells") are a special case. As discussed briefly in Chapter 3, in placental mammals, embryonic cells must remain in a totipotent state for several days between fertilization and implantation (4.5–5 days in the mouse, ~9 days in human). In life, following implantation, the entire embryo is formed from the descendants of these initially totipotent cells (epiblast cells). Thus in life the developmental trajectory of the descendants of the epiblast cells have nothing in common with the behavior of true stem cells, since they become progressively specified and do not indefinitely proliferate in an undifferentiated state. Ways have been found to commit epiblast cells to permanent replicating tissue culture, i.e., "embryonic stem cell" cultures. These lines of cultured ES cells can be induced to differentiate into numerous and diverse cell types in vitro by alteration of their culture conditions and addition of growth factor "cocktails". iPS cells and ES cell lines have obvious and enormous medical potential as avenues toward repair of damaged human tissues, and ES cells are hugely useful as well for the generation of reengineered mice. However, these experimental systems are highly artificial and even if various aspects of terminal differentiation can be reproduced using them, their behavior does not represent the genomic programs utilized in development. The genomic program for embryonic developmental process produces a deeply hierarchical series of diversifying regulatory states in space, which occurs canonically in almost the same way in every individual of the species. In contrast, the variable forms of ES cell differentiation in culture dishes short cut this entire program to induce terminal differentiation by environmental manipulations. The basic significance of the fact that beating cardiomyocytes or neurons, etc., can differentiate in ES cell cultures is for us the implication that differentiation codes are modular, and in general can be elicited as unit functions. This is related to their deployment in multiple developmental contexts, as discussed in the following section of this Chapter.

5. Modular Call-Up of Given Specification Processes in Multiple Developmental Contexts

Certain cell types are formed in many different spatial and developmental contexts. Body parts and their subparts, as we have seen in detail in Chapter 4, arise by institution of distinct, geometrically bounded spatial regulatory states. But the regulatory states of all body parts utilizing the same cell type must eventually come to activate the terminal cell type specification programs to be installed in them, and thus an apparent logic problem arises. If regional regulatory states are in fact definitively distinct from one another, then how could the same particular terminal differentiation program be installed in many such states? For example consider all the body parts of a vertebrate animal that require differentiation of skeletal or smooth muscle cell types. As we have seen before, these different muscle cells all express the same muscle differentiation gene cassette including their driver genes, such as *myod* and *myf5*. The conceptual issue is the level at which the regulatory coordination of these genes is organized in the genome. Three possibilities can be envisioned. The first is that each muscle effector gene contains *cis*-regulatory target sites that enable it to recognize all the diverse regulatory states in which muscle differentiation is required. The second is that the drivers of the muscle differentiation genes contain the *cis*-regulatory recognition systems that enable them to be expressed downstream of the regulatory states where muscle is required, and that the effector genes need only to respond to the drivers. The third is that the coordination of the muscle gene battery occurs at some higher level in the network. Since effector genes of given differentiation gene batteries can number into the hundreds, while the number of drivers is usually small, it is immediately apparent that a much larger amount of *cis*-regulatory sequence would be required for the first solution compared to the second. But both solutions occur prevalently in the genomes of modern animals. Across the Bilateria, the regulatory strategy used to solve the problem of response to diverse regulatory states devolves from the modularity of the control systems determining spatial expression of regulatory genes or effector genes. Some of these genes characteristically possess amazingly complex arrays of CRMs, responding respectively to the distinct regulatory states of all of the localities where expression of the differentiation gene batteries will be needed. Control architecture in skeletal muscle development in the mouse indeed provides a powerful and clear example.

5.1   Functionally mosaic organization of upper level control genes

As we saw in Section 3.3 of this chapter, skeletal muscle arises from undifferentiated precursors of somitic and head mesoderm origin, and muscle differentiation occurs in several waves of migratory cells that populate many structures all over the vertebrate body (Fig. 5.5). Muscle-specific differentiation pathways are imposed by a terminal set of four regulatory genes, in which the myogenic bHLH genes *myf5*, *mrf4*, *myogenin,* and *myod*, control muscle differentiation. The *cis*-regulatory control systems of the *myf5* and *mrf4* genes, which in vertebrates are linked, reveal the logic that accounts for the enormous variety of contexts in which the muscle specification program is activated during embryonic development. This logic is summarized in Fig. 5.7(A) (Carvajal et al., 2008). Here we see a map of the locus, in which each colored block represents a sequence of noncoding DNA containing a separate enhancer that drives expression in the indicated region of the body. The map represents a compilation of results obtained in more than a decade of studies carried out in transgenic mice (Hadchouel et al., 2003; Carvajal et al., 2008). This system activates *myf5* and *mrf4* in multiple precise spatial and temporal phases of development, and its internal control requires diverse promoters and other sequences that regulate internal traffic between the enhancers and promoters (cf. Chapter 2). The main point is that the explanation for the expression of the muscle specification program is the mosaic of CRMs that read the local regulatory states in all the places where this program is to be executed. The developmental complexity of program deployment is thus directly encoded in these modular sequence elements. Several of these enhancers have been analyzed in detail. For example, an enhancer located over 100-kb upstream from the *myf5* gene is responsible for activation of the myogenic program in myotomal regions of the somite, and then in intercostal, abdominal, and upper limb muscles. This enhancer utilizes sites at which Pax3 and Six1/2/4 bind, and as we saw earlier (Fig. 5.5(B2)), these transcription factors are known regulators of the trunk somitic muscle progenitor GRN (Daubas and Buckingham, 2013). The effector genes expressed in muscle cells in all these various locations in the body can now all respond to the same simple collection of driver regulatory factors, which includes members of the myogenic bHLH factors listed above (plus Srf and Mef2; for review, Davidson (2006)).

An example that concerns control of bone morphogenesis is summarized in Fig. 5.7(B) (Guenther et al., 2008). Bones are shaped by local growth patterns during their development, and this function is modulated by Bmp5. Consequently, the *bmp5* gene is expressed in numerous specific locations in skeletal precursors. Figure 5.7(B) summarizes the remarkable result of a large scale *cis*-regulatory analysis of this gene, which showed that even adjacent regions of the same bone are recognized by separate enhancers, thereby programming distinct growth patterns in each of these regions. The analyses illustrated show how separate enhancers control the two sides of the rib, thus by differential growth controlling the rib curvature; and how the exact shapes of the cartilaginous nasal turbinates is similarly controlled. Thus a modular logic is used to program the shapes of bones in multiple spatial domains of the vertebrate body, the principle of which is the regional activation of independently operating enhancers recognizing diverse spatial regulatory states. This strategy is obviously most significant from an evolutionary point of view: characters such as bone shape are heritable and species-specific, and we can perceive how regulatory evolution at these individual enhancers could readily account for morphological divergence.

5.2   Functionally mosaic organization at the effector gene level

Among the many local regulatory programs that are called into play in multiple regions of the body during development, some of the most important deploy morphogenetic cell biology functions such as regional growth control. The same regulatory logic is used for their developmental control as is evinced in the *myf5/mrf4* locus, i.e., mosaic *cis*-regulatory systems which constitute a sequence code for all the specific times and places where the function will be required. A perfect example is provided by the *string* gene of *Drosophila.* This gene encodes a Cdc25 phosphatase that induces cell division by activating CyclinB

Figure 5.7

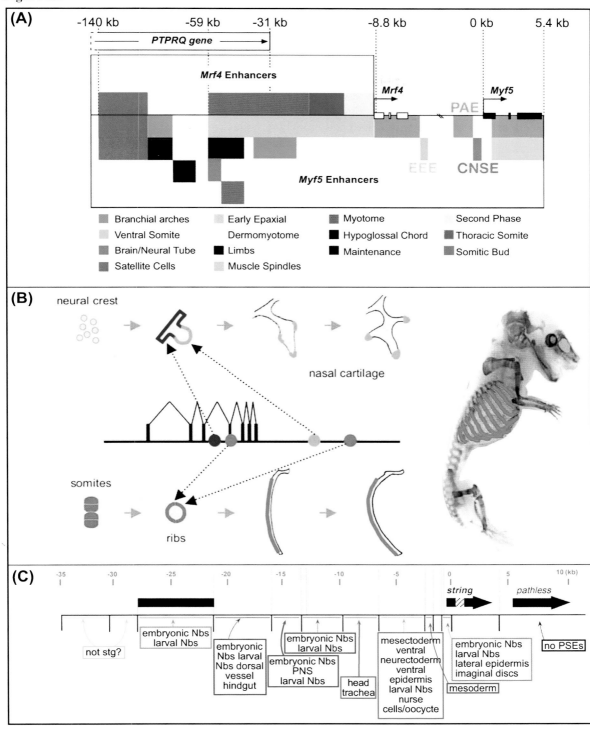

Figure 5.7 Multiple *cis*-regulatory modules account for deployment of given regulatory functions in the contexts of many diverse regulatory states. (A), Enhancers of the *myf5* and *mrf4* muscle regulatory genes *(from Carvajal et al. (2008))*. Enhancers of these genes are interdigitated and are scattered over 145 kb of genomic sequence; exons are indicated by black (*myf5*) and white (*mrf4*) boxes. Enhancers, which function at different stages of development in many regions of the body plan, are located within the color-coded regions. EEE, early epaxial enhancer; CNS enhancer; PAE, pharyngeal arch enhancer. (B), Diverse skeletogenic *cis*-regulatory modules of the *bmp5* gene *(from Guenther et al. (2008))*. Branching patterns in the developing nasal cartilages are controlled by *bmp5* expression driven by enhancers expressed at the base (blue) and branching nodes (orange) of the turbinates; see mouse embryo preparation at right. Other widely spaced enhancers (green and purple) control *bmp5* expression on opposite sides of the forming ribs. Genomic locations of these control modules are mapped in the center of the diagram. (C), *Cis*-regulatory modules of the *Drosophila string* gene *(from Lehman et al. (1999))*. The diagram maps approximate locations of enhancers over a 45-kb region of the genome (black rectangle indicated at position 0). These enhancers function to induce *string* expression and hence cell division to occur all over the embryo; nbs, neuroblasts.

(Cdc1). Its expression controls cell division all over the *Drosophila* embryo. Again, the mode of control accounting for this widespread expression is an enormous array of highly specific CRMs spread out over many kb of DNA sequence surrounding the *string* gene (Lehman et al., 1999). These account for expression in particular cell lineages at given times, in particular sets of neuroblasts, and in specific cell types (Fig. 5.7(C)). This example is akin to solution one above, in that *string* is an effector gene which by itself provides a replication checkpoint function. The diversity of spatial and temporal conditions by which it is activated is directly controlled in its own *cis*-regulatory system.

Another example of this mode of control is to be found in the analyses of an effector gene that encodes a vesicular transporter for glutamate, which is expressed in many different glutaminergic neurons in *C. elegans* (Serrano-Saiz et al., 2013). This gene is transcribed in many particular neurons widely distributed in the head and body of the worm, under control of distinct CRMs capable of reading their individual regulatory states. These modules recognize different particular combinations of 11 different homeodomain transcription factors. Thus again, the biological complexity of the organism is represented in *cis*-regulatory modular complexity, here at the effector gene level. Examples of this regulatory strategy at both this and upper control levels now abound.

Thus expression of effector genes in different developmental contexts can be seen to be controlled at both upstream and downstream levels, depending on the system. The strategy in which a small group of driver genes is expressed in many different developmental contexts due to their *cis*-regulatory modularity, thus allowing simple and monotonous effector gene control systems, would appear a particularly useful solution for large effector gene batteries. This is a simple and elegant control strategy from two points of view. First, it is maximally parsimonious in expenditure of regulatory sequence information, substituting hierarchy for redundancy. That is, instead of replicating local regulatory target site combinations in every effector gene, the mosaic response apparatus is encoded only in the few driver genes. Secondly, the genomic design by which effector gene batteries can be deployed is an almost infinitely expandable (i.e., evolvable) mechanistic solution: when a new spatial domain arises during the evolution of body part development, it is, as we have seen in Chapter 4, defined by institution of a new regional regulatory state. Effector gene cassettes and differentiation gene batteries can be deployed in any such region by the acquisition in the driver genes of dedicated enhancers that read these regional developmental regulatory states. We take up again in Chapter 7 the following fascinating thought: the arrays of CRMs in differentiation driver genes constitute an evolutionary genomic sequence record. Thus the fundamental evolutionary and developmental mosaicism of the organism is physically encoded in the *cis*-regulatory structure of its genome.

REFERENCES

Alberton, P., Popov, C., Prägert, M., Kohler, J., Shukunami, C., Schieker, M., Docheva, D., 2012. Conversion of human bone marrow-derived mesenchymal stem cells into tendon progenitor cells by ectopic expression of scleraxis. Stem Cells Dev. 21, 846–858.

Ando, T., Semba, K., Suda, H., Sei, A., Mizuta, H., Araki, M., Abe, K., Imai, K., Nakagata, N., Araki, K., Yamamura, K., 2011. The floor plate is sufficient for development of the sclerotome and spine without the notochord. Mech. Dev. 128, 129–140.

Bae, C.J., Park, B.Y., Lee, Y.H., Tobias, J.W., Hong, C.S., Saint-Jeannet, J.P., 2014. Identification of Pax3 and Zic1 targets in the developing neural crest. Dev. Biol. 386, 473–483.

Bajard, L., Relaix, F., Lagha, M., Rocancourt, D., Daubas, P., Buckingham, M.E., 2006. A novel genetic hierarchy functions during hypaxial myogenesis: pax3 directly activates Myf5 in muscle progenitor cells in the limb. Genes Dev. 20, 2450–2464.

Barembaum, M., Bronner, M.E., 2013. Identification and dissection of a key enhancer mediating cranial neural crest specific expression of transcription factor, Ets-1. Dev. Biol. 382, 567–575.

Barker, N., Bartfeld, S., Clevers, H., 2010. Tissue-resident adult stem cell populations of rapidly self-renewing organs. Cell Stem Cell 7, 656–670.

Barker, N., van Oudenaarden, A., Clevers, H., 2012. Identifying the stem cell of the intestinal crypt: strategies and pitfalls. Cell Stem Cell 11, 452–460.

Baron, M.H., 2013. Concise review: early embryonic erythropoiesis: not so primitive after all. Stem Cells 31, 849–856.

Baron, M.H., Isern, J., Fraser, S.T., 2012. The embryonic origins of erythropoiesis in mammals. Blood 119, 4828–4837.

Betancur, P., Bronner-Fraser, M., Sauka-Spengler, T., 2010a. Assembling neural crest regulatory circuits into a gene regulatory network. Annu. Rev. Cell Dev. Biol. 26, 581–603.

Betancur, P., Bronner-Fraser, M., Sauka-Spengler, T., 2010b. Genomic code for Sox10 activation reveals a key regulatory enhancer for cranial neural crest. Proc. Natl. Acad. Sci. U.S.A. 107, 3570–3575.

Betancur, P., Sauka-Spengler, T., Bronner, M., 2011. A Sox10 enhancer element common to the otic placode and neural crest is activated by tissue-specific paralogs. Development 138, 3689–3698.

Betancur, P., Simões-Costa, M., Sauka-Spengler, T., Bronner, M.E., 2014. Expression and function of transcription factor cMyb during cranial neural crest development. Mech. Dev. 132, 38–43.

Bhat, N., Kwon, H.J., Riley, B.B., 2013. A gene network that coordinates preplacodal competence and neural crest specification in zebrafish. Dev. Biol. 373, 107–117.

Biressi, S., Molinaro, M., Cossu, G., 2007a. Cellular heterogeneity during vertebrate skeletal muscle development. Dev. Biol. 308, 281–293.

Biressi, S., Tagliafico, E., Lamorte, G., Monteverde, S., Tenedini, E., Roncaglia, E., Ferrari, S., Cusella-De Angelis, M.G., Tajbakhsh, S., Cossu, G., 2007b. Intrinsic phenotypic diversity of embryonic and fetal myoblasts is revealed by genome-wide gene expression analysis on purified cells. Dev. Biol. 304, 633–651.

Blanco, J., Girard, F., Kamachi, Y., Kondoh, H., Gehring, W.J., 2005. Functional analysis of the chicken delta1-crystallin enhancer activity in *Drosophila* reveals remarkable evolutionary conservation between chicken and fly. Development 132, 1895–1905.

Brent, A.E., Braun, T., Tabin, C.J., 2005. Genetic analysis of interactions between the somitic muscle, cartilage and tendon cell lineages during mouse development. Development 132, 515–528.

Brent, A.E., Schweitzer, R., Tabin, C.J., 2003. A somitic compartment of tendon progenitors. Cell 113, 235–248.

Brent, A.E., Tabin, C.J., 2004. FGF acts directly on the somitic tendon progenitors through the Ets transcription factors Pea3 and Erm to regulate scleraxis expression. Development 131, 3885–3896.

Bronner, M.E., LeDouarin, N.M., 2012. Development and evolution of the neural crest: an overview. Dev. Biol. 366, 2–9.

Bryson-Richardson, R.J., Currie, P.D., 2008. The genetics of vertebrate myogenesis. Nat. Rev. Genet. 9, 632–646.

Buckingham, M., 2006. Myogenic progenitor cells and skeletal myogenesis in vertebrates. Curr. Opin. Genet. Dev. 16, 525–532.

Buckingham, M., Rigby, P.W., 2014. Gene regulatory networks and transcriptional mechanisms that control myogenesis. Dev. Cell 28, 225–238.

Buckingham, M., Vincent, S.D., 2009. Distinct and dynamic myogenic populations in the vertebrate embryo. Curr. Opin. Genet. Dev. 19, 444–453.

Cairns, D.M., Sato, M.E., Lee, P.G., Lassar, A.B., Zeng, L., 2008. A gradient of Shh establishes mutually repressing somitic cell fates induced by Nkx3.2 and Pax3. Dev. Biol. 323, 152–165.

Campbell, A.L., Eng, D., Gross, M.K., Kioussi, C., 2012. Prediction of gene network models in limb muscle precursors. Gene 509, 16–23.

Carvajal, J.J., Keith, A., Rigby, P.W., 2008. Global transcriptional regulation of the locus encoding the skeletal muscle determination genes *Mrf4* and *Myf5*. Genes Dev. 22, 265–276.

Chen, A.T., Zon, L.I., 2009. Zebrafish blood stem cells. J. Cell. Biochem. 108, 35–42.

Chen, M.J., Yokomizo, T., Zeigler, B.M., Dzierzak, E., Speck, N.A., 2009. Runx1 is required for the endothelial to haematopoietic cell transition but not thereafter. Nature 457, 887–891.

Ciau-Uitz, A., Liu, F., Patient, R., 2010. Genetic control of hematopoietic development in *Xenopus* and zebrafish. Int. J. Dev. Biol. 54, 1139–1149.

Ciau-Uitz, A., Pinheiro, P., Kirmizitas, A., Zuo, J., Patient, R., 2013. VEGFA-dependent and -independent pathways synergise to drive Scl expression and initiate programming of the blood stem cell lineage in *Xenopus*. Development 140, 2632–2642.

Cvekl, A., Duncan, M.K., 2007. Genetic and epigenetic mechanisms of gene regulation during lens development. Prog. Retin. Eye Res. 26, 555–597.

Daubas, P., Buckingham, M.E., 2013. Direct molecular regulation of the myogenic determination gene *Myf5* by *Pax3*, with modulation by Six1/4 factors, is exemplified by the -111 kb-Myf5 enhancer. Dev. Biol. 376, 236–244.

Davidson, A.J., Zon, L.I., 2006. The caudal-related homeobox genes *cdx1a* and *cdx4* act redundantly to regulate *hox* gene expression and the formation of putative hematopoietic stem cells during zebrafish embryogenesis. Dev. Biol. 292, 506–518.

Davidson, E.H., 2006. The Regulatory Genome. Gene Regulatory Networks in Development and Evolution. Academic Press/Elsevier, San Diego, CA.

Davidson, E.H., Erwin, D.H., 2006. Gene regulatory networks and the evolution of animal body plans. Science 311, 796–800.

Davis, J., Davis, D., Norman, B., Piatigorsky, J., 2008. Gene expression of the mouse corneal crystallin Aldh3a1: activation by Pax6, Oct1, and p300. Invest. Ophthalmol. Vis. Sci. 49, 1814–1826.

de Crozé, N., Maczkowiak, F., Monsoro-Burq, A.H., 2011. Reiterative AP2a activity controls sequential steps in the neural crest gene regulatory network. Proc. Natl. Acad. Sci. U.S.A. 108, 155–160.

de Jong, J.L., Davidson, A.J., Wang, Y., Palis, J., Opara, P., Pugach, E., Daley, G.Q., Zon, L.I., 2010. Interaction of retinoic acid and scl controls primitive blood development. Blood 116, 201–209.

Delassus, S., Titley, I., Enver, T., 1999. Functional and molecular analysis of hematopoietic progenitors derived from the aorta-gonad-mesonephros region of the mouse embryo. Blood 94, 1495–1503.

Della Gaspera, B., Armand, A.S., Sequeira, I., Chesneau, A., Mazabraud, A., Lécolle, S., Charbonnier, F., Chanoine, C., 2012. Myogenic waves and myogenic programs during *Xenopus* embryonic myogenesis. Dev. Dyn. 241, 995–1007.

Deneris, E.S., Wyler, S.C., 2012. Serotonergic transcriptional networks and potential importance to mental health. Nat. Neurosci. 15, 519–527.

Doitsidou, M., Flames, N., Topalidou, I., Abe, N., Felton, T., Remesal, L., Popovitchenko, T., Mann, R., Chalfie, M., Hobert, O., 2013. A combinatorial regulatory signature controls terminal differentiation of the dopaminergic nervous system in *C. elegans*. Genes Dev. 27, 1391–1405.

Drogat, B., Kalucka, J., Gutiérrez, L., Hammad, H., Goossens, S., Farhang Ghahremani, M., Bartunkova, S., Haigh, K., Deswarte, K., Nyabi, O., Naessens, M., Ferrara, N., Klingmüller, U., Lambrecht, B.N., Nagy, A., Philipsen, S., Haigh, J.J., 2010. Vegf regulates embryonic erythroid development through Gata1 modulation. Blood 116, 2141–2151.

Dzierzak, E., Philipsen, S., 2013. Erythropoiesis: development and differentiation. Cold Spring Harb. Perspect. Med. 3, a011601.

Eng, D., Ma, H.Y., Xu, J., Shih, H.P., Gross, M.K., Kioussi, C., Kiouss, C., 2012. Loss of abdominal muscle in Pitx2 mutants associated with altered axial specification of lateral plate mesoderm. PLoS One 7, e42228.

Evans, T., Felsenfeld, G., 1989. The erythroid-specific transcription factor Eryf1: a new finger protein. Cell 58, 877–885.

Ferreira, R., Spensberger, D., Silber, Y., Dimond, A., Li, J., Green, A.R., Göttgens, B., 2013. Impaired in vitro erythropoiesis following deletion of the Scl (Tal1) +40 enhancer is largely compensated for in vivo despite a significant reduction in expression. Mol. Cell. Biol. 33, 1254–1266.

Flames, N., Hobert, O., 2009. Gene regulatory logic of dopamine neuron differentiation. Nature 458, 885–889.

Fuchs, E., Chen, T., 2013. A matter of life and death: self-renewal in stem cells. EMBO Rep. 14, 39–48.

Galloway, J.L., Wingert, R.A., Thisse, C., Thisse, B., Zon, L.I., 2005. Loss of gata1 but not gata2 converts erythropoiesis to myelopoiesis in zebrafish embryos. Dev. Cell 8, 109–116.

Giordani, J., Bajard, L., Demignon, J., Daubas, P., Buckingham, M., Maire, P., 2007. Six proteins regulate the activation of Myf5 expression in embryonic mouse limbs. Proc. Natl. Acad. Sci. U.S.A. 104, 11310–11315.

Gouti, M., Briscoe, J., Gavalas, A., 2011. Anterior Hox genes interact with components of the neural crest specification network to induce neural crest fates. Stem Cells 29, 858–870.

Graf, T., Enver, T., 2009. Forcing cells to change lineages. Nature 462, 587–594.

Grandel, H., Brand, M., 2013. Comparative aspects of adult neural stem cell activity in vertebrates. Dev. Genes Evol. 223, 131–147.

Greenhill, E.R., Rocco, A., Vibert, L., Nikaido, M., Kelsh, R.N., 2011. An iterative genetic and dynamical modelling approach identifies novel features of the gene regulatory network underlying melanocyte development. PLoS Genet. 7, e1002265.

Grocott, T., Tambalo, M., Streit, A., 2012. The peripheral sensory nervous system in the vertebrate head: a gene regulatory perspective. Dev. Biol. 370, 3–23.

Guenther, C., Pantalena-Filho, L., Kingsley, D.M., 2008. Shaping skeletal growth by modular regulatory elements in the Bmp5 gene. PLoS Genet. 4, e1000308.

Göthert, J.R., Gustin, S.E., Hall, M.A., Green, A.R., Göttgens, B., Izon, D.J., Begley, C.G., 2005. In vivo fate-tracing studies using the Scl stem cell enhancer: embryonic hematopoietic stem cells significantly contribute to adult hematopoiesis. Blood 105, 2724–2732.

Göttgens, B., Broccardo, C., Sanchez, M.J., Deveaux, S., Murphy, G., Göthert, J.R., Kotsopoulou, E., Kinston, S., Delaney, L., Piltz, S., Barton, L.M., Knezevic, K., Erber, W.N., Begley, C.G., Frampton, J., Green, A.R., 2004. The scl +18/19 stem cell enhancer is not required for hematopoiesis: identification of a 5′ bifunctional hematopoietic-endothelial enhancer bound by Fli-1 and Elf-1. Mol. Cell. Biol. 24, 1870–1883.

Hadchouel, J., Carvajal, J.J., Daubas, P., Bajard, L., Chang, T., Rocancourt, D., Cox, D., Summerbell, D., Tajbakhsh, S., Rigby, P.W., Buckingham, M., 2003. Analysis of a key regulatory region upstream of the Myf5 gene reveals multiple phases of myogenesis, orchestrated at each site by a combination of elements dispersed throughout the locus. Development 130, 3415–3426.

Hammond, C.L., Hinits, Y., Osborn, D.P., Minchin, J.E., Tettamanti, G., Hughes, S.M., 2007. Signals and myogenic regulatory factors restrict pax3 and pax7 expression to dermomyotome-like tissue in zebrafish. Dev. Biol. 302, 504–521.

Harlan, S.M., Reiter, R.S., Sigmund, C.D., Lin, J.L., Lin, J.J., 2008. Requirement of TCTG(G/C) direct repeats and overlapping gata site for maintaining the cardiac-specific expression of Cardiac troponin T in developing and adult mice. Anat. Rec. Hob. 291, 1574–1586.

Hayashi, T., Xu, C., Carthew, R.W., 2008. Cell-type-specific transcription of prospero is controlled by combinatorial signaling in the *Drosophila* eye. Development 135, 2787–2796.

He, A., Kong, S.W., Ma, Q., Pu, W.T., 2011. Co-occupancy by multiple cardiac transcription factors identifies transcriptional enhancers active in heart. Proc. Natl. Acad. Sci. U.S.A. 108, 5632–5637.

He, S., Nakada, D., Morrison, S.J., 2009. Mechanisms of stem cell self-renewal. Annu Rev. Cell. Dev. Biol. 25, 377–406.

Hendershot, T.J., Liu, H., Clouthier, D.E., Shepherd, I.T., Coppola, E., Studer, M., Firulli, A.B., Pittman, D.L., Howard, M.J., 2008. Conditional deletion of Hand2 reveals critical functions in neurogenesis and cell type-specific gene expression for development of neural crest-derived noradrenergic sympathetic ganglion neurons. Dev. Biol. 319, 179–191.

Hobert, O., 2014. Development of left/right asymmetry in the *Caenorhabditis elegans* nervous system: from zygote to postmitotic neuron. Genesis 52, 528–543.

Hsiao, H.Y., Jukam, D., Johnston, R., Desplan, C., 2013. The neuronal transcription factor erect wing regulates specification and maintenance of *Drosophila* R8 photoreceptor subtypes. Dev. Biol. 381, 482–490.

Hsu, Y.C., Pasolli, H.A., Fuchs, E., 2011. Dynamics between stem cells, niche, and progeny in the hair follicle. Cell 144, 92–105.

Hutcheson, D.A., Zhao, J., Merrell, A., Haldar, M., Kardon, G., 2009. Embryonic and fetal limb myogenic cells are derived from developmentally distinct progenitors and have different requirements for beta-catenin. Genes Dev. 23, 997–1013.

Isern, J., He, Z., Fraser, S.T., Nowotschin, S., Ferrer-Vaquer, A., Moore, R., Hadjantonakis, A.K., Schulz, V., Tuck, D., Gallagher, P.G., Baron, M.H., 2011. Single-lineage transcriptome analysis reveals key regulatory pathways in primitive erythroid progenitors in the mouse embryo. Blood 117, 4924–4934.

Jin, H., Stojnic, R., Adryan, B., Ozdemir, A., Stathopoulos, A., Frasch, M., 2013. Genome-wide screens for in vivo tinman binding sites identify cardiac enhancers with diverse functional architectures. PLoS Genet. 9, e1003195.

Johnston, R.J., Otake, Y., Sood, P., Vogt, N., Behnia, R., Vasiliauskas, D., McDonald, E., Xie, B., Koenig, S., Wolf, R., Cook, T., Gebelein, B., Kussell, E., Nakagoshi, H., Desplan, C., 2011. Interlocked feedforward loops control cell-type-specific rhodopsin expression in the *Drosophila* eye. Cell 145, 956–968.

Junion, G., Spivakov, M., Girardot, C., Braun, M., Gustafson, E.H., Birney, E., Furlong, E.E., 2012. A transcription factor collective defines cardiac cell fate and reflects lineage history. Cell 148, 473–486.

Kamachi, Y., Uchikawa, M., Tanouchi, A., Sekido, R., Kondoh, H., 2001. Pax6 and SOX2 form a co-DNA-binding partner complex that regulates initiation of lens development. Genes. Dev. 15, 1272–1286.

Knezevic, K., Bee, T., Wilson, N.K., Janes, M.E., Kinston, S., Polderdijk, S., Kolb-Kokocinski, A., Ottersbach, K., Pencovich, N., Groner, Y., de Bruijn, M., Göttgens, B., Pimanda, J.E., 2011. A Runx1-Smad6 rheostat controls Runx1 activity during embryonic hematopoiesis. Mol. Cell. Biol. 31, 2817–2826.

Kumar, J.P., 2009. The molecular circuitry governing retinal determination. Biochim. Biophys. Acta 1789, 306–314.

Kumar, J.P., 2010. Retinal determination the beginning of eye development. Curr. Top. Dev. Biol. 93, 1–28.

Kusakabe, R., Kuraku, S., Kuratani, S., 2011. Expression and interaction of muscle-related genes in the lamprey imply the evolutionary scenario for vertebrate skeletal muscle, in association with the acquisition of the neck and fins. Dev. Biol. 350, 217–227.

L'honoré, A., Ouimette, J.F., Lavertu-Jolin, M., Drouin, J., 2010. Pitx2 defines alternate pathways acting through MyoD during limb and somitic myogenesis. Development 137, 3847–3856.

Lagha, M., Brunelli, S., Messina, G., Cumano, A., Kume, T., Relaix, F., Buckingham, M.E., 2009. Pax3:Foxc2 reciprocal repression in the somite modulates muscular versus vascular cell fate choice in multipotent progenitors. Dev. Cell 17, 892–899.

Lagha, M., Mayeuf-Louchart, A., Chang, T., Montarras, D., Rocancourt, D., Zalc, A., Kormish, J., Zaret, K.S., Buckingham, M.E., Relaix, F., 2013. *Itm2a* is a Pax3 target gene, expressed at sites of skeletal muscle formation in vivo. PLoS One 8, e63143.

Lancrin, C., Mazan, M., Stefanska, M., Patel, R., Lichtinger, M., Costa, G., Vargel, O., Wilson, N.K., Möröy, T., Bonifer, C., Göttgens, B., Kouskoff, V., Lacaud, G., 2012. GFI1 and GFI1B control the loss of endothelial identity of hemogenic endothelium during hematopoietic commitment. Blood 120, 314–322.

Le Grand, F., Rudnicki, M.A., 2007. Skeletal muscle satellite cells and adult myogenesis. Curr. Opin. Cell Biol. 19, 628–633.

Lehman, D.A., Patterson, B., Johnston, L.A., Balzer, T., Britton, J.S., Saint, R., Edgar, B.A., 1999. *Cis*-regulatory elements of the mitotic regulator, string/Cdc25. Development 126, 1793–1803.

Leung, A., Ciau-Uitz, A., Pinheiro, P., Monteiro, R., Zuo, J., Vyas, P., Patient, R., Porcher, C., 2013. Uncoupling VEGFA functions in arteriogenesis and hematopoietic stem cell specification. Dev. Cell 24, 144–158.

Li, Y., Jiang, Y., Chen, Y., Karandikar, U., Hoffman, K., Chattopadhyay, A., Mardon, G., Chen, R., 2013. Optix functions as a link between the retinal determination network and the dpp pathway to control morphogenetic furrow progression in *Drosophila*. Dev. Biol. 381, 50–61.

Liu, F., Walmsley, M., Rodaway, A., Patient, R., 2008. Fli1 acts at the top of the transcriptional network driving blood and endothelial development. Curr. Biol. 18, 1234–1240.

Liu, N., Olson, E.N., 2010. MicroRNA regulatory networks in cardiovascular development. Dev. Cell 18, 510–525.

Liu, S., Piatigorsky, J., 2011. Regulation of mouse small heat shock protein *αb-crystallin* gene by aryl hydrocarbon receptor. PLoS One 6, e17904.

Lu, C.P., Polak, L., Rocha, A.S., Pasolli, H.A., Chen, S.C., Sharma, N., Blanpain, C., Fuchs, E., 2012. Identification of stem cell populations in sweat glands and ducts reveals roles in homeostasis and wound repair. Cell 150, 136–150.

McGrath, K.E., Frame, J.M., Fromm, G.J., Koniski, A.D., Kingsley, P.D., Little, J., Bulger, M., Palis, J., 2011. A transient definitive erythroid lineage with unique regulation of the β-globin locus in the mammalian embryo. Blood 117, 4600–4608.

Medvinsky, A., Rybtsov, S., Taoudi, S., 2011. Embryonic origin of the adult hematopoietic system: advances and questions. Development 138, 1017–1031.

Mercer, E.M., Lin, Y.C., Murre, C., 2011. Factors and networks that underpin early hematopoiesis. Semin. Immunol. 23, 317–325.

Milet, C., Maczkowiak, F., Roche, D.D., Monsoro-Burq, A.H., 2013. Pax3 and Zic1 drive induction and differentiation of multipotent, migratory, and functional neural crest in *Xenopus* embryos. Proc. Natl. Acad. Sci. U.S.A. 110, 5528–5533.

Mishra, M., Oke, A., Lebel, C., McDonald, E.C., Plummer, Z., Cook, T.A., Zelhof, A.C., 2010. Pph13 and orthodenticle define a dual regulatory pathway for photoreceptor cell morphogenesis and function. Development 137, 2895–2904.

Moncaut, N., Cross, J.W., Siligan, C., Keith, A., Taylor, K., Rigby, P.W., Carvajal, J.J., 2012. Musculin and TCF21 coordinate the maintenance of myogenic regulatory factor expression levels during mouse craniofacial development. Development 139, 958–967.

Montarras, D., L'honoré, A., Buckingham, M., 2013. Lying low but ready for action: the quiescent muscle satellite cell. FEBS J. 280, 4036–4050.

Monteiro, R., Pouget, C., Patient, R., 2011. The gata1/pu.1 lineage fate paradigm varies between blood populations and is modulated by tif1γ. EMBO J. 30, 1093–1103.

Narula, J., Williams, C.J., Tiwari, A., Marks-Bluth, J., Pimanda, J.E., Igoshin, O.A., 2013. Mathematical model of a gene regulatory network reconciles effects of genetic perturbations on hematopoietic stem cell emergence. Dev. Biol. 379, 258–269.

Nitzan, E., Krispin, S., Pfaltzgraff, E.R., Klar, A., Labosky, P.A., Kalcheim, C., 2013. A dynamic code of dorsal neural tube genes regulates the segregation between neurogenic and melanogenic neural crest cells. Development 140, 2269–2279.

Ogilvy, S., Ferreira, R., Piltz, S.G., Bowen, J.M., Göttgens, B., Green, A.R., 2007. The SCL +40 enhancer targets the midbrain together with primitive and definitive hematopoiesis and is regulated by SCL and GATA proteins. Mol. Cell. Biol. 27, 7206–7219.

Ogino, H., Ochi, H., Reza, H.M., Yasuda, K., 2012. Transcription factors involved in lens development from the preplacodal ectoderm. Dev. Biol. 363, 333–347.

Pang, C.J., Lemsaddek, W., Alhashem, Y.N., Bondzi, C., Redmond, L.C., Ah-Son, N., Dumur, C.I., Archer, K.J., Haar, J.L., Lloyd, J.A., Trudel, M., 2012. Kruppel-like factor 1 (KLF1), KLF2, and Myc control a regulatory network essential for embryonic erythropoiesis. Mol. Cell. Biol. 32, 2628–2644.

Patterson, L.J., Gering, M., Eckfeldt, C.E., Green, A.R., Verfaillie, C.M., Ekker, S.C., Patient, R., 2007. The transcription factors Scl and Lmo2 act together during development of the hemangioblast in zebrafish. Blood 109, 2389–2398.

Pavan, W.J., Raible, D.W., 2012. Specification of neural crest into sensory neuron and melanocyte lineages. Dev. Biol. 366, 55–63.

Pignoni, F., Hu, B., Zavitz, K.H., Xiao, J., Garrity, P.A., Zipursky, S.L., 1997. The eye-specification proteins So and Eya form a complex and regulate multiple steps in *Drosophila* eye development. Cell 91, 881–891.

Pillay, L.M., Forrester, A.M., Erickson, T., Berman, J.N., Waskiewicz, A.J., 2010. The Hox cofactors Meis1 and Pbx act upstream of gata1 to regulate primitive hematopoiesis. Dev. Biol. 340, 306–317.

Pimanda, J.E., Göttgens, B., 2010. Gene regulatory networks governing haematopoietic stem cell development and identity. Int. J. Dev. Biol. 54, 1201–1211.

Pimanda, J.E., Ottersbach, K., Knezevic, K., Kinston, S., Chan, W.Y., Wilson, N.K., Landry, J.R., Wood, A.D., Kolb-Kokocinski, A., Green, A.R., Tannahill, D., Lacaud, G., Kouskoff, V., Gottgens, B., 2007. Gata2, Fli1, and Scl form a recursively wired gene-regulatory circuit during early hematopoietic development. Proc. Natl. Acad. Sci. U.S.A. 104, 17692–17697.

Plouhinec, J.L., Roche, D.D., Pegoraro, C., Figueiredo, A.L., Maczkowiak, F., Brunet, L.J., Milet, C., Vert, J.P., Pollet, N., Harland, R.M., Monsoro-Burq, A.H., 2014. Pax3 and Zic1 trigger the early neural crest gene regulatory network by the direct activation of multiple key neural crest specifiers. Dev. Biol. 386, 461–472.

Quan, X.J., Ramaekers, A., Hassan, B.A., 2012. Transcriptional control of cell fate specification: lessons from the fly retina. Curr. Top. Dev. Biol. 98, 259–276.

Relaix, F., Demignon, J., Laclef, C., Pujol, J., Santolini, M., Niro, C., Lagha, M., Rocancourt, D., Buckingham, M., Maire, P., 2013. Six homeoproteins directly activate Myod expression in the gene regulatory networks that control early myogenesis. PLoS Genet. 9, e1003425.

Relaix, F., Rocancourt, D., Mansouri, A., Buckingham, M., 2004. Divergent functions of murine Pax3 and Pax7 in limb muscle development. Genes. Dev. 18, 1088–1105.

Relaix, F., Rocancourt, D., Mansouri, A., Buckingham, M., 2005. A Pax3/Pax7-dependent population of skeletal muscle progenitor cells. Nature 435, 948–953.

Rinon, A., Lazar, S., Marshall, H., Büchmann-Møller, S., Neufeld, A., Elhanany-Tamir, H., Taketo, M.M., Sommer, L., Krumlauf, R., Tzahor, E., 2007. Cranial neural crest cells regulate head muscle patterning and differentiation during vertebrate embryogenesis. Development 134, 3065–3075.

Rodrigo, I., Bovolenta, P., Mankoo, B.S., Imai, K., 2004. Meox homeodomain proteins are required for Bapx1 expression in the sclerotome and activate its transcription by direct binding to its promoter. Mol. Cell. Biol. 24, 2757–2766.

Rogers, C.D., Phillips, J.L., Bronner, M.E., 2013. Elk3 is essential for the progression from progenitor to definitive neural crest cell. Dev. Biol. 374, 255–263.

Rohrer, H., 2011. Transcriptional control of differentiation and neurogenesis in autonomic ganglia. Eur. J. Neurosci. 34, 1563–1573.

Saint-Jeannet, J.P., Moody, S.A., 2014. Establishing the pre-placodal region and breaking it into placodes with distinct identities. Dev. Biol. 389, 13–27.

Sambasivan, R., Gayraud-Morel, B., Dumas, G., Cimper, C., Paisant, S., Kelly, R.G., Kelly, R., Tajbakhsh, S., 2009. Distinct regulatory cascades govern extraocular and pharyngeal arch muscle progenitor cell fates. Dev. Cell 16, 810–821.

Sambasivan, R., Kuratani, S., Tajbakhsh, S., 2011. An eye on the head: the development and evolution of craniofacial muscles. Development 138, 2401–2415.

Sato, T., Rocancourt, D., Marques, L., Thorsteinsdóttir, S., Buckingham, M., 2010. A Pax3/Dmrt2/Myf5 regulatory cascade functions at the onset of myogenesis. PLoS Genet. 6, e1000897.

Sauka-Spengler, T., Bronner-Fraser, M., 2008. A gene regulatory network orchestrates neural crest formation. Nat. Rev. Mol. Cell Biol. 9, 557–568.

Schütte, J., Moignard, V., Göttgens, B., 2012. Establishing the stem cell state: insights from regulatory network analysis of blood stem cell development. Wiley Interdiscip. Rev. Syst. Biol. Med. 4, 285–295.

Serrano-Saiz, E., Poole, R.J., Felton, T., Zhang, F., De La Cruz, E.D., Hobert, O., 2013. Modular control of glutamatergic neuronal identity in C. elegans by distinct homeodomain proteins. Cell 155, 659–673.

Shykind, B.M., Rohani, S.C., O'Donnell, S., Nemes, A., Mendelsohn, M., Sun, Y., Axel, R., Barnea, G., 2004. Gene switching and the stability of odorant receptor gene choice. Cell 117, 801–815.

Siatecka, M., Bieker, J.J., 2011. The multifunctional role of EKLF/KLF1 during erythropoiesis. Blood 118, 2044–2054.

Simoes-Costa, M., Bronner, M.E., 2014. Establishing Neural Crest Identity: A Gene Regulatory Recipe.

Simons, B.D., Clevers, H., 2011. Strategies for homeostatic stem cell self-renewal in adult tissues. Cell 145, 851–862.

Simões-Costa, M.S., McKeown, S.J., Tan-Cabugao, J., Sauka-Spengler, T., Bronner, M.E., 2012. Dynamic and differential regulation of stem cell factor FoxD3 in the neural crest is encrypted in the genome. PLoS Genet. 8, e1003142.

Singh, A., Tare, M., Puli, O.R., Kango-Singh, M., 2012. A glimpse into dorso-ventral patterning of the Drosophila eye. Dev. Dyn. 241, 69–84.

Skuntz, S., Mankoo, B., Nguyen, M.T., Hustert, E., Nakayama, A., Tournier-Lasserve, E., Wright, C.V., Pachnis, V., Bharti, K., Arnheiter, H., 2009. Lack of the mesodermal homeodomain protein MEOX1 disrupts sclerotome polarity and leads to a remodeling of the cranio-cervical joints of the axial skeleton. Dev. Biol. 332, 383–395.

Snippert, H.J., Clevers, H., 2011. Tracking adult stem cells. EMBO Rep. 12, 113–122.

Spensberger, D., Kotsopoulou, E., Ferreira, R., Broccardo, C., Scott, L.M., Fourouclas, N., Ottersbach, K., Green, A.R., Göttgens, B., 2012. Deletion of the Scl +19 enhancer increases the blood stem cell compartment without affecting the formation of mature blood lineages. Exp. Hematol. 40, 588–598 e581.

Spradling, A., Fuller, M.T., Braun, R.E., Yoshida, S., 2011. Germline stem cells. Cold Spring Harb. Perspect. Biol. 3, a002642.

Swamynathan, S.K., Davis, J., Piatigorsky, J., 2008. Identification of candidate Klf4 target genes reveals the molecular basis of the diverse regulatory roles of Klf4 in the mouse cornea. Invest. Ophthalmol. Vis. Sci. 49, 3360–3370.

Takahashi, K., Yamanaka, S., 2006. Induction of pluripotent stem cells from mouse embryonic and adult fibroblast cultures by defined factors. Cell 126, 663–676.

Takahashi, Y., Takagi, A., Hiraoka, S., Koseki, H., Kanno, J., Rawls, A., Saga, Y., 2007. Transcription factors Mesp2 and Paraxis have critical roles in axial musculoskeletal formation. Dev. Dyn. 236, 1484–1494.

Takahashi, Y., Yasuhiko, Y., Takahashi, J., Takada, S., Johnson, R.L., Saga, Y., Kanno, J., 2013. Metameric pattern of intervertebral disc/vertebral body is generated independently of Mesp2/Ripply-mediated rostro-caudal patterning of somites in the mouse embryo. Dev. Biol. 380, 172–184.

Takashima, S., Gold, D., Hartenstein, V., 2013. Stem cells and lineages of the intestine: a developmental and evolutionary perspective. Dev. Genes. Evol. 223, 85–102.

Taneyhill, L.A., Coles, E.G., Bronner-Fraser, M., 2007. Snail2 directly represses cadherin6B during epithelial-to-mesenchymal transitions of the neural crest. Development 134, 1481–1490.

Tasic, B., Nabholz, C.E., Baldwin, K.K., Kim, Y., Rueckert, E.H., Ribich, S.A., Cramer, P., Wu, Q., Axel, R., Maniatis, T., 2002. Promoter choice determines splice site selection in protocadherin alpha and gamma pre-mRNA splicing. Mol. Cell. 10, 21–33.

Templeton, J.P., Wang, X., Freeman, N.E., Ma, Z., Lu, A., Hejtmancik, F., Geisert, E.E., 2013. A crystallin gene network in the mouse retina. Exp. Eye Res. 116, 129–140.

Teslaa, J.J., Keller, A.N., Nyholm, M.K., Grinblat, Y., 2013. Zebrafish Zic2a and Zic2b regulate neural crest and craniofacial development. Dev. Biol. 380, 73–86.

Tober, J., Yzaguirre, A.D., Piwarzyk, E., Speck, N.A., 2013. Distinct temporal requirements for Runx1 in hematopoietic progenitors and stem cells. Development 140, 3765–3776.

Toro, R., Saadi, I., Kuburas, A., Nemer, M., Russo, A.F., 2004. Cell-specific activation of the atrial natriuretic factor promoter by PITX2 and MEF2A. J. Biol. Chem. 279, 52087–52094.

Tsachaki, M., Sprecher, S.G., 2012. Genetic and developmental mechanisms underlying the formation of the *Drosophila* compound eye. Dev. Dyn. 241, 40–56.

Tsai, S.F., Martin, D.I., Zon, L.I., D'Andrea, A.D., Wong, G.G., Orkin, S.H., 1989. Cloning of cDNA for the major DNA-binding protein of the erythroid lineage through expression in mammalian cells. Nature 339, 446–451.

Tzahor, E., Evans, S.M., 2011. Pharyngeal mesoderm development during embryogenesis: implications for both heart and head myogenesis. Cardiovasc Res. 91, 196–202.

Van Keymeulen, A., Rocha, A.S., Ousset, M., Beck, B., Bouvencourt, G., Rock, J., Sharma, N., Dekoninck, S., Blanpain, C., 2011. Distinct stem cells contribute to mammary gland development and maintenance. Nature 479, 189–193.

Vogeli, K.M., Jin, S.W., Martin, G.R., Stainier, D.Y., 2006. A common progenitor for haematopoietic and endothelial lineages in the zebrafish gastrula. Nature 443, 337–339.

Wall, L., deBoer, E., Grosveld, F., 1988. The human beta-globin gene 3' enhancer contains multiple binding sites for an erythroid-specific protein. Genes. Dev. 2, 1089–1100.

Wang, C., Cao, D., Wang, Q., Wang, D.Z., 2011. Synergistic activation of cardiac genes by myocardin and Tbx5. PLoS One 6, e24242.

Wang, Z., Wang, D.Z., Pipes, G.C., Olson, E.N., 2003. Myocardin is a master regulator of smooth muscle gene expression. Proc. Natl. Acad. Sci. U.S.A. 100, 7129–7134.

Weasner, B.M., Kumar, J.P., 2013. Competition among gene regulatory networks imposes order within the eye-antennal disc of *Drosophila*. Development 140, 205–215.

Wernet, M.F., Desplan, C., 2014. Homothorax and extradenticle alter the transcription factor network in *Drosophila* ommatidia at the dorsal rim of the retina. Development 141, 918–928.

Wernet, M.F., Mazzoni, E.O., Celik, A., Duncan, D.M., Duncan, I., Desplan, C., 2006. Stochastic spineless expression creates the retinal mosaic for colour vision. Nature 440, 174–180.

Wilson, N.K., Calero-Nieto, F.J., Ferreira, R., Göttgens, B., 2011. Transcriptional regulation of haematopoietic transcription factors. Stem Cell Res. Ther. 2, 6.

Xie, Q., Yang, Y., Huang, J., Ninkovic, J., Walcher, T., Wolf, L., Vitenzon, A., Zheng, D., Götz, M., Beebe, D.C., Zavadil, J., Cvekl, A., 2013. Pax6 interactions with chromatin and identification of its novel direct target genes in lens and forebrain. PLoS One 8, e54507.

Xie, T., 2013. Control of germline stem cell self-renewal and differentiation in the *Drosophila* ovary: concerted actions of niche signals and intrinsic factors. Wiley Interdiscip. Rev. Dev. Biol. 2, 261–273.

Yang, Y., Cvekl, A., 2005. Tissue-specific regulation of the mouse alphaA-crystallin gene in lens via recruitment of Pax6 and c-Maf to its promoter. J. Mol. Biol. 351, 453–469.

Yang, Y., Stopka, T., Golestaneh, N., Wang, Y., Wu, K., Li, A., Chauhan, B.K., Gao, C.Y., Cveklová, K., Duncan, M.K., Pestell, R.G., Chepelinsky, A.B., Skoultchi, A.I., Cvekl, A., 2006. Regulation of alphaA-crystallin via Pax6, c-Maf, CREB and a broad domain of lens-specific chromatin. EMBO J. 25, 2107–2118.

Yokomizo, T., Hasegawa, K., Ishitobi, H., Osato, M., Ema, M., Ito, Y., Yamamoto, M., Takahashi, S., 2008. Runx1 is involved in primitive erythropoiesis in the mouse. Blood 111, 4075–4080.

Zacharek, S.J., Fillmore, C.M., Lau, A.N., Gludish, D.W., Chou, A., Ho, J.W., Zamponi, R., Gazit, R., Bock, C., Jäger, N., Smith, Z.D., Kim, T.M., Saunders, A.H., Wong, J., Lee, J.H., Roach, R.R., Rossi, D.J., Meissner, A., Gimelbrant, A.A., Park, P.J., Kim, C.F., 2011. Lung stem cell self-renewal relies on BMI1-dependent control of expression at imprinted loci. Cell Stem Cell 9, 272–281.

Zhou, Q., Zhang, T., Jemc, J.C., Chen, Y., Chen, R., Rebay, I., Pignoni, F., 2014. Onset of atonal expression in *Drosophila* retinal progenitors involves redundant and synergistic contributions of Ey/Pax6 and So binding sites within two distant enhancers. Dev. Biol. 386, 152–164.

Chapter 6

On the Modeling of Developmental Gene Regulatory Networks

In this work we have considered the regulatory systems that control developmental process at the level at which they operate in bilaterian life. These systems are always composed of multiple genes engaged in multiple interactions in various causal conformations, and their scale has required continuous reliance on models which enable one to think about the structure of the control systems and about how they work. In the foregoing we considered results accruing from ordinary differential equation (ODE) models, topological models and Boolean models, without actually delving into the constraints and insights available from each of these types of gene regulatory network (GRN) model. This is now to be our subject. A GRN model is a formalization of a concept of how a developmental process is controlled by a genomically encoded regulatory system. It includes causal propositions that can be used to predict what is not yet known and that can be tested by experiment. Computational GRN models also constitute platforms for system analyses and redesign.

The structural features of developmental GRNs determine their various functionalities. Among these features are, at each network node, the identities of the regulatory inputs, the logic transactions by which these inputs determine the outputs, and the architecture of the gene interaction subcircuits into which these nodes are

Genomic Control Process
http://dx.doi.org/10.1016/B978-0-12-404729-7.00006-X

organized. Larger-scale features of GRNs include their hierarchy, the directionality of the regulatory transactions they encode, the signaling interactions among different GRNs, and their collective output in terms of spatial and temporal gene expression. Taken together, these features constitute the functional import of the GRN. The usefulness of any given modeling approach thus depends on the extent to which it captures the emergent properties of these GRN features. In this respect, the different types of model we discuss here are not equivalent.

Throughout this book we have approached our various subject matters with the idea that the different aspects of this field are best illuminated by consideration of particularly well-analyzed examples. We follow the same pathway in this chapter, where our objectives are the further insights into gene network biology and function that can be obtained by computational routes. We endeavor to provide neither a handbook of methods for modelers nor a comprehensive digest of the various technical issues encountered in every modeling approach. Rather our focus as always is on what these modes of analysis can contribute to understanding how genomic regulatory instructions give rise to spatially organized developmental patterns of gene expression.

The principle of model generation as we discuss it here starts with a mechanistic concept, and the model provides a means for experimentally testing the relevance and consistency of the concept. This approach is clearly distinct from attempts to "simulate" process output without particular concern for the underlying mechanism. That is, choosing a mathematical formulation primarily on the basis that its output "fits" experimental observation, without incorporating a plausible mechanistic idea, or without the constraints of biological reality, fails to illuminate the biological causality. Misconceptions are generated when, such simulations evoke the assumption that because they imitate the observed outcome, the mathematics used for the simulation can be taken to imply the physical nature of the underlying mechanism. However, recent history shows that purely phenomenological models fail to illuminate how a process actually works, and in the following we are concerned rather with models in which the mathematics attempts to capture more directly the assumed mechanisms of the process. Such models are particularly useful as they may demonstrate the extent to which given mechanistic assumptions explain the outcome of the process. Discrepancies between the model and the actual behavior of the system reveal where assumed mechanisms are incomplete, which phenomenological simulations cannot do. The primary imperative in designing the mechanistic concepts which will underlie given models is to take into account all available factual knowledge, and the more solidly a model is rooted in detailed knowledge the more useful it will be.

In the following we take up topological models, ODE models, and Boolean models, as each captures particular aspects of developmental process. The choice of modeling approach depends on the questions that need to be answered in order to access the biological issues at hand. Any regulatory network subcircuit can be subjected to topological analysis, that is, the attempt to make explicit the architecture of the subcircuit linkages, for which BioTapestry provides a suitable platform (Longabaugh et al., 2009). Or, the same subcircuit can be considered in dynamical terms, in order to rationalize quantitatively its kinetic behavior, and to establish predictively the effects of parameter values on its output, for which ODE analysis is required. Or, the logical structure of the information processing functions of the subcircuit can be formally abstracted, for which Boolean analysis is the method of choice. The usefulness of these modeling approaches is also dependent on the scale of the network considered. Dynamic ODE models rapidly become impractical for networks that exceed the size of typical subcircuits (i.e., less than ~10 genes) because of the formidable number of parameters that must be evaluated. On the other hand, topological models and Boolean logic models adapt well to large-scale networks. With respect to network analysis problems, dynamic ODE models aim at relating the kinetics of subcircuit outputs to the kinetics of inputs, and are thus ideally applied to temporal and quantitative interactions occurring within a given cell and its descendants. For this reason they have been prominently applied to the analysis of sequentially operating bacterial networks, and to animal systems generating physiological and biochemical transitions, including for example signal transduction systems. These are all fundamentally continuous processes. There have also been a number of applications to developmental processes, several of which we discuss below. In development, where the major output is discontinuous, as is disposition of regulatory state and cell fate choice in space, realistic modeling of the underlying genetic, logical, and morphological control events usually will require special approaches.

1. Topological Network Models

Topological network models provide graphical representations of the system of regulatory interactions and thus of the architecture of network circuitry. The graphs are literal representations of causal interactions among the regulatory genes in the system. These maps consist of the relevant regulatory genes (nodes or vertices), and they show explicitly the regulatory function these genes execute, i.e., they show for each gene how its outputs serve as inputs into other genes (linkages among genes, or "edges"). The inputs convey direct functional interactions between transcription factors and *cis*-regulatory DNA sequence, and thus they predict the regulatory sequence features that tie the network together. Topological models visualize and formally represent the genomic control system by relating individual regulatory DNA sequences to the circuitry specifying spatial regulatory states in development. These models do not only represent epistatic relationships, as do classic genetic interaction models, but rather functional, sequence-specific molecular interactions. The predicted functions of the transcriptional control system at any individual node of a topological model can therefore be directly tested by *cis*-regulatory analysis, in order to demonstrate the interaction between transcription factor and DNA sequence down to the single nucleotide level if desired.

Topological models uniquely capture major structural aspects of GRNs and they frequently serve as the starting point for other modeling approaches. First and foremost they explicitly indicate the identity of the interacting genes and the architecture of their interactions. They provide a telescopic view of the whole network: at the highest magnification they show the major *cis*-regulatory inputs, providing direct predictions that can be confronted by experimental *cis*-regulatory analyses. "Zoomed out", at the other extreme, they indicate overall aspects of network structure, such as layers of hierarchy, signaling interactions among GRNs, and the locations within GRN structure at which signals impinge. They reveal the modularity of the network and identify its peripheral as opposed to internal components. Beyond that they show where particular types of subcircuits are utilized within GRN hierarchy. Topological models are the presentations of choice for explanations of the genetic basis of spatial domain specification. The three features of GRNs that are not captured in topological models, or at best are only crudely indicated, are the dynamics of the encoded process, the functional characteristics of the circuitry, and the logic transactions executed by the driver inputs at the individual nodes. Thus additional modeling approaches are necessary to attain a predictive analysis of the functional output at subcircuit or GRN scale.

Topological models are static representations of GRNs which make directly accessible the circuit design features, on which the function of GRNs depends. A major strength of topological models for developmental networks is that they can be used to portray the particular network circuitries that operate within different spatial domains of the developing organism or body part. That is, they can represent a whole system of interacting GRNs, each encompassing the regulatory interactions functional in a specific spatial domain. This makes explicit the different circuitry operating within the nuclei of cells of diverse specification states. Furthermore, in topological models, interdomain interactions by means of signals are directly indicated by linkages into the specific regulatory genes whose activity is affected by these signals. Topological models may display qualitatively the sequential changes in circuit deployment as development progresses. Thus they are uniquely useful for regulatory process analysis at very different scales, from single nodes to large-scale networks, and from single differentially functioning cells to diversely specified territories, and thence to the overall shape of the network architecture. In our own experience, the iterative process of constructing topological models from experimental data is per se enlightening, in that it forces consideration of the regulatory system at all of these different scales. Many examples of topological models illustrating these features are included in the previous chapters, most of them presented in BioTapestry.

1.1 BioTapestry as a tool

BioTapestry is one of the widely used software platforms for presentation and analysis of topological GRN models. It was introduced with the initial presentation of the sea urchin endomesoderm GRN model

Figure 6.1

(A)

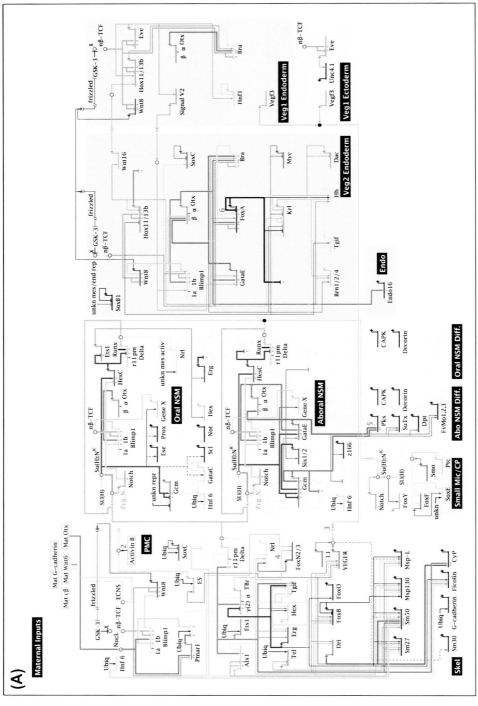

Figure 6.1 BioTapestry presentation of gene regulatory networks (GRNs). (A) The sea urchin endomesoderm GRN portrayed in BioTapestry (Longabaugh et al., 2009; Longabaugh, 2012). This is the largest experimentally solved, verified developmental GRN model so far available (Peter et al., 2012). It can also be seen in Fig. 3.5(I), to which the reader is referred for details; the symbolism employed is summarized in the following portions of this figure. (B) Graphic representation of regulatory features in BioTapestry

Figure 6.1

(B)

Components of topological GRN models as portrayed in BioTapestry

Node, Inputs, Output: a regulatory gene and its relevant *cis*-regulatory module(s), in which the inputs are the transcription factors encoded elsewhere in the GRN that bind to the *cis*-regulatory module(s) and the output is the transcription factor encoded by the gene at this node.

inputs

output

relevant *cis*-regulatory module

Positive regulatory linkage: interaction between two genes such that an upstream regulatory gene provides an activating input to a downstream target gene.

regulatory gene   target gene

Negative regulatory linkage: interaction between two genes such that an upstream regulatory gene provides a repressive input to a downstream target gene.

regulatory gene   target gene

Positive auto-regulation: activating input to a regulatory gene provided by the transcription factor that it itself encodes.

regulatory gene

Negative auto-regulation: repressive input to a regulatory gene provided by the transcription factor that it itself encodes.

regulatory gene

Intercellular signaling: stereospecific interaction occurs between extracellular ligand and membrane receptor, which when bound by the ligand engages intracellular signal transduction biochemistry. The result is to modify a dedicated transcription factor (response factor), thus functionally altering its regulatory effects on target genes.

Off-DNA
receptor modification of
ligand   response factor

response
factor

signal
transduction
biochemisitry

target gene

Janus signal response system: intercellular signal system in which the response factor acts as a default repressor for target gene(s), except in cells receiving signal. In these cells modification of the same response factor converts it to an activator for the same target gene(s).

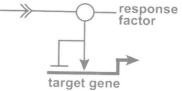

response
factor

target gene

Figure 6.1

(C)

General properties of BioTapestry GRN models

Time: sequential events are portrayed in BioTapestry such that earliest interactions are seen at the top and later interactions further down. Thus temporal order runs vertically downward, though the temporal representation is only roughly approximate, and there is no real time scale.

Space: spatial domains, distinguished by pastel background colors, are indicated in BioTapestry as rectangular areas within which regulatory linkages are portrayed; inter-domain signaling is indicated by horizontal ligand-receptor interactions crossing boundary between the domain from which the signal emanates and that in which the signal is received.

View from the Genome, VfG: BioTapestry diagram displaying all regulatory interactions in the GRN irrespective of when and where they occur during development; i.e., map of gene interactions encoded in *cis*-regulatory DNA sequence.

View from the Nucleus, VfN: BioTapestry diagram displaying the regulatory interactions in the GRN that take place in a given spatial domain within a given temporal interval, and indicating genes being transcribed and genes inactivated due to repression, within this interval. Commonly, multiple domains and large temporal intervals are portrayed to convey the overall GRN model, sometimes denoted "View from all Nuclei".

GRN hierarchy: Portrayed in BioTapestry as vertical depth, or stacking of sequential VfN transitions through time.

GRN periphery: Portrayed in BioTapestry in terminal rectangles containing effector genes and their circuitry (effector genes are genes encoding functional proteins other than transcription factors or signaling molecules).

and brief definitions of these features. (C) General properties of GRNs encompassed in BioTapestry representations.

(Davidson et al., 2002; Longabaugh et al., 2005). Thus BioTapestry has been specifically designed for the purpose of representing topological models of developmental GRNs. Its capacities and features have been continuously expanded (Longabaugh et al., 2009; Longabaugh, 2012). Here we briefly summarize how particular aspects of GRNs are specifically represented in BioTapestry. The current sea urchin embryo endomesoderm GRN model is shown in Fig. 6.1(A) as an example of a BioTapestry representation. The particular GRN features, molecular components and interactions that are symbolized in BioTapestry are summarized in Figs. 6.1(B) and (C). Here the inputs, outputs, regulatory genes, and representations of interactions are explicitly illustrated as shown in BioTapestry.

In addition to its visualization and analysis features, BioTapestry is designed to handle the multiple types of experimental data required for GRN analysis. These include quantitative perturbation data (e.g., QPCR, Nanostring, and RNAseq data showing the effects of blocking function of an upstream regulatory gene on activity of sets of target genes); spatial and temporal gene expression data; and *cis*-regulatory information including functional authentication of target sites and distinct functions of multiple modules. BioTapestry not only molds perturbation information into the rectilinear representation format according to the operator's design, but also stores tables of underlying experimental data of the types above. A BioTapestry GRN model is typically organized by the different spatial domains in which the GRNs of interest operate. In each domain, the set of expressed regulatory genes and any active regulatory interactions are explicit, providing a rational basis for internetwork comparisons.

The hierarchy of an overall developmental GRN, as discussed earlier in this book, can be instantly recognized by the depth of a BioTapestry model, that is, by the number of subcircuits stacked above one another upstream of the peripheral output of effector gene expression. These subcircuits represent the modular regulatory steps of the developmental process. Their sequential character is a manifestation of the fundamental directionality of development, and is ultimately a consequence of the unidirectional regulatory relation between transcription factor and downstream target gene (see Chapter 2). Thus any relevant representation of gene regulatory interactions must specify the unidirectional nature of the interactions, which is uniquely facilitated in BioTapestry. In network structure, this pertains to the output of an upstream subcircuit which serves as the input to the next subcircuit downstream. These relations typically are portrayed in the vertical dimension of a BioTapestry model. Similarly, the functional connections among regional GRNs due to intercellular inductive signaling are represented in BioTapestry models by horizontal interdomain linkages, as is the unidirectionality of these signaling interactions. Signaling is portrayed mechanistically (Chapter 3): the regulatory functions resulting in ligand gene expression are displayed as a feature of the GRN operating in the signal-sending domain; the immediate response transcription factor is in each case shown in the responding domain; and the specific target genes of this response factor are part of the GRN in the signal-receiving domain (Fig. 6.1(B)). Lastly, the basic modularity of GRN structure in terms of its subcircuit composition can easily be perceived in a BioTapestry presentation. Topological models are primarily designed to encompass the architecture of developmental regulatory networks and therefore they are most useful for analyses and conclusions based on architecture, such as for example evaluation of GRN hierarchy, subcircuit structure/function analyses, and considerations of evalutionary changes in network structure.

1.2 Comparative approaches to subcircuit topology

Our proposition is that subcircuit topology directly indicates subcircuit function in development. The concept of the GRN subcircuit derives from the observation that unit functions in development are performed by small sets of regulatory genes, usually 3–6, wired together by linkages of particular architecture, or topology. Furthermore, even in different development contexts, similar developmental jobs are frequently executed by subcircuits of similar topology, implying that linkage topology (the design of the activating and repressing interaction linkages) and not regulatory gene identity is the primary determinant of function. Thus comparative examination should reveal that subcircuits of similar architecture execute similar functions, even if composed of entirely different regulatory genes. In the following we utilize the topological networks presented earlier in this book for a meta-analysis of recurrent subcircuit architectures. Note that this comparison includes only networks the function of which is to execute spatial regulatory state specification in development.

In Fig. 6.2 is shown in BioTapestry format a compilation of instances representing eight different commonly occurring types of GRN subcircuit, almost all extracted from the networks discussed in chapters 3–5. These are positive feedback subcircuits, community effect subcircuits, coherent feedforward subcircuits, incoherent feed-forward subcircuits, signaling toggle switch subcircuits, mutual repression subcircuits, spatial exclusion subcircuits, and double-negative gate subcircuits. The theoretical impact of most of these subcircuits has been discussed earlier, in abstract terms (Mangan et al., 2003; Alon, 2007; Davidson, 2009; Peter and Davidson, 2009; Bolouri and Davidson, 2010;

Figure 6.2

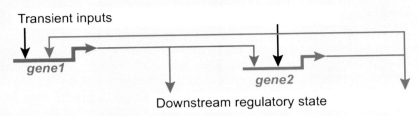

(A) **Positive feedback subcircuits**

Functions:
- **State stabilization**
- **Erase upstream variation**
- **Set level of downstream target gene expression**

(A1) **Sea urchin oral ectoderm**

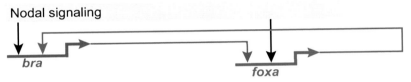

(A2) **Sea urchin aboral ectoderm**

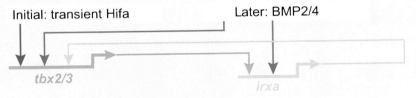

(A3) **Sea urchin non-skeletogenic mesoderm**

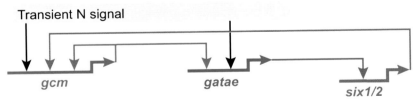

(A4) **Sea urchin skeletogenic mesoderm**

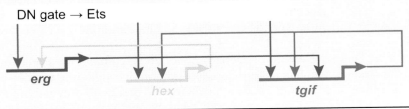

Figure 6.2

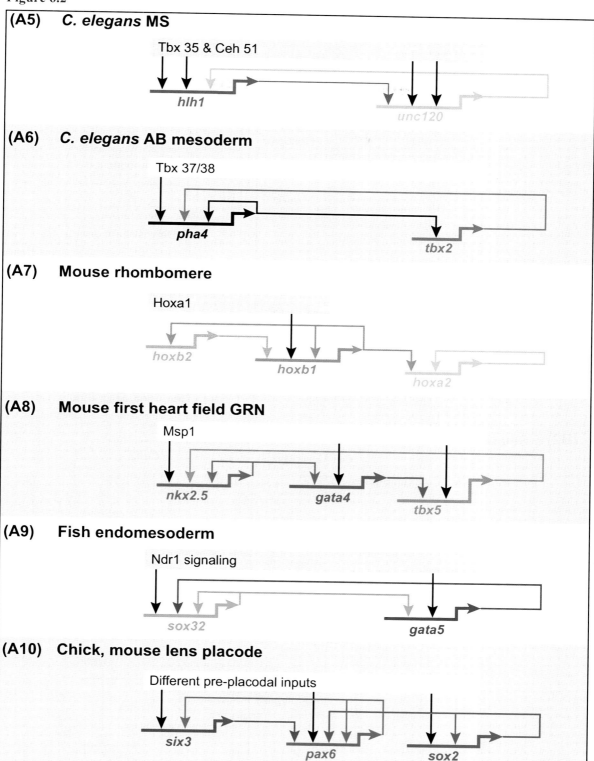

Figure 6.2

(B)　Community effect subcircuits

Functions:
- Intercellular communication among cells of a regulatory state domain required for maintenance of specific state
- Causes similar levels of expression of downstream target genes

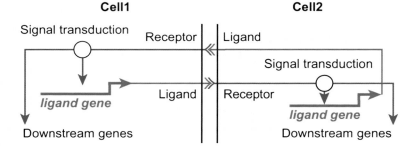

(B1)　Sea urchin oral ectoderm

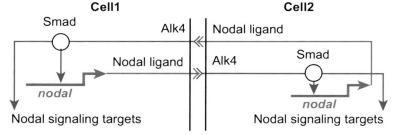

(B2)　Sea urchin skeletogenic micromeres

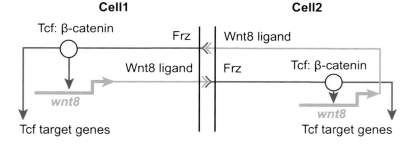

(B3)　Mouse limb bud

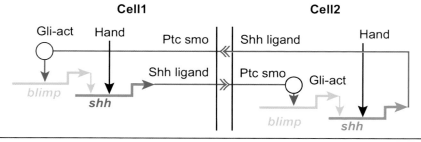

Figure 6.2

(B4) Mouse brain

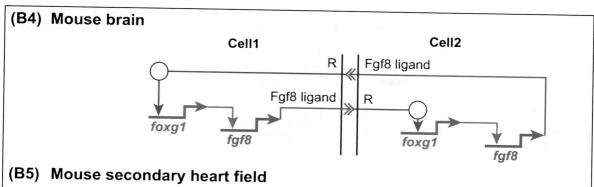

(B5) Mouse secondary heart field

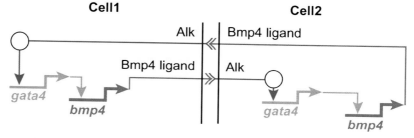

(B6) *Xenopus* **hemangioblast**

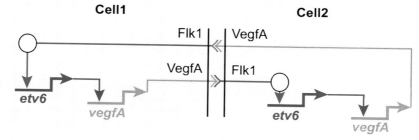

(C) Coherent feed forward

Functions:
- Device for high level expression of target genes
- Can cause temporal delay in target gene expression
- Integrates upstream spatial regulatory inputs to target gene
- Often used with multiple, similarly wired targets

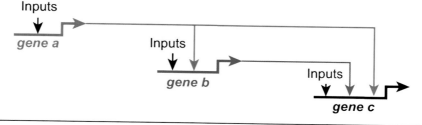

Figure 6.2

(C1) Sea urchin non-skeletogenic mesoderm

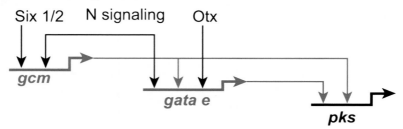

(C2) Sea urchin skeletogenic mesoderm

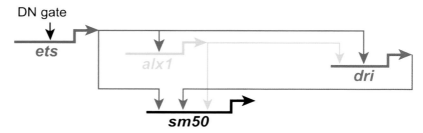

(C3) Sea urchin aboral ectoderm

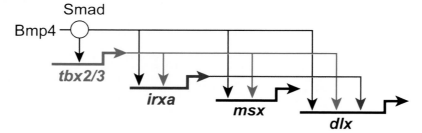

(C4) *Xenopus* blood cell

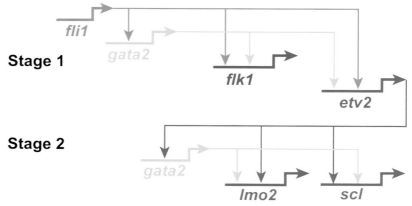

Figure 6.2

(C5) *C. elegans* **MS mesoderm**

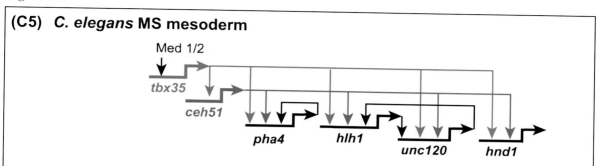

(C6) *C. elegans* **AB mesoderm**

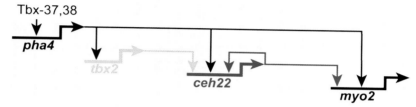

(C7) Mouse lens placode

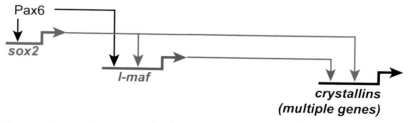

(C8) *Drosophila* **photoreceptor**

(D) Incoherent feed forward subcircuits

Functions:
• Spatial subdivision of domain defined by positional inputs
• Sequential cell fate decisions

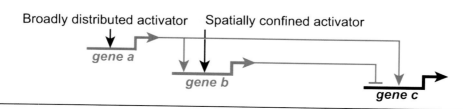

Figure 6.2

(D1) Sea urchin endoderm

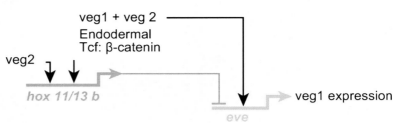

(D2) Sea urchin oral ectoderm

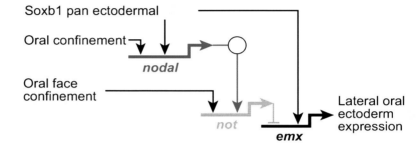

(D3)   Mouse secondary heart field

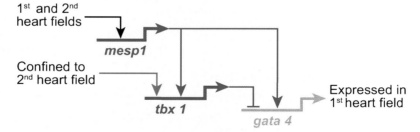

(D4) *Axolotl* endomesoderm

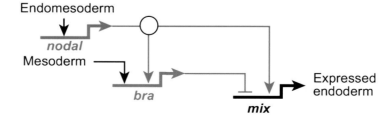

Figure 6.2

(D5) *Drosophila* **photoreceptor**

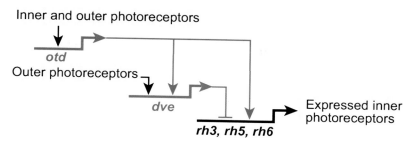

(E) **Signal mediated toggle switch circuitry**

Functions:
- Pre-existent response factor(s) act as default repressors on sequence-specific *cis*-regulatory targets
- In cells receiving signal, receptor mediated transduction events transform response factor into positive regulator for same *cis*-regulatory targets

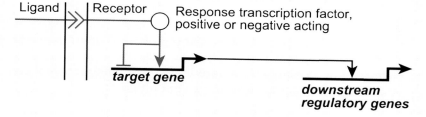

(E1) **Sea urchin endoderm**

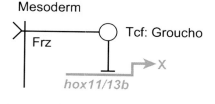
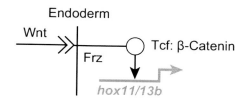

(E2) ***C. elegans* endomesoderm**

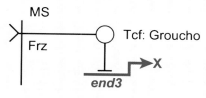
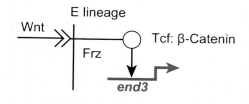

Figure 6.2

(E3)   Sea urchin mesoderm

Prospective endodermal veg2

Notch — Su(H): Groucho → gcm → X

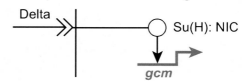

Prospective mesodermal veg2

Delta — Su(H): NIC → gcm

(E4)   Mouse limb bud

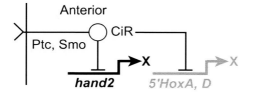

Anterior

Ptc, Smo — CiR — hand2 → X   5′HoxA, D → X

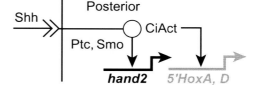

Posterior

Shh — Ptc, Smo — CiAct — hand2 →   5′HoxA, D →

(F)   Reciprocal repression subcircuits

Functions:
- **Irreversibly tips cell fate in one direction or another depending on relative inputs into each repressor**
- **Often found in single, non-parenchymal or mesenchymal cells where fate choices dependent on signals/cytokines encountered**

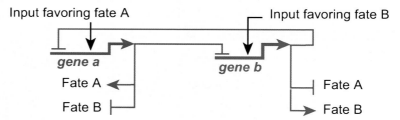

Input favoring fate A   Input favoring fate B

gene a   *gene b*

Fate A ←   Fate A
Fate B   Fate B →

(F1)   Mouse somitic cells

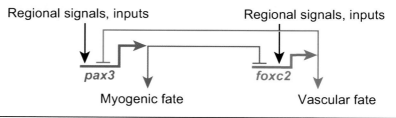

Regional signals, inputs   Regional signals, inputs

pax3   *foxc2*

Myogenic fate   Vascular fate

Figure 6.2

(F2) *Drosophila* **ocellus**

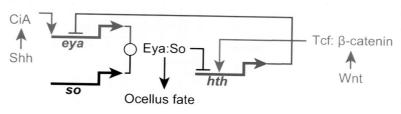

(F3) Mouse hematopoietic cells

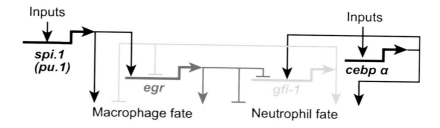

(G)   Spatial exclusion subcircuits

Functions:
- GRN establishing regulatory state includes transcriptional repression of key gene of alternative regulatory state
- Often used reciprocally in distinct regulatory state descendents from a common ancestor

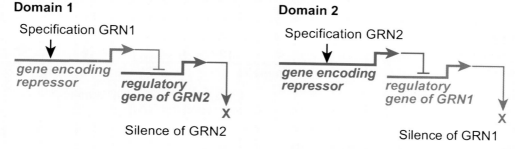

(G1) Sea urchin endoderm

Figure 6.2

(G2) Mouse rhombomeres

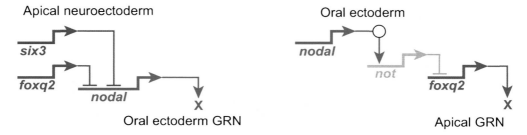

(G3) Sea urchin oral ectoderm

Apical neuroectoderm

Oral ectoderm

Oral ectoderm GRN

Apical GRN

(H) Double negative gate subcircuits

Functions:
- Spatial subdivision system ensuring genes are ON in a domain "A" and OFF elsewhere
- Repressors wired in tandem such that first repressor is expressed in A and transcriptionally represses second repressor, where latter is broadly expressed and controls set of downstream regulatory genes

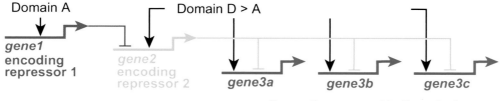

Genes 3 expressed in Domain A,
repressed in (D-A)

(H1) Sea urchin skeletogenic cells

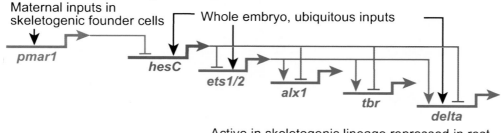

Active in skeletogenic lineage repressed in rest
of embryo

Figure 6.2

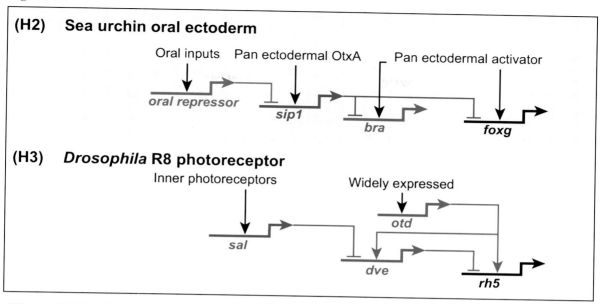

Figure 6.2 Recurrence of eight specific subcircuit architectures in developmental gene regulatory networks (GRNs). The canonical subcircuit types are presented diagrammatically using BioTapestry symbolism at the top of each section (A–H) of the figure, together with a brief bulleted legend indicating the functional role which that subcircuit plays in developmental GRNs. The instances that follow are drawn mainly from the GRNs reviewed earlier in this book, which serve as the database for most of this figure. To facilitate identification of the subcircuits in the parent GRNs, similar colors are used for the genes and linkages in this figure as in the respective parent GRNs. The identity of the GRN from which each example derives is given in the figure. (A) Positive feedback subcircuits: (A1) from GRN in Fig. 3.5(J); (A2) from Fig. 3.5(J); (A3) from Fig. 3.5(I); (A4) from Fig. 3.5(I); (A5) from Fig. 3.6(D); (A6) from Fig. 3.6(D); (A7) from Fig. 4.6(C); (A8) from Fig. 4.8(C); (A9) from Fig. 4.10(C); (A10) from Fig. 5.6(G). (B) Community effect subcircuits: (B1) from GRN in Fig. 3.5(J); (B2) from Fig. 3.5(I); (B3) from Fig. 4.2(C); (B4) from Fig. 4.7(E); (B5) from Fig. 4.8(C); (B6) from Fig. 5.4(D). (C) Coherent feed-forward subcircuits: (C1) from GRN in Fig. 3.5(I); (C2) also from Fig. 3.5(I); (C3) from Fig. 3.5(J); (C4) from Fig. 5.4(D); (C5) from Fig. 3.6(D); (C6) also from Fig. 3.6(D); (C7) from Fig. 5.6(G); (C8) from Fig. 5.2(F). (D) Incoherent feed-forward subcircuits; Gene A, activator expressed in a spatially broad domain compared to that in which gene B is expressed: (D1) from Fig. 3.5(I); (D2) from Fig. 3.5(J); (D3) from Fig. 4.8(C); (D4) from Fig. 4.10(A); (D5) from Fig. 5.2(F). (E) Signal-mediated switch subcircuits; "toggle" switch, in that response factor acts positively or negatively depending on presence of ligand/receptor complex: (E1) from Fig. 3.5(I); (E2) from Fig. 3.6(D); (E3) from Fig. 3.5(I); NIC, Notch intracellular domain; (E4) from Fig. 4.2(C). (F) Reciprocal repression subcircuits: (F1) from Fig. 5.5(C); (F2) from Fig. 4.9(B); (F3) from Laslo et al. (2006). (G) Spatial exclusion subcircuits: (G1) from Fig. 3.5(I); (G2) from Fig. 4.6(C); (G3) from Fig. 3.5(J). (H) Double-negative gate subcircuits; D>A indicates that spatial domain D is larger than and includes domain A: (H1) from Fig. 3.5(I); (H2) from Fig. 3.5(J); (H3) from Fig. 5.2(F).

Davidson, 2010; Shoval and Alon, 2010). Our focus here is on the role that these subcircuit topologies play in developmental contexts, and we consider the features of subcircuit architecture in respect to developmental process. While the particular collection of GRNs used as a source for the following compilation cannot be considered to be an unbiased database, the high level of recurrence of specific subcircuit types is still remarkable. For each of the subcircuit types, we begin in the following with a diagrammatic definition of the canonical linkage design and a brief list of the developmental functions that can be attributed to that subcircuit type. Examples drawn from the various GRNs follow.

Positive feedback subcircuits (Fig. 6.2(A))

Positive feedback subcircuits consist of regulatory genes which respond positively at the *cis*-regulatory level to their own or each other's gene products. The 10 cases of Fig. 6.2(A) portray direct mutual positive regulatory interactions that occur in the set of GRNs examined earlier in this book. The examples shown reveal a surprising incidence of particular structural features in addition to mutual cross-regulation. From an engineering point of view, it might be thought that the simplest form of circuitry that could accomplish feedback functions would require only a single gene with positive autoregulatory feedback. Irrespective of what a priori arguments might be made, the striking fact is that single genes displaying positive autoregulatory feedback are remarkably rare in the network models we have studied compared to multigenic feedbacks, except where autoregulatory wiring occurs in a regulatory gene also participating in a multigenic positive feedback subcircuit. One reason for this is that each of the genes participating in the multigenic feedback subcircuits that we show here provide inputs into different sets of downstream target genes, thereby coordinating their levels of expression. The multigenic feedback subcircuits thus have far greater combinatorial regulatory potentialities than if only a single autoregulatory gene were involved. It is also striking that all the cases observed here include either two or three mutually regulating genes, but never four or five. That positive feedback subcircuits consisting of either less than two or more than three regulatory genes are generally excluded, is not predicted simply in terms of the likelihood given by the prevalence of two- and three-gene feedback subcircuits. From this point of view it is interesting that a majority (seven out of ten) of the intergenic feedback subcircuits shown in Fig. 6.2(A) include at least one gene with positive autoregulation in addition to cross-regulation. Thus in the context of an intergenic positive feedback subcircuit, addition of an autoregulatory loop will not alter the Boolean behavior of the subcircuit and is allowed to occur. However, single gene positive auto-regulation is clearly disfavored in evolution.

As the 10 cases in Fig. 6.2(A) show, positive feedback subcircuits occur in developmental GRNs with great regularity. Most of the comprehensive GRN models included in this book in fact include such subcircuits, often one for each phase of specification. Our challenge is to understand in structure/function terms why this subcircuit design is so extremely useful. One way to approach this question would be consideration of the developmental contexts in which positive feedback subcircuits frequently appear. The cartoon at the top of Fig. 6.2(A) summarizes the contextual position of the feedback subcircuits portrayed in the following sections of the figure: these subcircuits are generally positioned immediately downstream of the transient inputs that initiate a specification process, and upstream of the GRN circuitry which ultimately executes the specification process. Another prominent though not invariable regulatory feature seen in these feedback subcircuits is that the same transient initial input directly regulates all of the interacting genes, as seen in about two-thirds of the examples in Fig. 6.2(A). The location of these subcircuits within GRN hierarchy reveals their fundamental functions. First, they convert transient inputs of regulatory significance, brief exposure to signals or briefly expressed transcription factors, into stable new regulatory states. These in turn provide inputs into the downstream specification program. Second, since the activities of the constituent genes come to depend on each other's function, positive feedback subcircuits act to erase the effects of variation in duration or intensity of the upstream transient inputs. Thus they act both as a regulatory memory device and as a mechanism that confers insensitivity to any variability of the initiating inputs in each phase of development. Third, thereby positive feedback subcircuits also set the levels of activity for their downstream transcriptional targets. Fourth, in dynamic terms, they can initiate changes of state with exponential kinetics.

Community effect subcircuits (Fig. 6.2(B))

The term "community effect" was introduced by Gurdon to denote the requirement for parenchymal cells to communicate to one another by signaling in order to maintain cell type-specific gene expression

(Gurdon, 1988). We here use this term for a specific subcircuit topology, which underlies reciprocal, symmetrical cell to cell signaling within given regulatory state domains (Bolouri and Davidson, 2010). The key architectural feature of community effect circuitry is that expression of the gene encoding the signaling ligand is driven by the signaling pathway activated by reception of the ligand. As can be seen at the top of Fig. 6.2(B), this circuit design institutes a positive feedback operating between cells, ensuring that they all express the same cohorts of downstream signal-dependent genes. Community effect signaling differs from more commonly discussed inductive signaling in that the cells expressing the ligand and the cells receiving the signal express the same regulatory state. Within a multicellular domain, intercellular feedbacks are established such that all cells express the ligand and also respond to the signal transduction system that it stimulates. Thus this subcircuit operates among cells of a domain, the boundaries of which are determined by other mechanisms usually involving repression. Community effect circuitry is not per se a pattern determining device, but contributes to homogenous intradomain regulatory function. An ODE analysis of the case portrayed in Fig. 6.2(B1) shows that community effect wiring can be expected to flatten out cell to cell differences in levels of expression of genes responding to the same signal transduction pathway (Bolouri and Davidson, 2010), a similar intercellular function as the intracellular function of the positive feedbacks discussed above.

It is remarkable that many different signaling systems are utilized in community effect wiring and this subcircuit constellation is thus in no way dependent on particular signal transduction pathways. Figure 6.2(B1) and (B5) show community effect circuits in which the ligands belong to the Tgfβ family (Nodal and Bmp4, respectively), Fig. 6.2(B2) concerns Wnt signaling, Fig. 6.2(B3) Hh signaling, Fig. 6.2(B4) Fgf signaling and Fig. 6.2(B6) Vegf signaling. In some of these subcircuits, ligand gene expression is directly regulated by the signal response transcription factor (open circles in Fig. 6.2(B)), as in the cases shown in Fig. 6.2(B1) and (B2), while in the other four cases shown, the signal response factor activates a regulatory gene which in turn controls transcription of the gene encoding the ligand; the effect is the same. Although relatively little is known of community effect signaling compared to what is known of inductive signaling, it is possible that this subcircuit topology is a defining feature of multicellular territories and tissues in which individual cells all express the same regulatory state. In GRN models, community effect subcircuits occur relatively high up in the hierarchy.

Coherent feed-forward subcircuits (Fig. 6.2(C))

Feed-forward subcircuits are defined by regulatory interactions among at least three genes such that an initiating gene A provides an input into a second regulatory gene B and both together regulate the expression of the target gene(s) C, as shown in the diagram at the top of Fig. 6.2(C). In "coherent" feed-forward subcircuits (Mangan et al., 2003; Alon, 2007), the regulatory import of the two pathways regulating expression of gene C is convergent, which is true for the following cases: (1) gene A activates gene C, gene A activates gene B, and gene B activates gene C; (2) gene A activates gene C, gene A represses gene B, and gene B represses gene C; (3) gene A represses gene C, gene A activates gene B, and gene B represses gene C; (4) gene A represses gene C, gene A represses gene B, and gene B activates gene C. The result of the first two cases is that the target gene(s) C are made to be expressed if A is expressed, while the result of the second two is that the target gene(s) are prevented from being expressed if A is expressed. By far the most commonly occurring of these four configurations is the first, in which all interactions are activating. This circuitry is very often used at the downstream periphery of developmental GRNs, where it drives the expression of effector genes. However, coherent feed-forward subcircuits are not confined to the periphery and are found throughout GRN hierarchy.

The frequency with which this subcircuit occurs raises the question of its functional significance for the operation of developmental GRNs. Previous considerations on the function of feed-forward motifs have focused on their dynamic behavior in the context of physiological response to external stimuli or metabolites (Mangan et al., 2003; Kalir et al., 2005; Alon, 2007). One conclusion is that feed-forward circuitry provides a temporal delay mechanism in respect to the initial input. However, the role of feed-forward subcircuits in the spatial transactions of the developmental process has in addition to be considered. Where feed-forward subcircuits are used for spatial patterning, they are most often used in the following way:

the inputs into gene A define a developmentally earlier and broader domain, for example a cell lineage or a regulatory state territory, while the inputs to gene B (which are different from those into gene A) define a subregion of the broad domain where gene A is expressed. If gene C integrates the inputs from genes A and B by AND logic, it is expressed only where and when both gene products are present. From an informational point of view, the temporal and spatial expression of gene C is therefore determined by that of gene B, and we might ask what the meaning is of the additional requirement of input from gene A. The significance of the input into gene C from gene A is that this input constitutes a regulatory memory of the developmental origin of the cells in which gene C is expressed, providing a safeguard which ensures that gene C will only be expressed in cells of this origin. Thus, gene B, like all constituent regulatory genes, might be used elsewhere for other purposes, but the innate combinatoriality of AND logic circuitry constrains expression of gene C to the particular developmental context defined by gene A.

Coherent feed-forward subcircuits are frequently found joined together in chains, for example in the subcircuits shown in Fig. 6.2(C3)–(C7). In addition, multiple feed-forward subcircuits converge on the same target gene, as in example Fig. 6.2(C2). Four out of the nine examples display the use of coherent feed-forward subcircuits to drive relatively high levels of expression of a downstream effector gene: here these are genes encoding polyketide synthase (pks; Fig. 6.2(C1)), the biomineralization protein Sm50 (Fig. 6.2(C2)), mouse lens crystallins (Fig. 6.2(C7)), and *Drosophila* rhodopsin rh3 (Fig. 6.2(C8)). An example is included of a coherent feed-forward loop that deploys repressive interactions. In Fig. 6.2(C8) we see that the *rh3* gene is activated through a type-2 circuit as defined above. Note that in all of these examples the consequence of the coherent feed-forward subcircuit is specifically to determine the spatial expression of the target gene.

Incoherent feed-forward subcircuits (Fig. 6.2(D))

While incoherent feed-forward subcircuits have many interesting dynamic properties (Alon, 2007), often applied in physiological contexts, we here focus on their function in the control of developmental spatial gene expression. The incoherent feed-forward subcircuit is defined by the property that its inputs into the terminal gene of the feed-forward architecture are of opposing import. Of the several configurations of activating and repressing interactions which conform to this property, the specific type of incoherent feed-forward subcircuit considered here is widely used for the specific purpose of subdividing a spatial domain. In this type of subcircuit, as shown in the cartoon at the top of Fig. 6.2(D), gene B receives a positive input from gene A but it also receives an additional input that confines its expression to a particular subregion. Gene B produces a repressor for the target gene C and although gene A provides the same positive memory input as in coherent feed-forward subcircuits, the repressor encoded by gene B is dominant. Thus gene C operates in the domain (A minus B), that is, in cells expressing gene A but not gene B. Temporally, these spatial subdivision mechanisms operate in such a way that the initially broad expression of gene C, which is activated by the input from gene A, precedes the confined expression of gene C determined by the dominant repressor encoded by gene B.

Each of the five examples in Fig. 6.2(D) follows the "A minus B rule". Thus gene A is expressed in a broad domain which is divided into two subregions expressing either gene B or gene C. Where gene B is expressed is determined by its *cis*-regulatory architecture, which integrates the activating input from gene A with a spatially confining input as indicated in the boxes in the examples of Fig. 6.2(D). Thus in the sea urchin endoderm, input A (Tcf/β-catenin) is present in the veg1 and veg2 lineages, whereas expression of gene B (*hox11/13b*) is restricted to veg2 and hence the target gene (*eve*) runs only in veg1 descendants (Fig. 6.2(D1)); in sea urchin oral ectoderm, input A (*soxb1*) is expressed throughout the ectoderm, whereas expression of gene B (*nodal*) is confined to oral ectoderm where it activates gene B' (*not*) in the oral facial ectoderm, and the target gene (*emx*), which is activated by Soxb1 and repressed by Not is consequently expressed in the lateral margins of the oral ectoderm (Fig. 6.2(D2)); in the mouse heart, gene A (*mesp1*) is expressed in the first and second heart field, whereas expression of gene B (*tbx1*) is confined to the second heart field and the target gene (*gata4*) is consequently expressed only in the first heart field (Fig. 6.2(D3));

in *Axolotl* endomesoderm, inputs A (nodal-like genes) are expressed in the endomesoderm, whereas gene B (*bra*) is expressed only in the mesoderm and consequently the target gene (*mix*) is only expressed in the endoderm (Fig. 6.2(D4)); in *Drosophila* photoreceptor specification, gene A (*otd*) is expressed in the inner and outer photoreceptors, whereas gene B (*dve*) is expressed only in the outer photoreceptors and consequently the target genes (*rh3*, *rh5* and *rh6*) are expressed only in the inner photoreceptors (Fig. 6.2(D5)).

Signal-mediated toggle switch subcircuits (Fig. 6.2(E))

Many developmental signaling systems display a 0/1 Boolean functional output. Even though the signal transduction biochemistry differs strikingly among these systems, their downstream effects are remarkably similar with respect to the GRNs in which they participate. Thus these systems have the quality that in cells receiving the signal they modify the activity of a transcription factor, termed here the "immediate response factor". In consequence of signal reception, this response factor changes its activity state from that of an obligate repressor for its sequence-specific target genes to that of a transcriptional activator or a factor that is permissive for activation of these same target genes by other regulators (Barolo and Posakony, 2002; see discussion in Chapter 3). This gives the response factor a Janus-like function, a switch which converts the advent of a signal into a change in downstream transcriptional activity. Inductive signaling in development always affects the regulatory state because some target genes of the response factor always encode transcription factors. Since the same small set of signaling systems is used over and over in bilaterian development, the expression of target genes is usually regulated combinatorially by the signal response factor plus transcription factors present as a function of the specification GRN up to that point, as shown by many recent examples (e.g., Trompouki et al., 2011).

In Fig. 6.2(E1)–(E4) similar Janus logic functions extracted from the foregoing GRN models are shown for Wnt signaling (with Tcf as the response factor), in sea urchin and *Caenorhabditis elegans* embryos; Notch signaling in the sea urchin embryo (with Suppressor of Hairless (Su(H)) as the response factor); and Hedgehog signaling in the mouse limb bud (with Cubitus Interruptus (CI) as the response factor). The alternative transcriptional functions of the response factors are regulated by different biochemistries. For Tcf, Wnt signaling causes the nuclear accumulation of the coactivator β-catenin, which replaces the corepressor Groucho in complex with Tcf. For Notch signaling, reception of the ligand causes the proteolytic cleavage of Notch, producing a polypeptide which interacts with Su(H), again replacing Groucho. For Hh signaling, the response factor CI contains more than one *trans*-regulating domain and is cleaved into either activating or repressive forms (Briscoe and Thérond, 2013). As with all classes of circuitry considered in this section, we see here completely distinct transcription factor identities, signal transduction biochemistries, and target genes, which however execute identical regulatory logic functions.

The overall regulatory functions of developmental inductive signaling is to link the activity of a GRN in a given domain which mediates ligand gene expression to the activity of another GRN in a second domain where the signal is received, and where it alters the regulatory state. Frequently, inductive signaling is used to subdivide spatial gene expression domains, where the Janus function embedded in this subcircuit ensures that the result of signal reception is unique. Thus, within a given gene expression domain, genes controlled combinatorially by tissue-specific sets of transcription factors and the signal response factor, will be activated only in cells receiving the signaling ligand and repressed in all other cells.

Reciprocal repression subcircuits (Fig. 6.2(F))

Results from several developmental systems have revealed a uniquely configured circuitry in which two genes encoding transcription factors mutually and directly repress one another. In the canonical diagram shown in Fig. 6.2(F), these factors are encoded by genes A and B. When either gene is expressed it shuts down transcription of the other. Which one of the two is expressed in any given developmental context depends on alternative activating inputs, frequently specification signals or cytokines, or inputs from other regulatory genes which respond to such signals. Thus immediately downstream of the specification inputs, this circuitry is used for cell fate decisions occurring within a progenitor cell that can give rise to two alternative developmental fates. Mutual repression circuitry can occur in regulatory chains, which mediate successive cell fate decisions in space and time. As indicated in the diagram in Fig. 6.2(F), expression of gene A results directly or indirectly in transcriptional promotion of fate A and transcriptional repression of fate B; correspondingly, expression of gene B results directly or indirectly in transcriptional promotion of fate B and transcriptional repression of fate A. Thus the product of gene A may have three different regulatory functions: the repression of gene B, the activation of regulatory genes required to generate fate A, and the repression of downstream regulatory genes required for fate B. The same applies to the product of gene B.

Several examples of this type of subcircuit are given in Fig. 6.2(F), drawn from different developmental contexts. In mice, as reviewed in Chapter 5, the cell fate decision between myogenic as opposed to vascular fate in the ventral somite is controlled by mutually exclusive expression of *pax3* and *foxc2*, such that *pax3* expressing cells become the precursors of skeletal myocytes, and *foxc2* expressing cells become the precursors of endothelial and smooth muscle cells of the aorta (Fig. 6.2(F1) and Fig. 5.5(C)). Within the somite and during their migration, these multipotential cells receive different positional signaling inputs, resulting in expression of either *pax3* and thereby repression of *foxc2*, or vice versa. Exclusive expression of *pax3* produces myogenic cell fate, while exclusive expression of *foxc2* results in vascular cell fate. In Fig. 6.2(F2), inter-ocellular versus ocellar fate in *Drosophila* can be seen to depend similarly on mutual repression between *eya* (and its partner *so*) and *hth* (Fig. 4.9). Downstream of the Eya:So complex are genes associated with ocellar fate. Upstream of *eya* is a positional input from the activating form of Ci as a result of Hh signaling, while upstream of *hth* is a positional input from Wnt signaling. Thus, in the inter-ocellar region Wnt signaling leads to the exclusion of ocellus formation by activation of the Hth repressor. The circuit in Fig. 6.2(F3) is utilized in postembryonic mouse hematopoietic lineage choice between macrophage and neutrophil fates from a common precursor (Laslo et al., 2006). Here, *egr* promotes macrophage fate and represses *gfi-1* and neutrophil fate, while *gfi-1* reciprocally promotes neutrophil fate and represses *egr* expression and macrophage fate. Thus we see all three functional outputs from each of the regulatory genes engaged in the mutual repression circuitry. Expression of *egr* is driven by PU.1, which responds to cytokine inputs, while expression of *gfi-1* is driven by C/EBPα in response to other cytokine inputs.

The mutual repression circuitry is a frequent subject of dynamic ODE models. However, the way this subcircuit is treated in these mathematical models differs from the way this subcircuit is often used in development. The presumption in these models is that initially the two mutually repressive transcription factors have the same chance of being expressed, but the consequence of a small nudge in either direction causes the regulatory system to assume either one or the other stable state. These models produce a series of specific predictions on the means by which such circuitry would effect cell fate decisions. Among these are that the mechanistic system underlying biological cell fate choice would be bi-stable in the sense that either direction in a given context is possible and of similar likelihood; that the transition to final state would be preceded by an unstable condition in which both repressors operate; that transcriptional stochasticity or "noise" could initiate an inequality in repressor levels and thus trigger the transition into one or the other state; and basically, that it would be the opposition between the mutually repressive transcription factors which causes the switch to function in one direction or the other. In development however, this type of circuitry operates in a different way. Most importantly, in development of the body plan, cell fate decisions occur in a nonstochastic, determinant manner with respect to the spatial component of the developing system in which the cells reside at the time of their specification. The expression of the two repressors at the heart of these subcircuits generally depend on distinct upstream control systems, which determinately favors expression of one or the other of the repressors in a context-dependent manner. That is, specific signals or previously generated transcriptional activators define the cell fate outcome in given spatial and temporal contexts. Thus the presence of these specific inputs is what determines the cell fate choice, and the mutual repression system ensures that only one cell fate will ensue. In each of the examples shown in Fig. 6.2(F) the determinant inputs driving the cell fate choices are shown in the boxes flanking the mutually repressive gene circuit.

Spatial exclusion subcircuits (Fig. 6.2(G))

Repression of alternative cell fates is an important and general feature of developmental process. The canonical underlying circuitry is the expression of a gene encoding a repressor that specifically targets a key regulatory gene in an alternative GRN. Specification GRNs often include genes engaged in exclusion subcircuits, the specific function of which is to ensure the silence of regulatory genes that if active would promote the specification of an alternative fate. Often this circuitry is utilized in situations where a regulatory state domain is subdivided into different specification states, whereupon each of the domain-specific GRNs represses genes essential for the operation of the alternative specification GRN. Sometimes the two GRNs initially express similar regulatory genes as a consequence of their regulatory history, or are exposed to similar signaling inputs. Typically, exclusion subcircuits are revealed experimentally when expression of the repressor is interrupted, leading to ectopic spread of the alternative regulatory state. Thus these circuits are not simply "redundant" aspects of "robustness" but rather act as functional determinants of regulatory state diversification. Their specific function is to set the boundaries of the regulatory state in which they are expressed. Since exclusion functions occur very generally, it is often found that such functions enforce both sides of a boundary between two adjacent specification domains.

Of the many cases that could have been chosen, Fig. 6.2(G) includes only a few to illustrate their function. The examples in Fig. 6.2(G1) and (G2) display exclusion functions expressed within given domains, while Fig. 6.2(G3) display such functions on both sides of the boundary separating two domains. In Fig. 6.2(G1) we see that a canonical gene of sea urchin anterior endoderm, *foxa*, specifically represses a key driver gene in aboral mesoderm, *gcm*. Anterior endoderm and aboral mesoderm have a common embryological origin (Chapter 3). In Fig. 6.2(G2) is portrayed an exclusion function required for the differential specification of mouse rhombomeres 3 and 5 vs rhombomere 4. The circuitry shown is essential for exclusion of *hoxb1* expression from rhombomeres 3 and 5 but it does not run in rhombomere 4 where it is essential that *hoxb1* be expressed (Chapter 4, Fig. 4.6). Figure 6.2(G3) shows the cross-boundary exclusion functions by which apical neuroectoderm and oral ectoderm are separated in the sea urchin embryo. Regulators of the apical neuroectoderm, *six3* and *foxq2*, exclude all oral ectoderm GRN expression by blocking transcription of the *nodal* signaling gene. In the oral ectoderm across the boundary, a repressor expressed downstream of nodal signaling, encoded by *not*, specifically excludes expression of the apical regulatory gene *foxq2*. Exclusion functions of similar design occur generally in cell fate specification GRNs.

Double-negative gate subcircuits (Fig. 6.2(H))

This type of subcircuit again features multiple genes encoding repressors, but in this case wired in tandem, such that the first of these repressors prevents the expression of a second repressor, and the second repressor prohibits transcription of a set of target genes. Double-negative gate subcircuits are used in development specifically for spatial control circumstances in which it is required that target genes be expressed exclusively in a specific subregion and that they be dominantly repressed elsewhere, despite the presence of their broadly distributed activators. As the cartoon at the top of Fig. 6.2(H) shows, the means by which this is accomplished depends on the spatial expression of the respective activating inputs into the two genes encoding the repressors. Thus the upstream repressor gene 1 is expressed in response to activators confined to region A, while the activators driving expression of repressor gene 2 function over a larger region D including region A, as do the activators driving the target genes 3. In consequence of this configuration, genes 3 can be expressed only in region A, in which repressor 2 is prevented from being expressed, and they are repressed elsewhere in the domain D (D minus A). This circuitry ensures global control of the target genes 3, both positive and negative, throughout domain D, and provides a more precise control system than one relying exclusively on the presence of an activator in domain A. Thus it combines activation and exclusion functions in one spatial specification subcircuit.

We have extracted three examples from the networks reviewed here. In Fig. 6.2(H1) is reproduced the double-negative gate responsible for installing the specific regulatory state of the skeletogenic lineage of the sea urchin embryo. Here gene 1 is *pmar1*, active in the founder cells of this lineage. Gene 2 is *hesc*, encoding a repressor of skeletogenic regulatory genes (gene 3). In this case, domain D (where *hesc* is expressed) is the entire embryo, while domain A (where *pmar1* is expressed) is the skeletogenic lineage. The consequence of the wiring is that the skeletogenic regulatory genes are expressed only in this lineage and are repressed throughout the remainder of the embryo (Chapter 3, Fig. 3.5). Figure 6.2(H2) shows another double-negative gate from the sea urchin embryo, this excerpted from the oral ectoderm GRN model. Here gene 1 encodes a repressor expressed exclusively in the oral face, gene 2 (*sip1*) encodes a repressor active throughout oral and aboral ectoderm, which targets the regulatory genes 3 (*bra*, *foxg*). The consequence again is expression of these target genes specifically in the oral face and repression in the rest of the ectoderm. Figure 6.2(H3) concerns photoreceptor specification in the *Drosophila* ommatidium. Gene 1 (*sal*) is expressed in inner photoreceptors, gene 2 (*dve*) is widely expressed in both inner and outer photoreceptors under control of *otd*, and the target gene 3 encodes rhodopsin 5. In consequence, *rh5* is expressed only in inner photoreceptors and specifically repressed in outer photoreceptors (Chapter 5, Fig. 5.2(F)). Double-negative gate circuitry is an example of a genomically encoded regulatory design specifically useful for the developmental processes of spatial specification and cell fate decision.

1.3   Significance of GRN subcircuits

A compelling conclusion from the foregoing comparative analyses is that the same topology of each of the subcircuits considered appears repeatedly in diverse developmental GRNs and executes similar network functions. But in each instance subcircuits of a given type are composed of entirely different regulatory genes. In fact, in these examples there is no case where the same regulatory genes are used more than once for a given type of circuit. This demonstrates that subcircuit function depends exclusively on topological design and is not determined by the specific biochemical nature of the constituent regulatory factors, except for their general properties as activating or repressing transcription factors. Perhaps most surprising is the fundamental similarity of inductive signaling logic despite the tremendous varieties of signal transduction pathways responding to diverse ligands. To summarize in one sentence, genomic regulatory functions in development depend on subcircuit logic functions and not on unique properties of particular individual transcription factors.

The necessary roles of the subcircuit repertoire considered here are attested to by the recurrence of each type in multiple GRNs, drawn from completely different branches of the bilaterian tree and completely different developmental episodes. It can only be concluded that feedback subcircuits, the various feedforward subcircuits, double-negative gate subcircuits, Janus signaling subcircuits, etc. are required constituents of GRNs for embryonic development. No doubt other subcircuit types will emerge as additional GRNs are solved, and indeed we consider more complex subcircuit configurations in the following section of this chapter. But the importance of the set of elemental subcircuits we have focused on here is obvious. The recurrent use of the same subcircuits implies that there may exist a finite complement of genomically encoded subcircuits which recur regularly in developmental GRNs, and from which these GRNs are constructed. This idea has deep implications for the evolutionary origins of the morphological programs that control bilaterian development.

The utilization of given subcircuit architectures is clearly not what would be expected on the basis of random occurrence of regulatory interactions. To give one striking example, as noted above, one might expect that isolated autoregulatory wiring would occur with equal frequency throughout the different regions of a GRN, that is, any gene might have a similar probability of engaging in autoregulation positive or negative. But what is observed is clearly different from this. Thus here we found among all these examples only three occurrences of isolated positive autoregulations but 10 occurrences of positive auto-regulation in the context of positive intergenic feedback subcircuits, even though there are many times more genes in these networks as a whole than are included in intergenic feedback subcircuits. Additional evidence that isolated positive autoregulation has been disfavored in evolution is the relatively high frequency of isolated negative autoregulation that we observe, 18 occurrences in the same set of networks, obviously none in the context of positive intergenic feedbacks. On a random basis, the likelihood of a positive and negative autoregulatory site should be quite similar, but in fact we see 3 versus 18 occurrences outside of intergenic feedback subcircuits. Subcircuit occurrence depends on requirement in GRN context for given subcircuit functions, which of course is not expected to be random. We can predict that as comparative subcircuit databases expand beyond the initial attempts we have made here, a large body of evidence will accumulate displaying the nonrandom occurrence of all the canonical subcircuit topologies we here consider.

2. ODE Models of Circuit Dynamics

2.1 Applying ODE models to developmental GRNs

Kinetic ODE models relate the rate of change in output of gene products to the rates of change in the inputs, according to the assumed mechanism. Since these mathematical models operate on microscopically incremental changes, they apply specifically to processes which behave continuously on a macroscopic scale. For example, the output of the physical–chemical processes of transcription indeed appear macroscopically continuous in time, and as we saw in Chapter 2, convenient ODEs can be applied to quantitative analysis and prediction of many aspects of transcription kinetics. These include *cis*-regulatory occupancy, transcription, transcript turnover, and simple gene cascades. Even here however, approaches differ widely with respect to how closely the mechanisms encompassed in the mathematics conform to the physical mechanics of the process. For example, as we discussed in Chapter 2, the transcriptional output of a target gene in respect to the concentrations and affinities of its input regulators can be modeled as a direct consequence of regulatory DNA occupancy using a probabilistic Boltzman distribution function. However, much more approximate mathematical approaches are frequently used in which, for example, transcription and translation are subsumed together rather than treated as separate processes, or transcription is computed using algorithms from enzyme kinetics, etc. These compromises are partly the result of the intrinsic difficulty of representing the statistical mechanics of complex processes, and are partly due to lack of knowledge of how parts of these processes work in detail.

ODE models that capture even simple transcriptional processes in mechanistic detail require evaluation of multiple parameters including relative equilibrium constants for transcription factor binding, cooperativity constants, polymerase translocation rates, RNA turnover rates, and protein turnover rates, among others. For models of real-life systems that include many genes and their interactions, a formidable number of parameters would be required. Models necessitating explicit treatment of realistic sets of participating genes can thus become intractable. The intractability is due to the virtual impossibility for animal systems, at our present level of technological capacity, to evaluate experimentally a large fraction of the physical–chemical parameters that would figure in any exact large-scale solution. When there are more than a small number of genes, parameter estimations from observed kinetics become dependent on one another. Therefore, in order to utilize the intellectual resolving power of kinetic analyses for regulatory processes, it becomes necessary to devise compromises in lieu of parameter evaluations. For example, parameters can be subsumed into one another or combined together into abstract terms which do not directly represent any physical entities; or intelligent guesses can be made about which few parameters are most important, ignoring the rest; or abstract scale-free models can be invoked. ODE analysis by these means can be used to predict the fundamental character of some processes, and in the following we visit several examples where ODE analyses of developmental processes have contributed fresh insights into the behavior of the underlying regulatory systems. But on the other hand, it is impossible to regard the results of abstract mathematical solutions literally in quantitative terms in respect to life processes. A common strategy employed to deal with large sets of ODEs and their attendant array of unknown parameter values is to use random variables to search the parameter field for solutions which reproduce the observed outcome. Such solutions may provide useful experimental targets. But simply because there exists a set of parameter solutions for the given mechanism assumed in the model, which produces the expected result, does not constitute evidence that both model and parameters represent the actual biological mechanism, despite the elegance of the mathematical solutions.

One significant problem in applying ODE models to real-life transcriptional cascades is how to capture *cis*-regulatory function, including the different types of logic operations by which *cis*-regulatory modules integrate their regulatory inputs. As reviewed in Chapter 2, *cis*-regulatory modules always operate on combinatorial inputs and produce outputs which are not related in any simple way to the instantaneous quantity of their multiple inputs. To add insult to injury, given genes frequently use alternative multiple *cis*-regulatory modules, the choice of which constitutes a regulatory transaction. Thus *cis*-regulatory inputs mediate a variety of interaction-dependent functions beyond concentration-driven activation or repression. None of these can be modeled by simple assumptions of singular inputs or functional equivalence among the inputs such as additive OR logic. For example, approximating the consequence of *cis*-regulatory AND logic, where the presence of two inputs is required, by multiplying their production rates, adding them in the case of OR logic, or applying subtraction or division to these rates in the case of repression, bear little or no relation to the mechanistic processes controlling *cis*-regulatory output. For treatment of combinatorial *cis*-regulatory logic functions, Boolean approaches provide a richer variety of possibilities that more exactly represent such functions where known, but of course at the expense of any quantitative description.

Finally and most importantly, the outcome of developmental process is fundamentally discontinuous. That is, body parts, their progenitor fields, spatial regulatory states, and cell types arise in discrete spatially bounded patterns. Mechanistically, transcriptional repression is causal in many specification events including boundary formation and cell fate decisions. Such repression is biochemically a multistep process that does not necessarily depend in a continuous manner on the concentration of the repressor. There are likely to be many different mechanisms of repression with different kinetic properties, but some result in irreversible silencing. In the initial stage, indeed, a transcriptional repressor reversibly binds to its *cis*-regulatory target site, but this is typically followed by a sequence of biochemical interactions with the transcription apparatus and then by enzymatically driven alterations of the chromatin environment, and sometimes even by the relocation of the repressed gene to a different region of the nucleus. These events may lead to the complete silencing of gene expression which can no longer be reversed simply by decrease of concentration of the initial repressor. Such repression thus actually behaves in a discontinuous way, and is frequently used for the specification of spatial domains of gene expression. Modeling discontinuous spatial gene expression using chains of continuous ODE functions thus in principle confronts developmental mechanism in a basically inconsistent way. In order to circumvent these problems and yet approximately encompass the behavior of developmental processes, ODE models often employ various mathematical devices to artificially mimic the outcome of the underlying molecular biology. Some such devices essentially "Booleanize" process output, for example the ad hoc use of high Hill coefficients, or using steep sigmoid functions to model gene expression boundaries. In the limit of any of these approximations are step functions which transform continuous ODE treatments into Boolean models, to which we return below.

2.2 "Gene circuit" models for the *Drosophila* gap genes

The gap gene transcriptional regulatory system of the thirteenth to fourteenth cleavage cycle *Drosophila* embryo has unique characteristics (see discussion in Chapter 3, Section 5, and Fig. 3.8) which lend themselves to a special ODE modeling approach. The most important of these characteristics are that there is a relatively small number of interacting genes in the system; that the gap gene products diffuse among adjacent nuclei, as the embryo is at this stage still syncytial, and the nuclei reside in a thin layer of superficial cytoplasm so that transcription factor diffusion is essentially two dimensional; that the initial inputs are maternal and some are spatially static in the embryo; and that it has been possible (as the result of a great deal of very careful experimental effort) to obtain exact relative quantitative spatial profiles of both the gap and maternal proteins during the time period considered (70 min). Furthermore, the broadly localized maternal inputs are activating, while the spatial pattern of gap gene expression depends on the repressive interactions among these genes. There have been three main objectives of the modeling approach that we briefly review here. First, the models have been used to derive the configuration of the network of gap gene interactions. The model results can then be compared with those deduced from genetic interaction data and the limited available *cis*-regulatory information. Second, these models provide a dynamic explanation for how the anterior-posterior gap gene expression patterns are established during this interval. Third, they illuminate the stabilizing effect of the network architecture on the spatial boundaries of the gap gene expression domains.

Many different kinds of modeling approach have been applied to the gap gene system over the years, some based on ODE dynamics, some on Boolean logic (see below) and earlier, mainly on diffusion kinetics (for review, Jaeger, 2009). Here we focus on "gene circuit" models (Jaeger et al., 2004; Perkins et al., 2006; Jaeger, 2009; Manu et al., 2009). In this approach, ODEs are used to compute the generation of gap gene products, i.e., transcription factor protein, in each nucleus along the anterior-posterior axis of the embryo (Fig. 6.3). The dynamics of production of each gap gene transcription factor is considered to depend upon: its rate of protein accumulation due to transcriptional regulation by other gap genes; its rate of protein accumulation due to regulatory inputs from other factors present in the embryo, the distribution of which does not depend on the gap genes, namely Bicoid, Caudal, Tailless, and Huckebein; the initial level of gap gene products at the twelfth cleavage cycle; diffusion of the transcription factors among adjacent nuclei; and the decay rates of the proteins. "Gene circuit" equations encompassing all these inputs are set up for *hunchback* (*hb*), *knirps* (*kni*), *giant* (*gt*), and *Kruppel* (*Kr*) genes. The input information used for the computation is the measured levels of Bicoid in each nucleus along the A/P axis, and measured levels of Tailless, Caudal, and Huckebein, which are all considered invariant in time over this interval. The regulatory impact of each of these inputs on each gap gene in each nucleus is derived by multiplying the relative concentrations of these input factors in that nucleus at a given time by variable interaction parameters. It is considered that expression of each gap gene might also depend on regulatory inputs from other gap genes if they are expressed in the given nucleus. For example, for the regulatory effect of gap gene *b* on the transcription of gap gene *a*, the instantaneous concentration of gap gene *b* product is multiplied by its interaction parameter with gap gene *a* (T^{ab}), where the protein concentration of gap gene *b* product at each time in each nucleus is derived from the computation. Thus the protein production rate of each gap gene is considered to depend on possible inputs from all other factors in the system. Each regulatory input can be activating, where the interaction parameter has a positive value, repressing, where the value of the interaction parameter is negative, or without consequence on gene expression, in which case the value is zero. These inputs are considered to be independent of one another and gene expression output is derived from the sum of all inputs. A major objective of the model is to derive computationally the gap gene interactions required to produce the observed expression patterns. This is done by finding the optimal set of interaction parameters representing the strength and sign of regulatory interactions between gap genes by fitting the computed concentrations of gap gene products over time and space to the observed gap gene expression data. As summarized in Fig. 6.3(A), the model is iteratively computed with free choice of the interaction

Figure 6.3

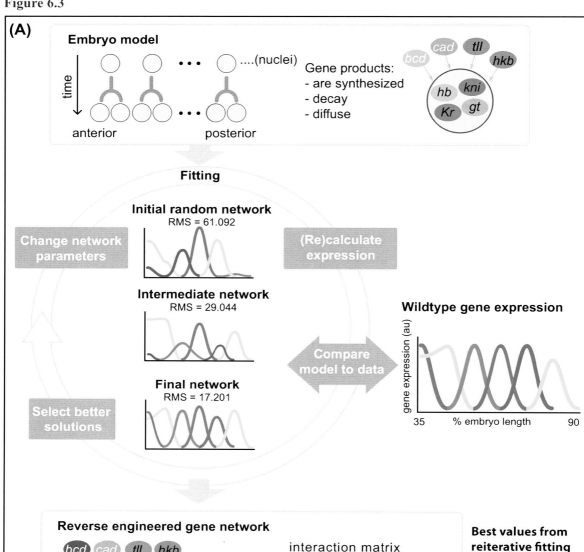

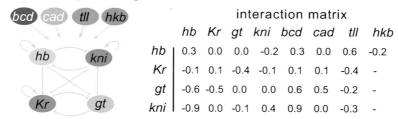

Figure 6.3

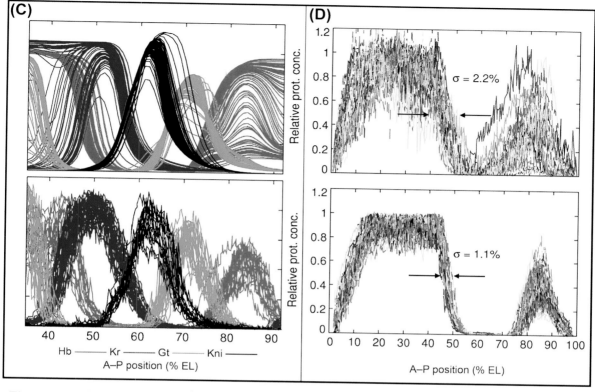

Figure 6.3 Gene circuit model for *Drosophila* gap genes. (A) Diagram of gene circuit modeling approach *(from Crombach et al. (2012))*. At the top is a schematic representing nuclei in the syncytial *Drosophila* embryo, undergoing a cleavage division; most of the action in the model in this figure occurs in the thirteenth and fourteenth cleavage cycles. The model concerns the regulatory interactions of the four gap genes (via their protein products) indicated on the right within the nucleus, viz. *hunchback* (*hb*), *giant* (*gt*), *Kruppel* (*Kr*), and *knirps* (*kni*). These interactions are affected kinetically by processes of synthesis, decay, and diffusion of gene products. The genes also receive external (i.e., to the gap gene interaction network) regulatory inputs from spatially static transcription factors encoded by the maternally expressed *bicoid* (*bcd*) and the maternally and zygotically expressed *caudal* (*cad*) genes, and from the zygotically expressed genes *huckebein* (*hkb*) and *tailless* (*tll*). Below are the iterative mathematical steps by which the key regulatory parameters of the model are varied, the spatial output calculated, and compared (by a least squares method) to measurements of gap gene expression. The process is repeated until no improvement in fit is attained by further parameter changes. The bottom panel displays on left the optimized network resolution, referred to as the "reverse engineered" network. Its configuration is determined by the best fit values of the regulatory "interaction matrix" shown at the right, which contains information on which genes interact positively (positive XY values), negatively (negative values), or not at all (zero values). (B)–(D) Mathematical formalization, model output and observations in normal and mutationally perturbed conditions *(from Manu et al. (2009))*. (B) Canonical gene circuit equation, left, and boundary sigmoid function, right. A gene circuit equation is formulated for each gap gene considered. Here i denotes a particular nucleus along the A/P axis, a indicates a particular gap gene, t is time, v is the amount of gap gene transcription factor, so the left side of the gene circuit equation gives the time rate of change in amount of that gene product (protein) at a given instant in a given nucleus. The right side of the equation contains three terms, the first representing the

several mechanisms that affect synthesis of the gap gene product; the second, diffusion of the protein into the ith nucleus from each of the adjacent nuclei (where D is a diffusion parameter varying inversely with (distance between nuclei)^2); and the third term gives the rate of decay of the factor, where λ is the first-order decay rate constant. In this model the synthesis rate of a given gap gene product depends on regulatory interactions with those other gap genes that interact transcriptionally with it, and with the "external" factors (Bcd, Hkb, Cad, Tll). The total regulatory effect, u^a, is modeled as a sum of the individual effects of these regulatory interactions, that is: $u^a = (h$, representing input from ubiquitous maternal factors; plus input from Bcd, based on measured Bcd spatial distribution (m^a is the regulatory input from Bcd on gene a and v_i^{BCD} is Bcd concentration in nucleus i); plus inputs from the spatially localized regulators β, viz. Tll, Cad, or Hkb (here $v_i(t)$ represents the measured instantaneous concentration of the gene product β in nucleus i and $E^{a\beta}$ is the regulatory input of gene product β on gene a); plus the regulatory input from the gap gene interactions per se). These last are modeled as the product of an interaction parameter T^{ab} which gives the effect of gap gene b product on transcription of gap gene a, as in the table in (A), multiplied by v_i^b, the concentration of the input gap gene b product in nucleus i. The sum of these various regulatory effects on total transcriptional output of gap gene a, u^a, is then used to compute the output of gap gene a in each nucleus i; at the boundary region for that gap gene expression domain, where the value of u^a goes from 0 to the maximum protein output R^a of gap gene a, this protein output is assumed phenomenologically to follow the sigmoid function g at the right of (B). As in (A), the values for T^{ab} are varied until the output profiles across all nuclei for all four gap genes statistically match the observed outputs as well as possible. (C) Computed runs (top) and observed outputs of gap gene gene regulatory network (GRN) (below), just prior to gastrulation. The observed profiles measured in 83 individual embryos are superimposed (color coding for individual genes is indicated). (D) Observed effects of mutationally altering the GRN on the accuracy of establishment of the posterior border of the *hb* anterior domain; top, 28 *Kr* and *kni* double mutant embryos observed between about 20 and 10 min prior to gastrulation; bottom, equivalent observations on wild-type embryos. Note that the variation of the border position (σ) is greatly suppressed in the wild-type circuitry.

parameters, until the spatial output as closely as possible matches the measured relative expression levels of each gap gene product. The "interaction matrix" in the table in Fig. 6.3(A) shows the computed optimum values for all possible interactions by which the four gap genes could be controlled. The topology of the interaction circuitry is thus derived by fitting the solutions of this system of ODEs to the expression data. In Fig. 6.3(B) we see the means by which the gap gene outputs are computed (see figure caption for terms and procedures).

Note that in this formulation RNA and protein synthesis are represented by a single parameter; inputs are considered to operate independently and their functions are simply added together; furthermore the regulatory output for each *cis*-regulatory input is treated as a linear function of input transcription factor concentration. In addition, the model employs an ad hoc sigmoid function which will sharpen the relation between regulatory input and spatial target gene expression, thus approximating the observed sharp boundaries between adjacent gap gene expression domains. These are technical issues from a mechanistic point of view, in that they abstract the processes of molecular biology, and the parameters in the equation do not represent measurable quantities. In addition, the parameters of the equations are correlated with one another (Ashyraliyev et al., 2008), and as these authors conclude, "the model cannot be used to infer quantitative regulatory weights". What is the most important result, however, is the qualitative topology of the regulatory interactions, which emerges reliably from this model (Ashyraliyev et al., 2008). This is the output that is portrayed in the topological diagrams in Fig. 3.8(E). The excellent agreement of the model output with the observed spatial distribution of the gap gene products is shown in Fig. 6.3(C). Furthermore, the linkages in the circuitry emerging from the model analysis are generally consistent with those found by genetics and molecular biology, which provides an important confirmation that indeed we have a system-level understanding of how this pattern forming system is encoded. Yet, as the authors point out, the model does not encompass certain features that have been observed experimentally, among which is failure to predict expression patterns in gap gene null mutants (Jaeger, 2009). Furthermore, this approach is perforce limited to systems consisting of only a small number of genes. Nonetheless, for this highly confined system, the gene circuit model successfully reveals the regulatory mechanisms for the dynamic changes in the spatial pattern of gap gene expression as it develops.

Gene circuit models have furthermore provided a means of rationalizing the way in which the gap gene GRN enforces fidelity and controls variability in the spatial specification process. In the absence of specific repressive linkages it is found experimentally that the variance in gap gene spatial expression is remarkably greater than in the wild-type circuit (Fig. 6.3(D)). The gene circuit models generate this effect for several specific linkages (Manu et al., 2009; Janssens et al., 2013; Surkova et al., 2013). Thus once again we see that the accuracy and reproducibility of the spatial expression control system in development is mediated by GRN circuitry, in which repressive as well as activating interactions are important.

2.3 A mathematical solution for the circuitry encoding emergence of hematopoietic stem cells in the embryo

Specification of hematopoietic stem cells (HSCs) occurs late in vertebrate embryogenesis and is driven by a well-defined GRN circuit supported by extensive cis-regulatory evidence (for review, Narula et al., 2010; Wilson et al., 2011), which is shown explicitly in Fig. 6.4(A) (also in Fig. 5.4(D2)). The circuit consists of three regulatory genes that are wired together with extensive positive feedback circuitry, scl, fli1, and gata2, referred to as the "triad", plus a "control module" composed of the smad1, smad6, and runx1 genes. The biological operation of this circuit is discussed extensively in Chapter 5 (Section 3.2). The control module provides both positive and negative feedbacks depending on developmentally presented signals that occur during HSC specification. The circuit as a whole operates in different states according to developmental circumstances. The triad runs at a relatively low level in hemogenic endothelial precursors. After receiving Notch and Bmp signals in the AGM, triad expression switches irreversibly to a high level and thereafter the positive feedbacks within the triad suffice to sustain the HSC state in the absence of any further signaling. Under these conditions, Smad6, which is a negative regulator of both smad1 and runx1, shuts down the control module which is now no longer necessary for triad expression. A glance at the circuitry in Fig. 6.4(A) illustrates the challenges to intuition that the behavior of this system under various circumstances presents. Here we review an ODE model that illuminates the diverse outputs of which the circuit is capable, and which provides an explanation of the functional significance of the circuit topology in the context of development (Narula et al., 2010, 2013).

The major features of this modeling approach are as follows. A chain of ODEs was constructed from which can be calculated the accumulation of each regulatory gene product, i.e., Scl, Gata2/3, Fli1, and Runx, and also the accumulation of Smad1, phosphorylated Smad1 (in response to Bmp4 signaling), the repressive Smad6, and of Smad1:Smad6 complex and Smad6:Runx complex. The accumulation of each gene product depends on its rates of production and turnover, while the rates of production depend on the cis-regulatory inputs to each gene from the others to which it is linked in the circuit in Fig. 6.4(A). The specific combinatorial cis-regulatory inputs are known from detailed gene transfer studies, which provide maps of the relevant inputs, as summarized in Fig. 6.4(B), and also measurements of their respective quantitative importance. In these models, the output of each gene was computed by assuming it to be directly proportional to the fractional occupancy of its cis-regulatory modules by the combinations of regulatory factors known to be required for its expression. To compute occupancy, the same probabilistic Boltzman partition function was utilized as discussed in Chapter 2 (Fig. 2.5 and attendant discussion). That is, the probability of productive occupancy is calculated as the sum of specific productive states of the cis-regulatory sites normalized

Figure 6.4

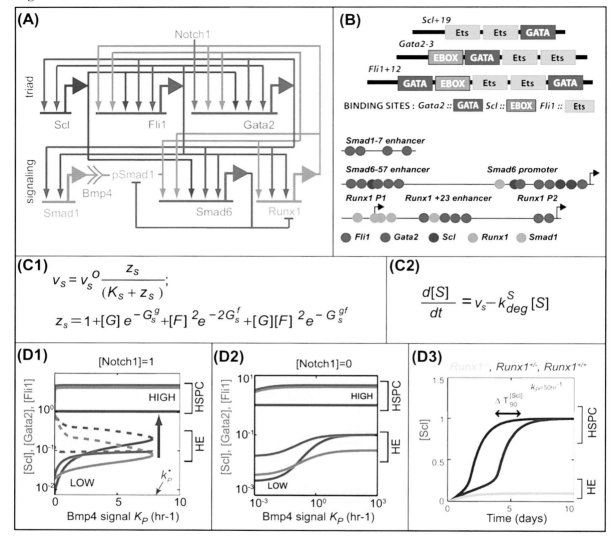

Figure 6.4 Probabilistic transcriptional model for developmental installation of hematopoietic stem cell (HSC) gene regulatory network (GRN) in mammalian embryogenesis. (A) Topological model of developmental GRN circuit, shown in BioTapestry *(from Narula et al. (2013))*. As described in text, this circuit is based on extensive *cis*-regulatory analysis, including measurements of relative quantitative inputs of interactions with individual *cis*-regulatory target sites. For discussion of the developmental origin of HSCs and functional interpretation of the circuitry, see Chapter 5. Note the almost saturated feedback linkages among the three genes *scl*, *fli1*, and *gata2* (the "triad"); the *smad1*, *smad6*, *runx*, subcircuit shown below functions as a signal-driven, developmental controller of triad gene expression, as described in text. Initial specification of precursors requires Notch and Bmp4 signaling and *runx* expression, but after the triad feedbacks are fully activated these become unnecessary. (B) Experimentally established *cis*-regulatory driver inputs into genes of the model in (A). Top, triad genes *(from Narula et al. (2010))*; bottom, genes of the control subcircuit *(from Narula et al. (2013))*. (C) Example of mathematical approach illustrated with equations for the *scl* triad gene

(from Narula et al. (2013)). The mathematical formulation of the model is described in text. (C1) Rate of Scl factor synthesis. The modified Boltzman partition function $z_s/K_s + z_s$ gives the probability of occupancy of the +19 enhancer of the *scl* gene (denoted by the subscript s), which *cis*-regulatory studies had shown to be activated by interactions with the products of the other triad genes, viz., at a Gata2 site and at two Fli1 sites ("Ets" binding sites in (B)). Thus (see Fig. 2.5, especially equations (4)–(7) and ancillary discussion), this fraction represents the ratio of the sum of the likelihood of occupancy of the Gata2 site (g), of the two Fli1 sites (f) and of all three sites (gf), i.e., all possible binding configurations, to the sum of all possible energetic configurations, i.e., the above plus a constant, K_s. For each site or site combination occupancy is the product of the instantaneous relative factor concentration (unitless) and the free energy of the interaction (G_s), a parameter which is equivalent and directly related to the equilibrium constants of Fig. 2.5 (see equation (3) of Fig. 2.5). Relative G values were measured in these studies as the effects of each type of interaction on quantitative output of *cis*-regulatory constructs. Here the constant K_s is taken to indicate a chromatin state energy on the basis of an ad hoc model of transcriptional activation, but as the authors point out, the mathematical form of the occupancy expression does not depend on this particular model and can be applied to more conventional images of transcriptional process where K_s would have a different interpretation; since the model is unitless, K_s simply serves as a representation of any energetic exchange not dependent per se on transcription factor binding. Thus for naked DNA with no binding, the value of G would be 0 and the equivalent K would be unity. In this model the occupancy fraction (probability) $z_s/K_s + z_s$ is simply multiplied by the maximum rate of transcription for the *scl* gene (in the absence of signal input), v_s^o, to obtain v_s, the instantaneous rate of Scl transcription factor protein production at a given point in time. These synthesis rates were then used to construct individual ODEs according to the interactions in the GRN shown in (A) for transcription of each of the triad genes in the network; and for the *smad1*, *smad6*, and *runx* genes. The *scl* ODE is shown as an example in (C2), where v_s is the synthesis rate and k_{deg}^S is the measured decay rate constant for the Scl factor protein. ODEs were also included in this model, based on many other measurements, to represent the rates of formation of phosphorylated Smad1, and of the repressive Smad1-Smad6 and Runx-Smad6 complexes (Narula et al., 2013). (D) Effects of the *smad1*-*smad6*-*runx* control circuit on triad gene expression; HE, hemogenic endothelial precursor; HSPC, HSC *(from Narula et al. (2013))*. (D1) Model output as a function of Bmp4 signal input in the presence of Notch signaling. Below the Bmp4 signaling rate of K^*p/h, two different levels of expression of Scl, Gata2, and Fli1 can be obtained in the model, heterogeneously representing both HE and HSC fates, but above this, exclusively high HSC levels are generated in the model solutions. This confirms experimental observation of the Bmp4-dependent developmental transition to the high, self-sustaining level of expression of the HSC triad feedback circuit. (D2) Requirement for Notch signaling. Even at >100× the Bmp4 signaling value of K^*p/h in (D1) heterogeneous low and high levels of triad gene output continue to be generated in silico in the absence of Notch signaling. This result explains the experimental observation that Notch signaling is necessary for transition to the permanent high-triad expression HSC expression state. (D3) Model prediction of effects of *runx* genotypes. Homozygous *runx* null and heterozygous genotypes were simulated in the model at the *runx* ODE, and the effects on Scl production were computed. The paradoxical effect observed experimentally that transition to high Scl output occurs earlier in the heterozygote than in wild type was reproduced, a consequence of attenuation of Runx-Smad6 repression in the heterozygote.

by the sum of all possible states. Here the probability of occupancy of each target site is given by the concentration of the regulatory factor and a parameter directly related to the equilibrium constant, as in equation (4) of Fig. 2.5, i.e., the negative exponent of the free energy of the transcription factor/target site interaction in equation (3) of Fig. 2.5. An example is the relations shown in Fig. 6.4(C) for the gene *scl*. A strength of this approach is that these free energies could be computed in relative terms from the results of mutational *cis*-regulatory experiments in which given target sites were destroyed and the *cis*-regulatory outputs measured. Thus this model directly utilizes experimental knowledge of *cis*-regulatory structure and function and appropriately applies this to probabilistic computation of *cis*-regulatory output using Boltzman partition functions. A technical issue, born of the necessity to escape the parameterization problem discussed above, arises in the highly abstract and simplified treatment in this model of the relationship between *cis*-regulatory occupancy and protein production. The ultimate protein production rate for each gene is computed simply as the product of the computed occupancy and the maximum production rate for that gene, thus escaping explicit consideration of any of the parameters for RNA and protein synthesis and turnover except for the previously measured half-lives of the triad proteins. The parameters in this model, other than the relative *cis*-regulatory measurements used to compute the expression functions, are unitless and abstract with respect to the physical–chemical parameters of transcription and translation.

The outcome of the model provides rational explanations for experimental observations which have been made in this intensely studied system over the years. First, the high level of expression of the triad genes is self-sustaining in the definitive HSCs and HSC precursors, and the underlying reason is the almost complete array of positive feedbacks among these genes. In addition, otherwise paradoxical features of the developmental process by which HSCs arise in the AGM are explained by the model. Three mechanistic aspects which the model illuminates are: the roles of Bmp4 and Notch signaling in the activation of the triad genes, the transient necessity of *runx* expression, and the outcomes of certain genetic perturbations. The first of these issues amounts to the mechanism by which Notch and Bmp4 signals kick the system into its irreversible high expression HSC state, which occurs as the cells transit from the hemogenic endothelial precursors to become the definitive HSCs (Fig. 5.3(B6)). Figure 6.4(D1) shows that in the presence of Notch signaling, which directly activates the *gata2* gene, there is a threshold value of Bmp4 signaling above which the circuit will exist only in the triad high-activity state, but this unitary state can never be achieved by Bmp4 signaling alone (Fig. 6.4(D2)), even though there are direct Bmp4 signaling inputs into all triad genes. The transient requirement for *runx* expression for the transition of hemogenic endothelial cells to the HSC state has been shown experimentally. However, Runx is no longer required after this transition has occurred, as demonstrated by conditional gene knockout experiments (for references, see Narula et al. (2013)). Simulated knockout of the Runx input at different stages in the model produce just this result, because even though two of the triad genes initially utilize direct Runx inputs, in the model these inputs are no longer required once the feedback circuitry is activated. An unexpected experimental observation is that in *runx* heterozygous mutant mice, HSCs emerge earlier than in homozygous wild-type mice (Cai et al., 2000; Lacaud et al., 2004). The model predicts this result, as shown in Fig. 6.4(D3), due to a negative feedback induced by Runx expression. That is, Runx directly activates the *smad6* gene, and Smad6 protein forms an inhibitory complex with phosphorylated Smad1, thus decreasing its positive input into the triad genes. Smad6 also causes degradation of Runx protein by forming a complex with it, and these off the DNA interactions are included in the model (Narula et al., 2013). Thus, the Smad6 negative feedback subcircuit functions to control the normal rate of HSC production.

In summary, the stringent attempts made in this model to encompass the critical regulatory mechanisms occurring at the *cis*-regulatory level, in the topological context of the circuit shown in Fig. 6.4(A), result in a qualitatively remarkable explanation of the behavior of this circuit under a variety of circumstances, irrespective of the abstract treatment of some aspects of transcription molecular biology.

2.4   Interpretation of morphogen gradients

The specification of discrete developmental domains in some particular instances relies on inputs which are present in graded concentrations, highest near the source. These inputs may consist of diffusible signaling ligands or, in the syncytial *Drosophila* embryo, of diffusing transcription factors. Graded inputs of both types are conventionally referred to as "morphogens". From a developmental regulatory point of view, the issue is the mechanism by which graded inputs result in discrete domains of gene transcription, spatially ordered with respect to these gradients. In the typical case, total removal of the graded input results in a failure of activation of these transcriptional domains. For several decades, the dominant view has been that concentration values along the morphogen gradient directly determine what the responding cells do, which has led to a definition of morphogens in terms of such a mechanistic relationship, for example: "As a minimal definition, a 'morphogen' is a molecule that diffuses away from a localized source to directly instruct cell identities in a concentration-dependent manner. [...] To qualify as a morphogen, it must form a graded distribution away from its source, directly act on cells at a distance [...], and induce the commitment of cells in the field, as a function of their distance from the morphogen source, to distinct developmental fates through the expression of different sets of genes" (Swarup and Verheyen, 2012). Prominent examples considered in detail from this point of view, where the transcriptional readouts are well defined, include in the syncytial *Drosophila* embryo the Bicoid gradient along the anterior-posterior axis and the Dorsal gradient along the dorsal–ventral axis; in developing *Drosophila* imaginal discs the Wg, Dpp and Hedgehog gradients; in developing spinal cords the ventral to dorsal Hedgehog gradient.

Quantitative measurements of unprecedented accuracy are now available, e.g., for Bicoid (Gregor et al., 2007), for Dorsal (Kanodia et al., 2009; Reeves et al., 2012), among others. The formation and dynamics of morphogen gradients are now realized to be more complex processes than previously assumed. In the following, however we are specifically only concerned with the regulatory mechanisms that determine transcriptional interpretation of inputs presented in graded clines. This subject has recently been illuminated by new system-wide experimental analyses that significantly revise traditional concepts of how morphogen gradients function. We briefly review results relevant to the major predictions of conventional morphogen theory in several systems and then focus on current propositions in which cellular interpretation of morphogen inputs is shown to depend on specific network architecture.

The abstract problem of understanding the causal relation between graded inputs and discrete outputs has a long history in developmental biology. Graded organization of developmental determinants in eggs was adduced more than a century ago (Child, 1900). The fundamental paradox in this problem is how a spatially continuous input function gives rise to spatially discrete Boolean developmental outputs. For much of the twentieth century the developmental functions of morphogens were treated mathematically as reaction-diffusion systems. New mathematical models of great elegance for the generation of biological pattern by graded morphogen inputs, which included self-organizing properties such as feedback, were invented by Hans Meinhardt and others (for a sympathetic historical overview see Roth (2011)). However, as they were essentially gene free, these dynamic mathematical models could not directly address the molecular mechanisms by which, in life, graded morphogenetic inputs affect the development of spatial gene expression domains.

The importance of morphogen gradient inputs for the spatial expression patterns of downstream genes was implied by experimental observations in which transcriptional responses in space were found to be altered when the quantitative dimensions of the gradient were manipulated. Furthermore, it was generally recognized that the sequence of transcriptional domains in space reflects the sequence of concentrations of the morphogen. Molecular biological models were generated which addressed the function of morphogen gradients in terms of transcriptional control of gene expression. The hard form of the resulting "morphogen theory" proposes that the absolute values of the morphogen concentration in space causally control the specific genes expressed in each spatial element, that is, that cells "read" gradient inputs quantitatively, and that each cell fate is the product of a particular morphogen concentration range. In *Drosophila*, where the morphogens are diffusing transcription factors, this lead to the idea that the spatial expression of morphogen target genes is determined directly and exclusively by *cis*-regulatory response to specific values of morphogen concentration. One mechanism that was proposed to underlie the response to Bicoid and Dorsal gradients is that the relevant *cis*-regulatory modules include target sites of different affinity for the morphogen transcription factor such that genes operating only close to the source of the morphogen contain only low affinity sites while genes operating at greater distance from the source, where morphogen concentration is lower, contain progressively higher affinity sites. Where the morphogen is a diffusible signaling ligand, bound by membrane receptors, a similar argument was made, such that the receptor/signal transduction system transmits its positive input to the transcriptional apparatus in accordance with the concentration of the external ligand. Since there are as many different types of signaling transduction system as there are families of ligand, this would require that a remarkable multiplicity of biochemistries produce the same type of quantitative transcriptional interpretation of external morphogen concentrations.

Experimental evidence

In the following, experimental challenges to particular aspects of this morphogen theory are briefly reviewed.

1. *Do absolute morphogen concentrations at different spatial locations directly determine differential transcriptional activity in individual nuclei?* A direct test of this proposition was carried out in *Drosophila* embryos, where the anterior to posterior Bicoid gradient was almost completely flattened by genetic interference with the maternal functions that position the Bicoid mRNA at the anterior end of the egg, and that ensure its clearance from the posterior region (Ochoa-Espinosa et al., 2005; Chen et al., 2012). Overall Bicoid levels were then changed by introducing different numbers of *bicoid* genes, thus producing embryos with different levels of almost uniformly distributed Bicoid transcription factor. The spatial expression of target gap genes was then monitored in these embryos. There were two main results, first that bounded gene expression domains are formed in the correct order despite the absence of a normal Bicoid gradient, and second, that the gap genes are expressed in these manipulated embryos at concentrations of Bicoid significantly lower than those obtaining in the nuclei where these genes are expressed in the wild-type embryo. This clearly demonstrates that it is not the absolute concentration of the Bicoid morphogen which determines transcriptional activity in space. Results of the same import have been obtained in a completely different system which utilizes a signal-dependent morphogen gradient. Thus, careful measurements of gene expression stripes forming dynamically in the ventral region of the vertebrate nerve tube in response to a sonic hedgehog gradient show that transcriptional identity fails to correspond simply to the strength of the intracellular signaling response to the morphogen both in wild-type and experimentally perturbed contexts (Balaskas et al., 2012). Thus in neither of these well-studied cases is the proposition supported that the transcriptional response is directly determined by morphogen concentration.

2. *Is there a direct and quantitative relation between signaling ligand concentration and transcriptional response?* A survey of the temporal relations between ligand presentation and transcriptional response for many different types of intercellular signal ligand shows that not only amplitude but also frequency and/or duration of signaling often determines transcriptional output (for general review, see Behar and Hoffmann (2010)). Thus in general, ligand concentration alone does not suffice to predict output. As knowledge of ligand presentation advances, we see that the morphogen gradients formed by signaling ligands change dynamically in space and time (in contrast to the static Bicoid gradient). This means that a simple relation between instantaneous concentration and the pattern of transcription is almost impossible to conceive, even more so in the developmental context, where the patterns of transcription in the end produce stable regulatory state domains. In the *Drosophila* wing, response to Hh signaling has been experimentally demonstrated to depend on the dynamic temporal behavior of the gradient, rather than simply on the absolute ligand concentration (Nahmad and Stathopoulos, 2009). The same conclusion was established experimentally for Hh signaling in the neural tube, where it was shown that the duration of the signaling determines the transcriptional response (Dessaud et al., 2010). Furthermore, attenuation of response to signals through negative feedbacks is a major feature of many inductive signal

transduction systems, including Hh (e.g., Dessaud et al., 2010; Briscoe and Thérond, 2013). By definition, in these cases ligand concentration is not directly reflected by transcriptional response through time, since this response is adaptively extinguished even though the ligand remains present. In a quantitative, mechanistic study of adaptation in TGF-β signaling, carried out in mammalian cells in culture, single cell transcriptional responses were monitored (Warmflash et al., 2012). The result was that even though the concentration of nuclear phosphorylated Smad2 mirrors the presence of external signaling ligand, the transcriptional response diminishes sharply over time. The mechanism is that phosphorylated Smad2 response factor complexes with Smad4, transits to the nucleus and activates transcription, but over time Smad4 is exported from the nucleus, extinguishing transcriptional activity. This process is accurately captured by an ODE model, for which the variables are the concentration of nuclear phosphorylated Smad2, Smad4, and the molecule responsible for export of Smad4 from the nucleus. In summary, morphogen gradients of signaling ligands are frequently dynamic, and thus signaling systems can generally not be treated as if transcriptional response depends directly on instantaneous ligand concentration.

3. *Is the spatial expression of morphogen target genes directly determined by cis-regulatory binding affinity?* As noted above, the morphogen theory proposes that the *cis*-regulatory modules of the target genes required to be expressed at the high end of the gradient have only low affinity-binding sites for the morphogen or its regulatory response factor, while genes expressed more distally have higher affinity target sites, and genes expressed at the far downstream end of the gradient have very high-affinity target sites. This proposition has been tested in a number of systems. In *Drosophila*, a direct assessment of the affinity of Bicoid target sites in genes expressed at different anterior-posterior positions along the whole length of the embryo, using a thermodynamic model for site evaluation, revealed no correlation whatsoever between position of posterior boundary of gene expression and the affinity of the target site for the Bicoid transcription factor (Segal et al., 2008). In the case of Dorsal-responsive genes, where a very large repertoire of *cis*-regulatory modules expressed in various ventral to dorsal positions has been recovered (Markstein et al., 2002; Stathopoulos et al., 2002; Hong et al., 2008), it is again true that differential affinity of target sites for the Dorsal transcription factor fails as the explanation for the spatial expression patterns of these genes. Genes expressed ventrally in the prospective mesoderm, where nuclear Dorsal concentrations are highest, do often contain multiple low affinity-binding sites for Dorsal and/or the product of its target gene *twist*. However, *cis*-regulatory studies show that genes operating in different positions along the dorsal–ventral axis depend on different activators in addition to Dorsal, for example Zelda in the dorsal-lateral regions (Liberman and Stathopoulos, 2009). In the most dorsal region other factors collaborate with Dorsal to provide repressive complexes. The most important conclusion is that specific repressors such as Snail, Vnd, Ind, SuH (in the absence of Notch signaling), and Capicua, are required to set both ventral and dorsal boundaries just as repressors are required for boundary formation in the anterior-posterior gap gene patterning system (Gray et al., 1994; Morel and Schweisguth, 2000; Cowden and Levine, 2003; Garcia and Stathopoulos, 2011; Ozdemir et al., 2014). Thus it is completely clear that the positioning of laterally expressed Dorsal target genes is not generally controlled by Dorsal target site affinity although the positioning of ventrally expressed Dorsal targets could be. The prediction that binding site affinities for the morphogen response factor determine spatial expression of target genes also fails for Hh control of neural tube specification. Here binding sites for the Gli transcription factor were identified in *cis*-regulatory modules of genes expressed at high and lower levels of the morphogen gradient, but were found to have target site affinities exactly opposite to those predicted on the basis of the morphogen theory, that is, the highest affinity sites were found in genes operating closest to the source of the morphogen (Oosterveen et al., 2012). It should be noted in this case that the response factor acts as a repressor as well as an activator, depending on the availability of the signaling ligand, and the situation is therefore not directly comparable. Most importantly, the proposition that morphogen gradients are interpreted by differential *cis*-regulatory affinity is not upheld in any of these cases.

Thus, a variety of current experimental information leads to the conclusion that the "morphogen theory" as defined in the beginning of this section does not represent mechanistic reality. As we discuss in the following, other explanations must obtain where graded inputs are required for specification of discrete spatial gene expression domains. It is at this point not even clear, however, whether per se the graded presentation of morphogens provides essential patterning information. Perhaps in some cases, the graded distribution of a required morphogen is simply a nonfunctional consequence of its diffusion from the source. For example, diffusion of one of the classic developmental morphogens, the Wnt signaling ligand (Swarup and Verheyen, 2012), turns out to be irrelevant for the complete normal development of *Drosophila*. In a recent study, the main Wnt ligand gene was reengineered so that the signaling ligand would remain tethered to the membrane of the cell producing it, with no consequence for developmental patterning (Alexandre et al., 2014). Thus, while the Wnt signaling ligand is essential for many regulatory processes, for example in wing development as analyzed in detail in the same paper, the graded distribution of the diffusible Wnt signaling ligand appears not to be functionally required.

An ODE model for the GRN circuitry operating the response to a sonic hedgehog morphogen gradient

A general conclusion that emerges from the analyses above is that encoded GRN circuitry and not singular morphogen inputs provide the machinery for transcription response to morphogen gradients in space. These circuits characteristically utilize repression systems to set the boundaries of the spatial domains that arise at different positions in respect to the morphogen gradient. Morphogens provide but one among multiple inputs into the *cis*-regulatory modules that control spatial expression of the genes encoding these repressors, and the topology of the circuit determines the genes that are expressed at given positions. We have seen this in the gap gene network shown in Fig. 6.3, and we now consider a different example in which ODE modeling has enabled at least a partial understanding of the dynamic positional changes in regulatory gene transcription (Balaskas et al., 2012). This is the patterning of the ventral neural tube, controlled by a Shh signal emanating from the notochord and ventral floor plate. The output is a ventral to dorsal series of transcriptional domains extending up the walls of the neural tube, each of which is characterized by expression of given regulatory genes (Oosterveen et al., 2012; Cohen et al., 2013). Ultimately, each domain gives rise to different subtypes of neuron. The following model concerns the operation of the circuitry controlling expression of three transcriptional repressors in the ventral region of the neural tube. The most ventral domain (p3 domain), closest to the Shh source, ultimately expresses *nkx2.2*. This domain is dorsally abutted by an *olig2* expression domain (pMN domain), dorsal to which is a *pax6* expression domain (p2 and p1 domains; Fig. 6.5(A)). Much data from genetics and molecular biology show that the network circuitry controlling formation of these domains is based on multiple reciprocal repressive interactions, executed by the products of genes that respond to the Shh signal, as shown by the topological model in Fig. 6.5(B). An important feature of the way the circuitry is proposed to operate is that the differential response to the morphogen gradient is independent of direct differential *cis*-regulatory interpretation of the morphogen input, and this feature is captured in the mathematical treatment.

The principle by which the elements of this circuit topology operate was elucidated earlier in consideration of the mechanism of response of *Xenopus* embryonic cells to an activin gradient (Saka and Smith, 2007). The circuitry explored in that context was simpler, consisting of only three elements. These were the morphogen and two genes encoding repressors, both of which respond to inputs from the morphogen and each of which transcriptionally represses the other. The output of this circuit is the specification of two distinct spatial domains exclusively expressing either one or the other repressor in response to modestly different levels of the morphogen. The Saka and Smith ODE model suggests how two different stable outcomes can arise from this circuit at different concentrations of

Figure 6.5

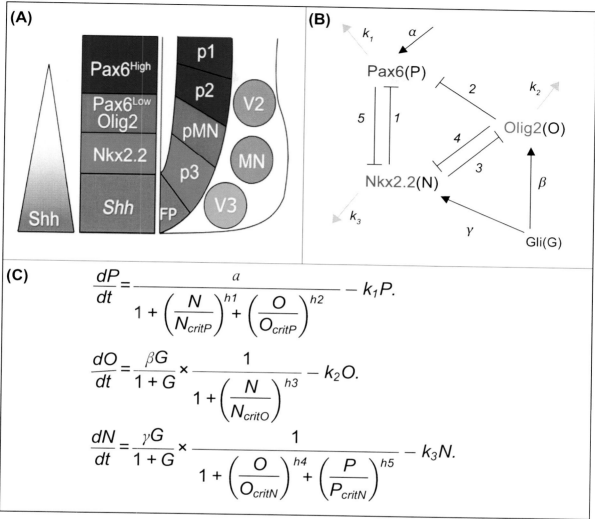

Figure 6.5 ODE model for Shh-dependent transcriptional patterning in ventral mouse neural tube *(from Balaskas et al. (2012)).* (A) Three transcriptional domains in the walls of the ventral neural tube, overlying floor plate cells secreting Shh. The ventral to dorsal Shh gradient is indicated at left. *nkx2.2* expressing precursors of the p3 domain give rise to V3 interneurons; *olig2* expressing precursors of the pMN domain give rise to motor neurons (MN); *pax6* expressing precursors of the p2 domain give rise to V2 interneurons. Additional regulatory genes are expressed in this, as well as in more dorsal regions of the neural tube, and the genes included here are only those to which the following model obtains. (B) Topological model of gene interactions based on genetic and molecular evidence (see text). The numerals 1–5 identify the individual Hill coefficients used in the mathematical model shown in (C); α represents the externally (to the model) controlled maximum rate of *pax6* expression; k_{1-3} are the respective transcription factor decay constants. γ and β are the maximum rates of expression of *nkx2.2* and *olig2*, respectively, which in the model are activated by the positive form of the Shh response factor, Gli_{act}. (C) ODE model of circuit shown in (B). In the equations, P,

Figure 6.5

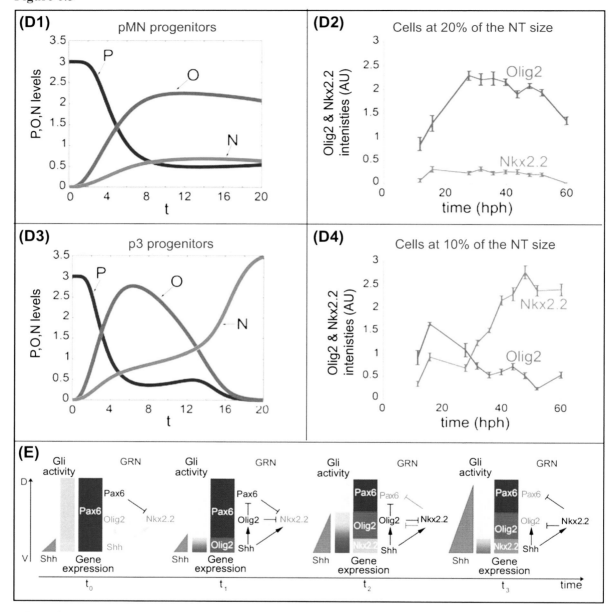

O, and *N* represent the levels of Pax6, Olig2, and Nkx2.2 factors. In this model activation of *olig2* and *nkx2.2* are due exclusively to input from Gli_{act} (*G*) set by the level of Shh signaling. The level of signal-driven activation of these genes is related to *G* in the model by a standard Michaelis–Menton form, the first term in the *O* and *N* ODEs. The circuit in (B) contains two pairs of reciprocal repressive interactions, and one unidirectional repression. Each of these repressions is modeled in the ODEs by an inverse relationship between the rate of activation and the instantaneous level of the respective repressors, raised to the power of the Hill coefficient for that interaction. The levels of the repressors are expressed as unitless ratios to the levels ("critical values") at which the target gene expression rate is reduced to half its maximum values (N_{critP}, N_{critO}, O_{critP}, O_{critN}, P_{critN}); i.e., the critical values could be taken as arbitrary equilibrium dissociation constants for the interactions between each particular repressor and each particular target gene cis-regulatory element. The behavior of these repression functions, i.e., the steepness of the relation between the amount of repression obtained as level of repressor changes depends on the dimensions of the Hill coefficients; however, the critical values determine the Hill coefficients needed for a given output, and vice versa (stronger repression occurs in the model at higher Hill coefficients and at lower critical values). Parameters were varied until the system behaved in a "biologically relevant" manner. As an example, for a given set of critical values (all set to 1), the Hill coefficients for Nkx2.2 repressions on *olig2* and *pax6* were required to be 5 and 6, while the reciprocal Hill coefficients were 1 and 2. (D) Comparison of model results for expression of *pax6* (P), *olig2* (O), and *nkx2.2* (N) at G profiles characteristic respectively of the pMN and p3 domains, vs measurements at these respective levels of the neural tube. (D1) Model results for intermediate levels of Shh signaling (G in model) as in the pMN domain (cf. (A)); t, arbitrary time. (D2) Measurements of Olig2 and Nkx2.2 levels at the pMN location. (D3) Model results for high levels of Shh signaling as in the ventral-most p3 domain. (D4) Measurements in the p3 region. (E) Summary of successive stages of the dynamic progression of Shh signaling, Gli$_{act}$ distribution, and spatial expression of the *nkx2.2*, *olig2*, and *pax6* genes. Note that only the Gli$_{act}$ gradient is shown, while as discussed in text a counteracting Gli$_{rep}$ gradient is also set up from dorsal to ventral, which affects expression of genes in the dorsal and midregions of the neural tube. The diagram shows the sequence in which the various gene interactions shown in (B) come into play.

morphogen despite a similar response of the two repressor genes to the morphogen. In this model, a small difference in the driver morphogen input can produce opposite results: either one repressor or the other is exclusively expressed. This model depends on a mathematical formulation which includes Hill coefficients set at a level which forcibly generates the binary outcome. The biological implication is that mechanistic differences in the effectiveness of the transcriptional repression functions mediated by the two genes in the repressive feedback circuit (modeled as differences in Hill coefficients) could result in the observed response to different morphogen concentrations.

This principle is built into the similar but more complex ODE model capturing the operation of the circuitry in response to the Shh gradient (Balaskas et al., 2012, Fig. 6.5(C)). Here we see three ODEs representing synthesis rates of the Pax6, Olig2, and Nkx2.2 gene products. In these equations, the rates of transcription factor expression vary inversely with the concentrations of their repressors raised to the power of Hill coefficients (h1–h5). In equation (1), for Pax6, an undefined activating function α, which is not dependent on the morphogen, is the parameter representing the activation of expression, and as shown by the circuitry in Fig. 6.5(B), the *pax6* gene is repressed by Nkx2.2 and Olig2. Equations (2) and (3) represent the morphogen-dependent synthesis of Olig2 and Nkx2.2, where activation of these genes depends on the concentration of Shh-dependent Gli$_{act}$ modeled according to a Michaelis–Menten form. These equations encompass the repression of the *olig2* gene by Nkx2.2 and the repression of *nkx2.2* by both Pax6 and Olig2, according to the circuit topology. Repression of *olig2* and *pax6* by Nkx2.2 defines the ventral-most domain according to the model, following transient early expression of Olig2. Repression of *nkx2.2* by both Olig2 and Pax6 defines the more dorsal pMN domain. The key mathematical aspect required for the differential spatial expression of these repressors with respect to the Shh gradient is again high Hill coefficients applied to the repression functions (as in the Saka and Smith model). To make the system work even on the assumption that the *olig2* and *nkx2.2* genes are activated equally by given levels of signaling (as opposed to assuming differentially sensitive cis-regulatory response elements), it is required that Nkx2.2 be a more powerful repressor than is either Olig2 or Pax6. This is modeled by higher values of the Hill coefficients for the repression of *olig2* and *pax6* by Nkx2.2 than for the other repressions in the model. It remains undefined what physical aspects of the transcriptional repression system these Hill coefficients might represent. Despite its very abstract construction with respect to the molecular biology of the transcriptional process, the model correctly captures several remarkable properties of this system. Thus it reproduces observed spatial effects of many mutational perturbations on these three genes; it provides an excellent simulation of the time course of expression of Olig2 and Nkx2.2 in given spatial locations (Fig. 6.5(D)); and with one exception it accurately represents the dynamic changes in the spatial pattern of gene expression with respect to the signaling process over time (Fig. 6.5(E)). The possible exception is that the model does not explain the maintenance of spatial gene expression as Gli_{act} levels decrease sharply at late times (Fig. 6.5(E), t_3). In essence, the topological model in Fig. 6.5(B) and the mathematical model in Fig. 6.5(C) propose a mechanism for the dynamic formation of one sharp boundary, that separating the Nkx2.2 domain from the Olig2 expression domain. A recent extension of this model includes an additional gene, *irx*, in the topological network model and relies on a more realistic treatment of repressor concentrations utilizing Boltzman partition functions to model *cis*-regulatory occupancy (cf. Chapter 2; Cohen et al., 2014). These ODE models illuminate how encoded network interactions might operate to produce bounded transcriptional domains in response to a morphogen gradient. This work now provides specific targets for mechanistic experimental examination.

The same principles might obtain throughout the neural tube, where additional repressors are regionally expressed. Studies on isolated Shh-responsive *cis*-regulatory modules from neural tube genes show that while *nkx2.2* does indeed respond to Gli_{act} in combination with a widespread coactivator, spatial control of expression of other genes depends on Gli_{rep} (Oosterveen et al., 2012). Thus in response to the Shh morphogen gradient, the signal response factor Gli generates two opposing gradients such that activator concentrations are highest near the source and repressor concentrations are highest distally, here in the dorsal neural tube. Yet again, the formation of spatial gene expression domains in response to Gli_{rep} requires additional transcriptional interactions with spatially specific repressors. As is true for response to Gli_{act}, different concentrations of Gli_{rep} are not interpreted directly by *cis*-regulatory binding site affinity. A prediction might be that circuitries of the form shown in Fig. 6.5(B) will ultimately pertain to the whole Shh response system of the neural tube.

* * *

When all is said and done we are left with two main conclusions about the mechanisms of morphogen gradient interpretation. First, the positions of the spatial gene expression domains formed in response to the morphogen gradient are not determined by absolute morphogen concentrations, but rather morphogen concentration differences may provide general vectorial inequalities which affect the relative order of the expression domains of some target genes, at least initially. Second, the actual spatial interpretive functions that result in discrete bounded transcriptional domains are generated by the operation of GRN circuitry, of which interdomain repression is a major feature. In these systems, the morphogen input is used over broad regions as an activator while the boundaries of the expression domains, as in other developmental spatial processes, are determined largely by the disposition of transcriptional repressors.

The very different examples reviewed in this section have in common that they use kinetic arguments and measurements to recreate computationally spatial as well as temporal transitions of regulatory state. Aside from the intrinsic difficulties of scale, parameterization, and the requirement for phenomenological mathematical approximations, ODE models are most naturally applied to temporal kinetic processes. In order to compute spatial processes in development, ODE models must be supplied externally with at least initial spatial inputs. Thus in the morphogen gradient problems, they are provided a priori with morphogen inputs that vary in space. Similarly, signals arriving from external locations "ex cathedra" may provide spatial inputs. On the other hand, ODE models utilize, as can no other computational approach, an extensive and increasingly available type of data, namely kinetic measurements of accumulation of specific regulatory molecules. These models relate changes in levels over time to regulatory circuitry and hence to developmental output. We turn now to a very different approach to modeling GRNs, one which is focused a priori on spatial output, and where it is the kinetics that must be supplied externally.

3. Boolean Models of Network Logic

3.1 Boolean models for developmental GRNs

The major feature of developmental processes captured in GRN models is the formation of spatial gene expression domains. Enormous collections of developmental gene expression data display discrete bounded spatial patterns of regulatory gene transcription, in both embryogenesis and adult body part formation. In space, the expression of regulatory genes is therefore of Boolean nature, as must be the mechanisms responsible for its control (see discussion in Chapter 2). These mechanisms are encoded in the networks of *cis*-regulatory logic functions by which bounded and discrete regulatory state domains are generated. For analysis of the question how successive regulatory transactions occur within a given developmental domain over time, as we have just seen, kinetic approaches may yield detailed insights. But for analysis of systems generating multiple discrete domains in development, what is needed is a modeling approach capable of dealing with the mechanisms that distinguish spatial regulatory states. This very general problem in capturing the developmental regulatory process has been successfully solved by application of Boolean logic models, in which the output of the computation is essentially the expression or silence of given genes, depending on the combinatorial logic by which their control systems operate.

Boolean models lend themselves to basic developmental processes for four fundamental reasons. First, as we have noted in chapters 3–5, the outputs of developmental GRNs are discrete patterns of gene expression, which in Boolean models can be captured as alternative states of given genes, ON or OFF. Second, these models adhere to the general character of developmental *cis*-regulatory modules: this is that their spatially exclusive activity is not a function of input level control, but rather depends on the presence or absence of particular combinations of regulatory factors, including repressors (as illustrated by many of the subcircuits shown in Fig. 6.2; see also Chapter 2). Third, the operations on inputs executed by *cis*-regulatory modules that control developmental gene expression resemble AND, OR and NOT logic processing functions. Fourth, these models do not become intractable as the scale or number of regulatory genes involved increases, and in fact their success depends on the completeness of the repertoire of genes included in the model.

Boolean models for developmental processes aim to include the following features. To start, they are constructed according to a topological GRN model established on the basis of molecular or genetic perturbation and expression data. The objectives of the Boolean model are to capture explicitly the logic transactions at each node of the GRN; to test whether the in silico computation executes the biological regulatory function; to determine the regulatory states that can be generated by a given circuitry structure; and to predict in silico the consequences of specific perturbations of GRN structure, such as can be imposed experimentally or as might occur in evolution. The foundation principle of Boolean models of developmental GRNs is that the rules for spatial expression of each gene determine the computation. In the models with which we are here concerned, these rules state the conditions upon which a given *cis*-regulatory module mediates activity or inactivity in terms of its inputs and information processing functions. That is, in the model as in life, the inputs are combinatorially processed according to the logic transactions executed by the *cis*-regulatory system. In this sense, each node of the model is constrained mechanistically. In life, the progression of regulatory gene expression and the formation of regulatory state domains emerge from the control functions mediated by genomic regulatory information, and node for node this is captured in Boolean models. Discrepancies between the operation of the Boolean model and the operation of the living system specifically identify missing regulatory functions in the topological GRN model, and thus Boolean models provide an exquisite tool for determining completeness of understanding.

3.2   Asynchronous Boolean logic models for diverse developmental processes

Three disparate applications of Boolean logic approaches to developmental GRNs follow. They have in common several features because they all are based upon what is known as asynchronous Boolean modeling, the initial principles of which were established by René Thomas (Thomas, 1991; Thomas et al., 1995). These models all include a set of logic rules which determine the conditions resulting in change of state of each component of the model (such as expression state of each gene) based on some prior knowledge encompassed in a topological GRN model. The inputs utilized by these rules are either initial conditions or the outputs of other components of the model. Essentially, the distinguishing feature of an asynchronous model is that the states of the system are updated for every individual sequential change of state of any of its components. Thus each change of state results in a next step in this computation irrespective of the real time in which the biological process underlying the change of state occurs. This allows processes, some of which occur rapidly, some slowly, e.g., enzyme reactions versus transcription cascades, to be combined in a sequential asynchronous computation. Computational methods have been devised which consider all possible state trajectories within this framework in order to arrive at the stable states that the system can give rise to (Gonzalez et al., 2006). In the examples that we touch on in the following, discovery of these stable states has been among the major objectives of the modeling exercise.

Logic models for the segment polarity network in *Drosophila*

Segmentation of the *Drosophila* embryo along the anterior/posterior axis is initiated with the formation of 14 domains, "parasegments". A complex GRN establishes positional values within each parasegment, in the form of stripes of expression of pair rule genes. A major important output of the pair rule gene network is the establishment of the parasegmental boundaries. These boundaries are defined functionally and morphologically by two exclusive regulatory state domains, each operating within a row of cells. The anterior border of each parasegment is defined by cells expressing *engrailed* (*en*) and across the boundary, the posterior border of the adjacent parasegment is marked by a row of cells expressing the Wnt signaling ligand gene *wg*. These "segment polarity genes" have been activated in the correct locations by the upstream pair rule gene network by the time of cellularization. Evolutionary comparisons discussed explicitly in Chapter 7 show that while the upstream pair rule GRN has evolved rapidly, its terminal output, definition of the segmental boundary by apposed rows of *wg* and *en* expressing cells is conserved across the Arthropods. After the formation of cellular boundaries the expression of *wg* and *en* is maintained by a different regulatory circuit. This depends in essence on intercellular signals exchanged across the boundary. The *en* expressing cells transcribe the *hedgehog* (*hh*) gene, secrete the Hh signaling ligand, and receive the Wg signaling ligand from the adjacent row of cells across the border. In turn the *wg* expressing cells secrete the Wg signal and receive the Hh signaling ligand in addition to responding to their own autocrine Wg signal. Much molecular and genetic evidence shows that these signals are causal to the maintenance of the two regulatory states. The circuitry in which this causality is embedded replaces the inputs from the upstream pair rule GRN, which becomes unnecessary once these signaling interactions are established.

The segment polarity cross-regulatory network has been the subject of several modeling attempts. As experimental information on this complex intercellular circuit has continued to accumulate, earlier models, both Boolean and ODE (von Dassow et al., 2000; Albert and Othmer, 2003; Ingolia, 2004), have become dated. A more recent logic model has utilized evidence from the literature and includes posttranslational modification in the signal transduction apparatus as well as gene regulatory interactions (Sanchez et al., 2008). This model is not strictly Boolean since for some of the components of the network three different levels of activity, 0, 1, and 2 were required rather than simply 0 and 1. As with all logic models of GRNs, this one begins with a topological model based on experimental evidence, in which, it should be noted, some linkages are more robustly supported than others. Logic statements indicating the conditionality of output and in some cases level of output were constructed. The computation was operated in steps rather than real time, as described above (the process being modeled actually occurs within about 100 min). The computation was directed at identifying stable states, that is, regulatory states which produce no further changes in the model. During the brief time period considered, the relatively fast changes in state due to protein modification events in the signaling cascade are more important than changes in state due to de novo protein synthesis and this was captured in the computational approach. The object of the model was to determine the interactions required to generate the biological output, which is the stable confrontation of *wg* and *en* expressing cells at the parasegment borders.

The results show that the network can assume five stable states, two of them corresponding to the ones present in the *en* and *wg* expressing cells. Targeted perturbations of the logic model also succeeded in deriving the stable states specifically resembling the effects of various genetic manipulations previously studied experimentally. Two mechanistic conclusions are derived from the model analysis. First, both paracrine and autocrine Wg signaling pathways are required for the wild-type stable state to be obtained. Thus in addition to the Hh signal from the adjacent *en* cells, *wg* expressing cells also depend on inputs from Wg signaling. Second, a mutual repression feedback between *en* and *sloppy-paired* (*slp*) genes operates exclusion functions in both cells. Slp is a driver of *wg* expression and its expression is excluded by En in *en* expressing cells. In turn, En is a driver of *hh* expression and its expression is excluded from *wg* expressing cells by Slp. In sum, the logic model illuminates the main overall features of the mechanism for border maintenance: the two rows of cells are locked in a positive intercellular signaling loop, such that in each cell reception of the signal directly or indirectly causes expression of the alternative signal, and in addition, within each cell, negative feedbacks guarantee the exclusion of the alternative regulatory state.

Logic model for the first and second mammalian heart field

A Boolean logic model was recently formulated from the GRN model accounting for the specification of the first and second heart field in mammalian development (Herrmann et al., 2012). The major objective of the computational analysis was to determine the stable states that this network would assume. The underlying topological GRN model is presented in BioTapestry format in Chapter 4, Fig. 4.8(C), together with a discussion of the molecular and developmental biology of early heart development. The Boolean model is constructed according to the same general principles as the segment polarity model just considered (i.e., Thomas, 1991 etc.). The experimentally derived inputs and outputs encompassed in the topological heart GRN model provide the basis for the rules governing expression of the core regulatory genes in the Boolean model. The model also incorporates evidence indicating the essential importance of Bmp and canonical Wnt signaling in the specification of the first and second heart fields (cf. Fig. 4.8C). The model was run stepwise and not in real time, although the order of some transitional regulatory states recovered computationally resembled the molecular sequence of gene expression observed during heart field development. The stable states this network could generate were computed by using all possible combinatorial conformations of the 15 network nodes as initial states (2^{15}). Indeed, the two significant stable states that emerged matched the expression profiles of the core genes of the first and second heart field GRNs, respectively. The model also sufficed to predict correctly the expression profiles observed in mice bearing several different mutations. This analysis shows that the interaction information assembled into the topological GRN model suffices to explain the generation and expression of the first and second heart field regulatory states, under normal and perturbed conditions.

Logic model for mammalian HSCs

Boolean analysis has also been applied to an experimentally well-supported topological network model for mammalian hematopoietic stem/progenitor cells (HSPCs) (Bonzanni et al., 2013). This network contains 11 regulatory genes interconnected by numerous positive and negative feedback loops. The interactions in the network are directly supported by extensive *cis*-regulatory analyses. HSPCs give rise to nine mature cell types (erythrocytes, monocytes, granulocytes, B-cells, NK cells, CD4 T-cells, CD8 T-cells, and activated CD4 and CD8 T-cells). The Boolean model was constructed based on logic statements for each regulatory gene, derived from the interactions controlling its expression. Again, the computation was operated in steps, each representing single changes in regulatory state. The computation revealed that the network predicts a canonical HSPC regulatory state, where all regulatory genes are expressed except one, and where this predicted stable state is composed of 32 interconnecting expression states. The diversity of internal expression states computed by the Boolean model suggests that the network may assume different states of activity, and indeed single cell analyses of HSPCs reveal a striking heterogeneity in expression of these same regulatory genes (Ramos et al., 2006). Despite its ability to capture potential heterogeneity of expression states in stem/progenitor cells, this network provides no pathways to the regulatory states of the mature blood cells unless an external stimulus is provided from outside of the network. The HSPC states are thus stable absent these external stimuli. The computation allowed delineation of the number of specific changes of gene expression along the pathways leading from the HSPCs to the diverse mature endpoint states. In addition, transitions back toward the HSPC states and transitions between states of mature cell type appear possible, although not observed in normal conditions, only in perturbed situations. Thus the Boolean analysis contributes a system-wide consideration of regulatory state transitions that can be generated by the topology of the HSPC network. From the outcome of this computation devolves a different and comprehensive view of the multipotentiality of HSPCs, compared for example to treatment of this problem in terms of serial binary lineage fate choices.

Boolean modeling is of great value for understanding the operation of developmental GRNs, particularly for larger networks. As the above examples and other Boolean models (e.g., Wittmann et al., 2009; Giacomantonio and Goodhill, 2010; Naldi et al., 2010) demonstrate, such models illuminate how *cis*-regulatory functions at the nodes of the network drive transitions from one regulatory state to the next; they illuminate the sequential dynamics of the transitions encoded in the network; and they provide predictions as to the different regulatory states that a given network topology can produce, in normal and perturbed developmental contexts. The models we have discussed in the foregoing all concerned GRNs of relatively modest dimensions. We now turn to a Boolean model analysis of a much larger experimentally established GRN. This model was constructed on entirely different and novel principles.

3.3   The Boolean model for the sea urchin endomesoderm GRN

Modeling large-scale GRNs presents novel challenges, due to the number of genes and interactions they include. Thus to generate a Boolean model of the sea urchin endomesoderm GRN shown in Fig. 6.1(A), which is the largest and most extensively analyzed model of its class, required the invention of new modeling approaches. Most importantly, in contrast to the models just briefly reviewed, the Boolean model that we now take up operates synchronously in real time. As discussed in Chapter 3, the sea urchin endomesoderm GRN model encompasses the interactions among about 40 regulatory genes plus signaling genes, and results in the generation of multiple spatial regulatory state domains over the first 30 h of development of this embryo. Because this Boolean model includes different modeling approaches that are easily applicable to other large-scale GRNs, we now discuss its principles of operation in more than usual detail.

The objectives of building a Boolean model for the sea urchin endomesoderm GRN were to test whether the regulatory linkages specified in the topological GRN model suffice to explain the observed expression patterns, and to compute the outcome of various perturbations (Peter et al., 2012). The Boolean model is a necessary tool for analysis of the behavior of such a complex network, the overall function of which lies beyond the range of intuition. It is important to note that this model is deeply grounded in the biology of the embryonic control system, and in perspective it is an abstraction of specific mechanistic biological features assembled into a computational automaton. Hence this model is constructed using unique strategies to capture key developmental functions such as interdomain signaling, transcriptional regulatory circuitry, embryonic geometry, and the measured dynamics of RNA synthesis and turnover in this embryo. A brief overview of the design features of the sea urchin endomesoderm Boolean model is as follows.

The fundamental operational principles in developmental GRNs are encoded in the *cis*-regulatory modules controlling regulatory gene expression. The sea urchin Boolean model is constructed on the basis of logic rules capturing the individual functions of these modules throughout the whole system. These rules are encompassed in vector equations that state the inputs to which the regulatory modules are capable of responding, and the logic functions that they apply to these inputs. Depending on the availability of its inputs, each vector equation can assume a state 0, representing absence of gene expression, or a state 1, indicating gene expression. In this model, signaling is treated as a mechanism causing transcription or silence of target genes, mediated by the signal response transcription factor (see discussion in Chapter 3). The vector equation for the response factor represents its activity state, which is then used just like the outputs of all other vector equations, as an input to the vector equations that pertain to signaling target genes. Computationally, the model is operated as a closed system, such that except for the initial inputs, there are no further external inputs. Instead, all inputs available to the vector equations in the model are generated as outputs of the vector equations at the previous computational step. A major objective of the model is to predict within this closed system, qualitative differences in activity states in various spatial domains. The four spatial domains of the sea urchin embryonic endomesoderm are considered in the model, each representing a group of equivalent cells. At every computational step, activity states are determined within each domain for all vector equations in the system. The activity states in separate domains may affect each other by means of interdomain signaling, by mechanisms we outline below.

This Boolean model is a computational representation of transcriptional activities at all the network nodes in the system through time. Interactions between upstream regulatory genes and downstream target genes occur with known dynamics in this embryo, and this provides a means of linking the steps of the computation to developmental real time. Relating the operations in this synchronous Boolean model to real time conveys major advantages for the analysis of complex developmental GRNs: (1) It allows an unlimited number of regulatory interactions to be computed in parallel, i.e., permitting multiple changes of activity states to be calculated in each computational step, rather than serially one change per computational step, as in asynchronous Boolean models. (2) It accommodates the different real-time kinetics of diverse kinds of regulatory process. (3) The computed activity state of all vector equations in each domain at a given time represents the actual intermediate regulatory state, and can be compared to measured gene expression in the experimental system in these domains at these times. Since development is driven by sequential regulatory states, all of which are transient, it is essential to be able to compute these transient states rather than to aim only at the stable states the system may assume. In development, stable regulatory states pertain only to the completed process. Models using asynchronous updating are well suited for serial processes such as occur in biochemical pathways and intracellular signaling cascades. But in contrast, regulatory networks simultaneously operate multiple combinatorial cross-regulatory interactions and a modeling framework that fits the conformation of the developmental process must be able to accommodate these.

Vector equations

The computed output of this Boolean model represents the transcriptional activity at each regulatory gene of the GRN in 0's and 1's. For each *cis*-regulatory module determining transcriptional expression of a regulatory gene in the system, a vector equation represents the functionality of that particular *cis*-regulatory module. The vector equations are based on a large amount of experimental data that permits deduction of the operational behavior of most of the nodes of the GRN. These data include direct *cis*-regulatory analyses, quantitative and spatial perturbation measurements, and detailed expression kinetics (e.g., Oliveri et al., 2008; Materna et al., 2010; Peter and Davidson, 2010, 2011). The regulatory inputs are stated qualitatively in the vector equations, identified according to the genes encoding them, irrespective of the number of binding sites contained in the respective *cis*-regulatory modules. The information processing of the inputs is expressed in terms of AND, OR, and NOT logic functions. These vector equations have the following form: for the case of a gene X which is activated only in the presence of both transcription factors A and B and which is repressed by transcription factor C, the vector equation computing the output of this gene would be "Gene X = 1 if gene A = 1 AND gene B = 1 AND NOT gene C = 1, else = 0". Vector equations of similar format are included for every relevant *cis*-regulatory module in the endomesoderm GRN model, amounting to about 65 such equations. Some genes respond to multiple *cis*-regulatory modules active in different developmental contexts. In these cases, a separate vector equation provides input processing function for each of the alternative modules, and for each such gene an additional statement is included specifying the contribution of these modules to the output of the gene. This permits attribution of hierarchical *cis*-regulatory functions, for example treatment of repressor functions which may control only specific modules or may dominantly extinguish gene output. In the model, the vector equations represent the relevant portions of the regulatory genome, and thus, like the DNA during development, they remain unchanged throughout the computation.

Signaling interactions

The Boolean model captures canonical features of signaling interactions in the developmental process, as we discussed in Chapter 3. Where inductive signaling occurs within the embryo depends on the expression of signaling ligand and receptor, and expression of both is directly encoded in the GRNs operating in the signal sending and receiving cells. In the Boolean model, the regulation of genes encoding signaling ligands and receptors is represented by individual vector equations. As we have discussed earlier, activation of the receptor by a developmental signaling ligand results in biochemical alteration of an already present transcription factor, the response factor, so as to change its transcriptional functionalities. Many common developmental signaling pathways display what we referred to above as "Janus behavior", in that in the absence of the ligand, the response factor acts as an obligate repressor of its target genes, while the alteration caused by the ligand-dependent activation of the receptor converts it to an activator of these same genes. Thus in the Boolean model, the Janus response factor is represented by a vector equation, the activity state of which is 0 if the Janus factor will act as a repressor, and 1 if it will act as an activator. The vector equation determining the state of activity of the target gene of a signaling interaction therefore includes as an input the Janus factor. In turn, the vector equation of the Janus factor depends on expression of the receptor and the expression of the ligand.

Spatial domains and geometry of embryonic territories

The endomesoderm GRN model (Fig. 6.1(A)) encompasses developmental events in almost half of the embryo up to gastrulation (30 h). This GRN model predicts the mechanism of specification of four distinct spatial regulatory state domains with distinct fates, the skeletogenic mesoderm, the nonskeletogenic mesoderm (only up to 18 h), the anterior and the posterior endoderm. During the first 30 h of development, the sea urchin embryo (*Strongylocentrotus purpuratus*) generates ~600 cells, about half of them included in the endomesoderm. It is important to note that this is a non-growing system and the total space occupied by the endomesoderm remains constant, being divided into an increasing number of smaller cells. Furthermore, after the earliest stages development, the whole system consists of a hollow sphere only one cell thick, and thus the disposition of its regulatory states can essentially be considered as a two-dimensional pattern. The embryonic domains included in the model encompass respectively about the same fraction of the total space throughout development, even though the number of cells within each increases. Within each domain, the regulatory state of all the cells is equivalent according to observation, and the project of the model is to determine the evolving regulatory states in each domain through time. In the Boolean model all four cell fate domains are represented throughout the 30 h (except the nonskeletogenic mesoderm), and at each step of the computation, the state of activity of every gene, as determined by the relevant vector equation, is assessed in each domain. The sum total of active genes in a given domain at a given time represents its regulatory state. In the embryo, early regulatory state domains subdivide to produce new cell fate domains, and some of these domains share common progenitor domains. In the Boolean model, the spatial subdivision of regulatory state domains is explicit in the divergence of the computed regulatory states in the separate domains, driven by mechanisms encoded in the vector equations.

The disposition of these domains within the embryo changes through time, as portrayed in Fig. 3.5(A), to which the reader is referred. Regulatory states expressed in separate domains can affect each other through signaling, that is, the signal is emitted from one domain and is received by another. In the sea urchin embryo, during the period considered, the relevant signals act only over short ranges. In order to compute the effects of signals, the specific neighbors of each domain must be explicit throughout time in the model. Thus information on the proximity of the cell fate specification domains throughout developmental time is available in the model computation as a lookup table, that is, the geometry of each embryonic domain is given by defining at each stage the proximal and distal neighboring domains. In addition, the model is given information on the potential range of action for each signaling ligand.

Dynamics

Computations in the Boolean model were executed in real time, such that each regulatory state pertains to an actual time point in sea urchin embryogenesis, in order to facilitate direct comparisons to the observed expression patterns. The key problem is how to consider the interval in real time between activation of a given regulatory gene and activation of its target gene(s). By this we mean specifically the interval between the time when the upstream gene begins to be transcribed, which in the model means that the vector equation for the upstream gene first turns from 0 to 1, and the onset of transcription of its target genes. This problem was addressed mathematically in Chapter 2 (Fig. 2.5). For the sea urchin embryo system operating at 15 °C, it was concluded from kinetic simulations using measured parameters that this interval (step time) is about 3 h. In other words, the 3 h interval is what is required on the average for transcription of a gene encoding a transcription factor, export of the mRNA to the cytoplasm, its translation, the accumulation of functional levels of transcription factor, transit of the transcription factor back into the nucleus, occupancy of the target *cis*-regulatory module, and initiation of target gene transcription. The assumption of a 3 h step time has been justified repeatedly by independent kinetic observations in this embryo. Since the Boolean model computes transcriptional activity, the step time is required to relate temporally the transcription of target genes to the transcription of the genes encoding their inputs. The 3 h step time was imposed throughout the Boolean model, with few exceptions where the expression kinetics indicated a slightly shorter step time. As we see below, the utilization of the uniform step time proved surprisingly successful. To insert this step time into the vector equations, we can use the same example as above: "Gene X=1 if AT-3 gene A=1 AND AT-3 gene B=1 AND NOT AT-3 gene C=1, else=0"; that is to say that genes A and B had to be transcribed at least 3 h earlier than the time when gene X is expressed. For signaling interactions, this step time is shorter, since signal transduction is a more rapid process than transcription.

Computation

In the sea urchin Boolean model, unchanging vector equations that represent the logic transactions encoded in the regulatory DNA are utilized to generate a changing set of spatial regulatory states. This is analogous to what happens in embryonic development. There are no "stable states" achieved during the processes of development, by definition. Thus, the model is focused on capturing transient regulatory states as they are generated progressively by the relevant positive and negative interactions. And just as in the embryo, where all genes are confronted by all the regulatory states obtaining in time and space, the computation presents each regulatory state to all the vector equations at once. Thus the status of all vector equations is updated synchronously. At intervals of 1 h, the output of every gene in the system is computed by applying to its vector equation the inputs available. For instance, for the vector equation for gene X given above which requires that both A and B inputs be present 3 h earlier in order for gene X to acquire a state of 1, suppose the activity state for six consecutive hours for gene A would be [0,1,1,1,1,1]; for gene B [0,0,1,1,1,1], then for gene X at each successive update time the output would be [0,0,0,0,0,1]. Similarly, the output of every vector equation is computed at each update time in a given domain based on the outputs produced at earlier times. These outputs are then utilized as inputs in the following intervals as the computation is iterated. The updating of gene expression states for all genes at every hour occurs in parallel in all four embryonic domains, as represented in silico. In this way, the spatial expression of every gene in the system is computationally predicted: thus for example if genes A and B are expressed in only one domain but not the others, the output of gene X would be 0 in all but that domain. In life, regulatory transactions and changes in regulatory states are occurring continuously in the different embryonic domains, and they are interrelated and affected by interdomain signaling. Given the resolution of experimental observation in time and the 3 h step time, the 1 h update interval utilized to separate computational steps is sufficiently fine scale. To begin the computation, initial conditions are specified in the model in each domain based on the measured maternal inputs of regulatory significance. Thereafter the computation operates as a regulatory automaton.

To evaluate the control system built into the vector equations of the Boolean model, it is necessary to devise a way to determine whether it suffices to regenerate computationally the observed patterns of expression. The spatial expression of regulatory genes in the endomesoderm of the sea urchin embryo during the first 30 h of development is abstractly summarized in Boolean form in Fig. 6.6(A). The computed output of the model is configured to resemble the format of Fig. 6.6(A), so that a direct comparison can be made between the computationally predicted and the observed gene expression for each gene at each point in time and in each domain. Since to define regulatory states it is just as important to know which genes are silent as to know which genes are expressed, models of the control system must predict both expression and absence thereof. Thus, in the Boolean model, the status of every gene, on or off, is computed in each domain throughout time.

Results: comparing model output to experimental data

A primary objective of the Boolean model analysis is to test whether the information captured in the endomesoderm GRN model is sufficient to explain the observed gene expression patterns through time and space. In the GeNeTool software developed to facilitate this comparison (Faure et al., 2013), the predicted output of the model is presented as a Boolean expression matrix, of exactly the same form as that generated from experimental observation as in Fig. 6.6(A). The comparison between the computed model output matrix and the sea urchin endomesoderm gene expression matrix is shown in Fig. 6.6(B). Summing over the four spatial domains and all the genes considered, between 0 and 30 h of development, produces over 2700 individual comparisons between computed and observed results. In Fig. 6.6(B), each rectangle of the matrix indicates the status of expression of one gene at a given time and in a given domain, as computed by the Boolean model. Active gene expression is shown in color while absence of expression is shown in gray. The extent to which this computed model output is in agreement with gene expression data is represented by the following symbolism: in the case of complete agreement, the cell is portrayed as a simple colored or gray rectangle, while discrepancies are indicated in the appropriate cell by open or closed black bars. As the time resolution of the underlying expression data is about 3 h, temporal discrepancies of less than 3 h might not be real and are represented as open bars. Other discrepancies, such as predicted expression of a gene in domains where it is not observed, or temporal discrepancies greater than 3 h are considered significant, and these are shown as closed bars (for additional symbolism, see caption).

The overall result is striking. First and foremost, the vast majority of cells in the matrix (98%) display agreement between computation and observation. This means that the information in the GRN model suffices to specify the spatial expression of almost all the genes and that the temporal dynamics of predicted expression within the resolution of the computation is largely accurate. Second, where major disagreements were observed, they identify the exact locations in the network where regulatory information is lacking. For example, in two places, genes were observed to be turned off at certain times in the embryo, while this is not predicted in the model (*delta* expression after 22 h in the skeletogenic cells and *foxa* expression in the veg2 endoderm after 24 h). In both these cases, the genes were turned off manually in the computation so as to prevent the consequences of these discrepancies from propagating (manual "overwrites" are shown by framed boxes). In the few cases where a gene is expressed in a domain where it should not be, for example *z13* in the Veg1 endoderm domain, the computed ectopic expression lasts for only a few hours. Third, the general assumption of the 3 h step time produced dynamics generally consistent with observation, indicating a surprising regularity in the transcriptional cascade kinetics of this system.

Figure 6.6

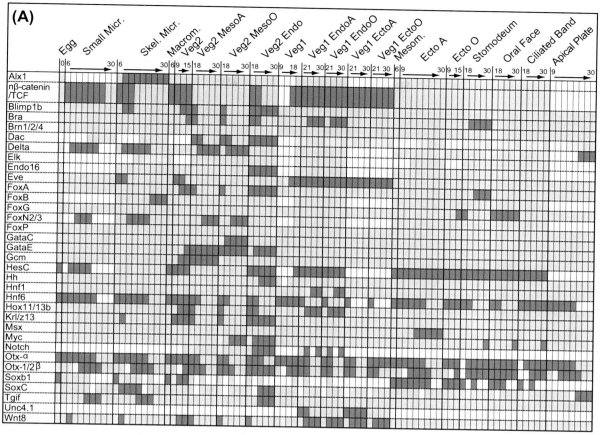

Figure 6.6 Outputs of a large-scale predictive Boolean computational model of the sea urchin endomesoderm gene regulatory network (GRN). (A) Boolean presentation of experimentally determined expression data for endomesodermal regulatory genes *(from Peter and Davidson (2011))*. Genes are listed alphabetically at left. Each vertical column represents a 3 h time interval. Spatial domains of the embryo are indicated across the top of the chart for the periods at which these domains obtain (hours post fertilization): Micr., micromeres; Skel., skeletogenic; MesoA., aboral nonskeletogenic mesoderm; MesoO., oral nonskeletogenic mesoderm; Endo., endoderm; EndoA., aboral endoderm; EndoO., oral endoderm; EctoA., aboral ectoderm; EctoO, oral ectoderm; Mesom., mesomeres. The output of the computational model was formatted similarly to show predicted expression of these genes at given times within spatial domains of the embryo, in order to facilitate exact comparison with experimental data. (B) Comparison of computed and observed expression patterns in real time, 0–30 h of development *(from Peter et al. (2012))*. The computational automaton was synchronously updated in silico, each step representing 1 h in developmental real time, and at each step every gene was assessed for the conditions determining its expression in all four spatial domains considered, according to the logic statements given to the model in its vector equations. See text for discussion of the structure of the model. Genes are listed across the top. The four embryonic spatial domains (skeletogenic micromere lineage, Veg2 mesoderm, Veg2 endoderm, Veg1 endoderm) are separately portrayed, each against 1 to 30 h timescales (Veg2 mesoderm only to 18 h). The results of the comparison are shown as follows: plain filled gray squares indicate that the computation and the data agree that the gene is not expressed in that time/space cell; plain colored filled squares indicate that computation and the data agree that the gene is expressed in that time/space cell. Where the computation

Figure 6.6

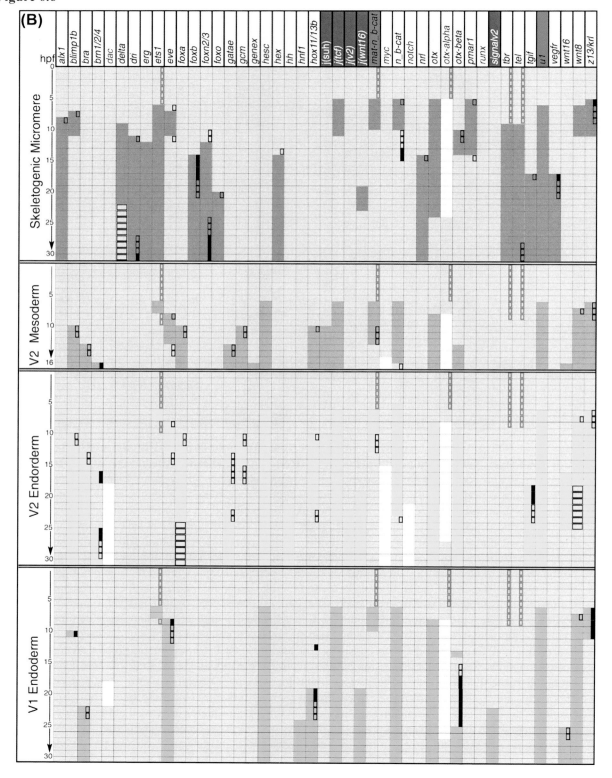

Figure 6.6

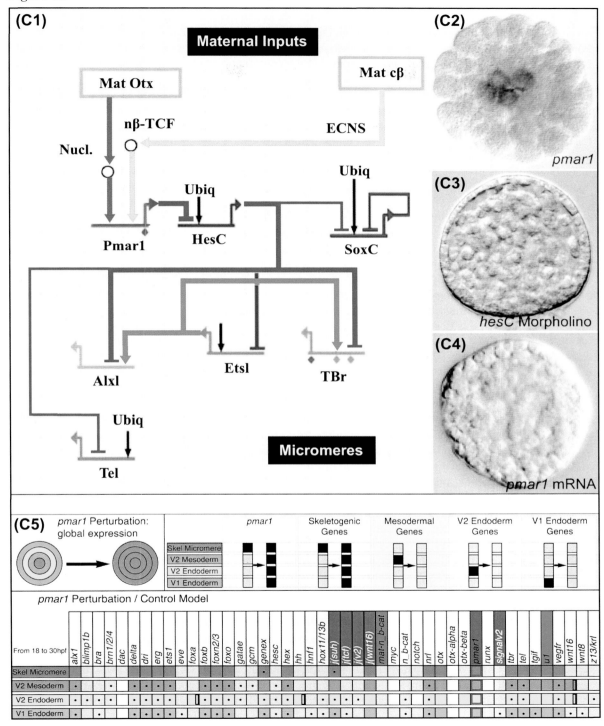

and the data do not agree black rectangles are superimposed on the time/space cells. Open black rectangles indicate discrepancies <3 h; as described in text, since the in situ hybridization observations have only 3 h resolution, these discrepancies may not be real. Discrepancies >3 h are indicated by solid black rectangles. For two signaling ligand genes (*delta* in skeletogenic mesoderm and *wnt8* in veg2 endoderm) expression is known to be extinguished after certain times, but the computation failed to produce these effects (i.e., the parent GRN lacks explanatory cause of the extinction of their expression), and these genes were turned off manually in the computation; similarly *foxa* continues to be expressed in veg2 endoderm at times beyond the point where the computation supported expression. The discrepant time/space cells are in these three cases marked as columns of cells with thick black borders. Narrow green striped bars at the top of the figure indicate maternal mRNAs that are not part of the zygotic GRN model, and are therefore used as initial conditions and not computationally predicted. For genes or regulatory operators shown on black backgrounds in the header, no observational data are available; for genes on white backgrounds, no upstream regulatory information is available, and the activation of these genes was not computed. Overall the comparison shows that in all but a very small fraction of the possible time/space cells the model computation correctly predicted the experimentally observed results. (C) Computational prediction by the Boolean model of the results of an experimental perturbation (Oliveri et al., 2008; Peter et al., 2012). (C1) Initiation of the skeletogenic GRN (i.e., expression of *alx1*, *ets1/2*, *tbr*, *tel*) by encoded double-negative gate GRN circuitry: ubiq, ubiquitous activator; mat, maternal; cβ, β-catenin; nβ-Tcf, nuclear β-catenin-Tcf complex; Nucl, nuclear (Oliveri et al., 2008; Damle and Davidson, 2011). For discussion, see Chapter 3, also Fig 6.2(H) and ancillary commentary. (C2–4) Experimental observations confirming circuit logic: (C2) the *pmar1* gene must be expressed specifically and exclusively in founder cells of skeletogenic lineage, as shown in this in situ hybridization; (C3) interference with *hesC* expression should convert the whole embryo into (skeletogenic) mesenchyme cells, visualized in an embryo treated with morpholino antisense vs hesC mRNA; (C4) the same result should be obtained if *pmar1* expression is forced to occur in all embryonic cells, as visualized in an embryo developing from an egg injected with *pmar1* mRNA (C2–C4 *from Oliveri et al. (2008)*). (C5) Boolean model computational result indicating that all domains of the embryo express skeletogenic genes if *pmar1* is "expressed" globally in silico (*from Peter et al. (2012)*). The top portion of (C5) represents the predicted expectations extrapolated from available data. The overall prediction is illustrated diagrammatically at upper left; radial regulatory state domains at blastula stage, viewed from the vegetal end in a normal embryo, are color coded as in (B), far left. From the experiments in (C2–C4) and other data (Oliveri et al., 2008), if Pmar1 is globally expressed, the skeletogenic regulatory state (purple) is expected instead to be expressed throughout the spatial domains normally giving rise to nonskeletogenic mesoderm and anterior and posterior endoderm. Pmar1 expression was enforced globally in silico in the model, by setting the state of this gene to 1, ab initio, in all domains of the embryo (indicated in white box at top). The expected result (purple rectangle), if the model includes enough information to behave as does a *pmar1* mRNA-injected embryo, would be for skeletogenic gene expression patterns to obtain in all domains. On the other hand, the normal domain-specific patterns of expression in cells of the other three prospective fates should be extinguished, as indicated diagrammatically, where black represents expression and gray represents absence of expression of the genes indicated on top (right three pastel boxes). The computational output of the in silico permutation is shown at the bottom: here the output of the computation under perturbed conditions was compared to the output of the computation under normal conditions (control). The matrix gives the expression results for the perturbed computational model. A gray square indicates no expression and a colored square indicates expression, with specific colors representing the embryonic domains as at upper left. Where the outputs of perturbed and control models are in agreement, the cell is plain, and where a change in expression has been caused by the perturbation, the cell is marked by a dot. The manually set changes used to mimic the experimental perturbations are indicated in red and red boxes around colored squares show where the *pmar1* gene was manually turned on. The result is that the model indeed installed skeletogenic expression in all domains, while leaving the normal pattern in the skeletogenic domain unchanged.

GeNeTool software makes it possible to redesign *cis*-regulatory function in silico, to add or subtract specific regulatory network linkages, to change regulatory gene expression patterns and to remove or alter signaling interactions. The specific experimental perturbations can be mimicked in silico and the response of the regulatory system can be predicted computationally and again compared to experimental results. An example is shown in Fig. 6.6(C). As we reviewed in Chapter 3, specification of skeletogenic cells depends on activation of a double-negative gate subcircuit, which is initiated by localized expression of a gene, *pmar1*, encoding a repressor. The logic of the circuit, which is shown in Fig. 6.6(C1), predicts that were *pmar1* expressed throughout the embryo, all cells would express skeletogenic functions at the expense of other specification states. This is indeed the dramatic result of injection of *pmar1* mRNA into the egg (Fig. 6.6(C4)). When in the Boolean model *pmar1* was turned on as an initial component in all domains, the result shown in Fig. 6.6(C5) was computed (Peter et al., 2012): in every domain, the same genes were expressed as in the skeletogenic domain, and genes representing other states of specification failed to be expressed. In sum, these results demonstrate that the model captured the essential regulatory relationships that control normal development, and also produces responses to alterations of GRN configuration which are consistent with those of the embryo. Additional examples of major perturbations that alter the course of development in vivo, and in silico, are to be found in Peter et al. (2012).

Three extensions into other areas now become accessible. This embryo has been utilized extensively for experimental exploration of the effects of developmental and molecular perturbations. The Boolean model serves as a tool with which to address the mechanisms underlying these effects, and with which to predict the outcome of future experiments. In addition, as synthetic circuit reengineering becomes more practical, the model provides a way to rationalize strategies and results. Perhaps most importantly, evolution proceeds essentially by perturbations of the developmental program, as we discuss in the next chapter, and finding the mechanisms responsible for evolutionary change will depend on understanding of how developmental networks operate. Logic models such as the one reviewed here may assist in the identification of the specific *cis*-regulatory functions underlying evolutionary change.

4. Conclusions

All these results lead to far-reaching conclusions. The predictive success of the computational Boolean model in its confrontation with detailed gene expression measurements provides a novel confirmation of the deep significance of GRN analysis. We see that the topological GRN model shown in Fig 6.1(A) indeed includes a sufficiently complete set of interaction relationships to explain causally the whole complex progression of regulatory states required to build half of an embryo up to gastrulation. This is in itself perhaps remarkable in that the topological model was assembled over a number of years from very disparate kinds of evidence. The evidence ranges from perturbation results to gene expression observations to *cis*-regulatory information, and the Boolean model demonstrates the existence of a coherent underlying principle. These different kinds of evidence correctly converge on what is indeed the focal plane where regulatory causality is to be found. Let us step back for a moment and consider the overall character of the genomic control system that this book is about. This consists essentially of arrays of *cis*-regulatory modules, physically separate from one another in the genome, executing independent functions, which when assembled together, produce the progressive programs for development. In constructing the Boolean model, we demonstrate that the unitary informational functions coded in the vector equations operate together to compute a similar outcome as does the genome in the developmental process. This directly illustrates that the genomic regulatory system in life consists essentially of a computational assembly of information processing modular units. It follows that this is the primary level of control in the developmental system and all other mechanisms that modulate gene regulation must occur downstream (see Chapter 1). By definition, developing systems function as automata. Stripped down to its abstract logic, the Boolean model reinforces the point that no information from outside is required following the initial circumstances, in that all inputs into the system are outputs from other parts of the system.

REFERENCES

Albert, R., Othmer, H.G., 2003. The topology of the regulatory interactions predicts the expression pattern of the segment polarity genes in *Drosophila melanogaster*. J. Theor. Biol. 223, 1–18.

Alexandre, C., Baena-Lopez, A., Vincent, J.P., 2014. Patterning and growth control by membrane-tethered wingless. Nature 505, 180–185.

Alon, U., 2007. Network motifs: theory and experimental approaches. Nat. Rev. Genet. 8, 450–461.

Ashyraliyev, M., Jaeger, J., Blom, J.G., 2008. Parameter estimation and determinability analysis applied to *Drosophila* gap gene circuits. BMC Syst. Biol. 2, 83.

Balaskas, N., Ribeiro, A., Panovska, J., Dessaud, E., Sasai, N., Page, K.M., Briscoe, J., Ribes, V., 2012. Gene regulatory logic for reading the sonic hedgehog signaling gradient in the vertebrate neural tube. Cell 148, 273–284.

Barolo, S., Posakony, J.W., 2002. Three habits of highly effective signaling pathways: principles of transcriptional control by developmental cell signaling. Genes Dev. 16, 1167–1181.

Behar, M., Hoffmann, A., 2010. Understanding the temporal codes of intra-cellular signals. Curr. Opin. Genet. Dev. 20, 684–693.

Bolouri, H., Davidson, E.H., 2010. The gene regulatory network basis of the "community effect," and analysis of a sea urchin embryo example. Dev. Biol. 340, 170–178.

Bonzanni, N., Garg, A., Feenstra, K.A., Schütte, J., Kinston, S., Miranda-Saavedra, D., Heringa, J., Xenarios, I., Göttgens, B., 2013. Hard-wired heterogeneity in blood stem cells revealed using a dynamic regulatory network model. Bioinformatics 29, i80–88.

Briscoe, J., Thérond, P.P., 2013. The mechanisms of Hedgehog signalling and its roles in development and disease. Nat. Rev. Mol. Cell Biol. 14, 416–429.

Cai, Z., de Bruijn, M., Ma, X., Dortland, B., Luteijn, T., Downing, R.J., Dzierzak, E., 2000. Haploinsufficiency of AML1 affects the temporal and spatial generation of hematopoietic stem cells in the mouse embryo. Immunity 13, 423–431.

Chen, H., Xu, Z., Mei, C., Yu, D., Small, S., 2012. A system of repressor gradients spatially organizes the boundaries of Bicoid-dependent target genes. Cell 149, 618–629.

Child, C.M., 1900. The early development of *Arenicola* and *Sternapsis*. Arch. Entwicklungsmech. Org. 9, 587.

Cohen, M., Briscoe, J., Blassberg, R., 2013. Morphogen interpretation: the transcriptional logic of neural tube patterning. Curr. Opin. Genet. Dev. 23, 423–428.

Cohen, M., Page, K.M., Perez-Carrasco, R., Barnes, C.P., Briscoe, J., 2014. A theoretical framework for the regulation of Shh morphogen-controlled gene expression. Development 141, 3868–3878.

Cowden, J., Levine, M., 2003. Ventral dominance governs sequential patterns of gene expression across the dorsal-ventral axis of the neuroectoderm in the *Drosophila* embryo. Dev. Biol. 262, 335–349.

Crombach, A., Wotton, K.R., Cicin-Sain, D., Ashyraliyev, M., Jaeger, J., 2012. Efficient reverse-engineering of a developmental gene regulatory network. PLoS Comput. Biol. 8, e1002589.

Davidson, E.H., 2009. Network design principles from the sea urchin embryo. Curr. Opin. Genet. Dev. 19, 535–540.

Davidson, E.H., 2010. Emerging properties of animal gene regulatory networks. Nature 468, 911–920.

Damle, S., Davidson, E.H., 2011. Precise *cis*-regulatory control of spatial and temporal expression of the alx-1 gene in the skeletogenic lineage of *S. purpuratus*. Dev. Biol. 357, 505–517.

von Dassow, G., Meir, E., Munro, E.M., Odell, G.M., 2000. The segment polarity network is a robust developmental module. Nature 406, 188–192.

Davidson, E.H., Rast, J.P., Oliveri, P., Ransick, A., Calestani, C., Yuh, C.H., Minokawa, T., Amore, G., Hinman, V., Arenas-Mena, C., Otim, O., Brown, C.T., Livi, C.B., Lee, P.Y., Revilla, R., Rust, A.G., Pan, Z., Schilstra, M.J., Clarke,

P.J., Arnone, M.I., Rowen, L., Cameron, R.A., McClay, D.R., Hood, L., Bolouri, H., 2002. A genomic regulatory network for development. Science 295, 1669–1678.

Dessaud, E., Ribes, V., Balaskas, N., Yang, L.L., Pierani, A., Kicheva, A., Novitch, B.G., Briscoe, J., Sasai, N., 2010. Dynamic assignment and maintenance of positional identity in the ventral neural tube by the morphogen sonic hedgehog. PLoS Biol. 8, e1000382.

Faure, E., Peter, I.S., Davidson, E.H., 2013. A new software package for predictive gene regulatory network modeling and redesign. J. Comput. Biol. 20, 419–423.

Garcia, M., Stathopoulos, A., 2011. Lateral gene expression in *Drosophila* early embryos is supported by grainyhead-mediated activation and tiers of dorsally-localized repression. PLoS One 6, e29172.

Giacomantonio, C.E., Goodhill, G.J., 2010. A Boolean model of the gene regulatory network underlying mammalian cortical area development. PLoS Comput. Biol. 6.

Gonzalez, A.G., Naldi, A., Sánchez, L., Thieffry, D., Chaouiya, C., 2006. GINsim: a software suite for the qualitative modelling, simulation and analysis of regulatory networks. Biosystems 84, 91–100.

Gray, S., Szymanski, P., Levine, M., 1994. Short-range repression permits multiple enhancers to function autonomously within a complex promoter. Genes Dev. 8, 1829–1838.

Gregor, T., Wieschaus, E.F., McGregor, A.P., Bialek, W., Tank, D.W., 2007. Stability and nuclear dynamics of the bicoid morphogen gradient. Cell 130, 141–152.

Gurdon, J.B., 1988. A community effect in animal development. Nature 336, 772–774.

Herrmann, F., Groß, A., Zhou, D., Kestler, H.A., Kühl, M., 2012. A boolean model of the cardiac gene regulatory network determining first and second heart field identity. PLoS One 7, e46798.

Hong, J.W., Hendrix, D.A., Papatsenko, D., Levine, M.S., 2008. How the Dorsal gradient works: insights from postgenome technologies. Proc. Natl. Acad. Sci. U.S.A. 105, 20072–20076.

Ingolia, N.T., 2004. Topology and robustness in the *Drosophila* segment polarity network. PLoS Biol. 2, e123.

Jaeger, J., 2009. Modelling the *Drosophila* embryo. Mol. Biosyst. 5, 1549–1568.

Jaeger, J., Blagov, M., Kosman, D., Kozlov, K.N., Manu, Myasnikova, E., Surkova, S., Vanario-Alonso, C.E., Samsonova, M., Sharp, D.H., Reinitz, J., 2004. Dynamical analysis of regulatory interactions in the gap gene system of *Drosophila melanogaster*. Genetics 167, 1721–1737.

Janssens, H., Crombach, A., Wotton, K.R., Cicin-Sain, D., Surkova, S., Lim, C.L., Samsonova, M., Akam, M., Jaeger, J., 2013. Lack of tailless leads to an increase in expression variability in *Drosophila* embryos. Dev. Biol. 377, 305–317.

Kalir, S., Mangan, S., Alon, U., 2005. A coherent feed-forward loop with a SUM input function prolongs flagella expression in *Escherichia coli*. Mol. Syst. Biol. 1. 2005 0006.

Kanodia, J.S., Rikhy, R., Kim, Y., Lund, V.K., DeLotto, R., Lippincott-Schwartz, J., Shvartsman, S.Y., 2009. Dynamics of the Dorsal morphogen gradient. Proc. Natl. Acad. Sci. U.S.A. 106, 21707–21712.

Lacaud, G., Kouskoff, V., Trumble, A., Schwantz, S., Keller, G., 2004. Haploinsufficiency of Runx1 results in the acceleration of mesodermal development and hemangioblast specification upon in vitro differentiation of ES cells. Blood 103, 886–889.

Laslo, P., Spooner, C.J., Warmflash, A., Lancki, D.W., Lee, H.J., Sciammas, R., Gantner, B.N., Dinner, A.R., Singh, H., 2006. Multilineage transcriptional priming and determination of alternate hematopoietic cell fates. Cell 126, 755–766.

Liberman, L.M., Stathopoulos, A., 2009. Design flexibility in *cis*-regulatory control of gene expression: synthetic and comparative evidence. Dev. Biol. 327, 578–589.

Longabaugh, W.J., 2012. BioTapestry: a tool to visualize the dynamic properties of gene regulatory networks. Methods Mol. Biol. 786, 359–394.

Longabaugh, W.J., Davidson, E.H., Bolouri, H., 2005. Computational representation of developmental genetic regulatory networks. Dev. Biol. 283, 1–16.

Longabaugh, W.J., Davidson, E.H., Bolouri, H., 2009. Visualization, documentation, analysis, and communication of large-scale gene regulatory networks. Biochim. Biophys. Acta 1789, 363–374.

Mangan, S., Zaslaver, A., Alon, U., 2003. The coherent feedforward loop serves as a sign-sensitive delay element in transcription networks. J. Mol. Biol. 334, 197–204.

Manu, Surkova, S., Spirov, A.V., Gursky, V.V., Janssens, H., Kim, A.R., Radulescu, O., Vanario-Alonso, C.E., Sharp, D.H., Samsonova, M., Reinitz, J., 2009. Canalization of gene expression in the *Drosophila* blastoderm by gap gene cross regulation. PLoS Biol. 7, e1000049.

Markstein, M., Markstein, P., Markstein, V., Levine, M.S., 2002. Genome-wide analysis of clustered Dorsal binding sites identifies putative target genes in the *Drosophila* embryo. Proc. Natl. Acad. Sci. U.S.A. 99, 763–768.

Materna, S.C., Nam, J., Davidson, E.H., 2010. High accuracy, high-resolution prevalence measurement for the majority of locally expressed regulatory genes in early sea urchin development. Gene Expr. Patterns 10, 177–184.

Morel, V., Schweisguth, F., 2000. Repression by suppressor of hairless and activation by Notch are required to define a single row of single-minded expressing cells in the *Drosophila* embryo. Genes Dev. 14, 377–388.

Nahmad, M., Stathopoulos, A., 2009. Dynamic interpretation of hedgehog signaling in the *Drosophila* wing disc. PLoS Biol. 7, e1000202.

Naldi, A., Carneiro, J., Chaouiya, C., Thieffry, D., 2010. Diversity and plasticity of Th cell types predicted from regulatory network modelling. PLoS Comput. Biol. 6, e1000912.

Narula, J., Smith, A.M., Gottgens, B., Igoshin, O.A., 2010. Modeling reveals bistability and low-pass filtering in the network module determining blood stem cell fate. PLoS Comput. Biol. 6, e1000771.

Narula, J., Williams, C.J., Tiwari, A., Marks-Bluth, J., Pimanda, J.E., Igoshin, O.A., 2013. Mathematical model of a gene regulatory network reconciles effects of genetic perturbations on hematopoietic stem cell emergence. Dev. Biol. 379, 258–269.

Ochoa-Espinosa, A., Yucel, G., Kaplan, L., Pare, A., Pura, N., Oberstein, A., Papatsenko, D., Small, S., 2005. The role of binding site cluster strength in Bicoid-dependent patterning in *Drosophila*. Proc. Natl. Acad. Sci. U.S.A. 102, 4960–4965.

Oliveri, P., Tu, Q., Davidson, E.H., 2008. Global regulatory logic for specification of an embryonic cell lineage. Proc. Natl. Acad. Sci. U.S.A. 105, 5955–5962.

Oosterveen, T., Kurdija, S., Alekseenko, Z., Uhde, C.W., Bergsland, M., Sandberg, M., Andersson, E., Dias, J.M., Muhr, J., Ericson, J., 2012. Mechanistic differences in the transcriptional interpretation of local and long-range Shh morphogen signaling. Dev. Cell. 23, 1006–1019.

Ozdemir, A., Ma, L., White, K.P., Stathopoulos, A., 2014. Su(H)-Mediated Repression Positions Gene Boundaries along the Dorsal-Ventral Axis of *Drosophila* Embryos. Dev. Cell. 31, 100–113.

Perkins, T.J., Jaeger, J., Reinitz, J., Glass, L., 2006. Reverse engineering the gap gene network of *Drosophila melanogaster*. PLoS Comput. Biol. 2, e51.

Peter, I.S., Davidson, E.H., 2009. Modularity and design principles in the sea urchin embryo gene regulatory network. FEBS Lett. 583, 3948–3958.

Peter, I.S., Davidson, E.H., 2010. The endoderm gene regulatory network in sea urchin embryos up to mid-blastula stage. Dev. Biol. 340, 188–199.

Peter, I.S., Davidson, E.H., 2011. A gene regulatory network controlling the embryonic specification of endoderm. Nature 474, 635–639.

Peter, I.S., Faure, E., Davidson, E.H., 2012. Feature article: predictive computation of genomic logic processing functions in embryonic development. Proc. Natl. Acad. Sci. U.S.A. 109, 16434–16442.

Ramos, C.A., Bowman, T.A., Boles, N.C., Merchant, A.A., Zheng, Y., Parra, I., Fuqua, S.A., Shaw, C.A., Goodell, M.A., 2006. Evidence for diversity in transcriptional profiles of single hematopoietic stem cells. PLoS Genet. 2, e159.

Reeves, G.T., Trisnadi, N., Truong, T.V., Nahmad, M., Katz, S., Stathopoulos, A., 2012. Dorsal-ventral gene expression in the *Drosophila* embryo reflects the dynamics and precision of the dorsal nuclear gradient. Dev. Cell. 22, 544–557.

Roth, S., 2011. Mathematics and biology: a Kantian view on the history of pattern formation theory. Dev. Genes Evol. 221, 255–279.

Saka, Y., Smith, J.C., 2007. A mechanism for the sharp transition of morphogen gradient interpretation in *Xenopus*. BMC Dev. Biol. 7, 47.

Sanchez, L., Chaouiya, C., Thieffry, D., 2008. Segmenting the fly embryo: logical analysis of the role of the segment polarity cross-regulatory module. Int. J. Dev. Biol. 52, 1059–1075.

Segal, E., Raveh-Sadka, T., Schroeder, M., Unnerstall, U., Gaul, U., 2008. Predicting expression patterns from regulatory sequence in *Drosophila* segmentation. Nature 451, 535–540.

Shoval, O., Alon, U., 2010. SnapShot: network motifs. Cell 143, 326–321.

Stathopoulos, A., Van Drenth, M., Erives, A., Markstein, M., Levine, M., 2002. Whole-genome analysis of dorsal-ventral patterning in the *Drosophila* embryo. Cell 111, 687–701.

Surkova, S., Golubkova, E., Manu, Panok, L., Mamon, L., Reinitz, J., Samsonova, M., 2013. Quantitative dynamics and increased variability of segmentation gene expression in the *Drosophila* Krüppel and knirps mutants. Dev. Biol. 376, 99–112.

Swarup, S., Verheyen, E.M., 2012. Wnt/Wingless signaling in *Drosophila*. Cold Spring Harbor Perspect. Biol. 4.

Thomas, R., 1991. Regulatory networks seen as asynchronous automata: a logical description. J. Theor. Biol. 153, 1–23.

Thomas, R., Thieffry, D., Kaufman, M., 1995. Dynamical behavior of biological regulatory networks. Bull. Math. Biol. 57, 135–153.

Trompouki, E., Bowman, T.V., Lawton, L.N., Fan, Z.P., Wu, D.C., DiBiase, A., Martin, C.S., Cech, J.N., Sessa, A.K., Leblanc, J.L., Li, P., Durand, E.M., Mosimann, C., Heffner, G.C., Daley, G.Q., Paulson, R.F., Young, R.A., Zon, L.I., 2011. Lineage regulators direct BMP and Wnt pathways to cell-specific programs during differentiation and regeneration. Cell 147, 577–589.

Warmflash, A., Zhang, Q., Sorre, B., Vonica, A., Siggia, E.D., Brivanlou, A.H., 2012. Dynamics of TGF-β signaling reveal adaptive and pulsatile behaviors reflected in the nuclear localization of transcription factor Smad4. Proc. Natl. Acad. Sci. U.S.A. 109, E1947–E1956.

Wilson, N.K., Calero-Nieto, F.J., Ferreira, R., Göttgens, B., 2011. Transcriptional regulation of haematopoietic transcription factors. Stem Cell Res. Ther. 2, 6.

Wittmann, D.M., Blochl, F., Trumbach, D., Wurst, W., Prakash, N., Theis, F.J., 2009. Spatial analysis of expression patterns predicts genetic interactions at the mid-hindbrain boundary. PLoS Comput. Biol. 5, e1000569.

Chapter 7

Evolution of Bilaterian Animals: Processes of Change and Stasis in Hierarchical Developmental Gene Regulatory Networks

Genomic Control Process
http://dx.doi.org/10.1016/B978-0-12-404729-7.00007-1

A simple but inescapable logic underlies the conceptual approach to the evolution of body plans that is elaborated in this chapter: since body plans are built by the operation of genomically encoded developmental gene regulatory networks (GRNs), evolutionary change in body plans must have been caused by evolutionary change in developmental GRNs. In general, evolution occurs by alteration of genomic sequence, and here we deal with the ways in which developmental GRNs have been heritably modified, so as to result in evolutionary change in developmental processes, and thus in their outcome, the body plan. The body plan as we use the term here refers specifically to the spatial disposition and morphological character of body parts and subparts, and general anatomical features such as segmentation or possession of a head.

Yet it is not just change in evolution that needs to be considered. The extreme conservation of distinct morphological features over immense geological periods is just the other side of the same coin. This requires consideration of how aspects of GRN structure which causally account for these conserved morphological features could have been stabilized through deep time. To view evolutionary process through the lens of change and stasis in encoded GRN structure introduces a series of concepts different from those of conventional "evo/devo". These considerations begin with the consequences of GRN hierarchy. As is instantly obvious, the effects of any given regulatory change in the genomic sequence included in a GRN will differ fundamentally according to where in GRN hierarchy this change has occurred, thus in one stroke destroying the traditional assumption that morphological evolution proceeds in a uniform fashion. We have three major goals in this chapter: to relate phylogeny to the comparative structure of developmental GRNs; to consider mechanistically, at least in principle, the modes of change in developmental GRNs at different hierarchical levels; and to understand conservation of GRN circuitry across deep time. Several of the specific arguments we take up in the following have been in part separately addressed earlier (Davidson and Erwin, 2009, 2010; Erwin and Davidson, 2009; Peter and Davidson, 2011).

1.   Introduction: Evolution by Genomic Change at Different Levels of GRN Hierarchy

As discussed in Chapter 2, the GRNs that determine the developmental process from embryo to body part formation are intrinsically hierarchical, just as is developmental process itself. Regulatory states, the specific combinations of transcription factors present in particular cellular domains, succeed one another in each of the progressively subdivided and subfunctionalized spatial domains of the organism. We can imagine this in the geometrical terms of the diagram in Fig. 7.1(A) (Peter and Davidson, 2011), where both time and developmental stage flow from top to bottom, and transcriptional causality within the hierarchy flows downward as well as laterally. At each point in time and embryonic space the

Figure 7.1

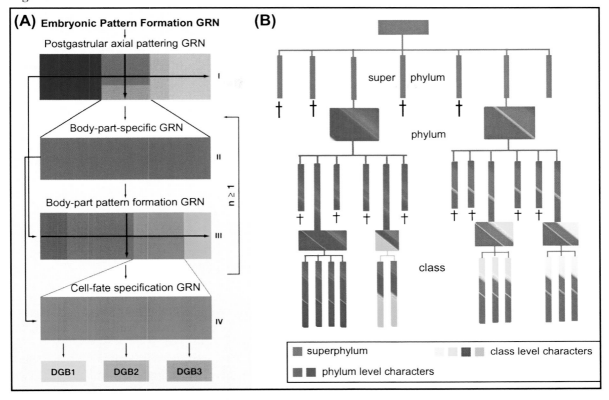

Figure 7.1 GRN hierarchy and bilaterian phylogeny. (A), Symbolic representation of the hierarchy of developmental GRNs. The diagram consists of a vertical stack of four rectangular boxes representing GRN circuitries proceeding from early (top) to late (bottom) in development. Box I symbolizes the control functions of early embryogenesis. The initial (i.e., pregastrular) embryonic GRNs establish regulatory states differentially throughout the embryo, as we discuss in Chapter 3, organized spatially with respect to the embryonic axes (axial organization and spatial subdivision are symbolized by orthogonal arrows and colored patterns). Progenitor fields for future adult body parts are later demarcated by signals plus local regulatory spatial information formulated in Box I, and given regulatory states are established in each such field by the earliest body part-specific GRNs. Many such progenitor fields are thus set up during postgastrular embryogenesis, and a GRN defining one of these is here symbolized as Box II. Subdivision of the progenitor field then establishes the different regional components of the body part at the regulatory level, and in each such region a discrete GRN operates. These "sub-body part" GRNs are symbolized by the oriented patterns of Box III. Some body parts are very complex and the subdivision process may be reiterated several times, symbolized by the upward arrow to right, extending from Box III back to Box II. Toward the termination of the developmental process in each region of the late embryo, the GRNs specifying the several individual cell types and deployed in each subpart of each body part, are symbolized for one subpart as Box IV. Generation of specific cell types in the adult from stem cells would also be considered a Box IV process. Activation of differentiation gene batteries ("DGB1, 2, 3") and of morphogenesis cassettes, constitutes the final regulatory output of each cell type *(from Peter and Davidson (2011))*. (B), Symbolic representation of concept that phylogenetically conserved GRN circuitry could account for the nested distribution of morphological characters (i.e., of developmental

processes) in animal phylogeny. Each colored field in the boxes represents a developmental GRN circuit which is conserved, as indicated by the color coding, in a nested phylogenetic pattern. Thus, all descendants of an ancestor in which the pattern originates inherit that pattern (for example, the blue pattern), while at each cladistic level certain descendants initiate novel patterns inherited by all of their descendants (for example, the red pattern) *(from Davidson and Erwin (2009))*.

regulatory state determines the succeeding regulatory state, as well as the states in adjacent territories, due to inter- and intradomain signaling. Therefore it must be the case that a given regulatory change at the upper levels of the GRN may have cascading, large effects on developmental outcome, while as spatial subdivision proceeds the effects of similar kinds of change in any of the more and more locally acting GRNs are more and more confined.

1.1   Evolution of regulatory linkages at different levels of developmental GRNs

Development and bilaterian phylogeny

In development, the function of the GRN circuits inhabiting the top level of the hierarchy in Fig. 7.1(A) is to control the basic spatial organization of the body plan in respect to the coordinates of the egg and early embryo. That is, such circuits define with respect to these coordinates the major parts of the organism, its head, the gut and its openings, the central nervous system, and the disposition of its mesoderm. These definitions of fate are executed by expression of sets of determinant transcription factors within distinct spatial domains. In the same general terms, the GRNs of the next level of the hierarchy in Fig. 7.1(A) set up the regulatory states that specifically define the respective progenitor fields for the organs, appendages if any, and other individual body parts that the animal will possess. The basic organization of the body plan and of its parts also constitutes the morphological characters that define animal Phyla and classes in the hierarchical phylogenies which underlie current evolutionary thought. Here we adopt a general cladistic approach in which monophyletic clades or taxa are based on shared derived characters, and phylogeny depends on the nested organization of sets of such characters. This concept logically implies the existence of developmental GRN circuitry that for example encodes phylum-level characters, which must at least functionally but to some extent also structurally be shared among all members of the phylum. Frequently, these shared features are defined early in postgastrular development, and thus it can be deduced that the underlying conserved circuitry will have to be encoded within the upper level of GRN hierarchy. Similarly the characters shared by members of each class should be encoded by a common GRN circuitry determining the layout of its body parts, which is deployed at somewhat later developmental stages. More locally acting developmental circuits should be shared respectively among all members of each order constituting each class. This is by no means to imply that all upstream GRN circuitry is conserved, particularly that controlling pregastrular development. However, we may conclude that circuitry which does control conserved characters at higher levels in phylogeny is more likely to occur at higher levels of GRN hierarchy (Davidson and Erwin, 2006, 2010; Erwin and Davidson, 2009). By this route we arrive at a basic theoretical precept of animal body plan evolution: the evolutionary cause of the cladistic distribution of body plan characters in Bilateria is the phylogenetically hierarchical, branching distribution of shared developmental GRN circuits, as diagrammed by color in Fig. 7.1(B). It is intriguing to realize that the intuitive hierarchical arrangement of animals generated by Linnaeus 200 years ago reflects patterns of evolutionary conservation of certain features of the underlying developmental GRNs.

Corollaries

From this relationship flow the corollaries that structure much of the following discussion. The first of these corollaries is that there must exist mechanistic aspects of hierarchical GRN structure that can account for the amazing stasis of phylum-level body plan characters in animals, which date from the Cambrian (Erwin et al., 2011). A majority of bilaterian class-level character complexes are also of Cambrian and Ordovician vintage, and almost all are Paleozoic (~80% of classes; Erwin et al., 2011). That is, most classes are at least 250 My old, as for example are all extant echinoderm and arthropod classes and their definitive characteristics. Understanding of animal body plan evolution requires that we identify a testable cause or causes that could explain this phylogenetic level-dependent stasis in terms of developmental GRN structure and function.

The paleontological record of some well-studied groups indicates that changes in more detailed anatomical structure, such as the morphological details of tetrapod appendages discussed in the following, have occurred again and again during evolution. These changes are due to alterations in the programs controlling embryonic development of particular body parts. Thus a second corollary is that there exist types of change in GRN structure responsible for such local alterations in body plan. We need to understand what are the evolutionarily flexible points in GRN structure, and the types of change which alter GRN linkages. But at the same time, an explanation is needed for the curious feature that once installed novel character sets typically persist unchanged for tens of millions of years. For example, the radiation of the extant mammalian order Carnivora from its common ancestral form occurred about 65 million years (My) ago and the same definitive body plan characters possessed by this ancestor persist in modern cats and dogs (Nyakatura and Bininda-Emonds, 2012).

A huge amount of classic and modern observation demonstrates continuous adaptive microevolution of the fine traits that distinguish different species belonging to the same genus, and even different variants belonging to the same species. Those changes generating visible species-specific characters, including color patterns on wings, skin or feathers, micromorphology of integument elements, minor differences in morphometric proportions, are also the product of control programs encoded in developmental GRNs. Unlike the upper level GRN components considered thus far, these traits are encoded at the GRN peripheries where differentiation gene battery deployment and function are programmed. In contrast to the extremely long evolutionary persistence of the developmental circuits that control the definitive characters of higher taxonomic levels, the duration of given species and their definitive characters is relatively brief in terms of geological time. Thus a third corollary emerging from the relation between GRN hierarchy and animal phylogeny is that at the GRN periphery continuous adaptive microevolutionary variation prevails, affecting the products and deployment of differentiation gene batteries and terminal morphology cassettes, including their immediate regulatory controllers. Such species-specific characters are essentially highly adaptive decorations superimposed on more widely utilized common body plans, as we will see in the following examples. There is a profound difference in the way these disparate parts of the GRN evolve. Since the functions of effector genes are terminal, what happens to their expression does not directly affect subsequent developmental processes. The stations at the end of the line are more free to change or come and go than those at internal exchange hubs where downstream traffic is controlled.

Thus, in summary, the obvious point that evolution of the body plan means evolution of encoded developmental GRNs leads to consideration of the direct relation between the intrinsic hierarchy of these GRNs and the intrinsic hierarchy of animal phylogeny. This relation in turn opens the way to a largely new view of the internal mechanisms of body plan evolution and morphological evolution in general, and to a largely new set of specific questions for which we seek answers. In retrospect it is evident that body plan evolution could not possibly have begun to be understood in mechanistic terms before GRN hierarchy was experimentally demonstrated.

1.2   Rapid *cis*-regulatory microevolution at the GRN periphery

In the following, our focus is on adaptive variation of regulatory functions at the GRN periphery. The function of these peripheral regulatory systems is to determine where, when and how intensely effector gene cassettes will be expressed in terminal developmental processes. Changes in the expression of such effector genes can be caused either by changes in the control system for effector gene transcription or by changes in the expression of their drivers (for multiple examples of effector gene control systems, see Chapter 5). The DNA sequences with which we are here concerned are thus either in *cis*-regulatory modules that control expression of individual differentiation genes; or one level upstream, in *cis*-regulatory systems of regulatory genes that control deployment of whole differentiation gene batteries or morphogenetic effector genes. This is of course not the only kind of evolutionary change that affects external downstream properties, as protein-coding changes in effector genes also can and do affect terminal characteristics in adaptive ways (Coyne and Hoekstra, 2007).

Cis-regulatory evolution of pigmentation patterns

Animal pigmentation patterns are of obvious adaptive significance, for mimicry, camouflage, and species-specific mate selection, and they evolve remarkably rapidly. This is evident from the prevalence of species-specific and intraspecific pigmentation differences, from insects to mammals. One reason for this is that relatively few genes control pigmentation pattern, and a major mechanism of evolutionary variation in this adaptive feature is *cis*-regulatory mutations which affect spatial and quantitative expression of such genes. In principle there are several levels at which such regulatory changes could occur: by controlling the expression of genes encoding the enzymes which generate the pigment from precursor compounds (Fig. 7.2(A)), by controlling the expression of genes encoding transcription factors that control pigmentation effector genes, or by controlling the expression of signaling molecules that determine where the transcriptional apparatus driving pigmentation is expressed. All are used to effect rapid evolutionary change in external pigmentation.

In *Drosophila*, the gene *ebony* provides a classic example of *cis*-regulatory control of pigment pattern at the effector gene level. As we see in Fig. 7.2(A) (Ferguson et al., 2011) *ebony* antagonizes black melanin formation leading instead to yellowish tan coloration. The intraspecific abdominal pigmentation variants in African *Drosophila melanogaster* may well be of adaptive significance, since the darker variants live at higher elevations. The causal events underlying this variation are specific *cis*-regulatory sequence changes which result in decreased *ebony* expression, and hence darker pigmentation (Rebeiz et al., 2009). Gene transfer rescue experiments into flies mutant for *ebony* directly demonstrate that the source of this difference is sequence variation in an *ebony cis*-regulatory control module, while the sequence of the Ebony protein remains the same. A key aspect of this evolutionary adaptation turns out to be the modularity of the *ebony cis*-regulatory system, in that expression in different parts of the body is controlled by different *cis*-regulatory modules. As Fig. 7.2(B) shows, the adaptive *cis*-regulatory mutations affect only abdominal expression of *ebony*, while expression of the same gene elsewhere remains unaffected. Similarly, in sister *Drosophila* species, namely the dark bodied *americana* and light bodied *novomexicana*, estimated to have diverged only 0.4 My ago, the interspecific pigmentation difference is due to *cis*-regulatory sequences that affect the expression levels of *ebony* and *tan* genes (see Fig. 7.2(A)), as directly demonstrated by allele specific measurements in species hybrids (Cooley et al., 2012). In both examples, adaptation of a peripheral feature has been achieved by changes within the *cis*-regulatory sequences of effector genes without affecting the regulatory state present in these cells. Only the response to these transcription factors is changed, leading to lowered expression of pigmentation enzymes and therefore brighter abdomens.

Evolutionary changes in *cis*-regulatory sequence controlling a pigmentation effector gene are also responsible for the distinct spatial patterns of wing spots in Diptera. The pigmentation pattern depends on how the landscape of spatial regulatory states in the developing wing is

Figure 7.2

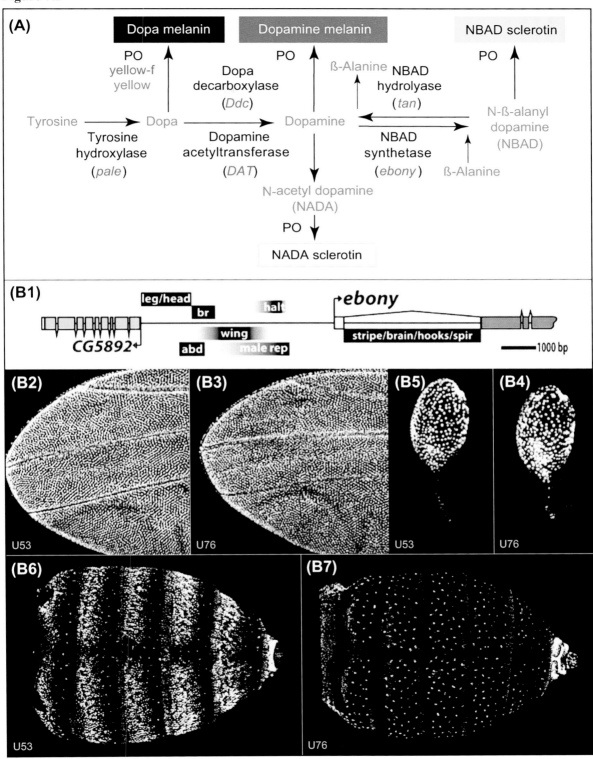

Figure 7.2

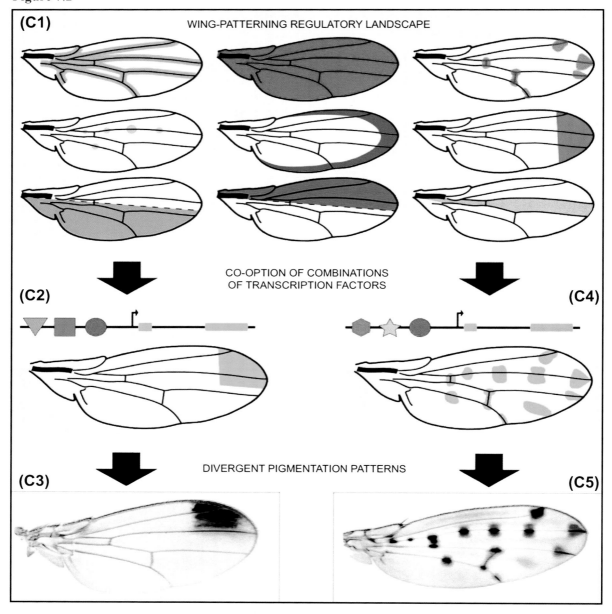

(C1) WING-PATTERNING REGULATORY LANDSCAPE

CO-OPTION OF COMBINATIONS
OF TRANSCRIPTION FACTORS

(C2)   **(C4)**

DIVERGENT PIGMENTATION PATTERNS

(C3)   **(C5)**

Figure 7.2 Evolution of color pattern at the GRN periphery in insects. (A), Pigment synthesis and sclerotinization pathway in *Drosophila (from Ferguson et al. (2011))*. Tyrosine and derivative intermediates, blue; enzymes, black; names of genes encoding these enzymes, red. PO, phenol oxidase, a cross-linking enzyme. The colors of the products are shown in boxes. (B), Intraspecific *cis*-regulatory evolution of the *ebony* pigmentation gene in *Drosophila (from Rebeiz et al. (2009))*. (B1), Modular structure of the genomic control system governing *ebony* expression in various tissues: halt, haltere; abd, abdomen; br, bristles; male rep, repression in males; stripe, repression in abdominal tergite stripe (cf. B6); spir, larval spiracles. Necessary enhancer regions (black bars) were mapped in gene transfer experiments, though boundaries are in some cases not defined (gray shading). (B2–7), Expression of

Figure 7.2

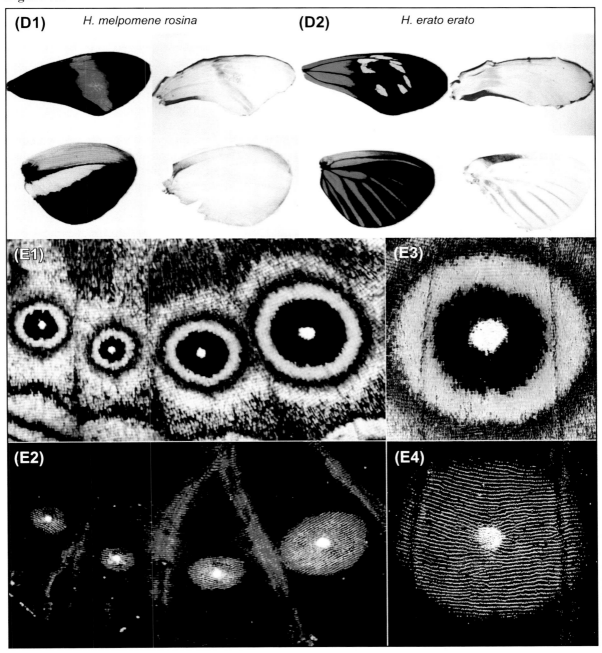

reporter constructs driving GFP (white) from *ebony cis*-regulatory sequences isolated from a light colored variant, U53, and from a dark colored variant, U76. (B2,3), Expression in wing blade; (B4,5), Expression in halteres; (B6,7), Expression in adult abdominal tergites. Only the *cis*-regulatory module driving this expression has diverged. (C), Evolutionary change in a *yellow* gene *cis*-regulatory module *(from Prud'homme et al. (2007))*. (C1), Diagrammatic representation of the conserved spatial distribution of nine transcription factors which constitute the regulatory landscape of the developing wing, and which offer downstream genes a large variety of regional combinatorial input patterns, according to their *cis*-regulatory target sites. (C2–5), Melanin pigment deposition patterns mediated by the *yellow* gene, such that the patterns would reflect co-option of different *cis*-regulatory target sites, interacting with different sets of the factors in (C1). (C2,3), "Spot pattern", combinatorially generated by the blue activator (expressed all over the wing) operating by AND logic with the red activator (expressed in the distal end), together with the dominant green factor, a posterior repressor (*viz.* the product of the *engrailed* gene). (C4,5), A different pattern generated by a different co-optive event; here the pattern is produced by the combination of two factors expressed differently along the veins plus the blue wing identifier. (C3,5), Respective pigment patterns generated by the *yellow cis*-regulatory modules pictured in (C2) and (C4), respectively. (D), Transcriptional regulatory basis of rapidly evolving pigment pattern in butterfly wings (*Heliconius* species). *Heliconius* species and races, as indicated, displaying various red pigment patterns, in each case foreshadowed exactly by the spatial expression of the *optix* gene *(from Reed et al. (2011))*. (D1,2), Forewings, top; hindwings, bottom; pigmentation patterns shown on mature butterfly wings, left; and in situ hybridization for *optix* mRNA on pupal wings, right. (E), Transcriptional regulatory basis of rapidly evolving eyespot mimicry in *Bicyclus anynana* butterflies *(from Brunetti et al. (2001))*. (E1), Four eyespots on ventral forewing surface; (E2) Patterns of expression of *engrailed* or *invected* (green) and *spalt* (purple) during pupation, which foreshadow these eyespots. (E3), High magnification of a forewing eyespot formed by colored scales; (E4), Earlier developmental expression of *engrailed* or *invected*, outer green ring; *spalt*, inner purple ring; both, center white spot.

read by the pigment gene control systems. This point has been illustrated in a study of the great diversity of pigmentation patterns in Diptera (Prud'homme et al., 2007). A key spatial regulator of these patterns turns out to be the transcriptional control system of the *yellow* gene (see Fig. 7.2(A); Gompel et al., 2005). The regulatory landscape to which this control system responds is shown in Fig. 7.2(C1). Detailed *cis*-regulatory analysis shows for example that evolution of the "spot" *cis*-regulatory pattern element found in the species *Drosophila biarmipes* (pattern in Fig. 7.2(C2,3)) occurred by acquisition of sites for certain of the activators and repressors present in the regulatory landscape. The same landscape is interpreted to give a different pattern in different species in which evolutionary sequence change has generated different *cis*-regulatory response elements controlling *yellow* gene expression (Fig. 7.2(C4,5)).

Pigmentation patterns in butterfly wings of the genus *Heliconius* are controlled at a different level (Fig. 7.2(D)). Here the striking interspecific and intraspecific variation of these patterns among species and races which diverged 12–25 My ago, maps to only one gene. This gene encodes a transcriptional regulator, Optix. Study of *optix* expression in pupal wings by in situ hybridization reveals that the spatial distribution of red patches on the adult wing is exactly foreshadowed by *optix* spatial expression domains (Reed et al., 2011). The amino acid sequence of the Optix protein has remained invariant, and the different patterns of *optix* expression are evidently the result of *cis*-regulatory evolution within the transcriptional control system of the *optix* gene, generating different *optix* expression patterns in response to the same regulatory landscape. It is also interesting that *optix* expression controls pointed scale morphology in all butterfly species examined, but only in the *Heliconius* species does it control red pigment formation. Thus an additional evolutionary change has been the *cis*-regulatory acquisition of Optix target site sequences by a red pigment gene. Similarly, the black pigment patterns of *Heliconius* wings are foreshadowed exactly by spatial patterns of *wntA* gene expression, and genetic mapping shows that this is the gene in which evolutionary *cis*-regulatory changes have been responsible for the diversification of that aspect of pattern (Martin et al., 2012). A broad range of butterflies produce "eyespots" of diverse color on their wings which are supposed to have adaptive functions in avoiding predators, and here again regulatory gene expression patterns foreshadow and developmentally organize the pigmentation design (Fig. 7.2(E); Brunetti et al.,

2001). Additional *cis*-regulatory changes in the linkages between the color pigmentation genes and their transcriptional regulators have to be invoked in accounting for the variations observed, because different colors appear downstream of the same regulators in different species. Similarly, returning to abdominal pigmentation patterns in *Drosophila* species, alterations in spatial expression of upstream regulators of *ebony* and *tan* have recently been implicated in the evolution of species-specific pigmentation pattern (Ordway et al., 2014).

Color patterning is a classic example of a developmental character in which multiple variations in the process controlling spatial and quantitative expression of genes are adaptively useful. Change in these processes is free to occur because they are peripheral; there is not the functional cost that would obtain were the evolutionary rewiring to affect causal regulatory linkages operating earlier in the development of the body plan. Similar evolutionary flexibility, operating by similar mechanisms, is likely to be revealed wherever color patterning systems are comparatively examined at the gene regulatory level. In mammals major aspects of color coat pattern have long been known to be controlled by the *agouti* gene system. This gene encodes a small paracrine signal molecule which is a dominant antagonist of a melanocortin receptor that in melanocytes determines the activity of pigment genes. The melanocortin receptor mediates cAMP activation of the regulatory gene *mitf*, a controller of pigment cell differentiation genes, and thus Agouti is a negative regulator of pigmentation, so that fur color becomes lighter in the presence of higher levels of Agouti (Otręba et al., 2012). Developmental changes in *agouti* expression have been experimentally shown to underlie adaptive intraspecific evolution of color patterning. For example, darkly pigmented deer mice (*Peromyscus polionotus*), originating as a population from a dark soil environment, rapidly evolved lighter dorsal camouflage pigmentation on migration to a sandy beach environment. This has been shown to be due to a "light allele" of *agouti* that encodes the same protein sequence but is regulated differently, so that during fetal life in beach mice this gene is expressed not only ventrally, but also more dorsally causing light pigmentation on the flanks and no ventral pigmentation. Experimental tests prove that indeed the light allele of *agouti* is able to downregulate melanocyte pigment when transfected into darkly pigmented mice (Manceau et al., 2011).

Multiple modes of adaptive *cis*-regulatory evolution at the GRN periphery

We have seen in the foregoing examples of adaptive evolutionary change within *cis*-regulatory sequences. Such functional change within regulatory sequence occurs in several entirely distinct ways. Many comparative studies of *cis*-regulatory structure and function have revealed an amazing flexibility in the organization of transcription factor target site placement, spacing, and multiplicity. As we reviewed in Chapter 2 (see Fig. 2.2 and ancillary discussion), a particularly striking example of this flexibility emerged from a functional- and sequence-level comparison of *evenskipped* enhancers in species of flies a few million to 100 My distant from *D. melanogaster* (Hare et al., 2008). The generally observed result is that functional output may remain essentially the same in comparisons of orthologous, evolutionarily diverged *cis*-regulatory modules, irrespective of gross changes in *cis*-regulatory structure. That is, the arrangement and organization of binding sites where transcription factors interact appears to have little effect on functional output, and what is important is that the identity of the inputs remains unaltered. Thus most attempts to define some rules or "grammar" by which the arrangement of binding sites within CRMs determines regulatory functionality have not been successful. The only frequent exceptions in this and another intensively studied *Drosophila cis*-regulatory module, *sparkling*, relate to the conservation of closely positioned target sites for factors that probably interact cooperatively on the DNA or bind overlapping sequences (Hare et al., 2008; Swanson et al., 2010, 2011; Fig. 2.2). Thus since so much change in sequence is tolerated without significant change of function, it becomes especially interesting to ask what kinds of structural alterations have occurred in cases of rapid, *cis*-regulatory evolution where it is of adaptive importance that function does indeed change. The answer is always multiple sequence changes, which often affect (or probably affect) the identities of the transcription factors these sequences respond to, rather than just their relative arrangement. The following examples show that such changes occur by a variety of mechanisms.

Comparative studies of *cis*-regulatory sequence evolution show that meaningful mutational sequence change occurs continuously and frequently over time. All but a few percent of genes in the genome are downstream effector genes expressed at the GRN periphery, and most such evolutionary change will affect this class of genes. It is of course their functions which adaptively confront the external environment. In addition to the few studies providing specific demonstrations of particular *cis*-regulatory sequence mutations that endow the corresponding CRM with new or altered functions, many descriptive and correlative observations indicate widespread and rapid *cis*-regulatory evolution at the effector gene level. For example, in rapidly diversifying African lake cichlid fishes, different species have sharply diverse adaptive sensitivity to light of different wavelengths, due to differential expression of various Opsins. Genetic analysis points to *cis*-regulatory sequence variation affecting transcription factor-binding sites as the cause (O'Quin et al., 2011). Indeed, enhancers per se turn over at a high rate. A genome-wide study of human, chimpanzee, and macaque DNAseI hypersensitive sites in primary skin fibroblasts and lymphoblastoid cell lines compared likely *cis*-regulatory sequences in the three species (Shibata et al., 2012). Though most such sites were conserved, hundreds of gains and losses in DNAse I hypersensitive sites were identified that had occurred in the lineages leading to human and chimpanzee, and these could be correlated with species-specific changes in gene expression.

A case of adaptive evolution where functional significance has been thoroughly examined at the sequence level is divergence in the *cis*-regulatory control systems of the *shavenbaby* (*svb*) regulatory gene of *Drosophila* (McGregor et al., 2007; Frankel et al., 2011). This regulatory gene directly controls expression of effector genes that cause epidermal cells to generate trichomes, which are noninnervated actin containing extensions of the epidermis (Chanut-Delalande et al., 2006). Spatial variation of the *svb* expression pattern results in species-specific differences in the distribution of trichomes. Unlike a number of other *Drosophila* species which produce dense lawns of dorsal and lateral trichomes on the posterior rows of epidermal cells of each segment ("quaternary" cells), in *Drosophila sechellia* these cells are covered by naked cuticle. Spatial expression of *svb* is controlled by multiple enhancers operating in different regions of the body. By transfection of *svb* enhancers of *D. sechellia* to *D. melanogaster*, it was demonstrated that cumulative changes in multiple enhancers underlie the loss of *svb* expression in quaternary cells. By testing the functional significance of the sequence differences in these enhancers between *D. melanogaster* and *D. sechellia,* it could be concluded that at least five single-base substitutions in one of the enhancers had phenotypic effects, which were complex and synergistic. Single-base substitutions are of course a continuous and reversible type of genomic change, and although the pathway toward spatial loss of function in the *D. sechellia* phenotype appears to have been unidirectional, as the authors point out, in principle it could be reversed. In another study, a scan of divergent expression patterns in 20 *Drosophila* genes (mainly downstream effector genes) across the same species subgroup and also a more distantly related subgroup displayed many evolutionary alterations in expression (Rebeiz et al., 2011). As might be expected, the most common form of derived change is just loss of expression, and heterochronic shifts were also frequent, but co-optive gains of novel spatial expression domains were rare. However, in one particularly variable gene encoding an extracellular metalloprotease (Neprolysin), a new domain of expression in optic neuroblasts of the brain had arisen in *Drosophila santomea*. This domain of expression is absent even in the sister species *Drosophila yakuba*, from which *D. santomea* diverged only ~0.5 My ago. However, every species of *Drosophila* examined retains a conserved expression pattern in the mushroom bodies of the brain. Mapping and functional tests showed that four mutations in an intronic enhancer of *neprolysin*, which is also required for expression in other tissues, were responsible for the evolutionarily novel optic lobe expression, together with other preexistent conserved sequences. In both these cases, then, a novel *cis*-regulatory function evolved by multiple mutations in a preexisting enhancer.

Point mutation in *cis*-regulatory sequence is not the only way adaptive changes in effector gene expression occur. A rapidly growing number of examples demonstrate adaptively useful regulatory change that has been mediated by transposable elements, internal sequences of which contain transcription factor-binding sites, as reviewed elsewhere (Kazazian, 2004; Gogvadze and Buzdin, 2009; Peter and Davidson, 2011). A particularly well-studied case is the expression of the *prolactin* (*prl*) gene in the uterine lining (endometrial decidua) in some but not all mammals, in addition to its usual pituitary locus of expression (Emera et al., 2012; Emera and Wagner, 2012). In elephants, mice, and humans (but not rabbits, pigs or dogs for example), *prl* is expressed in the endometrium during pregnancy, due to a second upstream transcriptional start site. This is located with its own enhancer within a DNA transposable element, remarkably, in each species a different type of transposon. But the mere presence of these particular transposable elements in the vicinity of the gene does not per se produce this regulatory adaptation; further sequence evolution subsequent to transposon insertion was required. In the human and great apes the adaptive feature of endometrial *prl* expression is apparently its role in controlling the deep interstitial invasiveness of placental cells, by activating uterine NK cells and thereby limiting invasion. Functional tests on the anthropoid transposon enhancer demonstrate the causal roles of evolutionary changes at five sites which act synergistically with an Ets site that was apparently carried in on the original transposon, evidently a key property of this transposon which potentiated its use by the *prl* gene.

1.3 Intraspecific and intrageneric morphological evolution by change affecting expression of signaling genes active in terminal developmental phases

The traits that distinguish bilaterian genera within the family, species within the genus, and sometimes heritable intraspecific variants, arise late in the developmental process. Relative morphometric dimensions, integument decoration and morphology, secondary sexual characteristics, feeding and locomotory structures, etc., then assume their final forms. Evolution may occur relatively rapidly at these taxonomic levels. Even cursory familiarity with the respective developmental processes predicts that a major mechanism underlying these kinds of evolutionary divergence should involve change affecting spatial, temporal, and quantitative transcriptional expression of developmental signaling ligands. Current evidence provides many notable examples based firmly on comparative sequence-level observations, and increasingly on functional and synthetic demonstrations.

The genetically defined adaptive variation in the shapes of beaks in the 14 species of Darwin's Galapagos finches (genus *Geospiza*), generated in embryonic craniofacial development, has been explained in just this way (Abzhanov et al., 2004, 2006; Mallarino et al., 2011). Species characterized by heavy beaks feed by crushing hard seeds, while species having long, slender beaks use them to recover nutrients from cactus flowers and fruits, and so forth; beak morphology is thus directly selective depending on the changing availability of various food sources (Grant, 1999); for examples, see Fig. 7.3(A). Several signaling systems affect beak shape and dimensions. The prenasal cartilage mass forms earlier in facial development, while the premaxillary bone ultimately generates the beak morphology. Variation in both underlies the variable morphology of the beak, which is part of the upper jaw. In situ hybridization revealed a strong correlation between beak morphology and Bmp4 expression in the mesenchyme that will form the prenasal cartilage during craniofacial development, such that the heavier beaks of ground finches displayed broader and earlier Bmp4 expression than in forms with gracile beaks. To get beyond correlation, Bmp4 expression was experimentally forced in mesenchyme of developing chick embryo beaks, using a retroviral expression vector. This resulted in significantly thicker and broader beaks formed by larger masses of skeletogenic cells (Abzhanov et al., 2004, 2006; Mallarino et al., 2011). The length of the prenasal cartilage mass is controlled independently, by Calmodulin, which mediates Ca^{2+} signaling, and this was similarly demonstrated experimentally by retroviral misexpression in chick embryos, as shown in Fig. 7.3(B) (Abzhanov et al., 2006). A different set of signaling molecules altogether determines the dimensions of the premaxillary bone. These signaling ligands bind to Tgfβ receptors which initiate an intracellular protein phosphorylation sequence that ultimately modifies transcription factor activity; additional signaling pathways involved in shaping this bone require the Wnt signaling mediator β-catenin, and a Dickkopf factor possibly also involved in Wnt signaling (Mallarino et al., 2011). Again it could be shown that forced expression of the relevant Tgfβ receptor, β-catenin, and the Dickkopf factor in developing chick embryos causes them to generate deeper and longer beaks (Fig. 7.3(C)). While none of the gene regulatory relationships between these signals and the genes determining bone differentiation are known, it is clear that in the diverse finch species these signals are expressed differently, more broadly in the species that generate larger beaks. At the sequence level, variation in expression of these signaling genes implies the evolutionary alteration of multiple *cis*-regulatory systems. Such sequence changes would account for the species-specific variation in the beak morphology of Darwin's finches, which includes independent changes in beak length, width, and breadth (Figs 7.3(D) and (E)). Similarly, in cichlid fishes, jaw morphologies have also rapidly and

Figure 7.3

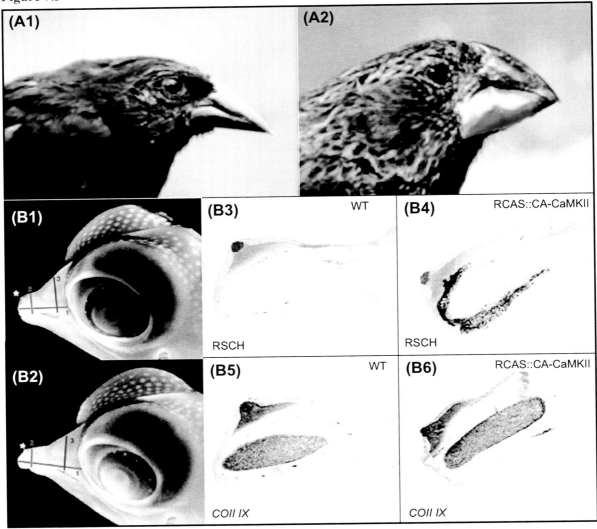

Figure 7.3

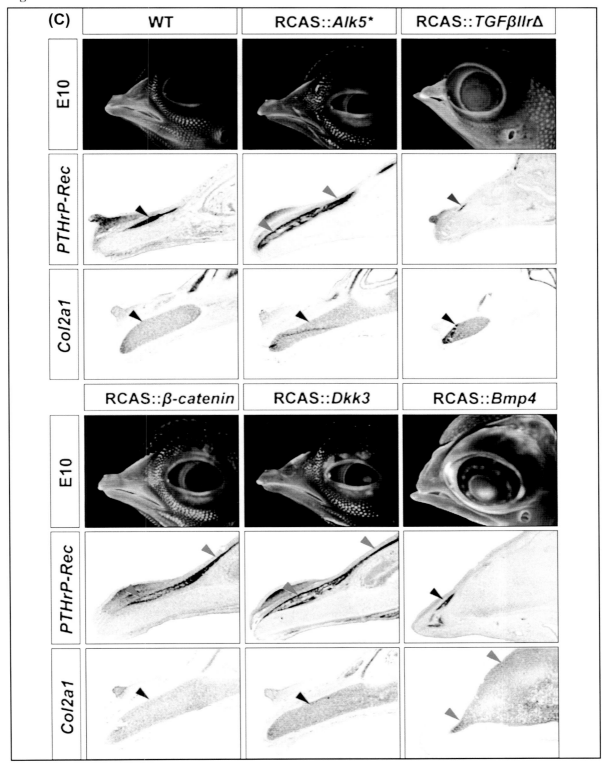

Figure 7.3

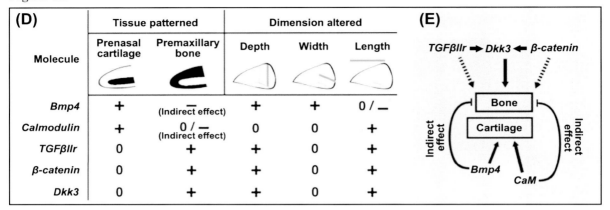

Figure 7.3 Mechanisms of selective variation in beak morphology in Darwin's finches. (A), Examples of divergent beak morphology *(from Grant and Grant (2009))*. (A1), *Geospiza scandens*, thin sharp beak; (A2), *Geospiza magnirostris*, heavy beak useful for crunching hard seeds. (B), Functional analysis of beak prenasal cartilage mass in locally transgenic chicken embryos *(from Abzhanov et al. (2006))*. (B1,3,5), Normal beak development; (B2,4,6) Development of beaks expressing constitutively active Ca-Calmodulin KinaseII from a retroviral vector (RCAS). (B1,2), Developing beak morphology, day 10; dimensions 1, 2, and 3 measure length, depth at base, and depth at tip, respectively. (B3,4), Expression of retroviral vector, monitored by in situ hybridization of a marker (RSCH). (B5,6), Expression of Coll IX indicating length of prenasal collagen mass. (B5), Control; (B6) greatly enhanced length of prenasal collagen mass (expression of Coll IX), resulting in longer beak seen in (B2). (C), Functional analysis of beak premaxillary bone development, a major determinant of beak dimensions, in locally transgenic chicken embryos *(from Mallarino et al. (2011))*. Chick embryos were infected in the appropriate areas with retroviral vectors bearing, as indicated, a constitutively active Tgfβ receptor (Alk5*), a dominant negative form of Tgfβ receptor (TgfβIIrΔ), activated β-catenin, chicken Dickkopf3 (Dkk3), which is of unknown function here, but is differentially expressed in finches of diverse beak morphology, and Bmp4. Top row, morphology of developing chicken embryo beaks; middle row, effects on premaxillary bone deposition, monitored by expression of a marker for osteogenesis (PTHrP-Rec); bottom row, effects on prenasal cartilage, monitored by expression of collagen2A1. Black marks denote control dimensions, red marks larger structure than control, blue marks, smaller structure than control. (D), Tgfβ signaling, Wnt signaling, DKK3, all positively affect the size and proportions of the premaxillary bone while Bmp4 (and Ca²⁺) signaling increase the length of the prenasal cartilage *(from Mallarino et al. (2011))*. Beak length, depth, and width effects are in any case independent of one another. (E), Summary of interactions; "indirect effects" reflect the developmentally earlier determination of the prenasal cartilage, which occurs independently of the patterning of the premaxillary bone *(from Mallarino et al. (2011))*.

adaptively diversified. Here again, according to genetic analyses, the differences in spatial expression of Bmp4 accounts for genus-specific distinctions between jaw morphologies that are used for biting and those used for sucking (Albertson et al., 2005).

In *Nasonia* wasps interspecific differences in wing size have been traced to regulatory sequences controlling expression of a signaling cytokine similar in sequence to *Drosophila unpaired* (*upd*), which controls cell proliferation via the JAK/STAT pathway. As in the other examples considered, multiple *cis*-regulatory regions act synergistically to control spatial expression of this gene in the prepupal *Nasonia* wing (Loehlin and Werren, 2012). That is, here is another example in which the *cis*-regulatory sequence controlling expression of a signaling molecule is the locus of adaptive evolutionary change in a downstream morphological trait. A similar argument relates to the *c-kit* gene, which encodes a signaling receptor tyrosine kinase: in humans and in stickleback fish regulatory mutations affecting expression of this gene in melanocytes account for rapidly occurring evolution of light skin color, and the authors of this discovery propose that pleiotropic genes with extensively modular regulatory systems are preferred loci for rapid adaptive evolutionary change (Miller et al., 2007).

Selective breeding for desired traits in dogs, an extreme form of "unnatural selection" is interesting in this context because it reveals the types of developmental process responsible for these traits, and the genomic features producing them. High-density mapping of genetic variation across domestic dog breeds and wild canids demonstrates that a small number of loci accounts for most of their canonical external phenotypes (Boyko et al., 2010). Protein sequence change in signaling molecules cause some of these phenotypic traits, but regulatory change that affects signal expression is also observed. For example, the very short legged phenotype characteristic of dachshunds and basset hounds, among other breeds, is caused by an *fgf4* retrogene (i.e., a DNA copy of *fgf4* mRNA present in the genome) which encodes a normal Fgf4 protein but which is embedded in transposable elements that probably cause its "atypical" expression in chondrocytes of the developing limbs (Parker et al., 2009).

As we see below, evolutionary redeployment of signaling systems has operated at many levels of developmental GRNs besides the peripheral kinds of function discussed here. But all the genomic mechanisms of change we have briefly traversed in these sections are similar in basic respects. (i) They have occurred recently and over the relatively short periods of time in which speciation within given genera occurs, or the even shorter periods in which variants within the species adaptively diverge. (ii) They affect properties arising relatively late or nearly at the end of the developmental process which begins with the formulation of the body plan in embryogenesis. (iii) They concern *cis*-regulatory control of effector genes of various kinds or of those regulatory gene functions that immediately control effector gene expression. Unlike changes in regulatory and signaling gene control systems higher up in the GRN hierarchy, change in peripheral functions is permissive because there are no feedback consequences for the rest of the control system. Thus the high rate at which adaptive evolutionary *cis*-regulatory change occurs in peripheral functions provides a kind of baseline against which can be measured the relative stasis of regulatory circuitry controlling more fundamental aspects of the body plan.

2. Evolution of the Body Plan by Co-Optive Alteration of GRN Structure

Over deep time many new body plans have evolved. In this section we try to address the kinds of mechanistic alteration in hereditary developmental programs that led to the large differences between the body plans shared by members of a given class and different from those of other classes of the same phylum, and of given orders compared to other orders of the same class. Typical examples of morphological change at this level are differences in disposition and character of appendages, differences in components and form of endo- or exo-skeleton and the muscular apparatus that serves them, identity and diversification of metameric structures, and disposition and organization of sensory organs. Developmentally, such differences begin to arise early in the formation of individual body parts, and are formulated during the processes of spatial specification within the body part. These are functions encoded in the upper and middle tiers of developmental GRNs (cf. Fig. 7.1(A)). From experimental developmental systems biology and from comparative genomics, informed by phylogenetic and paleontological identification of the polarity of events of body plan evolution, we are coming to be able to deduce what DNA-level changes accounted for given examples of body plan evolution. The bases of evolutionary change, and the guiding principles by which we can consider these events mechanistically, rest in the structure/function relations of developmental GRNs. That is, change in developmental GRN structure causes change in GRN function and hence change in developmental outcome (Peter and Davidson, 2011).

2.1   Co-option theory

Within homologous embryonic domains, the expression of some transcription factors might be lost completely, additional factors recruited, new downstream functions acquired or the same factors expressed at different times. Co-option, as used here, explicitly denotes a regulatory gain-of-function change such that a gene becomes expressed in a new spatial domain of a developing organism and/or in a new developmental context. Depending on the nature of the change, which as we see in the following can occur in a variety of ways, the new function may be added to the old or replace it (Fig. 7.4(A)). Co-optive events that affect developmental regulatory genes result in change in GRN linkages, and thus co-option is a major mechanism of GRN evolution (Peter and Davidson, 2011). The consequences of a co-optive event depend entirely on where in the GRN hierarchy it occurs. As in the foregoing, co-optive events affecting genes at the periphery of a GRN may range from redeployment of a single effector gene to redeployment of a differentiation gene battery to redeployment of a signal (Fig. 7.4(B)), and hence of all its downstream consequences. Co-option of regulatory genes and signaling functions operating at higher levels of a GRN can result in redeployment of larger patterning circuits (Davidson and Erwin, 2010).

    Considering the rapidity and frequency of co-optive *cis*-regulatory changes in genes operating at the GRN periphery, as we have just reviewed, co-option is potentially an extremely powerful mechanism for evolutionary revision of GRNs. Further considerations particular to regulatory gene co-option only increase the possible potency of this mechanism of evolutionary change in developmental GRNs. Thousands of laboratory experiments in which gain-of-function changes in regulatory gene expression have been deliberately introduced, particularly in *Drosophila* and also in mice, demonstrate that the forced ectopic expression of genes encoding transcription factors in embryonic development is generally haplosufficient; that is, expression of a single copy of the gene suffices to augment functionally the regulatory state. More generally, though there are exceptions (particularly in later development) we would not have all the scores of regulatory genes discovered over decades of recessive screens for developmental mutations in flies, worms, fish, mice, etc., were not all these genes haplosufficient for normal embryogenesis. Several authors have pointed out the evolutionary significance: a haploid co-optive event causing a novel domain of regulatory gene expression is likely to suffice for function, and for these genes homozygosity is not required (Ruvkun et al., 1991; Davidson and Erwin, 2010). Such a change will be inherited clonally. In summary, co-optive *cis*-regulatory change is the engine of GRN evolution, and an active engine it can be. Rather than wonder how something as intricate and interwoven as a hierarchical developmental GRN can change evolutionarily, given current knowledge of *cis*-regulatory lability, a moment's thought produces an opposite conundrum: how can GRNs remain stable?

    The following examples illuminate co-optive mechanisms at the *cis*-regulatory level, and the remainder of Section 2 of this chapter is devoted to evolutionary co-optive changes which affect upper levels of GRN hierarchy, and that have been analyzed at the gene regulatory level. We first consider the variety of co-optive changes in the regulatory DNA sequences of individual genes, then examples of evolutionary rewiring of GRN structure, and finally the largest scale instances of this, which involve redeployment of whole GRN subcircuits or even large portions of developmental GRNs.

2.2   *Cis*-regulatory signatures of co-option

Co-optive regulatory gene deployment and redeployment is a fundamental driver of the evolution of development. Co-option theory predicts that regulatory gene redeployment occurs by change in the *cis*-regulatory control system of that gene, thereby linking it into different GRN circuitry. The most common form of this is appearance of novel *cis*-regulatory modules with enhancer functions not required in other branches of phylogeny. The *pax6* gene provides a canonical example (Fig. 7.4(C); Kleinjan et al., 2006). In mammals this gene operates within diverse GRNs controlling development of pancreas, forebrain,

Figure 7.4

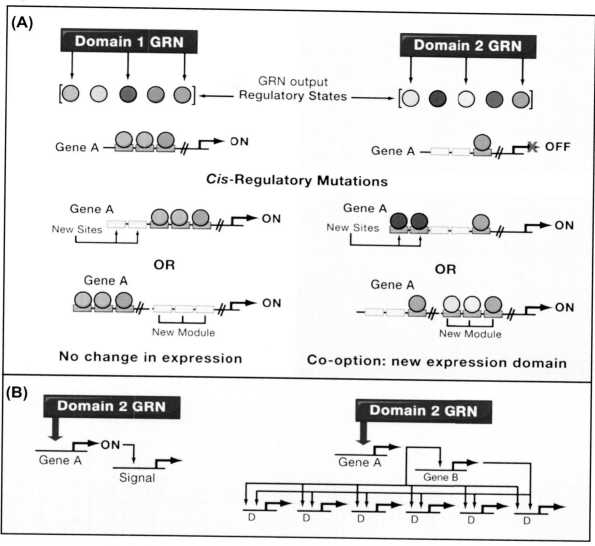

Figure 7.4 Co-optive gain of *cis*-regulatory function. (A), Diagrammatic illustration of co-option at the *cis*-regulatory level. The gene regulatory networks operating in spatial Domains 1 and 2 produce different regulatory states (colored balls, representing diverse transcription factors). A *cis*-regulatory module of Gene A, encoding a transcription factor, has target sites only for a set of factors present in Domain 1 but not Domain 2; hence Gene A and its downstream targets are only expressed in Domain 1. Two alternative types of potentially co-optive *cis*-regulatory mutations are portrayed: appearance of new sites within the module by internal nucleotide sequence change; and transposition into the DNA near the gene of a module from elsewhere in the genome bearing new sites. Although these gain-of-function changes do not affect the occupancy of the *cis*-regulatory sites of Gene A in Domain 1, by either route the new sites allow Gene A to respond to the regulatory state of Domain 2, resulting in a co-optive change in expression so that Gene A is now active in Domain 2. Note that in this example the upstream regulatory states of Domains 1 and 2 remain unchanged, i.e., are conserved, while the downstream regulatory state of Domain 2 has been altered by the addition of the Gene A transcription factor. (B), Possible consequences: if Gene A controls downstream expression of

Figure 7.4

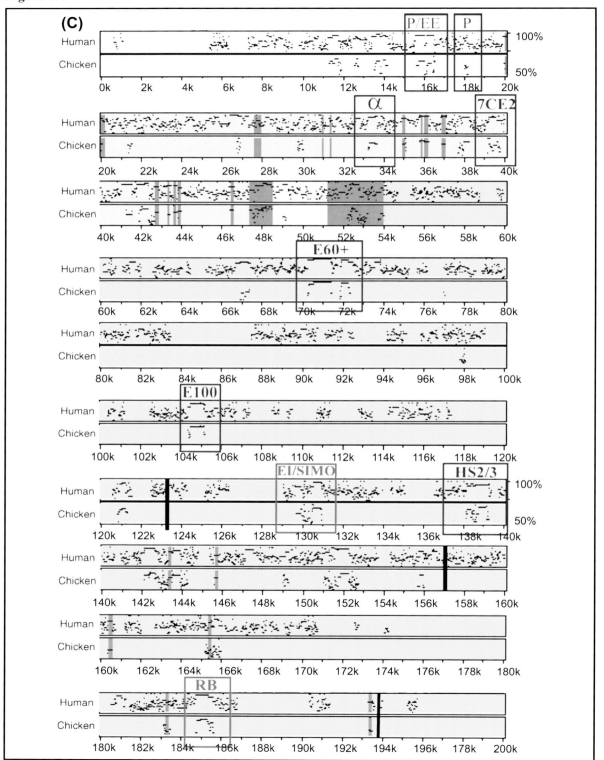

Figure 7.4

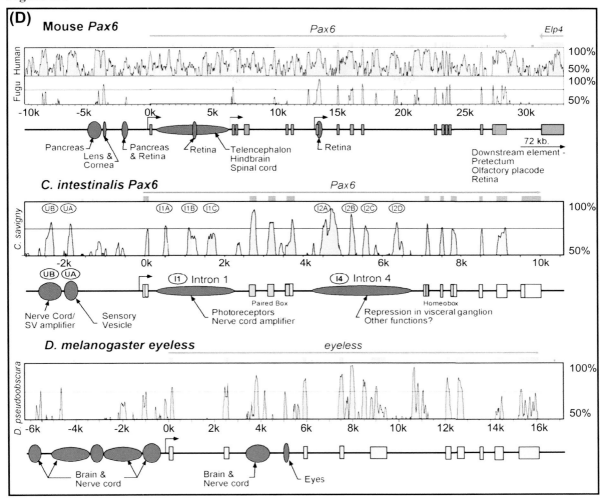

a signal ligand, this change could have the profound effect of altering expression of the signal ligand and its target subcircuit in Domain 2 (left); if it controls a driver of a differentiation gene battery (right) the effector function of that battery will now be expressed in a new domain of the organism ((A) and (B) *from Peter and Davidson (2011)*). (C), Multiple dispersed enhancers controlling *pax6* expression in amniotes *(from Kleinjan et al. (2006))*. 200 KB of chicken and human DNA are aligned, displaying conserved regions (black stipple, % identity), and at this evolutionary distance (250 My since divergence) these relatively conserved regions denote enhancers. Exons are shown as solid burgundy rectangles. The *pax6* transcription unit is shown as a yellow bar and in orange that of an adjacent gene, the ubiquitously expressed *elp4* gene, which is located downstream of *pax6* and transcribed in the opposite direction. Many of the *pax6* enhancers lie within the *elp4* transcription unit. Named enhancers have been functionally characterized by the spatial expressions they drive in transgenic mice: P/EE, pancreas/ectodermal enhancer; P, pancreas enhancer; α, intron 4 neural retinal enhancer; 7CE2, diencephalon enhancer; E60+, cerebellar enhancer; E100, pretectum, olfactory, neural retinal enhancer, EI/SIMO, lens, diencephalon, hindbrain enhancer; HS2/3, retinal enhancer; RB, forebrain enhancer. Some of these enhancers drive expression at different stages and in additional regions than those listed. (D), Structure/function evidence of multiple evolutionary co-option events in the *pax6* gene. A portion of the mammalian *cis*-regulatory region (up to about the E100 enhancer in (C)) is compared

Figure 7.4

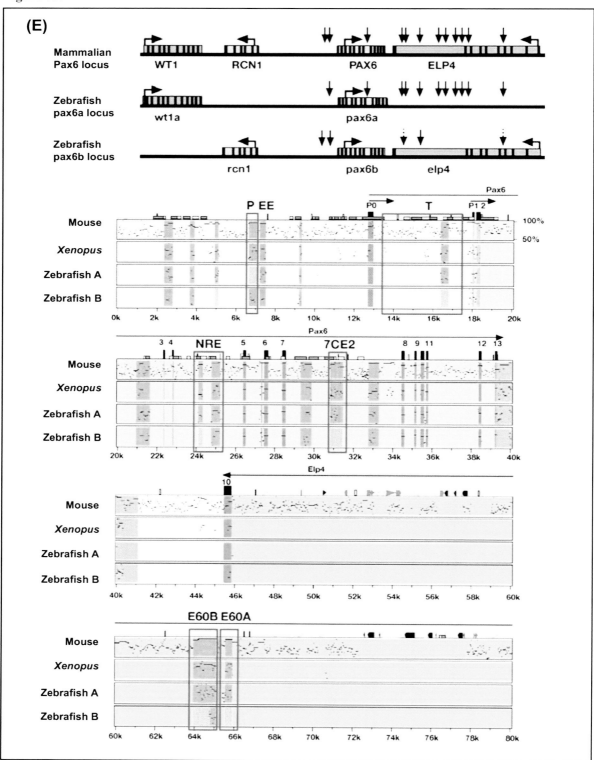

to *pax6* in *Ciona* and *Drosophila* genomes. Above each track is a conservation (Vista) plot in which mouse *pax6* is aligned with human and *Fugu pax6*; *Ciona intestinalis* was aligned with *Ciona savigny pax6*; and *Drosophila melanogaster pax6* (*eyeless*) was aligned with *Drosophila pseudoobscura pax6*. Needless to say no detectable sequence conservation among mouse, *Ciona*, and *Drosophila* survives. Exons are shown as colored rectangles usually displaying sequence conservation (gray peaks in Vista plot). Experimentally authenticated enhancers are indicated as purple ovals. Despite some underlying similarities noted in text the general import is that in each lineage novel *cis*-regulatory modules have evolved to control expression in body parts unique to that lineage, for example pancreas and lens in the mammal and visceral ganglia in ascidian. (E), Regulatory subfunctionalization of duplicated *pax6* genes in zebrafish *(from Kleinjan et al. (2008))*. At top are bird's eye maps of *pax6* gene synteny in mammals and zebrafish; enhancers are denoted by downward arrows. Mouse *pax6* is flanked by *wt1* and *rcn1* genes upstream and *elp4* gene downstream. The zebrafish *pax6a* and *pax6b* loci preserve alternate sets of flanking genes and enhancers. Below, in more detail, the region of the *pax6* gene upstream from the E60A enhancer (cf (C)) is compared between mammalian, amphibian, and the two zebrafish *pax6* genes. Conservation plots are shown as in (C) for mouse, *Xenopus*, and the two zebrafish genes; conserved noncoding regions are in green and known enhancers are indicated by the same names as in (C) (α-enhancer of (C) is NRE, neural retina enhancer here). Conserved exons are in red and conserved 3′ trailer regions in pink; *pax6* transcription unit yellow, *elp4* transcription unit in orange. Note the gray boxes which denote nonconserved sequence: zebrafish *pax6a* has lost the P enhancer while retaining others, while zebrafish *pax6b* has reciprocally lost 7CE2, E60B, and E60A enhancers while retaining others.

hindbrain, cerebellum and spinal cord, olfactory system, retina and lens of the eye. The *pax6* control system is extremely modular, and multiple enhancers have been discovered to control its expression in diverse developmental contexts, often overlapping in function, and spread out over 200 kb of genomic sequence, including within introns of an adjacent downstream gene (Kleinjan et al., 2006; McBride et al., 2011). Most but not all of the *pax6* enhancers are partially conserved in sequence throughout the vertebrates. But, *pax6* is a pan-bilaterian regulator, and we note that many bilaterians do not even possess most of the body parts where *pax6* is expressed in vertebrates. Thus in each branch of bilaterian evolution co-option to new developmental function has required installation of new *pax6 cis*-regulatory modules. This gene is an integral part of the underlying GRNs required to control development of the different body parts of these various species. In Fig. 7.4(D) a portion of the *pax6 cis*-regulatory system is compared in mouse, *Ciona*, and *Drosophila* (Irvine et al., 2008). There are some fascinating similarities in regulatory organization. For example, in all three species the *pax6* gene has enhancers driving expression in photoreceptors and nerve cord. These enhancers can be regarded as pleisiomorphic functional aspects of the *pax6* regulatory system in that they have evidently survived from the common ancestor which utilized the *pax6* gene for operating photoreceptors and for nervous system development. But as can be seen even in this comparison of a small part of the *pax6* control system, multiple vertebrate-specific enhancers have appeared, such as those enabling *pax6* participation in the GRNs controlling telencephalon, retina, and pancreas development. These enhancers thus are signatures of evolutionary co-option events which have enabled the participation of *pax6* in the development of these "novel" vertebrate-specific body parts. A particular example is afforded by the *pax6* lens enhancers, as the type of lens in the vertebrate eye does not even exist in other clades. Nonetheless, in mammals *pax6* has apparently gained control of large numbers of eye lens effector genes, as shown in Fig. 5.1(A) (Wolf et al., 2009). Thus, while some regulatory functions and some part of the expression pattern of *pax6* has been conserved, this gene has been co-opted multiple times independently to different GRNs which control the development of nonhomologous body parts. Looking back at the overall map of *pax6* enhancers in Fig. 7.4(C), we can now see that some are of more ancient or pleisiomorphic vintage than others, and as a whole, the array of enhancers provides a structural map of some of the pathways of vertebrate evolution. Recent studies expand this map and identify even more vertebrate-specific *pax6* enhancers expressed in body parts that only vertebrates possess (Bhatia et al., 2014).

The human *gli3* gene provides another clear example in which a regulatory gene was co-opted to participation in the development of an evolutionarily more modern structure, evidently mediated by the appearance of a new enhancer (Abbasi et al., 2010). This gene has a highly modular *cis*-regulatory system, much of which is conserved from fish to man. By testing in mouse and chick transgenes, an enhancer was identified which drives expression in the evolutionarily older long bones of the forming limb but not in the evolutionary more recent digits. However, a different enhancer works during development of the digital extensions (see below for discussion of vertebrate limb evolution). Enhancers disappear during evolution as well as appear: Fig. 7.4(E) shows two duplicated zebrafish *pax6* genes deriving from a single ancestor which have subfunctionalized, reciprocally having lost enhancers present in the ancestor, as seen by comparison to out-groups.

We now have many examples of regulatory genes expressed in multiple phases of development, at multiple places and times, each controlled by locally active, distinct *cis*-regulatory modules, sometimes overlapping in function (Davidson, 2006). The arrays of *cis*-regulatory modules characteristic of such genes (and of many others that do not encode regulatory proteins) comprise the genomic paleontology of the co-option events that the gene underwent as it was drawn into GRNs controlling development of evolutionarily novel structures.

2.3   Evolution of GRNs

In a few cases orthologous GRNs descendant from a common ancestor have been worked out, and cannot only be directly compared, but also used to derive a model for the GRN of the common ancestor. Thus the changes, or presumed changes, along each lineage become explicit, and we can directly visualize the processes by which GRN structure evolves. In order for such exercises to be informative the organisms must be close enough to the common ancestor so that some basic pleisiomorphic aspects of GRN structure remain visible (i.e., aspects that were also present in the common ancestor). The insects *Tribolium castaneum* and *D. melanogaster*, which derive from lineages thought to have diverged at least 250 My ago, provide examples of informative GRN comparison, which we now briefly consult.

Evolution of segmentation gene networks upstream of *engrailed* and *wingless*

In *Drosophila* embryos, as has long been known, the regulatory gene *engrailed* (*en*) is expressed in cells at the anterior boundary of each parasegment, directly confronting cells at the posterior boundary of the preceding parasegment which express the *wingless* (*wg*) signal ligand gene (see discussion of segment polarity model in Chapter 6). The boundaries between the 14 parasegments mark the initial metameric units of the thoracic and abdominal portions of the body plan (the definitive future segments are composed of the posterior half of one parasegment plus the anterior half of the next parasegment). The initial metameric organization is established by complex, transient GRNs, consisting of regulatory genes initially expressed in a two-segment periodicity along the anterior/posterior axis. These "pair rule" genes are expressed in various stripes of cells along the anterior/posterior axis, and given genes execute mutual spatial repressions as well as activations within each parasegmental domain. An ultimate output of the metamerism GRNs is determination of the location of the juxtaposed *en* and *wg* expressing boundary cells. The GRNs are differently structured in even-numbered and odd-numbered segments, following initial institution of expression of the gene *evenskipped* in each odd-numbered parasegmental domain. In *Tribolium* the same *en* and *wg* genes are expressed by the anterior and posterior parasegment boundary cells as in *Drosophila*, and this is general for arthropods (Peel et al., 2005; Damen, 2007). However, the *Drosophila* and *Tribolium* GRNs which generate the upstream inputs to *en* and *wg* are remarkably different (Choe and Brown, 2009), and thereby hangs our tale.

The segmentation GRNs of *Drosophila* and *Tribolium* are summarized in the parasegmental spatial interaction diagrams in Fig. 7.5(A) and (B) (Choe and Brown, 2009). Here the *Drosophila* network summarizes a great deal of prior observation; the *Tribolium* network is based mainly on the RNAi study cited here, plus prior observations. A simple ancestral segmentation circuit in Fig. 7.5(C) (Choe and Brown, 2009) is inferred from comparative expression studies which show that juxtaposed parasegmental expression of *en* and *wg* at every boundary is more basal and general in arthropods than is pair rule (two-segment periodicity) expression of genes upstream in the segmentation network. The pleisiomorphic circuitry controlling *en* and *wg* expression may thus have been the same in every metameric segment. In insects such as *Drosophila* and *Tribolium* expression of initial pair rule genes in every parasegment is achieved only at the second stage of the metamerization process, after the "secondary" segmentation genes are activated, when for example the additional secondary stripes of *eve* and *prd* are expressed (marked with 2° in Fig. 7.5(A) and (B)). The inferred ancestral metameric circuit (Fig. 7.5(C)) is based on two observations. First, it is a fact that the most basal and widespread expression of *eve* in arthropods occurs in every metameric segment, rather than in every other one (Liu and Kaufman, 2005); and second, part of the proposed primitive mechanism is still preserved at the anterior borders of the odd-numbered parasegments in both *Drosophila* and *Tribolium*. At these borders (see Fig. 7.5(A)–(C)) *prd* expression extends anteriorly across the boundary and is used to drive both *en* and *wg* expression; furthermore, as can still be seen in *Tribolium*, direct *eve* repression dominantly prevents *wg* expression in the anterior border cells (Fig. 7.5(A)). We can now see in this three-way comparison the types of co-optive change in the upstream GRNs that evolution has wrought since the divergence of the Diptera (flies, mosquitoes) and the Coleoptera (beetles), even though the *Tribolium* GRN, which is less studied, may turn out to be more complex than portrayed here. Two major kinds of alteration in GRN structure have occurred, assuming the pleisiomorphic metamerization circuit of Fig. 7.5(C), or something equivalently simple. These are the insertion of additional upstream circuitry at the anterior borders of the odd segments between *prd*, the driver of *en* and *wg*, and *eve*, the gene at the top of the subcircuit; and the insertion of altogether different circuitry at both the posterior borders of the odd parasegments and the anterior borders of the even-numbered parasegments. These changes amount to the evolutionary institution of the pair rule mechanism. Before the divergence of Coleoptera and Diptera, the *slp* regulatory layer was intercalated, though it plays partially different roles in the two systems. While in *Tribolium en* is directly activated by *eve* as well as *prd*, in *Drosophila*, *en* is activated in the odd parasegments by *prd*, while *eve* is used to repress *slp*, which here encodes the repressor of *en*. Even though in both systems *eve* supports expression of *en*, this is encoded as direct activation in *Tribolium* and as double repression in *Drosophila*. The pair rule drivers of the even parasegments evolved differently since divergence: in *Tribolium* the secondary *runt* stripe plus *eve* are upstream of *en* activation while in *Drosophila* this role is played by *ftz*. As detailed in the caption to Fig. 7.5 numerous additional changes in the regulatory wiring including co-options of additional genes such as *opa* also occurred. These many alterations in GRN linkages mean that with a few exceptions the *cis*-regulatory systems of almost every gene incorporated in Figs 7.5(A) and (B) must differ in the sets of inputs they respond to. Change of upstream circuitry while retaining a required output is an essential concept in considering processes of GRN evolution: it allows for addition of diverse machinery and loss of regulatory interactions once no longer necessary, both without endangering the basic outcome underpinned by the pleisiomorphic circuit (such as in Fig. 7.5(C)). Here the basic required outcome is of course the expression of *en* and *wg* in juxtaposed cells at the parasegment boundaries.

Other changes have occurred upstream, even within Diptera, in the linkages from gap genes into primary pair rule genes such as *eve*. A study of the seven primary *eve* stripes with respect to gap gene expression patterns in a basal dipteran, *Clogmia albipunctata*, indicates that while *eve* stripes 1, 2, and 3 could be regulated the same as in *Drosophila*, this cannot be the case for the posterior stripes (García-Solache et al., 2010). Gap genes which repressively determine some posterior *eve* stripe boundaries in *Drosophila* are not expressed at the right times and places to perform these functions in

Figure 7.5

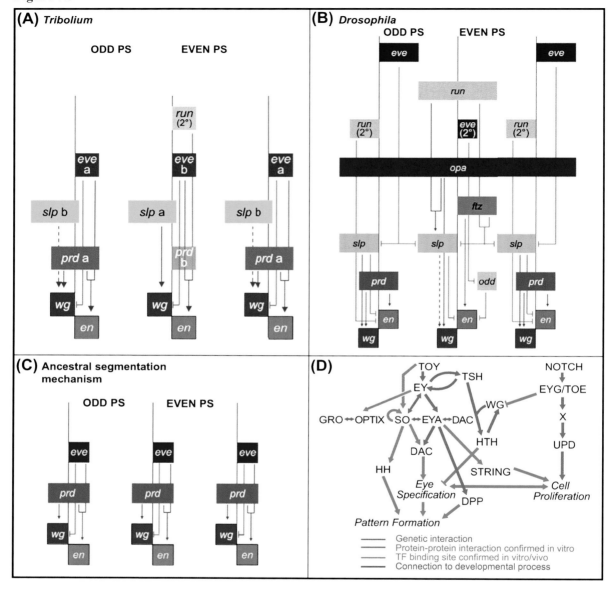

Figure 7.5 Divergent pair rule segmentation circuitry in insects generating conserved boundary gene expression domains, and *Drosophila* retinal determination GRN model. GRN organization in *Tribolium* and *Drosophila* upstream of expression of wingless (*wg*) and engrailed (*en*) in cells respectively anterior (left) and posterior (right) of parasegmental borders, which are indicated as thin vertical black lines *(from Choe and Brown (2009))*. (A), *Tribolium* pair rule gene network, from RNAi and gene expression studies. (B), *Drosophila* pair rule gene network assembled from multiple literature sources. (C), Hypothetical basal arthropod segmentation network based on comparative observations in other arthropods, and features conserved between *Drosophila* and *Tribolium*, as reviewed in text. Despite their different upstream structures, both the *Drosophila* and *Tribolium* GRNs output *wg* and *en* expression on either side of the parasegment borders. But the wiring is different. For example in *Tribolium eve* directly prevents *wg* expression at the anterior boundary of both even and odd parasegments, while in *Drosophila*

it is *en* that is repressed at the posterior boundary of each parasegment, by different mechanisms in even and odd parasegments. In the even parasegments *en* repression depends on expression of *slp* plus *runt*, and in the odd parasegments *en* repression depends on *runt*-driven *slp* expression, while *eve* expression has nothing to do with repression of *wg*. Instead in *Drosophila eve* operates two different double-negative gates that allow *en* to be expressed at the anterior boundaries of the parasegments: in the odd parasegments the primary (pair rule) *eve* stripes repress *slp* which would otherwise repress *en*; and in the even-numbered parasegments the secondary *eve* stripes repress *odd*, which would otherwise repress *en*. Nor is the role of *slp* the same in *Tribolium* and *Drosophila*: while *en* repression in *Drosophila* depends on *slp*, in *Tribolium slp* does not repress *en*. In addition the activating roles of *Drosophila ftz* on *en* in even-numbered parasegments and *opa* on *wg* in odd-numbered parasegments are co-options not seen in *Tribolium*, and many other specific differences can be adduced upstream of expression of *en* and *wg* at the parasegment boundaries. (D), Retinal determination GRN model initiating eye development in *Drosophila (from Kumar (2009))*; see text for discussion. As indicated by the color coding, the data supporting these interactions include *cis*-regulatory observations, genetic inferences, and observed protein–protein interactions.

Clogmia. For example, while adjacent high-level *knirps* expression causes the repression of *eve* and generates the posterior boundary of *eve* stripe 4 in *Drosophila*, in *Clogmia* both boundaries of this stripe lie within the middle of the broad *knirps* domain. Though there is yet no functional information, the implication is that the *cis*-regulatory modules controlling posterior *eve* stripe expression in *Clogmia* respond to very different inputs than in *Drosophila*.

Evolutionary change in insect retinal determination GRNs

The GRN controlling retinal determination in the compound eye primordia of *Drosophila* is by now supported by much direct *cis*-regulatory molecular biology as well as by extensive genetic and misexpression data (Kumar, 2009; Gehring, 2012). At the top of the core of this GRN are two functionally nonredundant *pax6* genes, *eyeless* (*ey*), and *twin of eyeless* (*toy*), of which *toy* is activated first and provides a direct input into *ey*. A powerfully recursive circuit is then established in which *toy* and *ey* both activate a *six* class gene, *sine oculis* (*so*), generating a feed-forward circuit, and *so* maintains its own activity by auto-activation and then feeds back directly on *ey* (Fig. 7.5(D)). The *ey* gene also activates the *eyes absent* (*eya*) gene, which encodes a transcriptional coactivator that binds to the So transcription factor, forming an active heteromeric complex. The final member of the core circuit is the *dachshund* (*dac*) gene, enhancers of which are activated directly by the Eya:So complex plus Dpp signaling (Pappu et al., 2005). In addition, protein:protein complexes as well as direct transcriptional inputs and feedbacks cross-link this GRN (Datta et al., 2011). In turn the core circuit provides regulatory inputs downstream, into the proneural gene *atonal*, which is needed for photoreceptor cell type specification (cf. Chapter 5). Other regulatory outputs of the core circuit include *optix*, activated by Ey, which acts as a repressor of nonretinal circuitry via its cofactor Groucho (Anderson et al., 2012); and the *hh* gene, activated by So, which is involved in downstream eye pattern formation. Surrounding the retinal determination core circuit are other regulatory genes and several signaling inputs that are used to position the GRN in eye-antenna disc development by activating or repressing it: these include *teashirt* and its paralog *tiptop*, and the Wingless pathway which via its target *homothorax* (also downstream of *tsh*) represses eye formation (cf. Fig. 5.2).

In evolution this GRN, like the embryonic segmentation GRN just considered, has also proved subject to evolutionary change by processes of co-optive rewiring. In addition to the compound eyes there are two other light-sensitive ocular systems in *Drosophila*, the larval eyes or Bolwig organs, and the three dorsal ocelli (specification of ocellar versus inter-ocellar and peri-ocellar cuticle is discussed in Chapter 4;

Fig. 4.9). Ocelli (median eyes) as well as compound eyes occur throughout the arthropods, and the larval eyes have a common developmental origin with the adult eyes (Friedrich, 2006). Comparisons between the GRNs underlying the specification of ocelli and Bolwig organs or larval eyes, to the extent they are known, to the retinal determination GRN of the adult compound eyes, display this malleability. A major difference in the mechanism underlying ocellus development, compared to the canonical compound eye retinal determination circuitry, is that surprisingly the *pax6* ortholog *ey* plays no role in ocellus development, by either genetic or expression evidence (Blanco et al., 2010). Activation of *so* depends instead on Eya or an Eya:Toy complex. Thus the feed-forward subcircuit by which *so* is activated in the retinal determination GRN (*toy* > *ey* > *so*; *toy* > *so*) is replaced in the ocellar GRN. Furthermore, rather than activation of *eya* by *ey*, *eya* is also activated differently, by two inputs, *toy* and (directly or indirectly) *otd*. Even more remarkably, neither *toy* nor *ey* are even necessary for Bolwig's organs to form (reviewed by Friedrich, 2006). One constant, however, is the expression in all three *Drosophila* visual systems of the genes immediately downstream of *toy* and *ey* in the retinal determination GRN, i.e., *so* and *eya*. In *Tribolium*, in contrast to *Drosophila*, larval eye development is strongly dependent on expression of *toy* and *ey*, and *toy* may operate redundantly with *ey* rather than being epistatic to *ey* (Yang et al., 2009a). In *Tribolium* *so* and *eya* are coexpressed in both larval and adult eyes and are essential to their formation, just as in the *Drosophila* visual system. However, although *dac* seems also to be utilized for normal eye patterning in *Tribolium*, it is clearly wired differently into the adult eye GRN since loss of *dac* function does not seem to result in complete eye loss as in *Drosophila* (Yang et al., 2009b).

Thus in a way analogous to the evolutionary changes in the segmentation GRN shown in Fig. 7.5(A)–(C), changes in the retinal determination GRN have occurred upstream of the unchanging circuit output, in segmentation expression of *wingless* and *engrailed* at the parasegment boundaries, and in visual system development expression of *so* and *eya* in the primordia. An interesting way to think about the pathway by which these changes occur is that in both of these cases, the evolutionary process has an inverse trajectory from the developmental process giving rise to the structure: in development, the upstream circuitry precedes the downstream output, while in these evolutionary processes, the downstream output has been conserved, that is, is older, while novel linkages have been inserted in the upstream GRN, a kind of inverse evolution. In segmentation, the downstream output is activation of the differentiation processes that produce the metameric patterns of denticles and other integumentary structures (Chanut-Delalande et al., 2006). In the visual system the downstream output is differentiation of photoreceptors.

In the evolution of both the segmentation and visual GRNs, each retained a similar set of regulatory genes while rewiring their interactions. Remarkably, orthologous regulatory genes also function in GRNs underlying the development of analogous structures that have evolved independently, and one clue as to how this might have occurred is given by the multiple functions of *pax6* in unrelated visual system GRNs. For example, even though the development of insect compound eyes and vertebrate eyes evolved independently, *pax6* plays a crucial upstream role in both developmental GRNs, which apart from sharing some common regulatory genes are structured very differently. The key to this conundrum is that Pax6 and other transcription factors in the retinal determination GRNs are expressed in differentiated photoreceptor cells both in arthropods and vertebrates, and this is the ancestral conserved relationship between the *pax6* gene and visual systems. In the independent evolution of the transcriptional patterning apparatus giving rise to the diverse morphologies of vertebrate and arthropod eyes, the *pax6* gene acquired further functions higher up in the hierarchy of the developmental GRNs, since it was already expressed in the domain where visual cells developed. This has been adduced as a canonical example of intercalary evolution, where regulatory genes acquire additional regulatory functions in the upper hierarchy of the same GRN in which they execute a conserved downstream function (Gehring and Ikeo, 1999; Davidson, 2006). Intercalary evolution may account for the amazing diversity of eye morphologies, though all utilize homologous downstream photoreceptor genes, and *pax6*- and *eya*-related drivers (Gehring and Ikeo, 1999; Suga et al., 2010; Graziussi et al., 2012).

2.4   Evolutionary co-option of whole subcircuits

In evolution of gene regulatory circuits, if something can happen, it sooner or later will. The hierarchical structure shown in Fig. 7.1(A) suggests that a co-optive change in developmental address occurring at the top of a regulatory subcircuit could result in redeployment of the whole subcircuit to a new location in a GRN, which could equate to a new use for an already encoded genetic spatial control system. There is increasing evidence that this was a major pathway by which large changes in body plan occurred deep in evolutionary time (for reviews, Peter and Davidson, 2011; Monteiro, 2012). Similarly, relatively few negative changes at the top of a regulatory subcircuit can remove it from operation in a given regulatory context, with again a large effect on body plan. Evolutionary alteration of developmental GRN organization by redeployment of whole patterning subcircuits is a potent consequence of the innate modularity and hierarchy of these GRNs.

Most of the data in the following examples are descriptive, in the form of spatial regulatory gene expression patterns. The persuasiveness of their argument lies in the recurrence of spatial expression patterns generated by the same set of genes, in the same relative positions. The underlying patterning GRN circuitry that causes expression of these genes in entirely different developmental contexts may thus have been co-opted from a pleisiomorphic use, to another use of the particular spatial pattern of regulatory states that this subcircuit generates. This is a priori a possible explanation of evolutionary conundrums of certain kinds, one of which is the origin of proximal-distal appendage patterning (Lemons et al., 2010).

Co-option of an axial patterning subcircuit

The precambrian last common bilaterian ancestor almost certainly had no appendages, while its Paleozoic descendants, including arthropods and vertebrates, were equipped with appendages that are structured in entirely different ways. Study of modern representatives shows that these respective appendage types arise by very different developmental processes. Yet at the earliest stages of progenitor field formation, similar specific small sets of regulatory genes are expressed in both arthropod and vertebrate appendage fields along the future proximal-distal appendage axes (for review, Lemons et al., 2010). It can hardly be coincidence that a number of regulatory genes are expressed in the same respective patterns in these two basically unrelated yet functionally analogous processes. These same regulatory genes turn out to be used to set out positional anterior to posterior domains in the embryonic neuroectoderm of arthropods and vertebrates, as reviewed in Chapter 4 (cf. Fig. 4.4). As illustrated diagrammatically in Fig. 7.6(A) the neuroectoderm anterior-posterior patterning sequence and the *Drosophila* leg proximal-distal patterning sequence are basically similar. Thus arises the theory that the proximal-distal appendage patterning networks of both arthropods and vertebrates was co-opted from a common, ancestral, neuroectoderm patterning GRN originally expressed in the bilaterian ancestor along the anterior-posterior axis.

Co-option of skeletogenic circuitry in echinoderms

Another example comes from interclass comparisons of development in echinoderms. A definitive characteristic of the "modern" sea urchins (subclass Echinoidea) is the distinctive larval skeleton they build during embryogenesis. This is the sole output of a special lineage of cells segregated in early cleavage, the micromeres (cf. Chapter 3). As we have seen, the GRN controlling the specification of this lineage is well known (Fig. 3.5(I)). Micromeres also have essential signaling functions in early development. The top portion of their specification GRN consists of the circuitry by which their particular regulatory state is initiated by means of the double-negative gate discussed in Chapter 3, which also accounts for signal ligand transcription. The remainder of the encoded regulatory apparatus, the majority of the GRN linkages, is composed of circuitry producing expression of the drivers of skeletogenic biomineralization and cell biology. Since no other echinoderm class performs the function of creating a precociously specified embryonic/larval skeleton, this is clearly a derived echinoid feature, but how did such a complex structure as the skeletogenic GRN get assembled in evolution? An easy guess is that the circuitry preexisted and was co-opted as a whole by placing it under the control of a new embryonic address. The adult forms of all echinoderms, as a definitive phyletic character, generate a calcite biomineral endoskeleton (Bottjer et al., 2006), and so the skeletogenic circuitry must have been present in their common Cambrian ancestor. We determined whether every gene of the embryonic skeletogenic GRN is expressed during development of the adult skeleton, which can be studied in late larvae, and mapped those that are and those that are not onto the embryonic GRN model (Gao and Davidson, 2008). The outcome is shown in Fig. 7.6(B). Here we see that indeed almost the whole of the regulatory state established downstream of the embryonic double-negative gate is also utilized for adult skeletogenesis, while the micromere initiation and signaling machinery is not. Inspection of the skeletogenic circuitry shows that it will run stably because of a mutual feedback set up in the tier of interactions immediately downstream of the several double-negative gate target genes. What would have been required for co-option of the whole of the circuitry and thus of a whole new structure within the larval body plan, is merely to emplace a few genes of the skeletogenic GRN under the control of a new embryonic spatial address, in this case the double-negative gate. Note in addition that there are individual regulatory genes of the embryonic skeletogenic GRN that are not utilized for adult skeletogenesis and that were co-opted individually. An example is *t-brain* (*tbr*) which in other echinoderm classes is expressed in the embryonic archenteron (Minemura et al., 2009). This gene provides a dramatic example of co-optive developmental change within the evolution of a single phylum (Fig. 7.6(C)); in sea urchins *tbr* is required for embryonic skeleton formation, while in sea stars it is required for archenteron formation (Hinman et al., 2007; Oliveri et al., 2008). Thus in sea urchins, absence of *tbr* expression causes failure of skeletogenesis, while in sea stars absence of *tbr* expression blocks archenteron development.

Regulatory origins of the gnathostome jaw

As the following examples show, the requirement in this kind of analysis is the comparison of modern descendants representing evolutionary states before and after a given change in body plan. A case of this nature that has drawn much attention is the origin of the jaws of gnathostome vertebrates, given the prior existence of jawless bony fish which date back to the Cambrian; a cartoon of the relevant phylogeny is shown in Fig. 7.6(D1). The common ancestor of bony fish with jaws existed about 450 My ago. The upper and lower oral jaws of gnathostomes develop respectively from dorsal and ventral components of the first branchial arch, the intermediate domain of which gives rise to the hinge joint of the jaw.

Figure 7.6

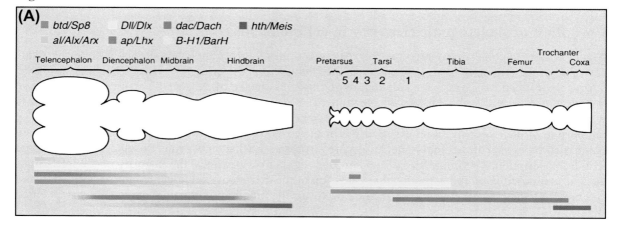

Figure 7.6

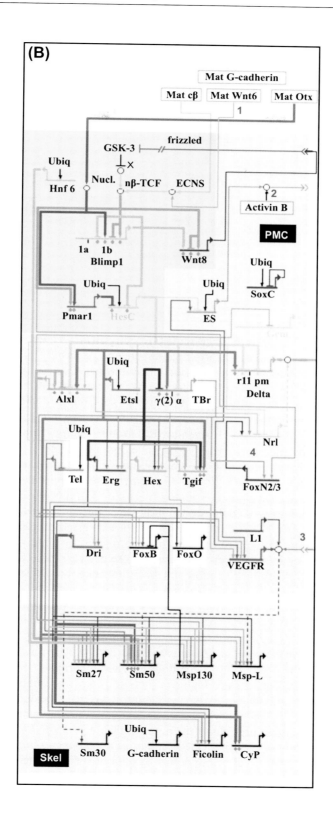

Figure 7.6

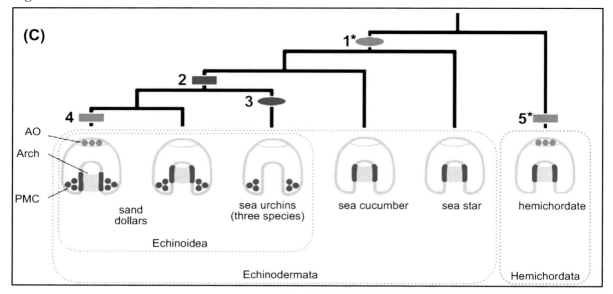

Figure 7.6 Evolutionary co-options of large-scale GRN circuitry. (A), Putative co-option of anterior-posterior head patterning system to distal-proximal appendage patterning system *(from Lemons et al. (2010))*. Left, generalized chordate brain, anterior left, with conserved anterior to posterior patterns of neuroectodermal expression of seven regulatory genes color coded, as indicated. Similar relative expression domains are seen in distantly related bilaterians (see Chapter 4). Right, very similar relative distal to proximal expression domains of the same regulatory genes in a developing *Drosophila* leg. (B), Evidence for co-option of a large skeletogenic GRN used in the formation of test plates and spines of the adult sea urchin body plan to an embryonic developmental address *(from Gao and Davidson (2008))*. The experimentally established GRN controlling specification and initial differentiation of the pregastrular skeletogenic micromere linage of *S. purpuratus* is reproduced (Fig. 3.5(I)). Almost every gene in this network was tested to determine if it is used in adult skeletogenesis. The results were superimposed on the network as three color-coded indicators: in red are genes that are also utilized in formation of adult skeletal elements; in blue are genes not expressed in adult skeletogenesis that are utilized for initial specification of the micromere lineage regulatory state very early in cleavage; in green are genes not expressed in adult skeletogenesis that are used for micromere signaling to adjacent embryonic cells. Thus, except for specification and signaling functions which have embryonic but no intrinsic skeletogenic significance, virtually the whole regulatory apparatus appears to have been co-opted from an adult skeletogenic GRN to the micromere lineage GRN, except for the four genes shown on light background. These are not expressed in adult skeletogenesis and were evidently individually co-opted into the skeletogenic embryonic network. (C), Co-optive history in echinoderm evolution of one of these genes, *t-brain* (*tbr*) *(from Minemura et al. (2009))*. The phylogenetic tree shows three classes of echinoderm plus the sister group of echinoderms, hemichordates (these two phyla are together known as the Ambulacraria). The class Echinoidea includes both sea urchins and sand dollars. Blue shading denotes *tbr* expression in the embryonic archenteron (Arch); blue dots denote *tbr* expression in the skeletogenic descendants of the embryonic skeletogenic micromere lineage (primary mesenchyme cells, PMC); red dots denote *tbr* expression in the neurogenic apical organ (AO). Co-optive gains, bars, and losses, ovals, of *tbr* expression are indicated: 1, loss of apical expression in the common ancestor of the free-living echinoderms or of all echinoderms; 2, gain of skeletogenic micromere lineage which expresses *tbr* (cf (B)); 3, loss of *tbr* expression in the archenteron of regular sea urchins; 4, gain of apical *tbr* expression in some but not all sand dollar lineages; 5, gain of *tbr* expression in the apical

Figure 7.6

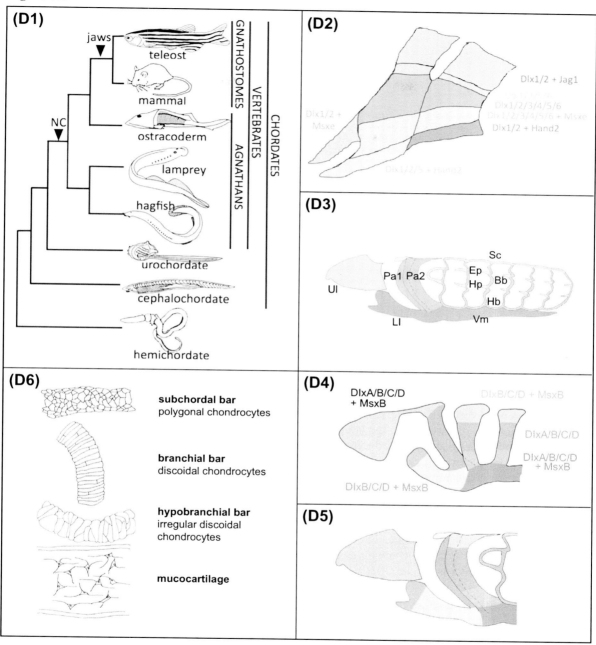

domain of hemichordate embryos. Note that all echinoids shown here have an embryonic skeletogenic lineage expressing *tbr* and that all Ambulacraria have embryonic archenterons expressing *tbr* except for the regular sea urchins (three species have been examined). (D), Apparent evolutionary co-option of the agnathan dorsal-ventral branchial arch patterning network to the dorsal-ventral jaw patterning network in gnathostomes. (D1), Chordate phylogeny, indicating jawed and jawless vertebrates, i.e., gnathostomes and agnathans respectively, and nonvertebrate chordates; NC, origin of neural crest. (D2), Components of dorsal-ventral regulatory states in first and second branchial arches of (36 h)

Figure 7.6

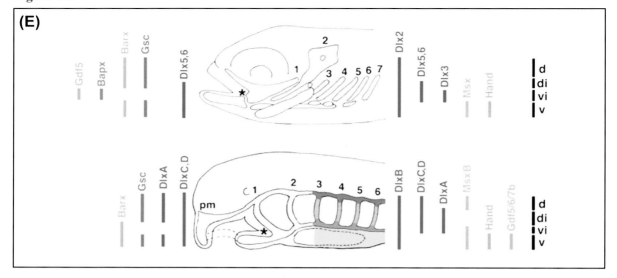

zebrafish embryo; green, dorsal; light and dark blue, ventral; gray, maxillary prominence. Regulatory genes expressed in these respective domains are indicated by color. (D3), Cartilagenous structure of a developing agnathan (lamprey) pharyngeal "basket" 35 days after fertilization. UI, upper lip; LI lower lip; Vm, Ventral pharynx; PA1, PA2, pharyngeal or branchial arches; Sc, subchordal bar; Hb, hypobranchial bar; a mesh of fused cartilaginous rods, the branchial bars and their extensions, extends posteriorly in the pharynx supporting the remaining arches. (D4,D5), Dorsal-ventral regulatory states in the developing lamprey pharyngeal basket at 13 days (D4) and 35 days (D5); color coding as in (D4). Only the maxillary region and the first three pharyngeal arches are included in these drawings (cf. (D3)). Regulatory genes expressed in the respective dorsal to ventral regions are indicated by color. (D6), Dorsal-ventral types of cartilage in the lamprey pharyngeal basket (cf. (D3)). The different regulatory states evidently activate different effector gene cassettes ((D1–D6) are *from Medeiros and Crump (2012)*). (E), Comparison of dorsal-ventral regulatory states in 48 h zebrafish pharyngeal skeleton, above, and developing lamprey pharyngeal skeleton, below. Regulatory state components specific to the first arch are shown to left, those more posterior to right. In the lamprey these give rise to the ventral to dorsal succession of cartilage types shown in (D6), the anatomical position of which can be seen by the color coding: yellow, mucocartilage; green hypobranchial cartilage; light blue, branchial bar cartilage; dark blue, subchordal bar cartilage. The arches are numbered. Pm, premandibular mesenchyme. The asterisk marks the position of the jaw hinge in the gnathostomes: only they express *gd5* and *bapx* genes in this position *(from Cerny et al. (2010))*.

Extensive observations have revealed that the dorsal-ventral positions in this structure express distinct regulatory states (Fig. 7.6(D2)), generated by GRNs that are differentially activated by regional signal inputs (Medeiros and Crump, 2012). Vertebrate jaw patterning GRNs include many demonstrated regulatory interactions. The skeletal elements forming the jaws and hinge are built by immigrant neural crest mesenchyme cells, which under the influence of certain *dlx* regulatory genes that they express, form localized aggregations, and execute chondrogenesis (Gordon et al., 2010). Since the gnathostomes descend from jawless ancestors, comparative observations on regulatory gene expression in the anterior branchial arches of extant jawless fishes may indicate whether these jaw patterning GRNs were entirely novel evolutionary assemblages. In modern jawless fish, exemplified by the lamprey, the branchial arches form a cartilaginous "basket"-like structure (Fig. 7.6(D3)). The patterns of expression of many key genes of the gnathostome jaw-forming GRN have been examined in lamprey larvae (Cerny et al., 2010; Medeiros and Crump, 2012). Strikingly, dorsal, intermediate, and ventral regulatory states largely equivalent to those of the gnathostome zebrafish were observed (Fig. 7.6(D2,4,5)). Thus much of the encoded spatial patterning apparatus is older than is jaw morphogenesis. In the lamprey these different regulatory states are apparently used to drive different batteries of effector genes than in gnathostomes, generating diverse types of cartilage (Fig. 7.6(D6)). The spatial regulatory state comparison is summarized in Fig. 7.6(E). Here we see two things: First, the gnathostomes largely retain the whole dorsal/ventral patterning system of the anterior branchial arches in jawless fish, i.e., possibly much of the GRN generating it. This is an illustrative example of use of a large preexistent spatial subdivision circuit for generation of a new structure by modifying its downstream morphogenetic output, i.e., its connections to effector gene cassettes. Second, a basically new gnathostome feature is the hinge formed from the intermediate domain of the first arch, and two genes are expressed in that location in the gnathostomes that are not expressed in the equivalent region in the lamprey, of which a particular Tgfβ paralog, Gdf5, may be key (Fig. 7.6(E)). This gene is expressed in the lamprey in other cells which synthesize "mucocartilage", a loosely packed cartilage more similar to that of the jaw joint (Fig. 7.6(D6)) and it is speculated that perhaps this differentiation program was co-opted for the hinge joint (Medeiros and Crump, 2012).

Jaws are used to support teeth, a major facilitator of the gnathostome predatory habit, and of gnathostome dominance, but teeth also predate jaws. They originated within the posterior pharynx in early jawless fish, and some modern fish still produce posterior pharyngeal teeth. Here hangs yet another tale of whole GRN co-option (Fraser et al., 2009). Some modern teleosts, here African cichlids, have pharyngeal jaws as well as oral jaws, and these bear teeth just as do the oral jaws. There is a striking difference in the cellular environment in which the teeth develop: the oral teeth are constructed by neural crest mesenchyme in an ectodermal environment, while the pharyngeal teeth are constructed by neural crest mesenchyme in an endodermal environment. But despite this, the GRN utilized for patterning teeth is the same, as shown by observations on over a dozen regulatory and signaling genes operating within the known GRN circuitry, and it can be concluded that the vertebrate tooth GRN was invented only once (Fraser et al., 2009). It was first used in a posterior pharyngeal location, then co-opted to the oral jaws, then co-opted again to pharyngeal jaws as well. The example demonstrates the fundamental consequences for evolution of the modular hierarchy of GRN structure, wherein whole patterning subcircuits can be plugged in to diverse regions of the program for body plan development.

The progenitor fields for embryonic regions in Bilateria, and thence for their body parts, are organized from the beginning of embryonic development to its end with respect to vectorial geometrical axes, the anterior/posterior and dorsal/ventral axes of the body per se, and the proximal/distal, as well as anterior/posterior and dorsal/ventral axes of appendages. This requires vectorial patterning systems that possess direction and polarity, and ultimately give rise differentially to regulatory states along these axes. Two major types of vectorial patterning systems are utilized repeatedly in bilaterian development. The first of these are vectorial signal ligand gradients, in which continuous concentration clines are interpreted spatially by Booleanizing GRN circuitry that generates sequences of regulatory gene expression domains, as we discuss in Chapter 6. The others are vectorially expressed regulatory gene clusters, of which by far the most prominent and broadly applied is the bilaterian *hox* gene system.

2.5 *hox* genes and GRN circuit deployment

hox genes encode transcription factors that are expressed in nested patterns along the anterior/posterior axis of the bilaterian body plan during development, except for the anterior structures of the head and brain (cf. Fig. 4.4). *hox* genes are also expressed in the orthogonal trunk appendages in vertebrates (see Chapter 4). The pleisiomorphic genomic organization of *hox* genes in Bilateria is a single contiguous gene cluster such that the genes at the 3′ end of the cluster are expressed most anteriorly, and those at the 5′ end most posteriorly.

In vertebrates, the anterior boundaries of *hox* gene expression domains occur in the same sequence along the anterior-posterior body axis as the order of the respective *hox* genes in their chromosomal clusters. This gene organization can still be observed in amphioxus (Garcia-Fernández and Holland, 1994; Minguillón et al., 2005). However, major alterations in chromosomal organization of *hox* genes have often occurred in bilaterian evolution: in many clades the cluster has been reorganized by breakage and translocation, as in *Drosophila*, or by inversion and translocation as in sea urchins, or the cluster has been eradicated by fragmentation and gene deletion, and the remaining *hox* genes dispersed in the genome, as in *Caenorhabditis elegans* and *Ciona*; yet in all these animals anterior to posterior expression is still observed for the respective *hox* genes. Thus, where it exists, the cluster organization or surviving elements of it may be used for vectorial expression, but the genes are individually equipped with local *cis*-regulatory control systems that read the diverse regulatory states along the body axis. Major genomic reorganizations with respect to the pleisiomorphic *hox* gene cluster have occurred by whole, paralogous cluster duplication in vertebrates. Mammals have 4 clusters, bony fish 7, and following the duplications by which these arose, individual *hox* gene loss and subfunctionalization has to some extent made each surviving cluster unique. The patterns of *hox* gene expression are extremely complex, as these genes are used in multiple tissues at multiple times, and thus they are serviced by a plethora of *cis*-regulatory modules. Several nonexclusive regulatory mechanisms militate toward physical survival of *hox* gene clusters per se. One structural aspect that can possibly explain the general conservation of clustered organization of the *hox* genes is that their *cis*-regulatory modules are not uncommonly found in or on the other side of the next *hox* gene of the cluster, a prominent feature that became evident with the mapping of *hox* gene enhancers in *Drosophila*. Furthermore, in vertebrates, multiple linked genes are to some extent controlled by potent distant enhancers located far outside the clusters per se, and the position in the cluster of individual genes relative to these control sequences is essential to their correct function, as we briefly discuss in a particular evolutionary context below.

A quarter century of intensive observation and experimentation on *hox* gene expression and function has produced several fundamental generalizations. First, within the Bilateria the order in which *hox* genes are expressed in respect to the anterior-posterior axis has been preserved, despite massive changes in the genomic organization and number of *hox* genes, and in their regulation and downstream functions. Thus most importantly, *hox* genes may function as a vectorial patterning system by which the sequential order of expression along an axis is controlled. Second, within each phylum, the location of given body parts correlates closely with the particular combination of *hox* genes expressed where these body parts form. This implies that within the phylum, there are stable linkages between given *hox* genes and the GRNs that control development and morphogenesis of given body parts. Third, the precise position and spatial extent of given *hox* gene expression domains is evolutionarily flexible, as seen for example in comparisons of diverse arthropod or vertebrate classes and orders (for review, Davidson, 2006). Fourth, both experimental gain and loss of *hox* gene function have produced overwhelming evidence that expression of given *hox* genes is required for development of given body parts, and this is their dominant developmental function; exemplary cases are discussed in the following section. As is now abundantly clear, body part formation is mechanistically the outcome of progressive spatial regulatory state patterning controlled by specific GRN circuitry (Chapter 4). Thus, this dominant function of *hox* genes can only mean that their vectorial spatial coordinate system is utilized to deploy the GRN circuitry that actually programs the development of given body parts. This circuitry is deployed at clade-specific locations on the axes along which the relevant *hox* genes are expressed. In this sense one of the roles of *hox* genes in GRNs would be to function as switches which turn on or off GRN subcircuits (Erwin and Davidson, 2009; Peter and Davidson, 2011). This of course could not be as frequent and general a function as it is, were developmental GRNs not fundamentally modular and hierarchical in structure. Thus we can consider the *hox* gene system as a vectorial device for recruiting large GRN circuits to be activated at the right positions in the body plan.

hox gene deployment of developmental GRNs controlling body part formation

Experimental deletion and rearrangement of *hox* gene regions in mice provides causal evidence of their ultimate effects on development of the body plan. A typical, and also revealing example is reproduced in

Fig. 7.7(A), which illuminates the roles of *hox6* and *hox10* paralog groups in axial skeleton formation (Wellik and Capecchi, 2003; Wellik, 2009; Mallo et al., 2010). As we see here, the structure of mammalian vertebrae and ribs varies along the anterior-posterior axis, correlated with the sequence of *hox* genes expression (Fig. 7.7(A1)); the thoracic vertebrae produce ribs, the cervical and lumbar vertebrae do not, while the sacral vertebrae again generate small rib-like projections. Fig. 7.7(A2–6) (Mallo et al., 2010) shows the consequences for vertebral specialization of spatial gain of function (forced ectopic expression) and loss of function (deletion) for *hox6* and *hox10* paralog group genes (see Fig. 7.7(A2) for genomic map). These experiments show that *hox6* promotes rib formation and can do so ectopically (Fig. 7.7(A3,5)), while *hox10* suppresses rib formation (Fig. 7.7(A3,4,6)). In addition, *hox11* is required for sacral vertebral formation, counteracting *hox10* rib suppression in the sacral region (Wellik and Capecchi, 2003). As we discuss in Chapter 5, ribs are generated from the sclerotomal (mesenchymal) portions of the somites according to a morphogenetic regulatory GRN, the linkages of which are not yet defined experimentally but which is known to include many regulatory and signaling genes (Vinagre et al., 2010). The *hox* genes sit astride the patterning networks that control these development functions. The *hox6* and *hox10* genes function in the mesodermal cells adjacent to the sclerotome, evidently controlling signal inputs into the *myf5* and *myf6* regulators at the top of the network circuit that controls rib morphogenesis. In short, they act to deploy these networks spatially.

Similarly, as Fig. 7.7(B) shows, *hox10* and *hox11* also deploy the programs for generation of the long bones of forelimb and hindlimb (Wellik and Capecchi, 2003). Loss of *hox11* function obliterates the lower limb long bones (zeugopod) of both forelimb and hindlimb (Fig. 7.7(B1,2,3,5)); loss of *hox10* obliterates the upper long bone (stylopod) of the hindlimb (Fig. 7.7(B3,4)). Domains of *hox* gene expression required for the operation of limb morphogenetic programs throughout the limbs for the whole set of posterior *hox* genes are summarized in Fig. 7.7(B6,7).

The segment-specific morphology of the whole appendage in the crustacean *Paryhale* is also dependent on specific *hox* gene expression (Liubicich et al., 2009; Pavlopoulos et al., 2009). As illustrated in Fig. 7.7(C1–5), uniform expression of the *Ubx hox* gene causes transformations of what should be various head segment appendages to thoracic appendages of quite different construction, due directly to the ectopic Ubx input and indirectly to Ubx repression of the anterior *hox* gene *Scr*. Conversely, interference with *Ubx* expression causes transformation of walking legs to anterior thoracic appendages. Thus here again, the developmental program deployed is controlled by, i.e., switched on or off by, *hox* gene expression in given domains along the anterior-posterior axis of the animal.

The same general relation between *hox* gene inputs and developmental GRNs that control spatial development of given body parts can be deduced from experiments on many *hox* genes and many body parts done over the last two decades. An additional example shown in Fig. 7.7(D) is the role of *Abd-A* in heart and aorta development from the dorsal vessel in *Drosophila* (Lo et al., 2002). Here we see that *Antp*, *Ubx*, *Abd-A*, and *Abd-B* are expressed in that order anterior to posterior in the aorta and heart proper, such that normally only *Abd-A* is expressed throughout the heart proper, except for the posterior valve. But if *Abd-A* is expressed in the whole structure, the aorta developmental program is suppressed, and only the heart developmental program is operated, while the reverse happens in an *Abd-A* mutant, where only the aorta program is operated.

Figure 7.7 illustrates multiple cases of Boolean switch functions, in which the GRNs underlying formation of given body parts do or do not operate in given locations depending on expression of given *hox* genes. This feature of network wiring means that redeployment by *hox* gene switches of circuitry controlling development of given body parts can be used in evolution to effect large-scale spatial changes in morphological outcome, such that the unit of change is the whole body part. Thus in vertebrate classes, the proportions of cervical- versus thoracic-type vertebrae varies as expression of key *hox* genes varies, and these variations lie specifically in the relevant *cis*-regulatory control sequences of given *hox* genes (Belting et al., 1998; Anand et al., 2003). Within the phylum, where the metameric organization of clades of related animals can be aligned and homologized, as in arthropods, it can be seen that *hox* gene expression

Figure 7.7

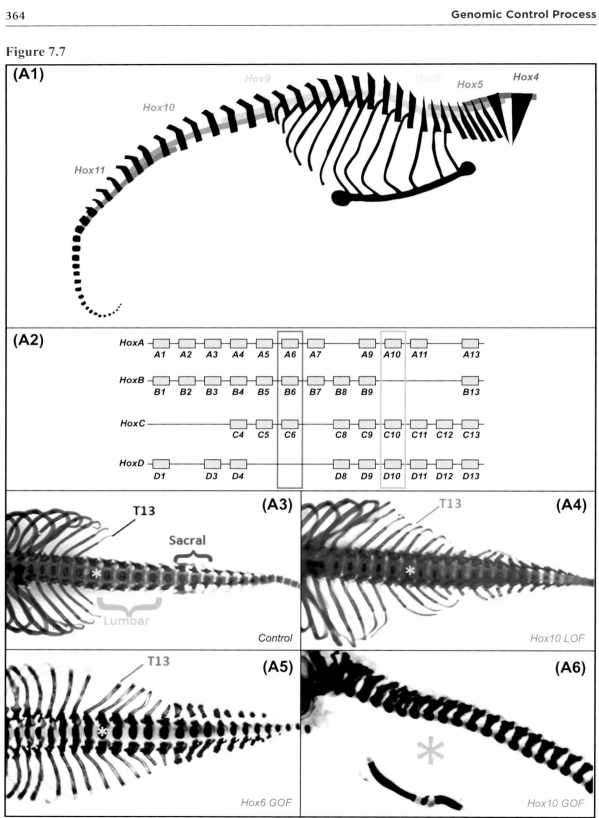

Figure 7.7

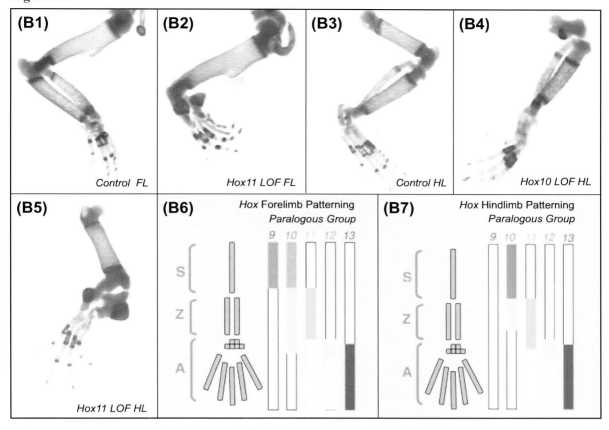

Figure 7.7 *hox* genes as switches controlling developmental patterning circuitry. (A), Mammalian axial skeletal patterning. (A1), Mouse axial skeleton, anterior to right. Colored bars show overlapping regional requirements for *hox* gene expression, based on paralog group mutational analysis *(from Wellik (2009))*. (A2–A6), Axial effects of gain and loss of function of *hox6* and *hox10* genes *(from Mallo et al. (2010))*. (A2), The four mouse *hox* gene clusters, the three *hox6* and *hox10* genes boxed in red and green respectively. (A3), Normal posterior skeleton showing thoracic rib bearing vertebrae, lumbar vertebrae, and sacral vertebrae; T13 is normally the most caudal rib followed by the first lumbar vertebra, asterisk. (A4), Effects of loss of function mutation of *hox10* genes, resulting in ectopic caudal ribs (cf. (A2)). (A5), Effects of gain of function of *hoxb6* gene, also resulting in ectopic caudal ribs (cf. (A2)). (A6), Effects of gain of function, here precocious activation of *hoxa10* gene in somitic mesoderm, resulting in complete loss of ribs (large green asterisk). These and many other experiments show that *hox10* genes prevent execution of rib patterning functions, and *hox6* has the opposite role of deploying this developmental function. (B), Posterior *hox* gene requirements for forelimb (FL) and hindlimb (HL) long bone morphogenesis *(from Wellik and Capecchi (2003))*. (B1), Control forelimb skeleton of mouse. (B2), Effects on forelimb of deletion of *hox11* genes (cf. (B1)), resulting in absence of forearm (zeugopod) long bones. (B3), Control hindlimb skeleton. (B4), Effects on hindlimb of deletion of *hox10* genes resulting in almost complete loss of long bone of the upper limb (stylopod; cf. (B3)). (B5), Effects on hindlimb of deletion of *hox11* genes resulting in absence of zeugopod. (B6), Summary of *hox* paralog group functions in forelimb regions, S, stylopod; Z, zeugopod, A, autopod. (B7), Summary of *hox* paralog group functions in hindlimb regions. Operation of the forelimb stylopod patterning GRN requires *hox9* and *hox10* inputs; the forelimb zeugopod GRN requires *hox11* input; the autopod

Figure 7.7

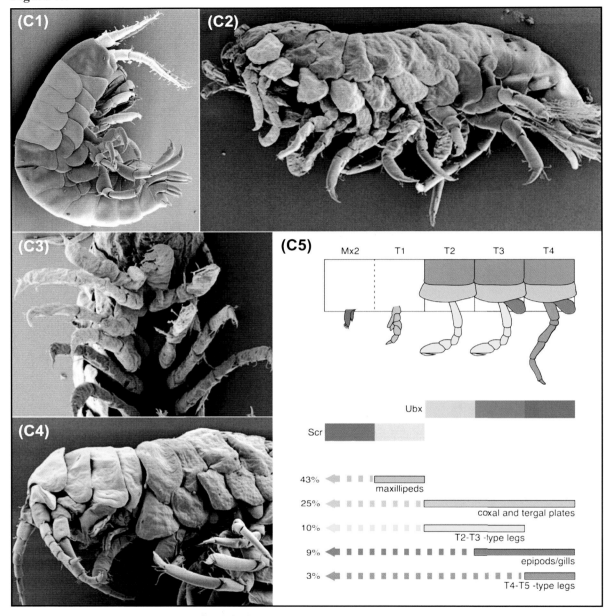

patterning GRN requires *hox12* or *13* (discussed later in this chapter), and similarly for the hindlimb skeletal patterning systems, except that *hox9* genes are not expressed in the hindlimb. (C), Effects of gain of function of the *Ubx hox* gene on appendage patterning in the crustacean *Paryhale (from Pavlopoulos et al. (2009)).* In (C1, C5) appendages of *Paryhale* are shown according to following color code: antennae, white; maxillary (MX) head feeding organ, blue: T1 thoracic maxilliped (MXP; feeding and locomotory organ), green; T2–T3 maxillipeds yellow, T4–T5 legs magenta. The dorsal (tergal) and ventral (coxal) thoracic plates are shown in light and dark orange. (C1), Color-coded SEM of control animal. (C2–C4), Transformations of appendages caused by ectopic misexpression of *Paryhale Ubx* during development, using a heat-inducible transgenesis vector, highlighted in turquoise. (C2),

Figure 7.7

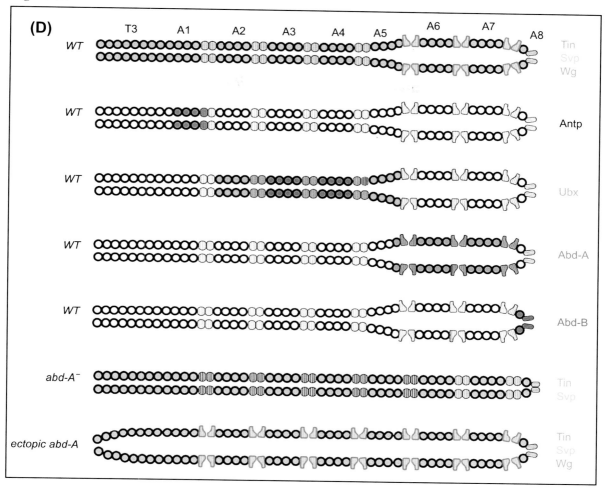

Transformation of MX and MXP toward thoracic legs. (C3), MX to thoracic leg, MXP to thoracic leg and T2/T3 to T4/T5. (C4), Ectopic tergal thoracic plates formed in normally unsegmented head region. (C5), Summary of misexpression transformations, and normal correlation between domains of *Ubx* and *Scr hox* gene expression and appendage type, anterior left. At top segmental morphological appendage characters and tergal and coxal thoracic plates are shown using same color coding as in (C1). Intensity of normal expression of *Ubx* and *Scr* is indicated by shading. The normal locations of each morphological feature are indicated by the colored bars, and the transformations caused by ectopic *Ubx* expression in the indicated percentages by dashes. (D), Regional requirement for AbdA function in developmental specification of the *Drosophila* heart *(from Lo et al. (2002))*. The dorsal vessel is shown at top in a late embryo: T, thoracic segment; A, abdominal segment. The heart proper is the enlarged chamber at the posterior end (right), preceded by the aorta, which extends forward into the thorax. The *tinman* gene (orange) is expressed throughout; *svp* (yellow) and *wg* (green) expression marks pairs of special cardioblasts that form intake valves or ostia in the heart. Below are the normal domains of expression of the *hox* genes *Antp*, *Ubx*, *Abd-A*, *Abd-B*, expressed colinearly in the heart structures, anterior to posterior. In the absence of Abd-A only the aorta is formed, and in Abd-A overexpression, only heart chamber and intake valves are formed.

patterns are correlated with position of homologous body parts and that these expression domains differ extensively among orders and classes, accounting for differences in position of the homologous body parts (Abzhanov and Kaufman, 1999, 2000; Abzhanov et al., 1999; Mansfield and Abzhanov, 2010).

A paradigmatic example of exactly how a *hox* gene controls the presence or absence of a body part has been obtained for the control of haltere formation in *Drosophila*. Expression of the *Ubx hox* gene (a paralog of chordate *hox 6-8* genes) is required for the dorsal imaginal disc of the third thoracic segment (T3) to produce the haltere, a small balancing organ (the mechanism is considered below). The developmental program underlying haltere formation is quite different at its final stages from that by which the wing is formed at the same time from the dorsal imaginal disc of the T2 segment. *Ubx* expression is also required for the more ventral imaginal disc of the T3 segment to generate the third leg. Normally *Ubx* expression, at the time important for these developmental pathways, occurs in the T3 but not T2 segment. Total loss of *Ubx* function generates Ed Lewis' famous four-winged fly (Fig. 7.8(A)), in which the dorsal T3 discs have developed into a pair of wings, normally the product only of the T2 dorsal imaginal discs (Bender et al., 1983). The *Ubx*⁻ phenotype predicts directly that in normal flies Ubx executes a negative switch-like effect on wing development in the T3 disc, allowing it to give rise instead to a haltere. The upper level functions of *Ubx* expression in assigning T3 vs T2 imaginal disc specification identity are no doubt as old as the dipteran imaginal disc program, at least several hundred million years.

Molecular investigations on how the wing developmental program is affected by *Ubx* were carried out over a decade ago (Weatherbee et al., 1998; Galant et al., 2002). Extensive experiments using mosaic flies in which clones of wing cells were made to express *Ubx* ectopically revealed the Ubx repression targets. These experiments demonstrated that ectopic Ubx expression causes repression of several key genes of the wing patterning GRN (Fig. 7.8(B)). Thus it is clear that in the normal T3 haltere disc where *Ubx* is expressed, the Ubx gene product blocks the wing program from functioning at multiple genes (Weatherbee et al., 1998; Galant et al., 2002; Hersh and Carroll, 2005). Furthermore, as shown in the *cis*-regulatory experiments of Figs. 7.8(C,D) for the key wing circuit regulatory genes *spalt* (*sal*) and *knot*, these are direct *cis*-regulatory repressions, and this is also demonstrated for *vestigial* (see Fig. 7.8(B) for the roles of these genes; Hersh and Carroll, 2005; Galant et al., 2002). But Ubx also directly activates haltere-specific genes in this disc, as well as repressing wing-specific genes.

Target gene sets of the Ubx transcription factor and its cofactor Homothorax (Hth) have been studied on a genome-wide scale in *Drosophila*. In one large-scale set of observations, normal T3 haltere and leg imaginal discs were isolated, and all Ubx- and HtH-bound sites were identified by ChIP and analyzed for target sites at the sequence level (Slattery et al., 2011). Distinct sets of genes encoding transcription factors were prominently recovered from the target gene datasets for each imaginal disc, including genes already known as specific Ubx targets, and also other *hox* genes. Also prominent were Wg, Notch, and Dpp signaling pathway genes. However, there were in addition many targets encoding downstream cell biology and cellular morphogenesis functions. Of genes overexpressed in ectopic *Ubx* expression (i.e., direct plus indirect target genes), about half, ~200, display Ubx binding at putative *cis*-regulatory sites. In another study (Pavlopoulos and Akam, 2011), an inducible system was used to switch on *Ubx* expression in developing wing blade epithelium of T2; ectopic expression of *Ubx* in the wing arrests the wing developmental program and causes transformation to the distal blade of the haltere. The affected tissue was isolated and analyzed for gene expression very shortly after induction at different stages, in order to focus on direct up- and downregulated targets. Several hundred largely stage-specific genes were thus identified, again prominently including regulatory genes and signaling components but also effector genes involved in cuticle formation, growth, cell adhesion, and cytoskeletal functions as well. Many of these genes were affected only modestly, displaying on average less than twofold-level differences as a result of the induced *Ubx* expression compared to controls.

Though we have focused on upper level switch-like functions of *hox* genes so far, it becomes apparent that they provide inputs at many levels of the GRN hierarchy. Whole genome studies of targets of other *hox* genes have recently been carried out using microarrays, ChIP-seq, or transcriptome sequence analysis, both in vertebrates and in *Drosophila*. An example is an overexpression study of genes up- and

Figure 7.8

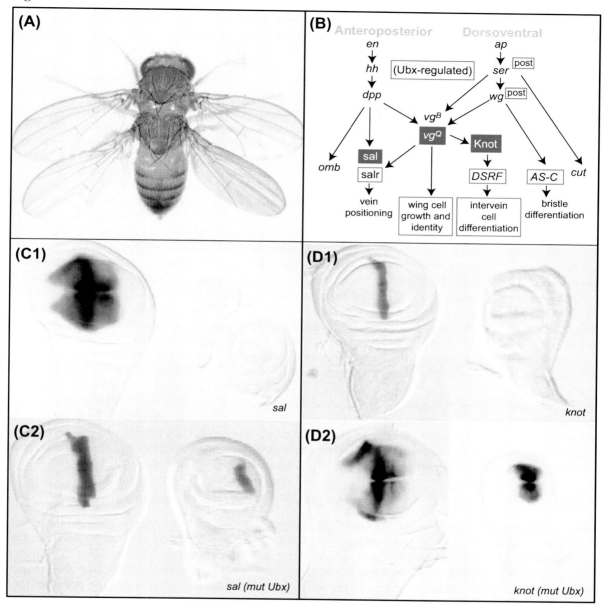

Figure 7.8 Cis-regulatory repression of wing patterning transcriptional circuitry by Ubx, in the *Drosophila* haltere disc. (A), Four-winged fly resulting from total loss of *Ubx* function, the consequence of absence of Ubx repression of wing patterning in the haltere discs *(image provided by Ed Lewis; from Bender et al. (1983))*. The four-winged fly results from execution of the same wing developmental program in the second and third dorsal thoracic discs. (B), Elements of the anterior-posterior and dorsal-ventral patterning GRN of the wing disc, displaying in red the direct regulatory targets of Ubx repression in the haltere disc, thereby normally prohibiting wing development there *(from Weatherbee et al. (1998), Galant et al. (2002))*. Open red boxes indicate genes which are repressed in the wing disc in mosaic clones ectopically expressing *Ubx*, and which are ectopically activated in the haltere disc in mosaic clones lacking *Ubx* expression. Solid red boxes denote direct *cis*-regulatory evidence for *Ubx*

repression. For example VgQ is the "quadrant enhancer" of the vestigial gene, expressed widely in the wing blade, and as the name suggests required for wing outgrowth and normal-sized wings. VgQ cis-regulatory constructs are silent in the haltere disc but are derepressed in haltere clones lacking *Ubx* expression (Weatherbee et al., 1998). (C), *Cis*-regulatory evidence for direct Ubx repression of the *spalt* (*sal*) gene (cf. (B); *from Galant et al. (2002)*). (C1), A *sal cis*-regulatory lacZ construct expresses normally in wing only; note relatively small size of developing haltere to right. (C2), Mutation of seven Ubx-binding sites in this construct results in its derepression in the haltere. (D), *Cis*-regulatory evidence for direct Ubx repression of the *knot* gene (cf. (B); *from Hersh and Carroll (2005)*). (D1), Normal expression of *knot cis*-regulatory lacZ construct along the anterior-posterior axis of wing disc. This construct is silent in haltere, as shown to right. (D2), If the 10 Ubx sites in the *knot* construct are mutated, expression now extends to the haltere as well.

downregulated after injection of *hoxb1b* mRNA in zebrafish eggs (van den Akker et al., 2010). In zebrafish this particular *hox* gene is normally expressed first in embryonic ventral/lateral mesoderm. Scores of genes were directly or indirectly affected by early gastrula stage, probably representing immediate targets, including terminal downstream effector genes that function in cell adhesion, metabolism, and cell motility. A main result, however, was many-fold overrepresentation of genes encoding transcription factors. In mouse embryo fibroblasts, the apparently direct targets of *hoxc8* also include genes involved in cell adhesion, cell proliferation, and other cell biology functions (Lei et al., 2006). Analysis of putative direct targets of Hoxa2 and Hoxa2-Pbx in mouse embryonic development by ChIP-seq revealed hundreds of putative targets, of which Wnt signaling pathway components were shown to be coexpressed in the specific domain of this *hox* gene (Donaldson et al., 2012). In *Drosophila*, a *cis*-regulatory study of an essential specific head enhancer of the *dpp* gene, which causes expression in a stripe at the lateral margin of the eye-antennal disc, demonstrates direct control of this signaling gene by the *labial hox* gene, together with cofactors (Stultz et al., 2012).

We can conclude that at least some *hox* genes intervene directly, if subtly, at all levels of the developmental GRN hierarchy. Thus upstream patterning and signaling genes as well as effector genes that operate in the final construction and specification of the body part, all may be regulated directly by *hox* genes. This could explain the rigidity and conservation within phyla of the functional relations between expression of given *hox* genes and the operations of GRNs for the formation of given body parts.

Regulatory evolution of *hoxd* cluster genes in vertebrate appendages

The evolution of tetrapod limbs beginning in the Devonian (419–359 My ago) has presented a perfect challenge for understanding the mechanisms by which major changes in body plan may occur. The fossil record of the transition from lobe-finned (Sarcopterygian) fish to tetrapod is increasingly fine grained (e.g., Shubin et al., 2006), and so it has become clear that the basic long bone structure of the tetrapod forelimb and hindlimb is already present in the paired fins of these fishes; what is missing is the autopod (hand or foot) with its digits. Tetrapod autopod structures appeared in a great variety of forms in the succeeding geological period (Carboniferous, 359–299 My ago), circumscribing the time of their evolutionary invention. Experimental exploration of the evolutionary process of fish fin to tetrapod limb transition is facilitated by the comparative examination of the different forms of paired appendages in certain groups of living fish, in which developmental studies can also be carried out. In addition, as reviewed in Chapter 4, we know a good amount about the GRN underlying the earlier stages of appendage formation, particularly the establishment of the anterior-posterior, dorsal-ventral and proximal-distal regulatory state axes of the limb buds from which form the appendages. Only the gnathostomes possess two sets of paired appendages. A single pair of pectoral fins was present in the now extinct armored jawless fish (Osteostracans) which had appeared earlier, in the Silurian (455–419 My ago). Probably during the initial evolution of paired fins the posterior *hox* genes became expressed in the appendages more or less orthogonally to the main body axis of the animal, but in the same order, a clear example of co-opted use of this vectorial patterning system (Shubin et al., 1997), more of which below. As we have seen (Fig. 7.7(B)), *hox* gene expression in the elongating limb is required for deployment of the programs controlling long bone morphogenesis.

Normal autopod development absolutely requires expression of posterior *hoxd* cluster genes, particularly of *hoxd13* at the margin of the growing limb bud, in a second phase of posterior *hox* gene expression. This has been established in an extensive series of targeted mutations and deletions in mice following detailed earlier observations of *hox* gene expression in developing limb buds (reviewed by Zakany and Duboule, 2007). The genomic basis of the control system causing this second phase of *hoxd13* expression is in some way related to autopod evolution. However, the relation is not simple, for fish which do not have autopod digits nonetheless express posterior *hoxd* genes across the limb bud margin in patterns grossly similar to tetrapods. Furthermore, a highly conserved murine *hoxd13* enhancer active in promoting this second phase of expression also produces distal expression in transgenic zebrafish, and the conserved zebrafish enhancer promotes autopod expression in transgenic mice (Schneider et al., 2011); these results together with a summary of extinct and living limb bone morphologies are reproduced in Fig. 7.9(A). This figure supports several conclusions: first, in Fig. 7.9(A1) we see that the enhancer CsB which generates robust *hoxd13* expression in the autopod of the developing amniote limb bud, displays remarkably conserved sequence all the way from mouse to skate. When the skate and mouse version of this enhancer are tested in transgenic mice, the enhancer is shown to run in the autopod region, although only the mouse enhancer drives expression distally where the digits are going to form; thus the function of this conserved *hoxd13* enhancer has changed from fish to mouse (Fig. 7.9(A2)). The morphological structures of the appendages of transitional forms evolving from sarcopterygian fish, represent the stepwise evolution of the tetrapod autopod. Since the primitive fish *Polyodon* also expresses *hoxd13* at the distal end of the limb bud but does not produce an autopod, it is clear that the evolution of the hand required extensive additional patterning circuitry.

In mice, the *hoxd* cluster, itself about 100 kb in length, is flanked by two gene deserts, one on the centromeric (5′) side extending from the *evx2* and *lunapark* genes near *hoxd13* for about 600 kb, until the next gene up; and on the telomeric (3′) side of the cluster another gene desert extends from genes immediately flanking the *hoxd* cluster for about 780 kb. Both of these gene deserts contain arrays of conserved regulatory sequences which control certain aspects of expression of the *hoxd* genes (for review, Tschopp and Duboule, 2011). Fig. 7.9(B) shows diagrammatically that the first phase of *hoxd* gene expression, required for forearm and upper arm long bone development, is mediated by enhancers located on the 3′ telomeric side. Large deletions that bring the posterior *hoxd* genes closer to these enhancers cause earlier expression the time when the genes are activated in limb development is colinear with gene position. Similarly, on the 5′ side, distant enhancers located in the flanking gene desert control the second phase of expression, which is required for autopod development. Here the proximity of the gene to these external enhancers determines their levels of expression. Both results mean that the positional sequence of the genes in the *hoxd* cluster has functional significance. An interesting evolutionary simplicity emerges (Tschopp and Duboule, 2011). The most ancient aspect of *hox* gene function, the one shared primitively across all Bilateria, is colinear expression along the main anterior-posterior axis of the body, whether or not the body includes appendages. As summarized in the cartoon of Fig.7.9(C), this aspect of *hoxd* function is controlled by enhancers within the cluster, no doubt local enhancers determining individual aspects of *hox* gene function such as are well characterized in both *Drosophila* and vertebrates (for a detailed example of this, consider the *hox* gene *cis*-regulatory systems active in rhombomere specification discussed in Chapter4). Thus the process of appendage evolution in vertebrates involved the appearance of new, externally located control elements which made use of the order of the genes in the cluster. Chromatin capture studies have shown decisively that these distant regulatory elements activate the genes of the *hoxd* cluster by looping to their promoters (Montavon et al., 2011), which makes perfect mechanistic sense of observations that relate transcriptional activity to position with respect to the distant enhancers.

Figure 7.9

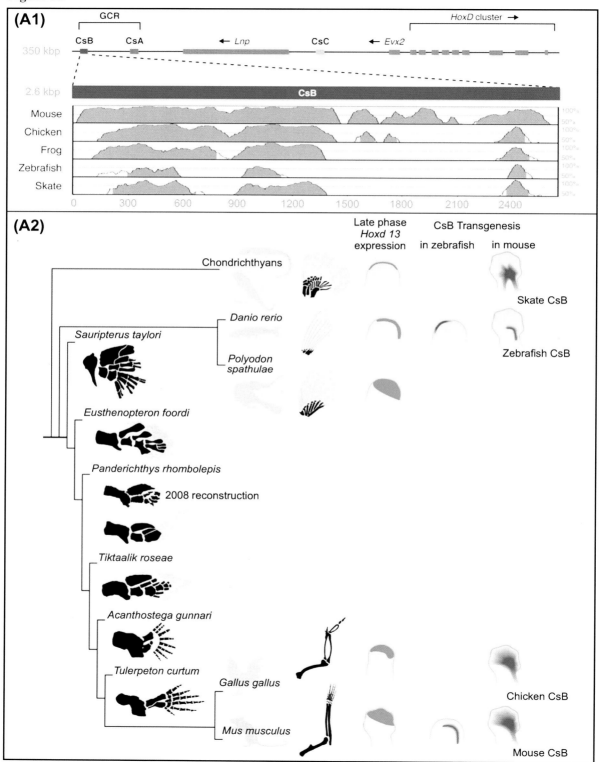

Figure 7.9

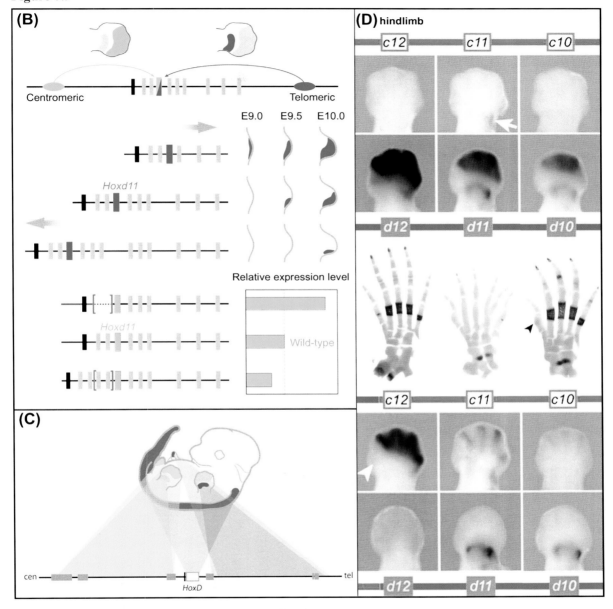

Figure 7.9 Posterior *hox* genes and the evolution of the tetrapod autopod. (A), Paleontological transition from fin to limb, and limb bud *cis*-regulatory function of an external 5′ *hoxd* control sequence *(from Schneider et al. (2011))*. (A1), Map of a proximal portion of the centromeric gene desert flanking the *hoxd* cluster with Vista plots below displaying conservation of regulatory CsB sequence from amniotes to cartilaginous fish. In mice CsB *cis*-regulatory constructs generate autopod expression: GCR, "global control region", *evx* and *lnp*, *evenskipped* and *lunapark* genes; Mouse CsA enhancers express only axially; CsB constructs express axially in neural tube, as well as in genital bud and heavily in autopods; CsC also expresses in autopods. (A2), Paleontological evidence for stepwise transition of skeletal morphology from lobe-finned fish fin to limb with autopod. The diagram is phylogenetically organized, from the fossil record. Chondrychthians, cartilaginous fish, here skate; *Danio rerio*, the teleost zebrafish; *Polyodon*, paddlefish, a primitive finned

Figure 7.9

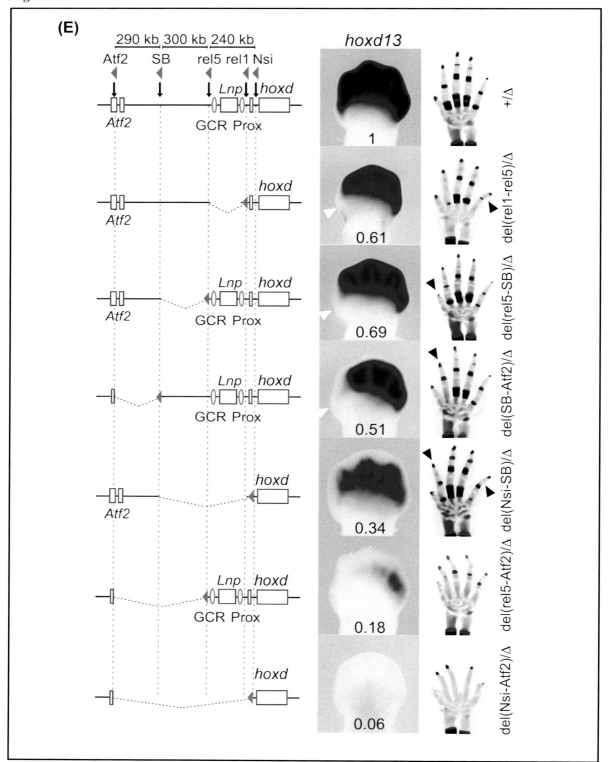

bony fish. To right at top of diagram, skeletal structures at the base of the fin rays of these fish are shown in black; nonetheless, as portrayed in green, *hoxd* expression patterns in the portion of the limb bud from which in tetrapods the autopod forms, particularly in *Polyodon*, are directly comparable to the amniote tetrapod patterns shown at bottom (mouse and chicken). The purple patterns to right show expression in transgenic zebrafish and mice, as indicated, of CsB *cis*-regulatory constructs in the relevant region of the limb bud. Thus evolutionarily the *hoxd* expression pattern across the top of the limb bud predates its co-option to the developmental processes required for autopod morphogenesis, since this feature of *hoxd* spatial regulation already exists partially even in skate. However, in transgenic mice, the skate CsB construct does not express up to the distal edge where the digits develop, but only to the wrist and digit bases, unlike the mouse and chicken CsB constructs shown at bottom. (B), Spatial genomic position of *hoxd11* with respect to the distant external enhancers affects expression in the autopod *(from Tschopp and Duboule (2011))*. At top is a map of the *hoxd* complex, *hoxd11* in color, with roles of distant external enhancers on telomeric (green arrow) and centromeric (yellow arrow) sides indicated: the telomeric enhancers generate early phase expression in the elongating limb in cells the descendants of which will give rise to proximal structures (green patch); in late phase expression the centromeric enhancers generate expression in the autopod (yellow patch). Experiments in which deletions or insertions were used to change the position of *hoxd11* (green bar) with respect to the telomeric enhancers are illustrated diagrammatically below; wild type, middle row. Closer proximity to the telomeric enhancers causes premature early phase activation and more anterior late phase expression, and the converse occurs when the gene is moved further away. Similar experiments in which *hoxd11* (yellow bar) is moved closer to the centromeric enhancer by means of deletion of intervening genes (red brackets) or further away by means of duplication of these genes, are shown at bottom; wild-type position again in middle row. As indicated by the histogram, the closer the gene is to the external enhancer the greater its activity, and the further away the less. (C), Overall regulatory organization of the *hoxd* locus *(from Tschopp and Duboule (2011))*. Color coding for the more recently evolved phases of expression in the paired appendages is as in (B), i.e., the centromeric external enhancers, location and effect in yellow, determine late phase autopod expression, while telomeric enhancers, location and effect in green, determine early phase expression affecting upper limb outgrowth. The *hoxd* complex is also primordially expressed in the anterior-posterior axis of the body, and the enhancers that control this evolutionarily ancient expression are located within the cluster (indicated in blue). (D), Demonstration that *hoxc* genes are functionally equivalent to *hoxd* genes in digit development if expressed similarly *(from Tschopp et al., 2011)*. *hoxc12, c11,* and *c10* genes are not expressed in the distal hindlimb buds of the wild-type mouse as shown in top row of panels (except for some proximal *hoxc11* transcripts, white arrow), while the *hoxd* paralogs are, as shown in second row of panels. The newborn skeletal structure of the wild-type foot is shown on the left; center, failure of formation of cartilage rods and delayed ossification in digits results from absence of *hoxd* function, but in right skeletal preparation, this phenotype is almost completely rescued if *hoxc12, c11,* and *c10* are placed under the control of the *hoxd* centromeric enhancer system. In the experiment where the centromeric enhancers are exchanged between the *hoxc* and *hoxd* clusters, as the embryonic limb bud in situ hybridizations in bottom panels show, the *hoxc* genes are now expressed in the autopod much like the *hoxd* genes normally are, except for failure of expression at the far anterior side, and hence development of the thumb digit is not rescued (white and black arrows; third and forth row of panels). Expression of the *hoxd* genes under control of the *hoxc* centromeric enhancer fails to extend into the autopod (fifth row of panels). (E), Demonstration of multiple dispersed control elements throughout the centromeric gene desert by successive deletion (from Montavon et al., 2011). Loxp sites were inserted as indicated by red arrowheads within the enormous centromeric gene desert which extends up to the *atf2* gene; GCR, Prox, previously studied external enhancers (cf. (A)), *lnp*, *lunapark* gene flanking *hoxd* complex. Successive deletions were as shown (dotted lines, sequences deleted), and resulting patterns of expression of *hoxd13* are illustrated for each by in situ hybridization; numerals indicate relative levels of *hoxd13* expression (QPCR). The resulting digital phenotypes are shown at right; white arrowheads indicate position of future thumb, black arrowheads indicate skeletal abnormalities. The last two deletions show that much regulatory apparatus is located distal to GCR and Prox, and that only when the entire gene desert is deleted is expression of *hoxd13* in the autopod almost completely lost.

In the second of the two whole genome duplications believed to have occurred in the vertebrate stem, the *hoxd* and *hoxc* clusters were formed from a common ancestor. These clusters were ultimately sub-functionalized. In mammals for example, *hoxc13* is needed for development of hair follicles because it controls *foxn1*, the gene responsible for the nude mouse phenotype (Potter et al., 2011), and the *hoxc13* gene is not expressed in the autopod; on the other hand, *hoxd13* is, as we have seen, essential for autopod morphogenesis. This is wholly a difference due to the regulatory apparatus, not to the genes themselves. Thus in a recent experiment the whole *hoxc* cluster was reciprocally translocated into the chromosomal position of the *hoxd* cluster, and thus placed under control of the external regulatory system resident in the *hoxd* centromeric gene desert (Tschopp et al., 2011). *hoxc12* expression now occurs ectopically in the autopod and this suffices to rescue the phenotypic effects of absence of posterior *hoxd* expression, which include morphogenetic failure of autopod ossification (Fig. 7.9(D)). That is, the difference between *hoxc* and *hoxd* function lies to a large extent within their regulatory control systems. This strengthens the argument that autopod evolution depended in part on assembly of the external *hoxd* regulatory apparatus on the centromeric side, by excluding the possibility that alteration of the structure of the HoxD protein itself is somehow responsible for specificity of its autopod function.

The vertebrate-specific external control systems of the *hoxd* complex, located within the centromeric gene desert, which drives expression throughout the developing autopod, is of previously unsuspected complexity. It consists of multiple individual modular elements. In the autopod these modules physically interact in specific ways with each other, and with the *hoxd* genes, while these interactions do not occur in a control tissue, in this case brain (Montavon et al., 2011). At least some of these modular elements are conserved across vertebrates although certain regulatory sequences common to tetrapods are lacking in fish, as shown in Fig. 7.9(A1) (Schneider et al., 2011). What is most striking is that active regulatory elements are distributed across the whole gene desert, as demonstrated by progressive deletion experiments reproduced in Fig. 7.9(E) (Montavon et al., 2011). Note, however, that even in the deletion of the whole distal control region five digits are still formed (see also Fig. 7.9(D)), though they are not ossified and are malformed, each containing only two rather than three phalanges. The GRN operating downstream of the *hoxd* genes in the autopod is only part of the circuitry required to generate the five digits, a result that follows from the mechanism for digit specification reviewed in Chapter 4.

The multiplicity of the elements constituting the external *hoxd* autopod control system in vertebrates, no one of which is apparently individually essential, renders this a powerfully "evolvable" system (Montavon et al., 2011). As each tetrapod clade diversified, new control elements could be added or older ones changed, without danger of loss of overall functionality, and the more elements, the more opportunity for fine scale regulatory change. By extension, this could be true of the gene deserts with conserved noncoding sequence elements which surround all of the vertebrate *hox* gene clusters (Lee et al., 2006).

2.6   Precis: Processes of upper level GRN evolution

In this section we have consulted data which show or imply the particular mechanisms by which developmental GRNs were functionally altered in deep geological time, so as to result in evolution of the body plan, sometimes generating class- or order-specific characters. Most of these changes have remained fixed in their various clades ever since, though some, such as digit morphology, continue to evolve, within limits. It should be noted here that there is one class of upper level hierarchical change in GRNs which can occur rapidly and frequently: this is mutational change that simply destroys GRN circuitry by pulling the plug on a key regulatory gene function. The most clearly analyzed example is the repeated adaptive evolution of pelvic spine loss in lake bound stickleback fish, which has been proved to be due to deletion of a key enhancer in the *pitx1* gene (Shapiro et al., 2006; Chan et al., 2010). Reductive evolution, i.e., destructive change, generally occurs rapidly where it is of selective advantage (for review, Peter and Davidson, 2011). However, this type of process is scarcely a model for how circuitry was built to perform novel developmental functions in the evolutionary diversification of body plans.

We see in this chapter that the basis of the genomic processes by which large-scale body plan diversification has occurred is to a large extent regulatory co-option, at different levels of GRN structure. Co-option can be intercalary, where regulatory genes adopt new functions higher up within the same GRN. Thus co-option may result in the inverse process of adding regulatory functions above a certain developmental step that remains fixed because it is necessary to the continuity of the clade. By this means novel spatial regulatory state patterns are generated, leading to different developmental processes, and giving rise to different programs of morphogenesis. Co-option may also move whole patterning subcircuits to new locations in developmental GRNs, a major mechanism probably very often exploited in evolution. Such large-scale co-option is potentiated by the modular organization of developmental circuits. Another common mechanism of morphological evolution is alteration in signal ligand presentation causing the redeployment of large-scale GRN circuitry underlying morphogenetic programs such as bone formation to new spatial locations. Signal apparatus thus becomes a primary locus where major morphological change can be caused by relatively little *cis*-regulatory sequence alteration. That is, the regulatory apparatus of the signal ligand gene, which determines when and where it will be expressed, is a frequent target of evolutionary change. Often too, complex circuitry can be triggered by expression of one or two initial activators (Chapter 3), the regulatory systems of which may thereby act as potent evolutionary levers. The *hox* gene clusters constitute an especially widespread class of co-opters of developmental GRNs (except in the anterior head where these genes are not expressed at all). In evolution, *hox* genes affect body plan by deploying developmental GRNs to new locations. Finally, at least in vertebrates, the unusual structure of the external *hox* gene regulatory systems, which both account for and capitalize on the colinearity of their vectorial expression patterns, have permitted extensive regulatory evolution resulting in morphological novelty.

In the first section of this chapter we saw how rapid and continuous is *cis*-regulatory evolution when it occurs at the GRN periphery, where the only feedback penalties are environmental. In this section we see mechanisms by which changes in organization of upper level GRN structure have been occurring throughout the Phanerozoic (the period beginning with the Early Cambrian boundary at about 541 My ago and extending up to the present). These are large, not microscale changes, accounting for large changes in body plan. Such changes do not occur continuously; instead they took place and then were fixed for enormous geologic periods. We turn now to the problem of explaining stasis in developmental programs, as this is as much an aspect of the evolutionary properties of GRN structure as are the means of change.

3. GRN Stasis and Phylogeny

At the outset of this chapter the geological antiquity of phyla and classes was noted in relation to the hierarchy of developmental GRNs. Here we return to the point that definitive properties of phyla are installed in the developing organism by upper level circuitry operating relatively early in embryonic life; think for example of the embryological origins of the dorsal nerve chord in phylum Chordata or of metamerism in phylum Arthropoda. Similarly, characters distinguishing classes appear in mid-development, but still far up in the symbolic diagram of GRN hierarchy in Fig. 7.1(A), for instance specification of the type and placement of jointed appendages in arthropod classes, or the development of the ray-finned appendages of class Actinopterygii, which includes teleost fish. As pointed out above, a corollary of this relationship is that those aspects of GRN structure concerned with phylum-, class-, or order-specific structures have remained unchanged for as long as these clades have existed, or if now extinct for as long as they did exist. Thus the higher in the hierarchy of the developmental GRN, the older some circuitry to be found there may be. This is a more inclusive view than that represented by the cliché known as the "phylotypic stage", for it encompasses not just one level of shared characters and one stage of development, but rather characters built early in development that define all members of a phylum, characters built later that are shared by all members of a class, and the same for orders, and on down. The result, as summarized above (Fig. 7.1(B)), is that the conservation, change, or redeployment of particular developmental GRN circuitry at the phylum, class, and order level underlie and explain branching, hierarchical cladistics. This is a profound relation and it is worth taking a closer look, which can only be afforded by illumination available in the real-time fossil record, organized according to molecular phylogeny.

3.1   Quantification of stasis in body plan characters from the fossil record

In Fig.7.10(A) a convincing recent compilation of phylogenetic and fossil evidence is reproduced (Erwin et al., 2011; Erwin and Valentine, 2013). The phylogram is calibrated by fossil ages in real time marked by open circles. For instance, the paleontological date for the divergence of mammals from birds and reptiles, marked in Fig. 7.10(A) by an open circle (here at the divergence branch between *Anolis* and *Gallus*), is about 320 My ago in the Carboniferous (C). This affords a real-time calibration for the differences in sequence of orthologous proteins extracted from modern birds and mammals. Such calibrations are available for many branches of bilaterian evolution as the distribution of open circles in Fig. 7.10(A) shows. By using these rates for protein evolution across the Bilateria in the Phanerozoic, where we have good fossil records, the dates for the last common ancestors in each branch can be extrapolated. As the figure shows (see caption), most phyla, represented at the top by images and colored bars, had by the Early Cambrian achieved crown group characteristics, that is they possessed the total set of features defining each phylum (closed colored circles in Fig. 7.10(A)). But according to this extrapolated calibration, the last common ancestor of all Bilateria (black circle in Fig. 7.10(A)) lived much earlier, in the Cryogenian, before the last Snowball Earth episode; and the last common ancestor of Bilateria and their sister group Cnidaria lies even deeper. The main usefulness of this diagram in the present context is to date the real time Phanerozoic divergences within phyla in the best calibrated branches, e.g., in the vertebrate and the arthropod portions of the phylogeny. Thus we see that divergence between the tetrapod classes of vertebrates represented by *Xenopus*, *Anolis*, and *Gallus* occurred stepwise between 350 and 250 My ago, while the vertebrates as a whole originated in the Early Cambrian and the last common ancestor between the ray-finned fish (here *Danio* and *Tetraodon*) and the lobe-finned fish that gave rise to the tetrapods lived over 400 My ago. Even the two orders of ray-finned fish in this calibration are at least 150 My apart. In the Arthropods, divergences within class Insecta are hundreds of millions of years deep and the Arthropod classes had been established by the Cambrian and Ordovician. Enough details: the take home lesson is that throughout the calibrated phylogeny there is an almost unbelievably long lasting real-time stasis in those developmental programs directing earlier developmental processes. What is the explanation, given the rate at which *cis*-regulatory apparatus can and does change?

A finer look at the dynamics of intraphyletic evolutionary change and stasis can be seen in Fig. 7.10(B), which shows the divergence and subsequent history of some extinct and of all surviving Echinoderm classes (Bottjer et al., 2006). Here the change in rate of appearance of class-level evolutionary novelty through time is very dramatic, as supported by an excellent fossil record. New echinoderm classes arose throughout the Cambrian following divergence of echinoderms from their sister phylum, the hemichordates (at the node labeled "stereom", which refers to the unique endoskeletal biomineral made by all echinoderms). The five modern echinoderm classes arose in the Late Cambrian and Ordovician (490–440 My), and since then the body plan characters defining these echinoderm classes have remained static. Consistent with the freezing of developmental programs that this implies, is the lack of intermediate class- and order-level characters. As often noted, this is a general and deep-seated feature of animal phylogeny that is in direct contrast to the continuous variation seen in species-specific characters such as those discussed in the first section of this chapter.

3.2   GRN kernels and hierarchical patterns of body plan stasis

Kernels are phylogenetically conserved GRN circuits dedicated to specific developmental processes, such as for example setting up a spatial regulatory state defining the progenitor field for a given body part (Davidson and

Figure 7.10

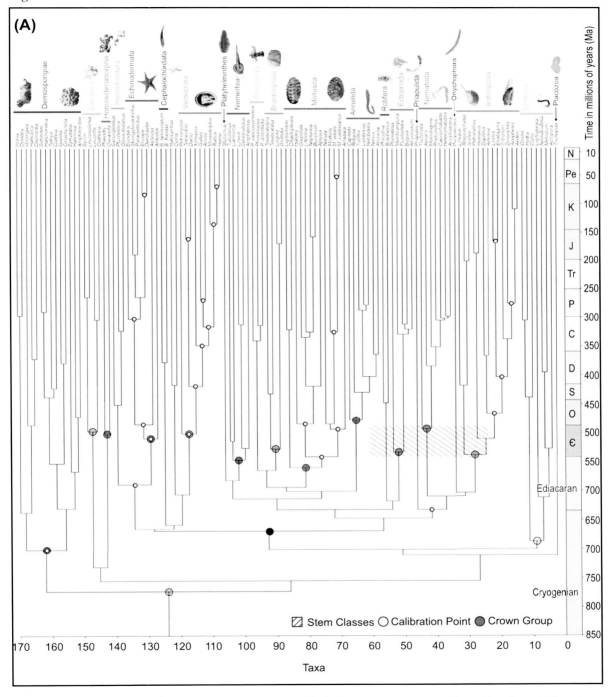

Figure 7.10 Stasis and diversification of bilaterian clades. (A), Origins and diversification of bilaterian phyla and classes, calibrated in real time against the fossil record *(from Erwin et al. (2011))*. Crown group phylum-level characters (solid circles, see key at bottom) had arisen for all phyla more than 500 My ago, and little diversification in upper level body plan characters has since occurred; cf. discussion in text.

Figure 7.10

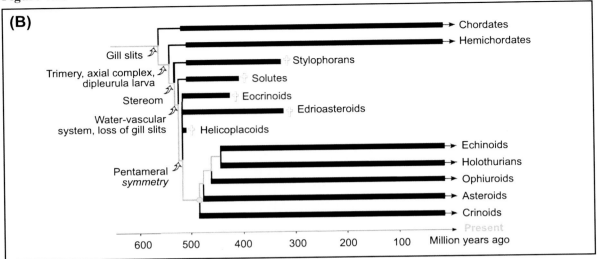

Phyletic crown group origins are color matched to the bars representing each phylum at top. Geologic periods are shown on a real timescale at right: Є, Cambrian; O, Ordovician; S, Silurian; D, Devonian; C, Carboniferous; P, Permian; Tr, Triassic; J, Jurassic; K, Cretaceous; Pe, Paleogene; N, Neogene. Divergence times were computed by calibration of sequence change in a common set of housekeeping genes against the paleontological record of evolutionary divergence of given lineages (calibration points, open circles). Using the molecular clock rate thus established beyond the Cambrian boundary, the Bilateria would have originated >150 My before this boundary in the Cryogenian ("Snowball Earth" Period), prior to the Marinoan Glaciation (solid black circle). The sister group to the Bilateria, the cnidarians, are shown to the right, the last common ancestor with the Bilateria occurring deeper in the Cryogenian; and divergence of the sponges is shown on the left of the Bilateria. (B), Discontinuous rates of evolution in echinoderms *(from Bottjer et al. (2006))*. Echinoderm evolution at the class level is shown for living and extinct (white crosses) forms; the five extant echinoderm classes are at bottom. The earliest Cambrian echinoderms (Stylophorans) possessed stereom, the definitive biomineral of echinoderms, but were not pentamerally symmetric, had no water vascular systems, and still retained gill slits. Pentameral symmetry appeared during the Cambrian (Eocrinoids and Edrioasteroids). All stem group forms became extinct by the Carboniferous period. The phylogeny is shown against a real timescale. In magenta at left the acquisition of definitive characters in the stem group forms can be seen to have occurred relatively rapidly compared to stasis of crown group forms ever since the Ordovician (yellow circle). Class-level crown groups appeared by the later Paleozoic.

Erwin, 2006). Not only are the orthologous identities of the regulatory and signaling gene components of the subcircuit conserved, but so also are at least major aspects of the wiring design of the circuit conserved within the clade where the kernel is found. The clade could be all Bilateria; for example there is good evidence for a kernel directing heart territory specification shared from *Drosophila* to mouse (Davidson and Erwin, 2006). Or it could be all members of a phylum, or all members of a class or all members of an order, barring secondary simplifications. In other words we can imagine that kernels are phylogenetically distributed as are the circuits indicated by the different colors in Fig. 7.1(B). The existence of kernels implies that indeed the deep time conservation of GRN circuitry is possible despite the rapidity of *cis*-regulatory change. Since GRN kernels control upstream or midlevel developmental processes, they could be causally responsible for the distribution of those body plan characters used to define phylogenetic clades. Kernel conservation can thus be regarded as one of the mechanisms that account for the paleontological record of cladistic stasis that we see in Fig. 7.10(A).

Kernel conservation

Even in the absence of experimental evidence of kernels, it would still be a reasonable prediction that phylogenetic hierarchy must ultimately depend on shared network circuitry encoding development of the nested sets of body plan characters that define clades at each level. But we do have increasing experimental indication of unusually highly conserved network circuitry that performs reproducible, phylogenetically distributed developmental tasks. Comparative observations reveal that modules of very conserved structure may be surrounded by entirely nonconserved circuitry, highlighting the significance of the conservation (for example, Hinman et al., 2003). Evidence, of various degrees of completeness, has been reported for kernels that contribute to many aspects of the body plan. We have discussed many of these kernels and putative kernels at the circuit level in chapters 4 and 5. These include kernels that establish the coordinates of vertebrate limb buds; kernels that set up the regulatory state axes of insect wings (Tomoyasu et al., 2009); that control hindbrain specification in vertebrates (Davidson, 2006); that control tooth bud development (Fraser et al., 2009); that control aspects of hematopoiesis (Pimanda et al., 2007); that control mesoderm development in vertebrates (Swiers et al., 2010); that control early medio-lateral patterning in the CNS, as well as anterior-posterior CNS regulatory states (Lowe et al., 2003; Tessmar-Raible et al., 2007; Steinmetz et al., 2010); that control endoderm specification in gnathostome vertebrates (Tseng et al., 2011). What we would like to know is how kernels remain conserved, and what they may have to do with overall stasis in developmental GRNs over geologic time periods.

The most obvious and basic explanation for conservation of dedicated kernel circuitry is implicit in the hierarchical structure of developmental GRNs. Given the roles in GRNs that kernels would have to play if they are to account for the presence of body parts definitive for major clades, they must act at the beginning of the processes in which these body parts are formed, high up in the hierarchy. This is indeed where many examples of which we have some knowledge function. The development of given body parts and of their major characteristics depends specifically on installation of progressively subdivided progenitor field regulatory states (Chapter 4; Fig. 7.1(A)). Collapse of the circuitry generating the progenitor field regulatory state would mean loss of the body part in development, usually a catastrophe. Once formed, the progenitor field is subdivided into regions that will give rise to the different future components of the body part. But since in development the regulatory state of each progressively defined subregion of the body part depends on inputs from the prior regulatory state, loss of even some of these inputs could mean downstream loss of that subregion, also a catastrophe. The circuits that set up the progenitor fields at the top of each developmental process are absolutely essential to each body plan, and they are constrained from below, so to speak, by the subdivision circuits which depend on their inputs. This is a general explanation for the extreme conservation of certain upper level developmental circuitry. The deployment of the progenitor field which determines where the body part will form is, however, a major source of variation of the body plan and as we have seen, this deployment depends on further upstream developmental switches and signaling, both of which are evolutionarily malleable. As the previous section of this chapter indicates, when it did happen, evolutionary divergence at these levels occurred by various mechanisms, be they in any given case intercalary evolution, co-option of whole regulatory state patterning subcircuits for new uses, duplication and subfunctionalization, or redeployment of morphogenetic circuits due to regulatory alteration in expression of *hox* genes. All these processes have in deep time resulted in generation of new upper level spatial regulatory state patterns, but once connected to preexistent circuitry for further spatial subdivision, and for effector gene expression, the circuitry that produced the novel upstream regulatory state will be locked in, resulting in evolutionary stasis.

Internal design features of developmental GRNs that lead to evolutionary stability

It could be predicted that GRNs have built-in stabilizing features in that so much of the structure of modern developmental GRNs seems designed to obviate any chance of system failure (Chapter 6). For example, a subcircuit design we have encountered in the best known large developmental GRN, the sea urchin endomesoderm GRN (Fig. 3.5(I)), is a "vertical feedback latch", a direct mechanism for ensuring constraint from below. In the skeletogenic portion of this GRN the initial regulatory state is set up by lineage-specific release of four regulatory genes from repression. By various routes these all contribute to activation of the next downstream circuit feature, which is a triple positive feedback loop composed of three other regulatory genes. Thence other regulatory genes are activated downstream, using as inputs the outputs of the feedback loop genes, as well as those of the initial four genes. But the output of the triple feedback circuit also goes back "upstairs" to provide a permanent positive input to the most important of the four initial genes. As might be predicted from this architecture, loss of expression of any of the three genes in the feedback subcircuit causes total loss of skeletogenesis downstream (Oliveri et al., 2008). In addition, all three regulatory genes within the skeletogenic positive feedback circuitry and the upstream regulatory gene they control are expressed in the skeletogenic centers of juvenile sea urchins and starfish as well. It is very possible that the positive feedback wiring between these regulatory genes is indeed a conserved feature of echinoderm skeletogenic GRNs used for the formation of the adult endoskeleton (Gao and Davidson, 2008). In general it could be said that since positive feedbacks are regulatory state stabilization devices, this kind of circuitry will act in evolution to increase gene interdependence within the circuit. In Chapter 6 we discussed the remarkable prevalence of positive feedback circuitry within developmental GRNs. The signature functional characteristics of these feedback subcircuits is that experimental loss of function of any of the genes involved causes collapse of the whole subcircuit. Considering the connections which typically link the feedback subcircuit to the downstream developmental program, we can see that loss of operation of the subcircuit may well result in the loss of the developmental program. This would operate as a powerful enforcer of design conservation, and thus preclude functional rewiring other than addition of new layers of positive feedbacks and occasional loss of no longer necessary linkages.

To summarize, we have concluded that the hierarchical phylogeny of Bilateria is an external manifestation of the structural hierarchy of developmental GRNs. Furthermore, the striking evolutionary stasis of phylum-, class-, and even order-level body plan characters can be predicted to depend on the extreme conservation of particular spatial regulatory state patterning circuitry, i.e., of network kernels. The enormous contrast between this stasis and the continuous active evolutionary change at the GRN periphery brings us back to where we started: evolution of the body plan has to be viewed through the lens of GRN hierarchy and its mechanistic significance. That is why consideration of evolution in nondevelopmental systems, or of evolution of species-specific characters encoded at the GRN periphery, cannot per se provide an explanation of body plan evolution.

Since cis-regulatory sequence variation will certainly occur randomly and continuously at all levels of the system, how could kernel linkages be conserved for tens or hundreds of millions of years; how could phylogenetically definitive regulatory state specification patterns survive such variation in every major evolutionary lineage? The general answer to this is closely related to a fundamental aspect of body plan development: its Boolean character. As we discuss at length earlier in this book (Chapters 2 and 6), regulatory states in embryonic space are essentially Boolean; a regulatory gene is or is not expressed in a given domain at a given time. Close analysis of transcription kinetics in embryos, furthermore, shows that for the good physical chemical reasons touched on in Chapter 2, the function of transcriptional gene cascades in developmental GRNs is generally insensitive to quantitative variation in cis-regulatory inputs, unlike processes of terminal development or physiological responses. In Chapter 6 we reviewed our recent study (Peter et al., 2012) which directly demonstrated the system-wide, predictive success of a Boolean automaton model of the sea urchin embryo GRN. While cis-regulatory mutations resulting in total loss of modules or essential sets of target sites may have catastrophic loss of function effects, in which case if homozygous the bearer might be dead, mutations leading to quantitative variation of gene output levels over a wide range will not in general alter the Boolean patterns of spatial regulatory gene expression. In addition to level insensitivity of embryonic gene cascades, many aspects of network circuit design contribute to the "Booleanization" of spatial regulatory circuit output, including the multiple roles of spatial repression circuitry as is discussed in detail in chapters 4–6. As long as the Boolean spatial expression output of the regulatory system upon which development essentially depends is maintained, multiple variations at the *cis*-regulatory level are tolerated. Examples such as that in Fig. 2.2(B) show that precise spatial function can persist in evolution despite massive sequence-level variations in *cis*-regulatory module structure. A type of GRN structure that has the specific effect of damping the significance of variations in the activity of *cis*-regulatory modules is again the ubiquitous positive feedback subcircuit. Once such subcircuits are activated they obviate the significance of variation in strength of all inputs upstream, as they mutually determine their own outputs and reset the activities of their downstream targets. Thus, in general, developmental control networks have evolved stability, and hence evolutionary stasis, by Booleanizing the effects of *cis*-regulatory variation in the upstream circuitry that deploys spatial regulatory state. Inverting this argument, the requirement for stasis and program stability is likely to have been the evolutionary force that led to the discontinuous or Boolean character of spatial regulatory gene expression patterns, and thereby to the developmental process per se.

4. *Trans*-Phyletic Conservation of Cell Type-Specific Regulatory States

Morphological character states (and also genomic sequence divergence states) are phylogenetically distributed in Bilateria, in the discontinuous hierarchical way that we have been discussing, but many differentiated cell types are not phylogenetically distributed and instead occur broadly across phylum and superphylum boundaries. Of course there are indeed some phylum-specific differentiated cell types, as for example the skeletogenic cell types that produce the calcite biomineral (stereom) specific to echinoderms. On the other hand, bilaterians of all clades share a variety of muscle cell types expressing differentiation gene batteries consisting of widely distributed contractile myosin isotypes and muscle actins, driven by orthologous sets of bHLH, Mef2, and Srf drivers. Bilaterians of all sorts also share several types of neuron, again expressing a set of orthologous transcription factors driving orthologous effector genes. The conserved linkages between these differentiation genes and their regulators across Bilateria suggest that these very same linkages were already encoded in the genomes of the common bilaterian ancestor. The monophyletic origin of bilaterian striated muscle, for example, is highlighted by evidence that striated muscle of the sister group of the bilaterians, the cnidarians (Fig. 7.10(A)) evolved independently from that shared by all Bilateria (Steinmetz et al., 2012). On the other hand both at the effector and regulatory gene levels cnidarian neurogenesis looks similar to bilaterian neurogenesis though it is deployed differently in development (Hayakawa et al., 2004; Layden et al., 2012; Nakanishi et al., 2012). Thus the terminal differentiation of cell types might indeed be very broadly conserved, even if the developmental process which leads to the differentiation of these cells and which locates them within the context of specific body parts might be entirely different from clade to clade.

Other cases of what may be pan-bilaterian differentiation gene batteries in which we know (some of) the regulatory drivers as well as specific effector genes have recently come to light. For example the deuterostome zebrafish and the lophotrochozoan polychaete annelid *Platynereis* share non-ocular neurosecretory forebrain cells that produce both Vasotocin-Neurophysin and Opsin (Tessmar-Raible et al., 2007). These cells similarly coexpress orthologous *nkx2.1*, *rx*, and *otp* homeodomain regulatory genes plus *mir-7* microRNA. In zebrafish it is known that *rx3* expression is specifically required for Vasotocin-Neurophysin synthesis. Chemosensory neurosecretory cells that generate RF-amide, also expressing *mir-7*, are found as well in the brains of these same two very distantly related bilaterian organisms, Platynereis and zebrafish. Other examples are ocular photoreceptor cell types which provide much evidence of pan-bilaterian conservation both at the effector gene and the regulatory driver gene level. There are two very widely distributed types of ocular photoreceptors, those in which the photoreceptor structures derive from cilia, including our rods and cones, and those in which they derive from microvilli, rhabdomeric receptors.

Yet all photoreceptors express various forms of the visual protein Opsin (Feuda et al., 2012), which on absorption of light causes changes in the state of ion channels via G-protein activation (for a recent comparative review of Opsin-based signal phototransduction, see Fain et al., 2010). While vertebrates utilize only ciliary-type photoreceptors and their Opsins, both types are found in at least some protostomes (Passamaneck et al., 2011). In addition to opsin, many other effector genes are widely expressed from flies to mammals in these various ocular photoreceptor cell types such as transducins, phosphodiesterases, arrestins, guanine-binding proteins, etc., (Ranade et al., 2008). What is most striking is the remarkably conserved dedication of a small set of regulatory genes to the direct transcriptional control of *opsin* and other photoreceptor effector genes in these photoreceptors (Corbo et al., 2010; Swaroop et al., 2010; Emerson and Cepko, 2011). These regulatory genes encode specific transcription factors: homeodomain factors of the K50 family, e.g., in mammals Otx2 at the commitment stage, and Crx in terminal differentiation; NeuroD1; specific proteins of the nuclear receptor family; and some additional factors that are expressed specifically in terminal rods or cones (Swaroop et al., 2010). Furthermore, rhabdomeric photoreceptors expressing *opsin* genes are also to be found in non-ocular locations, such as in light-sensitive cells of the *Platynereis* ventral nerve cord and in the appendages of this animal, but without *pax6* expression; and also in the tubefoot appendages of sea urchins, where *pax6* is expressed (Agca et al., 2011; Ullrich-Lüter et al., 2011). Thus there exists a conserved general light-sensing machine which is deployed in the various photoreceptor cells across Bilateria, utilizing similar regulatory and effector genes. These photoreceptors appear to descend from a primordial ancestral bilaterian cell type (Ranade et al., 2008).

All these pan-bilaterian, cell type-specific differentiation gene batteries, and many other less intensively studied cases that could be cited, tell us something about the last common bilaterian ancestor. Clearly these very specific, functionally dedicated regulatory and effector gene cassettes are shared by the disparate modern bilaterian clades due to descent from this common ancestor. Thus in their antiquity such cassettes predate the phylogenetic bilaterian morphological divergence that has been our subject thus far in this chapter. But this may only be the tip of the iceberg: most unexpectedly, given driver regulatory states have also been *trans*-phyletically conserved in cells of similar function, while the effector genes that carry out the function are not conserved. We now look briefly at two examples, integumentary wound healing, and adaptive immune response.

4.1   Repair of the utterly different integuments of mice and flies

Ecdysozoans such as *Drosophila* and *C. elegans* have cross-linked chitin-based integuments, while in us the external barrier (stratum corneum) is composed of entirely different components. It is built by linked terminal keratinocytes that are filled with keratin, have lost their nuclei, and as they rise toward the surface of the epidermis die and become covered with cornified membrane proteins, peptides, and lipids. Although completely diverse sets of effector genes build the integument in insects such as *Drosophila*, and in vertebrates such as mice or ourselves, from an abstract functional point of view they do similar jobs for the organism: they generate and maintain the organism's barrier against the outside world. Whether chitin or stratum corneum based, if the external integuments are breached by a wound, they must be repaired, which requires a complex sequence of functions, in each case beginning as soon as possible after the wound. The molecular nature of the wound repair apparatus required to heal breaches in the integument in mammals is of course not at all the same as that required in insects. Yet a truly amazing homology exists at the regulatory level: regulatory genes encoding proteins of the same unusual DNA recognition family respond to wounding and direct the expression of integument repair effector genes in *Drosophila* as in mouse (Mace et al., 2005; Ting et al., 2005). The *Drosophila* version of these genes is known as *grainyhead* (*grh*). There are several orthologous *grh* genes in mammals, of which *grh2* and *3* are relevant here; note also that *grh* plays several other completely different roles in *Drosophila*, including in early embryonic development.

The important point for our purposes is the evidence that in wound healing and epidermal barrier development, the orthologous regulatory genes directly activate completely distinct, though in the end functionally similar processes of integument repair. Figure 7.11(A1–3) illustrates the expression at a *Drosophila* wound site of effector genes encoding Dopa decarboxylase (*ddc*) and Tyrosine hydroxylase (*ple*), which are required to produce quinones that in turn work as the active agents in cross-linking cuticular chitin in wound repair (Pearson et al., 2009). *Cis*-regulatory studies show that activation of these and other target effector genes require direct inputs from Grh as well as from AP-1 factors (Fig. 7.11(A4,5); Pearson et al., 2009). According to microarray analyses, *Drosophila grh* is also required for activation of a large variety of effector genes that function in chitin metabolism as well as in wound healing and other related functions, such as triggering genes of the antibacterial innate immune system (Paré et al., 2012). Curiously, related genes also function in formation of the chitinous fungal cell wall, and the linkage between *grh* genes and chitin may be anciently descendant from the fungal/animal ancestor. In *Drosophila*, the specific wound repair functions of Grh are triggered by phosphorylation, mediated by local, extracellularly induced ERK activity (Fig. 7.11(B); Kim and McGinnis, 2011). In mice, where the requirement for *grh3* expression in wound healing is dramatic (Fig. 7.11(C)), a primary effector gene function needed to rebuild the integumentary barrier is executed by the protein cross-linking enzyme *Trans*-glutaminase. The gene encoding this enzyme is a direct target of Grh3 (Ting et al., 2005), while several protein constituents of intercellular apical epithelial junctions including Claudin and E-cadherin are directly activated by Grh2 (Fig. 7.11(D); Werth et al., 2010). Grh3 may also directly target a great many other effector genes required to build the stratum corneum (Yu et al., 2006).

In terms of evolutionary process it would seem in this case that the developmental GRNs produce regulatory states for rebuilding the integument after injury which from flies to mice have retained a striking

Figure 7.11

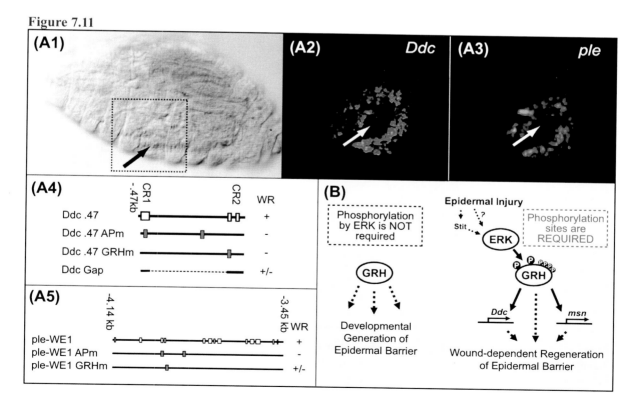

Figure 7.11

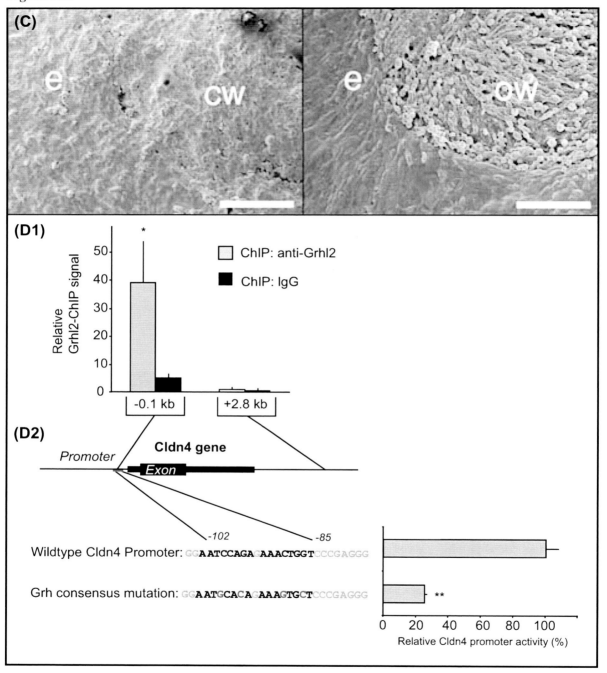

Figure 7.11 Pan-bilaterian regulatory signature in integumentary wound healing despite complete turnover in effector genes. (A), Regulatory control of integumentary wound healing in *Drosophila (from Pearson et al. (2009))*. (A1), DIC image of 17h *Drosophila* embryo showing region wounded by microneedle puncture (arrow). (A2), Transcriptional induction of *dopa decarboxylase* (*ddc*) gene

within 30 min of wounding, visualized by fluorescent in situ hybridization; area imaged is indicated in (A1). (A3), Transcriptional induction of *tyrosine hydroxylase* (*ple*) gene within 30 min of wounding, visualized by fluorescent in situ hybridization. (A4), Requirement for AP1 and Grainyhead (Grh) sites in *cis*-regulatory system of *ddc* gene. White boxes indicate sequences conserved in other drosophilid genomes, red boxes mutated sequences, and expression results (wound response, WR) are shown at right. (A5), Similar *cis*-regulatory experiment with *ple* gene. (B), Control of wound healing response by phosphorylation of Grh *(from Kim and McGinnis (2011))*. In normal development Grh activates genes required to build the chitinous integument without phosphorylation (left). The wound repair response, however, is stimulated by ERK, a kinase activated by an extracellular signal, which functions by phosphorylating specific sites on Grh; *msn* is another wound repair gene (also encoding a kinase) that like *ddc* is transcriptionally controlled by Grh. (C), Requirement for *grh3* in wound healing in mice *(from Ting et al. (2005))*. SEMs of hindlimb amputation sites in day-12 embryos are shown after 24 h; left, wild-type, right homozygous *grh3* mutant: e, epidermis; cw, closed wound in wild type; ow, open wound in *grh3* mutant. (D), Direct *cis*-regulatory control of mammalian epithelial differentiation by *grh2 (from Werth et al. (2010))*. (D1), Grh binding to proximal 5′ sequence of the *claudin4* gene, which encodes a protein of the epithelial tight junction, demonstrated by ChIP. (D2), Site-specific mutation experiment using a *claudin4* luciferase *cis*-regulatory construct, demonstrating requirement for Grh2 target site.

homology, centered on the unique *grh* regulatory genes. But downstream of this there has been a total switch of effector genes. Probably we are here seeing only one example of something that might have happened repeatedly, since animal integuments come in great variety and directly confront diverse environments. Because of this adaptive force, and the peripheral position of these effector genes in the GRN, the regulatory systems of these genes can be expected to alter rapidly. Thus it is fascinating that what has not altered ever since the bilaterian ancestor is what could be described as the genomic definition of "integument" in terms of regulatory state.

4.2 Cell type conservation and effector gene divergence in the adaptive immune systems of jawed and jawless vertebrates

Across bilaterian phylogeny immune effector genes are famously different from one another. Genome sequences reveal these gene sets to be among the most clade-specific of all classes of gene, and model systems from diverse branches of the phylogenetic tree, such as flies, snails, annelids, sea urchins, cephalochordates, and jawless and jawed vertebrates, have been found to utilize remarkably unique sets of immune effector proteins (Messier-Solek et al., 2010). Yet where it has been examined, the gene regulatory apparatus activating the immune effector responses seems much more conserved. For example, the innate immune response to infection in *Drosophila* features the synthesis of special antimicrobial peptides, but this reaction is mediated transcriptionally by NF-κB just as is our own very different innate immune response (Hoffmann, 2003). Similarly, sea urchins possess an innate immune system which depends on an effector gene repertoire of hundreds of specific innate immune receptors of several different classes (Hibino et al., 2006; Messier-Solek et al., 2010; Smith et al., 2010). Yet adult sea urchin coelomocytes and embryonic immune cells expressing various of these receptors utilize a set of regulatory genes very familiar to those who study mammalian hematopoietic cells (Hibino et al., 2006; Rast et al., 2006; Messier-Solek et al., 2010; Smith et al., 2010). These include *scl*, a *pu.1*-like gene, *e2a*, *runx*, *gata1/2/3*, *ikaros*, etc. The most striking comparison, however, is between the adaptive immune systems of jawed and jawless vertebrates, because these are in unique ways so similar in their functional components and processes, yet so different in the nature of their effector genes (Herrin and Cooper, 2010; Boehm et al., 2012).

The adaptive immune systems of gnathostomes such as ourselves, and of jawless fish or agnathans such as lampreys (Fig. 7.12), both deploy cells that generate a huge variety of antibodies which specifically bind respective antigens. In both, generation of a specifically reactive antibody stimulates proliferation and terminal differentiation of the cells producing the effective antibodies, followed by secretion of antibody (McCurley et al., 2012). In us the antibodies are immunoglobulins (Igs), but in lampreys and hagfish they are assembled from variable leucine-rich repeat sequences (VLRs). The functionally homologous antibody secreting cells of gnathostomes and agnathans are respectively known as B cells and VLRB$^+$ cells. Both gnathostomes and agnathans also generate another type of adaptive immune cell. These functionally homologous cells use specific antigen-reactive proteins which are retained on the cell surface and not secreted. In us these are the T cells and the recognition proteins are T cell receptors. In agnathans these specific receptors are also assembled from VLR family sequences, deriving from germ line VLR sequence clusters, which generate different sets of VLRA's and the cells are denoted VLRA$^+$ cells (Herrin and Cooper, 2010). The overall strategies of the agnathan and gnathostome adaptive immune responses are obviously very similar, despite the completely different receptor protein families constituting their adaptive immune systems, their different recombination molecular biology, and the anatomically different immune tissues in the gnathostome and agnathan body plans. The agnathan-specific manner in which germ line VLR gene sequences are copied by gene conversion and joined stepwise by homologous recombination to generate mature antigen-binding proteins is summarized in the diagrams of Fig. 7.12(D). Because there are hundreds of VRLA and VLRB sequence elements in each of the respective clusters, and because they can be copied and combined in a great many different orders it is calculated that the potential diversity of effector proteins is of about the same magnitude as the diversity available in mammalian T cell receptors and B cell antibodies (Boehm et al., 2012). In addition, a whole new set of reassembling VLR genes, the VLRC locus, representing a second T cell like lineage, has been discovered in lamprey genomes, adding to the potential complexity of their immune response (Das et al., 2013; Hirano et al., 2013).

Even though the proteins mediating epitope recognition are completely different, the functionally homologous cell types express similar sets of immune stimulatory and other effector genes. This remarkably illuminating evolutionary comparison is enriched by the details. Fig. 7.12(A) and (B) shows schematically the interactions of VLRA$^+$ and VLRB$^+$ immune cells. Note that there are so many molecular parallels between these and gnathostome T and B cells respectively that their common vertebrate origin cannot be doubted (McCurley et al., 2012): like T cells, VLRA$^+$ cells express CD45, as well as a chemokine receptor, CCR9, and the IL8 receptor; they also secrete IL17. VLRB$^+$ cells produce IL8 and mount the IL17 receptor, evidence for a possible functional interaction with VLRA$^+$ cells. VLRB$^+$ cells express several Toll receptors just as do B cells, as well as other proteins specific to various aspects of B cell function (Herrin and Cooper, 2010; Boehm et al., 2012). Lampreys have no thymus proper but instead they have tissue patches located in the gill tips (Fig. 7.12(C)) which express a regulatory gene also expressed in thymic epithelium in gnathostomes, *foxn1*, as well as *vlra* genes; the *delta* signal ligand gene is expressed as well in these patches (Bajoghli et al., 2011), just as it is in the thymic stroma in mice where it is required for T cell development (Rothenberg, 2011). Thus both mammalian T cells and VLRA-expressing cells of agnathans develop in unique environments. Most striking, the regulatory state of VLRA$^+$ cells includes the T cell transcription factors Gata2/3; Bcl11b, c-Rel, and an AhR (aryl hydrocarbon receptor) bHLH factor. But, despite the detailed functional similarities of these adaptive immune systems, as well as the orthologous use of many ancillary genes in agnathan and gnathostome adaptive immunity, the adaptive effector proteins are totally unrelated.

How can we imagine the evolutionary history of these adaptive immune systems? It would seem clear that the last vertebrate common ancestor of gnathostomes and agnathans possessed an adaptive immune system which utilized both a T cell-like and a B cell-like functionality, although it cannot be excluded that this is a character originating much deeper in the vertebrate stem lineage. Some evidence exists for

Figure 7.12

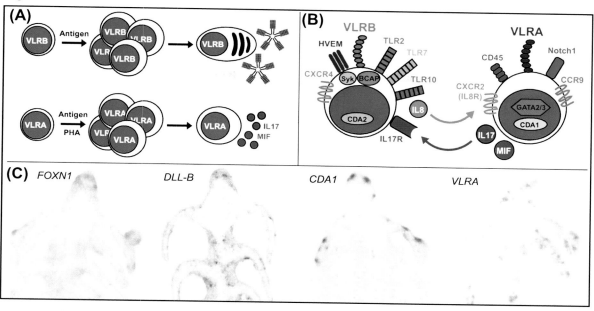

Figure 7.12 Effector genes of the agnathan adaptive immune system and its evolutionary relation to ours. (A), Diagrammatic summary of B cell like (VLRB$^+$) and T cell like (VLRA$^+$) lamprey adaptive immune effector cells, and their interaction *(from Herrin and Cooper (2010))*. Antigen stimulates clonal proliferation of VLRB$^+$ and VLRA$^+$ immune cells. Top, VLRB$^+$ cells differentiate into VLRB antibody secreting cells (pentagonal red structures); bottom, VLRA$^+$ cells differentiate into pro-inflammatory effectors which do not secrete their receptors but produce cytokines such as IL17. (B), Both cell types express many genes expressed by B cells and T cells respectively in gnathostomes (see text) as well as their VLR receptors. The expression of IL17 by VLRA$^+$ cells and of IL17 receptor by VLRB$^+$ cells suggests their functional interaction. (C), Molecular identification of lamprey "thymoids" at tips of gill filaments: in situ hybridization shows these cells coexpress the epithelial regulatory genes *foxn1* and *distalless-B* together with the lymphocyte-specific *vlra* genes and the *cytidine deaminase1* gene *(from Bajoghli et al. (2011))*. (D), Assembly and structure of VLR genes *(from Boehm et al. (2012))*. Multiple variable leucine-rich repeat (LRR) encoding genes (each internal module is 24 amino acids long) are located next to the germ line VLR gene: LRR NT, N-terminal LRR; LRR CT, C-terminal LRR; LRR1, first module after NT; CP, connecting peptide; LRR-Ve, last (3′) LRR. The mature VLR proteins (bottom) are assembled from either direction by gene conversion mediated by cytidine deaminases. Below is shown a comparison of the structures of germ line VLR genes in lamprey and hagfish. (E), A possible evolutionary scenario for the diversification of the effector gene systems of jawed and jawless vertebrates *(from Boehm et al. (2012))*. Here the vertebrate common ancestor possessed both VLR (ancestrally a GPI transmembrane-anchored receptor) and Ig-based B cell receptor and T cell receptor types of immune response genes. This ancestor also had a Cytidine Deaminase conversion system and functional division of B cell and T cell like lymphocytes, as well as much ancillary molecular biology shared today between the agnathan and gnathostome systems (as, e.g., in (B)). The RAG recombination system would have entered the gnathostome receptor diversification mechanism later, and both systems evolved further independently following separation; WGD, whole genome duplication.

Figure 7.12

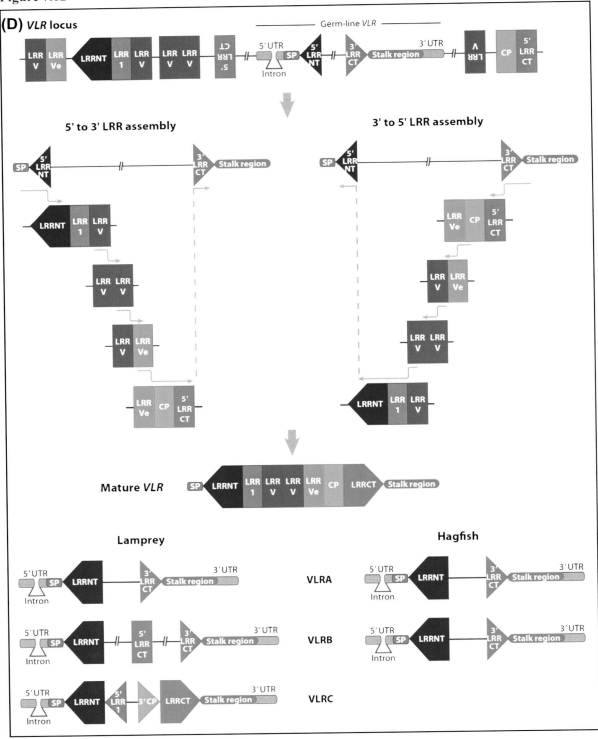

Figure 7.12

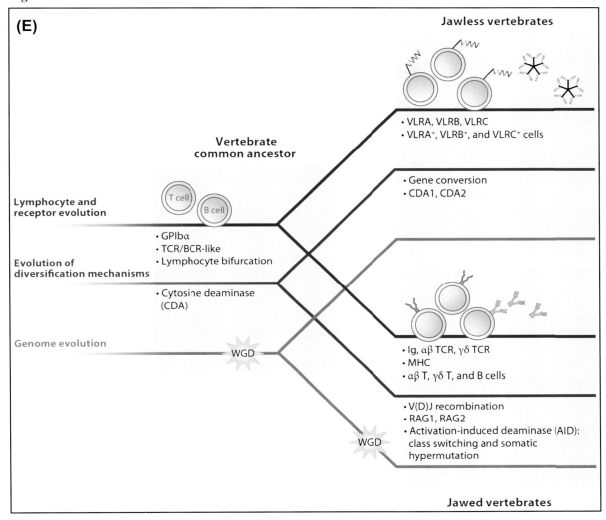

(E)

Jawless vertebrates

Vertebrate
common ancestor

• VLRA, VLRB, VLRC
• VLRA⁺, VLRB⁺, and VLRC⁺ cells

• Gene conversion
• CDA1, CDA2

Lymphocyte and
receptor evolution

T cell

B cell

• GPIbα
• TCR/BCR-like
• Lymphocyte bifurcation

Evolution of
diversification mechanisms

• Cytosine deaminase
(CDA)

Genome evolution

WGD

• Ig, αβ TCR, γδ TCR
• MHC
• αβ T, γδ T, and B cells

• V(D)J recombination
• RAG1, RAG2
• Activation-induced deaminase (AID):
class switching and somatic
hypermutation

WGD

Jawed vertebrates

the presence in agnathan genomes of Ig genes of the appropriate structure to have served as ancestral molecular stock for the diversification of the gnathostome Ig receptors (Pancer et al., 2004). Thus it may have been the case that both types of effector gene were ancestrally available, and that the two branches of the vertebrates capitalized on them alternatively; this speculation is shown in Fig. 7.12(E) (Boehm et al., 2012). Consistent with these observations, this model requires that the regulatory state powering the operation of the different descendant systems was assembled only once. Judging from the regulatory gene repertoire utilized in sea urchin innate immune cells, a similar regulatory state could have been associated with immune function at the dawn of the deuterostome radiation.

In both the wound healing and adaptive immune system examples, an anciently evolved regulatory state was dedicated to a given specific biological function in an ancestral form. While the regulatory state is conserved in the divergent descendants of this ancestor and applied to the same function, this function is now executed by entirely distinct sets of effector genes. Thus we come to the realization that bilaterian GRNs generate anciently evolved regulatory "packages" assigned to cell type-specific jobs. Fascinatingly, the visible external coat of effector functions executed by every portion of the developed organism, that which confronts the environment, may change ceaselessly, unlike some aspects of the underlying regulatory hierarchy. Like paleontological evidence, these features indicate deep time ancestral characters of bilaterian developmental GRNs.

5.   Bilaterian Evolution

This chapter is focused on the phylogenetic diversification of the Bilateria throughout the Phanerozoic. For crucial epochs of the Phanerozoic we have an extremely informative, well-interpreted fossil record, which is continuously being enhanced by new discoveries. By the Early Cambrian virtually all major bilaterian phyla are represented in the fossil record as are most extant classes. Paleontologically calibrated extrapolations from phylogenomic divergence estimates indicate the depth of origin of the bilaterian "superclades" to which these phyla belong. These superclades are the ecdysozoans (to which belong current arthropods, nematodes, rotifers), the lophotrochozoans (to which belong mollusks, annelids, brachiopods), and the deuterostomes (to which belong chordates, echinoderms, hemichordates). The bilaterian superclades all appear to have originated deep in the Ediacaran (630–541 My), while the last common bilaterian ancestor is of even deeper, Neoproterozoic vintage (Erwin et al., 2011). Thus if treated comprehensively, discussion of bilaterian evolution would have to encompass the pre-Phanerozoic origins of these superclades, and primordial bilaterian relations with their sister groups, indicated as well in Fig. 7.10(A). A detailed theoretical discussion of this speculative subject has been presented elsewhere, taking into account both the relatively sparse and often enigmatic Ediacaran fossil record, and what deductions seem possible on how bilaterian GRNs might have been assembled from less hierarchical precursors (Davidson and Erwin, 2009). Although there is paleontological evidence of various degrees of incompleteness for Ediacaran bilaterian forms, it is too discontinuous to provide the kinds of logical boundaries for specific questions of evolutionary change and stasis as does the Phanerozoic fossil record. Hence we have chosen to deal here with basic though accessible mechanistic issues, in the contexts of known phylogenetic relationships that emanate from at least approximate real time divergence points; that is, to focus on the Phanerozoic events of bilaterian body plan evolution. Our method in this chapter has been to combine insights from what has been learned about structure and function in developmental GRNs, with deductions from evolutionary comparisons that have been analyzed mechanistically, and are also well constrained in phylogenetic and/or paleontological terms. Perhaps it is now possible to erect a series of conceptual interpretations of the processes by which bilaterian body plans have evolved. For throughout the Phanerozoic, according to comparative observation, phylogeny, and paleontology, these body plans have indeed been evolving, at different rates, times, and levels of structure. Such interpretations, our proposed explanations, if they are to be in the least satisfactory, must specifically illuminate evolutionary processes at all levels of the body plan, and must also provide a framework for understanding the overall output of the evolutionary process, the cladistic structure of the Bilateria.

5.1   Interpretation of the hierarchical cladistic phylogeny of the Bilateria

Developmental GRNs are hierarchical, such that generation of regulatory states for the spatial coordinates of early pre- and postgastrular embryos is encoded at their top, followed by circuitry encoding spatial regulatory states that define the body parts and their subparts, and ultimately by circuitry defining cell types, followed at the periphery by control apparatus for terminal effector functions, differentiation and local morphogenesis. These GRN "layers" are causally related, since each provides the regulatory inputs for the succeeding layer. But the morphological aspects of the body plan that each layer developmentally programs, are those that in parallel succession give rise to the nested shared characters that define phylum, class, order. Thus the encoded (postgastrular) characters that block out the body plan are phylogenetically most widely shared, i.e., phylum-level characters; those particular to given body parts are in general class-level characters within each phylum. Hence GRN hierarchy gives rise to the branching pattern of shared characters seen in bilaterian phylogeny. This extends the Haeckelian concept of the developmental "phylotype" to the further levels of the "classotype" and the "ordertype", i.e., from one given "type" per phylum to nested character hierarchy at all cladistic levels.

5.2 Interpretation of evolutionary stasis of bilaterian phyla, class, and order character sets

The clear implication is that character sets retained by all members of a phylum or class are encoded by very highly conserved GRN circuitry, such that the conserved circuit elements are distributed phylogenetically. Since conservation means here lack of functional change in developmental process, this would account for phylogenetic stasis. As discussed above, some such highly conserved modular subcircuits, termed network "kernels", have been discovered and many more are likely to be discovered in the near future. Predicted correlations between the patterns of phylogenetic conservation and the roles of such kernels in development will afford the means of testing this and related interpretations.

5.3 Interpretation of the phenomenon of modular subcircuit output conservation

Explanations are required for the mechanisms underlying subcircuit conservation in light of the demonstrated lability of individual *cis*-regulatory control systems. We have addressed this issue at three levels: at the level of GRN hierarchy, at the level of subcircuit design, and at the level of *cis*-regulatory structure/function relations. In hierarchical GRNs the inputs responsible for installation of each temporal/spatial regulatory state depend on the prior network circuitry; i.e., the output of a given circuit provides the regulatory state animating the next circuits below it. Thus a given prior network circuitry specifically constrains the sequential circuitry ("canalization"), and so forth down the GRN hierarchy, generating a huge potential penalty for upstream circuit change. Additionally, the internal properties of a variety of developmental GRN subcircuits act to damp the effects of variation and to Booleanize spatial regulatory state output in development. The overall effect is to obviate gradual change in basic body plans due to fluctuation in *cis*-regulatory sequence within circuits, and thus to preserve spatial developmental regulatory output.

5.4 Interpretation of co-option, as the basic mechanism of evolutionary change in upper level regulatory circuitry

During the Phanerozoic, diversification events have occurred in each major bilaterian lineage that affect upper level characters of the body plan, generating new superclasses (such as the gnathostomes or tetrapods), classes, and orders, even given the mechanisms that normally impose stasis. We know of the existence of these events from the fossil record and from the branching topology of the phylogenetic tree. The explanation of how this type of change could happen at the DNA level begins with co-option of regulatory gene expression by *cis*-regulatory sequence change, such that the gene now responds to novel inputs causing it to be expressed in new spatial domains during development. This is the mechanism by which GRNs are restructured by adding new functional linkages. Three essential aspects of co-optive processes are: that the historical signature of such events is the appearance of novel *cis*-regulatory modules in given genes with respect to their pleisiomorphic regulatory apparatus; that co-optive gain of regulatory gene function events are likely to be haplosufficient in respect to function, increasing the likelihood of significant change; and that because of GRN hierarchy, a given co-option event can affect many downstream regulatory genes, resulting in co-optive redeployment of whole developmental regulatory circuits.

5.5   Interpretation of how upper level co-optive changes have occurred in developmental GRNs

The Phanerozoic fossil record proves that significant divergence in bilaterian body plans (e.g., fin to limb transition) has taken place in the midst of hierarchical GRNs which encoded the development of what were already complex earlier forms. It is important to realize this, in contrast to the concept that such changes can occur only at early evolutionary stages. We know this from the fossils of already complex stem group bilaterian forms, at the earliest Phanerozoic, forms which later gave rise to other different, but similarly complex forms, as in Cambrian echinoderms and arthropods. The examples in the foregoing provide several answers as to how changes can occur within upper level GRN structures without destructive consequences on all subsequent developmental process. Inverse evolutionary processes including intercalary evolution provide the means of preserving crucial downstream functions of regulatory genes while recruiting these or other regulatory genes into the upstream morphogenetic program. In the vertebrate lineage whole genome duplications, or in any lineage regulatory gene duplications, provide additional opportunities of utilizing preexisting circuitry to deploy new circuitry without catastrophic loss of function. Most powerful in their effects are redeployment of whole patterning subcircuits, preserving the original, but by new regulatory wiring, connecting such subcircuits to novel portions of the developmental GRN. A particularly flexible point is deployment of signaling apparatus, which because of the way developmental signals are typically wired into the regulatory GRN, can and has caused activation of developmental programs in additional novel locations. Many evolutionary changes involve loss as well as gain of morphogenetic function, but this is relatively easy to understand, as ruining circuit functionality is far simpler than building it.

5.6   Interpretation of correlations between evolutionary changes in *hox* gene expression and divergence in body plan

Numerous and illuminating loss of function *hox* gene mutations, and gain-of-function spatial alterations of *hox* gene expression in flies and mice have produced experimental effects on development of the body plan, some atavistic. Contrary to the usual cliché, *hox* genes do not per se build the structures of the developing organism, but they act like switches to deploy or prevent deployment of the GRN subcircuits that set up the regulatory states required for morphogenesis of given body parts. It is these subcircuits which actually do the work of building the structures of the organism. *hox* genes determine subcircuit function by providing crucial transcriptional inputs to these subcircuits, or transcriptional repression at crucial nodes of these subcircuits, or by mediating signals which act in turn to deploy these subcircuits. But while this provides an interpretation of the prominent homeotic functions of *hox* genes, many additional mechanistic aspects of *hox* gene function in evolution have come to light: (1) *hox* genes participate at every level of the GRN, including activation of many downstream morphogenetic and cell biology effector genes. (2) Their deep entwinement in the GRNs for specific body parts accounts for the persistent conservation within phyla of the requirement for given *hox* genes in the development of given body parts, though between phyla, the body parts are different as are the relations of *hox* genes with their respective developmental programs. (3) Since *hox* gene control systems organize their spatially vectorial expression with respect to one another, they also provide sequential positional transcriptional inputs within given structures. Their subcircuit deployment functions mean that their own *cis*-regulatory systems are effective loci of evolutionary change that alters developmental GRN output. (4) Though broadly pleiotropic, *hox* gene clusters are not used in all aspects of bilaterian body plan formation; e.g., they do not as a cluster contribute to transcriptional specification of pregastrular anterior/posterior or dorsal/ventral axes of the embryo, nor in later development to anterior head structures, for instance in vertebrate midbrain and forebrain. (5) The many phases of expression of each *hox* gene during development are mediated by respective *cis*-regulatory modules, meaning that evolutionary alteration in any particular one need not affect all the other functions of that *hox* gene, allowing gain or loss of switch functions in context of otherwise fully operative developmental GRNs. The evolutionary appearance of complex, multimodular, distributed, external control systems in gene deserts flanking vertebrate *hox* clusters adds even further possibilities for subtle upper level evolutionary change in GRN patterning circuit deployment.

5.7 Interpretation of speciation in terms of GRN structure

Cis-regulatory microevolutionary change occurs continuously in effector gene output, quantitatively, spatially, and temporally, generating selectively significant heritable differences between and within species. The pace of such change is extremely rapid, compared to the long periods of stasis in major aspects of phylogenetically definitive body plan characters. This is because intra- and interspecific evolutionary change takes place most easily in developmental functions encoded at or near the periphery of GRNs. Therefore such changes are constrained only by their selective consequences, not by any downstream effects on GRN function and not by feedback control except between organism and environment. Genes and gene batteries operating at the downstream GRN periphery are at the termini of the developmental control system in each portion of the organism. Species-specific selective negative changes that cause loss of function of patterning systems operating at higher levels of the GRN also occur very rapidly, because destruction is easily achieved. But we can observe the direct adaptive consequences of continuous regulatory microevolution only near the GRN periphery, that is, in development only in processes occurring relatively late in embryogenesis. However, since these processes affect deployment of effector genes, and since effector genes constitute over 90% of the protein coding genes in the genome, adaptive changes resulting in species-level divergence indeed occur in almost the whole of the coding genome and those portions of the regulatory genome which function in the control of effector genes.

5.8 Interpretation of nonphylogenetically distributed circuitry

Some differentiation gene batteries are pan-bilaterian or extend across whole bilaterian superclades. These, including their dedicated transcriptional regulatory drivers, are to be considered conserved characters inherited from the bilaterian common ancestor. Cell biological pathways which we share with yeast, for example, are much older. As reviewed above, we also must account for the unexpected *trans*-phyletic conservation of regulatory drivers dedicated to given effector functions, where the effector genes have, however, become phylogenetically distributed. This property again illuminates the contrast between the GRN periphery, where change is indeed endless and selective, and the regulatory apparatus per se. The effector functions required in animals may vary adaptively, while the regulatory drivers producing each qualitative set of functions, may be pleisiomorphic and not confined to given branches of animal phylogeny.

* * *

Thus we come to where we began this discussion. The genomically encoded GRNs that control development of the body plans of modern Bilateria, are the product of Phanerozoic evolution within their brachiating clades. We can understand the processes of development of the bilaterian body plan, or the processes of evolution of the bilaterian body plan, only in light of the properties of the encoded developmental GRNs. In both explanatory domains, the most fundamental general aspect of the structure of the bilaterian genomic control system is its deep causal hierarchy.

REFERENCES

Abbasi, A.A., Paparidis, Z., Malik, S., Bangs, F., Schmidt, A., Koch, S., Lopez-Rios, J., Grzeschik, K.H., 2010. Human intronic enhancers control distinct sub-domains of Gli3 expression during mouse CNS and limb development. BMC Dev. Biol. 10, 44.

Abzhanov, A., Kaufman, T.C., 1999. Homeotic genes and the arthropod head: expression patterns of the labial, proboscipedia, and deformed genes in crustaceans and insects. Proc. Natl. Acad. Sci. U.S.A. 96, 10224–10229.

Abzhanov, A., Kaufman, T.C., 2000. Crustacean (malacostracan) Hox genes and the evolution of the arthropod trunk. Development 127, 2239–2249.

Abzhanov, A., Kuo, W.P., Hartmann, C., Grant, B.R., Grant, P.R., Tabin, C.J., 2006. The calmodulin pathway and evolution of elongated beak morphology in Darwin's finches. Nature 442, 563–567.

Abzhanov, A., Popadic, A., Kaufman, T.C., 1999. Chelicerate Hox genes and the homology of arthropod segments. Evol. Dev. 1, 77–89.

Abzhanov, A., Protas, M., Grant, B.R., Grant, P.R., Tabin, C.J., 2004. Bmp4 and morphological variation of beaks in Darwin's finches. Science 305, 1462–1465.

Agca, C., Elhajj, M.C., Klein, W.H., Venuti, J.M., 2011. Neurosensory and neuromuscular organization in tube feet of the sea urchin strongylocentrotus purpuratus. J. Comp. Neurol. 519, 3566–3579.

Albertson, R.C., Streelman, J.T., Kocher, T.D., Yelick, P.C., 2005. Integration and evolution of the cichlid mandible: the molecular basis of alternate feeding strategies. Proc. Natl. Acad. Sci. U.S.A. 102, 16287–16292.

Anand, S., Wang, W.C., Powell, D.R., Bolanowski, S.A., Zhang, J., Ledje, C., Pawashe, A.B., Amemiya, C.T., Shashikant, C.S., 2003. Divergence of Hoxc8 early enhancer parallels diverged axial morphologies between mammals and fishes. Proc. Natl. Acad. Sci. U.S.A. 100, 15666–15669.

Anderson, A.M., Weasner, B.M., Weasner, B.P., Kumar, J.P., 2012. Dual transcriptional activities of SIX proteins define their roles in normal and ectopic eye development. Development 139, 991–1000.

Bajoghli, B., Guo, P., Aghaallaei, N., Hirano, M., Strohmeier, C., McCurley, N., Bockman, D.E., Schorpp, M., Cooper, M.D., Boehm, T., 2011. A thymus candidate in lampreys. Nature 470, 90–94.

Belting, H.G., Shashikant, C.S., Ruddle, F.H., 1998. Modification of expression and *cis*-regulation of Hoxc8 in the evolution of diverged axial morphology. Proc. Natl. Acad. Sci. U.S.A. 95, 2355–2360.

Bender, W., Akam, M., Karch, F., Beachy, P.A., Peifer, M., Spierer, P., Lewis, E.B., Hogness, D.S., 1983. Molecular genetics of the Bithorax complex in *Drosophila melanogaster*. Science 221, 23–29.

Bhatia, S., Monahan, J., Ravi, V., Gautier, P., Murdoch, E., Brenner, S., van Heyningen, V., Venkatesh, B., Kleinjan, D.A., 2014. A survey of ancient conserved non-coding elements in the PAX6 locus reveals a landscape of interdigitated *cis*-regulatory archipelagos. Dev. Biol. 387, 214–228.

Blanco, J., Pauli, T., Seimiya, M., Udolph, G., Gehring, W.J., 2010. Genetic interactions of eyes absent, twin of eyeless and orthodenticle regulate sine oculis expression during ocellar development in *Drosophila*. Dev. Biol. 344, 1088–1099.

Boehm, T., McCurley, N., Sutoh, Y., Schorpp, M., Kasahara, M., Cooper, M.D., 2012. VLR-based adaptive immunity. Annu Rev. Immunol. 30, 203–220.

Bottjer, D.J., Davidson, E.H., Peterson, K.J., Cameron, R.A., 2006. Paleogenomics of echinoderms. Science 314, 956–960.

Boyko, A.R., Quignon, P., Li, L., Schoenebeck, J.J., Degenhardt, J.D., Lohmueller, K.E., Zhao, K., Brisbin, A., Parker, H.G., vonHoldt, B.M., Cargill, M., Auton, A., Reynolds, A., Elkahloun, A.G., Castelhano, M., Mosher, D.S., Sutter, N.B., Johnson, G.S., Novembre, J., Hubisz, M.J., Siepel, A., Wayne, R.K., Bustamante, C.D., Ostrander, E.A., 2010. A simple genetic architecture underlies morphological variation in dogs. PLoS Biol. 8, e1000451.

Brunetti, C.R., Selegue, J.E., Monteiro, A., French, V., Brakefield, P.M., Carroll, S.B., 2001. The generation and diversification of butterfly eyespot color patterns. Curr. Biol. 11, 1578–1585.

Cerny, R., Cattell, M., Sauka-Spengler, T., Bronner-Fraser, M., Yu, F., Medeiros, D.M., 2010. Evidence for the prepattern/cooption model of vertebrate jaw evolution. Proc. Natl. Acad. Sci. U.S.A. 107, 17262–17267.

Chan, Y.F., Marks, M.E., Jones, F.C., Villarreal Jr., G., Shapiro, M.D., Brady, S.D., Southwick, A.M., Absher, D.M., Grimwood, J., Schmutz, J., Myers, R.M., Petrov, D., Jonsson, B., Schluter, D., Bell, M.A., Kingsley, D.M., 2010. Adaptive evolution of pelvic reduction in sticklebacks by recurrent deletion of a Pitx1 enhancer. Science 327, 302–305.

Chanut-Delalande, H., Fernandes, I., Roch, F., Payre, F., Plaza, S., 2006. Shavenbaby couples patterning to epidermal cell shape control. PLoS Biol. 4, e290.

Choe, C.P., Brown, S.J., 2009. Genetic regulation of engrailed and wingless in Tribolium segmentation and the evolution of pair-rule segmentation. Dev. Biol. 325, 482–491.

Cooley, A.M., Shefner, L., McLaughlin, W.N., Stewart, E.E., Wittkopp, P.J., 2012. The ontogeny of color: developmental origins of divergent pigmentation in *Drosophila americana* and *D. novamexicana*. Evol. Dev. 14, 317–325.

Corbo, J.C., Lawrence, K.A., Karlstetter, M., Myers, C.A., Abdelaziz, M., Dirkes, W., Weigelt, K., Seifert, M., Benes, V., Fritsche, L.G., Weber, B.H., Langmann, T., 2010. CRX ChIP-seq reveals the *cis*-regulatory architecture of mouse photoreceptors. Genome Res. 20, 1512–1525.

Coyne, J.A., Hoekstra, H.E., 2007. Evolution of protein expression: new genes for a new diet. Curr. Biol. 17, R1014–R1016.

Damen, W.G., 2007. Evolutionary conservation and divergence of the segmentation process in arthropods. Dev. Dyn. 236, 1379–1391.

Das, S., Hirano, M., Aghaallaei, N., Bajoghli, B., Boehm, T., Cooper, M.D., 2013. Organization of lamprey variable lymphocyte receptor C locus and repertoire development. Proc. Natl. Acad. Sci. U.S.A. 110, 6043–6048.

Datta, R.R., Cruickshank, T., Kumar, J.P., 2011. Differential selection within the *Drosophila* retinal determination network and evidence for functional divergence between paralog pairs. Evol. Dev. 13, 58–71.

Davidson, E.H., 2006. The Regulatory Genome. Gene Regulatory Networks in Development and Evolution. Academic Press/Elsevier, San Diego, CA.

Davidson, E.H., Erwin, D.H., 2006. Gene regulatory networks and the evolution of animal body plans. Science 311, 796–800.

Davidson, E.H., Erwin, D.H., 2009. An integrated view of precambrian eumetazoan evolution. Cold Spring Harb. Symp. Quant. Biol. 74, 65–80.

Davidson, E.H., Erwin, D.H., 2010. Evolutionary innovation and stability in animal gene networks. J. Exp. Zool. B. Mol. Dev. Evol. 314, 182–186.

Donaldson, I.J., Amin, S., Hensman, J.J., Kutejova, E., Rattray, M., Lawrence, N., Hayes, A., Ward, C.M., Bobola, N., 2012. Genome-wide occupancy links Hoxa2 to Wnt-β-catenin signaling in mouse embryonic development. Nucleic Acids Res. 40, 3990–4001.

Emera, D., Casola, C., Lynch, V.J., Wildman, D.E., Agnew, D., Wagner, G.P., 2012. Convergent evolution of endometrial prolactin expression in primates, mice, and elephants through the independent recruitment of transposable elements. Mol. Biol. Evol. 29, 239–247.

Emera, D., Wagner, G.P., 2012. Transformation of a transposon into a derived prolactin promoter with function during human pregnancy. Proc. Natl. Acad. Sci. U.S.A. 109, 11246–11251.

Emerson, M.M., Cepko, C.L., 2011. Identification of a retina-specific Otx2 enhancer element active in immature developing photoreceptors. Dev. Biol. 360, 241–255.

Erwin, D.H., Davidson, E.H., 2009. The evolution of hierarchical gene regulatory networks. Nat. Rev. Genet. 10, 141–148.

Erwin, D.H., Laflamme, M., Tweedt, S.M., Sperling, E.A., Pisani, D., Peterson, K.J., 2011. The Cambrian conundrum: early divergence and later ecological success in the early history of animals. Science 334, 1091–1097.

Erwin, D.H., Valentine, J.W., 2013. The Cambrian Explosion. Roberts and Company, Greenwood Village, Colorado.

Fain, G.L., Hardie, R., Laughlin, S.B., 2010. Phototransduction and the evolution of photoreceptors. Curr. Biol. 20, R114–R124.

Ferguson, L.C., Maroja, L., Jiggins, C.D., 2011. Convergent, modular expression of ebony and tan in the mimetic wing patterns of Heliconius butterflies. Dev. Genes. Evol. 221, 297–308.

Feuda, R., Hamilton, S.C., McInerney, J.O., Pisani, D., 2012. Metazoan opsin evolution reveals a simple route to animal vision. Proc. Natl. Acad. Sci. U.S.A. 109, 18868–18872.

Frankel, N., Erezyilmaz, D.F., McGregor, A.P., Wang, S., Payre, F., Stern, D.L., 2011. Morphological evolution caused by many subtle-effect substitutions in regulatory DNA. Nature 474, 598–603.

Fraser, G.J., Hulsey, C.D., Bloomquist, R.F., Uyesugi, K., Manley, N.R., Streelman, J.T., 2009. An ancient gene network is co-opted for teeth on old and new jaws. PLoS Biol. 7, e31.

Friedrich, M., 2006. Ancient mechanisms of visual sense organ development based on comparison of the gene networks controlling larval eye, ocellus, and compound eye specification in *Drosophila*. Arthropod. Struct. Dev. 35, 357–378.

Galant, R., Walsh, C.M., Carroll, S.B., 2002. Hox repression of a target gene: extradenticle-independent, additive action through multiple monomer binding sites. Development 129, 3115–3126.

Gao, F., Davidson, E.H., 2008. Transfer of a large gene regulatory apparatus to a new developmental address in echinoid evolution. Proc. Natl. Acad. Sci. U.S.A. 105, 6091–6096.

Garcia-Fernández, J., Holland, P.W., 1994. Archetypal organization of the amphioxus Hox gene cluster. Nature 370, 563–566.

García-Solache, M., Jaeger, J., Akam, M., 2010. A systematic analysis of the gap gene system in the moth midge *Clogmia albipunctata*. Dev. Biol. 344, 306–318.

Gehring, W.J., 2012. The animal body plan, the prototypic body segment, and eye evolution. Evol. Dev. 14, 34–46.

Gehring, W.J., Ikeo, K., 1999. Pax 6: mastering eye morphogenesis and eye evolution. Trends Genet. 15, 371–377.

Gogvadze, E., Buzdin, A., 2009. Retroelements and their impact on genome evolution and functioning. Cell. Mol. Life Sci. 66, 3727–3742.

Gompel, N., Prud'homme, B., Wittkopp, P.J., Kassner, V.A., Carroll, S.B., 2005. Chance caught on the wing: *cis*-regulatory evolution and the origin of pigment patterns in *Drosophila*. Nature 433, 481–487.

Gordon, C.T., Brinas, I.M., Rodda, F.A., Bendall, A.J., Farlie, P.G., 2010. Role of Dlx genes in craniofacial morphogenesis: dlx2 influences skeletal patterning by inducing ectomesenchymal aggregation in ovo. Evol. Dev. 12, 459–473.

Grant, P., 1999. The Ecology and Evolution of Darwin's Finches. Princeton University Press, Pricneton, N.J.

Grant, P.R., Grant, B.R., 2009. The secondary contact phase of allopatric speciation in Darwin's finches. Proc. Natl. Acad. Sci. U.S.A. 106, 20141–20148.

Graziussi, D.F., Suga, H., Schmid, V., Gehring, W.J., 2012. The "eyes absent" (eya) gene in the eye-bearing hydrozoan jellyfish *Cladonema radiatum*: conservation of the retinal determination network. J. Exp. Zool. B Mol. Dev. Evol. 318, 257–267.

Hare, E.E., Peterson, B.K., Eisen, M.B., 2008. A careful look at binding site reorganization in the even-skipped enhancers of *Drosophila* and sepsids. PLoS Genet. 4, e1000268.

Hayakawa, E., Fujisawa, C., Fujisawa, T., 2004. Involvement of Hydra achaete-scute gene CnASH in the differentiation pathway of sensory neurons in the tentacles. Dev. Genes. Evol. 214, 486–492.

Herrin, B.R., Cooper, M.D., 2010. Alternative adaptive immunity in jawless vertebrates. J. Immunol. 185, 1367–1374.

Hersh, B.M., Carroll, S.B., 2005. Direct regulation of knot gene expression by Ultrabithorax and the evolution of *cis*-regulatory elements in *Drosophila*. Development 132, 1567–1577.

Hibino, T., Loza-Coll, M., Messier, C., Majeske, A.J., Cohen, A.H., Terwilliger, D.P., Buckley, K.M., Brockton, V., Nair, S.V., Berney, K., Fugmann, S.D., Anderson, M.K., Pancer, Z., Cameron, R.A., Smith, L.C., Rast, J.P., 2006. The immune gene repertoire encoded in the purple sea urchin genome. Dev. Biol. 300, 349–365.

Hinman, V.F., Nguyen, A., Davidson, E.H., 2007. Caught in the evolutionary act: precise *cis*-regulatory basis of difference in the organization of gene networks of sea stars and sea urchins. Dev. Biol. 312, 584–595.

Hinman, V.F., Nguyen, A.T., Cameron, R.A., Davidson, E.H., 2003. Developmental gene regulatory network architecture across 500 million years of echinoderm evolution. Proc. Natl. Acad. Sci. U.S.A. 100, 13356–13361.

Hirano, M., Guo, P., McCurley, N., Schorpp, M., Das, S., Boehm, T., Cooper, M.D., 2013. Evolutionary implications of a third lymphocyte lineage in lampreys. Nature 501, 435–438.

Hoffmann, J.A., 2003. The immune response of *Drosophila*. Nature 426, 33–38.

Irvine, S.Q., Fonseca, V.C., Zompa, M.A., Antony, R., 2008. *Cis*-regulatory organization of the Pax6 gene in the ascidian *Ciona intestinalis*. Dev. Biol. 317, 649–659.

Kazazian, H.H., 2004. Mobile elements: drivers of genome evolution. Science 303, 1626–1632.

Kim, M., McGinnis, W., 2011. Phosphorylation of grainy head by ERK is essential for wound-dependent regeneration but not for development of an epidermal barrier. Proc. Natl. Acad. Sci. U.S.A. 108, 650–655.

Kleinjan, D.A., Bancewicz, R.M., Gautier, P., Dahm, R., Schonthaler, H.B., Damante, G., Seawright, A., Hever, A.M., Yeyati, P.L., van Heyningen, V., Coutinho, P., 2008. Subfunctionalization of duplicated zebrafish pax6 genes by *cis*-regulatory divergence. PLoS Genet. 4, e29.

Kleinjan, D.A., Seawright, A., Mella, S., Carr, C.B., Tyas, D.A., Simpson, T.I., Mason, J.O., Price, D.J., van Heyningen, V., 2006. Long-range downstream enhancers are essential for Pax6 expression. Dev. Biol. 299, 563–581.

Kumar, J.P., 2009. The molecular circuitry governing retinal determination. Biochim. Biophys. Acta 1789, 306–314.

Layden, M.J., Boekhout, M., Martindale, M.Q., 2012. *Nematostella vectensis* achaete-scute homolog NvashA regulates embryonic ectodermal neurogenesis and represents an ancient component of the metazoan neural specification pathway. Development 139, 1013–1022.

Lee, A.P., Koh, E.G., Tay, A., Brenner, S., Venkatesh, B., 2006. Highly conserved syntenic blocks at the vertebrate Hox loci and conserved regulatory elements within and outside Hox gene clusters. Proc. Natl. Acad. Sci. U.S.A. 103, 6994–6999.

Lei, H., Juan, A.H., Kim, M.S., Ruddle, F.H., 2006. Identification of a Hoxc8-regulated transcriptional network in mouse embryo fibroblast cells. Proc. Natl. Acad. Sci. U.S.A. 103, 10305–10309.

Lemons, D., Fritzenwanker, J.H., Gerhart, J., Lowe, C.J., McGinnis, W., 2010. Co-option of an anteroposterior head axis patterning system for proximodistal patterning of appendages in early bilaterian evolution. Dev. Biol. 344, 358–362.

Liu, P.Z., Kaufman, T.C., 2005. Even-skipped is not a pair-rule gene but has segmental and gap-like functions in *Oncopeltus fasciatus*, an intermediate germband insect. Development 132, 2081–2092.

Liubicich, D.M., Serano, J.M., Pavlopoulos, A., Kontarakis, Z., Protas, M.E., Kwan, E., Chatterjee, S., Tran, K.D., Averof, M., Patel, N.H., 2009. Knockdown of Parhyale Ultrabithorax recapitulates evolutionary changes in crustacean appendage morphology. Proc. Natl. Acad. Sci. U.S.A 106, 13892–13896.

Lo, P.C., Skeath, J.B., Gajewski, K., Schulz, R.A., Frasch, M., 2002. Homeotic genes autonomously specify the anteroposterior subdivision of the *Drosophila* dorsal vessel into aorta and heart. Dev. Biol. 251, 307–319.

Loehlin, D.W., Werren, J.H., 2012. Evolution of shape by multiple regulatory changes to a growth gene. Science 335, 943–947.

Lowe, C.J., Wu, M., Salic, A., Evans, L., Lander, E., Stange-Thomann, N., Gruber, C.E., Gerhart, J., Kirschner, M., 2003. Anteroposterior patterning in hemichordates and the origins of the chordate nervous system. Cell 113, 853–865.

Mace, K.A., Pearson, J.C., McGinnis, W., 2005. An epidermal barrier wound repair pathway in *Drosophila* is mediated by grainy head. Science 308, 381–385.

Mallarino, R., Grant, P.R., Grant, B.R., Herrel, A., Kuo, W.P., Abzhanov, A., 2011. Two developmental modules establish 3D beak-shape variation in Darwin's finches. Proc. Natl. Acad. Sci. U.S.A. 108, 4057–4062.

Mallo, M., Wellik, D.M., Deschamps, J., 2010. Hox genes and regional patterning of the vertebrate body plan. Dev. Biol. 344, 7–15.

Manceau, M., Domingues, V.S., Mallarino, R., Hoekstra, H.E., 2011. The developmental role of Agouti in color pattern evolution. Science 331, 1062–1065.

Mansfield, J.H., Abzhanov, A., 2010. Hox expression in the American alligator and evolution of archosaurian axial patterning. J. Exp. Zool. B Mol. Dev. Evol. 314, 629–644.

Martin, A., Papa, R., Nadeau, N.J., Hill, R.I., Counterman, B.A., Halder, G., Jiggins, C.D., Kronforst, M.R., Long, A.D., McMillan, W.O., Reed, R.D., 2012. Diversification of complex butterfly wing patterns by repeated regulatory evolution of a Wnt ligand. Proc. Natl. Acad. Sci. U.S.A. 109, 12632–12637.

McBride, D.J., Buckle, A., van Heyningen, V., Kleinjan, D.A., 2011. DNaseI hypersensitivity and ultraconservation reveal novel, interdependent long-range enhancers at the complex Pax6 cis-regulatory region. PLoS One 6, e28616.

McCurley, N., Hirano, M., Das, S., Cooper, M.D., 2012. Immune related genes underpin the evolution of adaptive immunity in jawless vertebrates. Curr. Genomics 13, 86–94.

McGregor, A.P., Orgogozo, V., Delon, I., Zanet, J., Srinivasan, D.G., Payre, F., Stern, D.L., 2007. Morphological evolution through multiple cis-regulatory mutations at a single gene. Nature 448, 587–590.

Medeiros, D.M., Crump, J.G., 2012. New perspectives on pharyngeal dorsoventral patterning in development and evolution of the vertebrate jaw. Dev. Biol. 371, 121–135.

Messier-Solek, C., Buckley, K.M., Rast, J.P., 2010. Highly diversified innate receptor systems and new forms of animal immunity. Semin. Immunol. 22, 39–47.

Miller, C.T., Beleza, S., Pollen, A.A., Schluter, D., Kittles, R.A., Shriver, M.D., Kingsley, D.M., 2007. cis-Regulatory changes in Kit ligand expression and parallel evolution of pigmentation in sticklebacks and humans. Cell 131, 1179–1189.

Minemura, K., Yamaguchi, M., Minokawa, T., 2009. Evolutionary modification of T-brain (tbr) expression patterns in sand dollar. Gene Expr. Patterns 9, 468–474.

Minguillón, C., Gardenyes, J., Serra, E., Castro, L.F., Hill-Force, A., Holland, P.W., Amemiya, C.T., Garcia-Fernàndez, J., 2005. No more than 14: the end of the amphioxus Hox cluster. Int. J. Biol. Sci. 1, 19–23.

Montavon, T., Soshnikova, N., Mascrez, B., Joye, E., Thevenet, L., Splinter, E., de Laat, W., Spitz, F., Duboule, D., 2011. A regulatory archipelago controls Hox genes transcription in digits. Cell 147, 1132–1145.

Monteiro, A., 2012. Gene regulatory networks reused to build novel traits: co-option of an eye-related gene regulatory network in eye-like organs and red wing patches on insect wings is suggested by optix expression. Bioessays 34, 181–186.

Nakanishi, N., Renfer, E., Technau, U., Rentzsch, F., 2012. Nervous systems of the sea anemone Nematostella vectensis are generated by ectoderm and endoderm and shaped by distinct mechanisms. Development 139, 347–357.

Nyakatura, K., Bininda-Emonds, O.R., 2012. Updating the evolutionary history of Carnivora (Mammalia): a new species-level supertree complete with divergence time estimates. BMC Biol. 10, 12.

O'Quin, K.E., Smith, D., Naseer, Z., Schulte, J., Engel, S.D., Loh, Y.H., Streelman, J.T., Boore, J.L., Carleton, K.L., 2011. Divergence in cis-regulatory sequences surrounding the opsin gene arrays of African cichlid fishes. BMC Evol. Biol. 11, 120.

Oliveri, P., Tu, Q., Davidson, E.H., 2008. Global regulatory logic for specification of an embryonic cell lineage. Proc. Natl. Acad. Sci. U.S.A 105, 5955–5962.

Ordway, A.J., Hancuch, K.N., Johnson, W., Wiliams, T.M., Rebeiz, M., 2014. The expansion of body coloration involves coordinated evolution in cis and trans within the pigmentation regulatory network of Drosophila prostipennis. Dev. Biol. 392, 431–440.

Otręba, M., Rok, J., Buszman, E., Wrześniok, D., 2012. Regulation of melanogenesis: the role of cAMP and MITF. Postepy Hig. Med. Dosw (Online) 66, 33–40.

Pancer, Z., Mayer, W.E., Klein, J., Cooper, M.D., 2004. Prototypic T cell receptor and CD4-like coreceptor are expressed by lymphocytes in the agnathan sea lamprey. Proc. Natl. Acad. Sci. U.S.A. 101, 13273–13278.

Pappu, K.S., Ostrin, E.J., Middlebrooks, B.W., Sili, B.T., Chen, R., Atkins, M.R., Gibbs, R., Mardon, G., 2005. Dual regulation and redundant function of two eye-specific enhancers of the Drosophila retinal determination gene dachshund. Development 132, 2895–2905.

Parker, H.G., VonHoldt, B.M., Quignon, P., Margulies, E.H., Shao, S., Mosher, D.S., Spady, T.C., Elkahloun, A., Cargill, M., Jones, P.G., Maslen, C.L., Acland, G.M., Sutter, N.B., Kuroki, K., Bustamante, C.D., Wayne, R.K., Ostrander, E.A., 2009. An expressed fgf4 retrogene is associated with breed-defining chondrodysplasia in domestic dogs. Science 325, 995–998.

Paré, A., Kim, M., Juarez, M.T., Brody, S., McGinnis, W., 2012. The functions of grainy head-like proteins in animals and fungi and the evolution of apical extracellular barriers. PLoS One 7, e36254.

Passamaneck, Y.J., Furchheim, N., Hejnol, A., Martindale, M.Q., Lüter, C., 2011. Ciliary photoreceptors in the cerebral eyes of a protostome larva. Evodevo 2, 6.

Pavlopoulos, A., Akam, M., 2011. Hox gene Ultrabithorax regulates distinct sets of target genes at successive stages of *Drosophila haltere* morphogenesis. Proc. Natl. Acad. Sci. U.S.A. 108, 2855–2860.

Pavlopoulos, A., Kontarakis, Z., Liubicich, D.M., Serano, J.M., Akam, M., Patel, N.H., Averof, M., 2009. Probing the evolution of appendage specialization by Hox gene misexpression in an emerging model crustacean. Proc. Natl. Acad. Sci. U.S.A 106, 13897–13902.

Pearson, J.C., Juarez, M.T., Kim, M., Drivenes, O., McGinnis, W., 2009. Multiple transcription factor codes activate epidermal wound-response genes in *Drosophila*. Proc. Natl. Acad. Sci. U.S.A. 106, 2224–2229.

Peel, A.D., Chipman, A.D., Akam, M., 2005. Arthropod segmentation: beyond the *Drosophila* paradigm. Nat. Rev. Genet. 6, 905–916.

Peter, I.S., Davidson, E.H., 2011. Evolution of gene regulatory networks controlling body plan development. Cell 144, 970–985.

Peter, I.S., Faure, E., Davidson, E.H., 2012. Feature Article: predictive computation of genomic logic processing functions in embryonic development. Proc. Natl. Acad. Sci. U.S.A. 109, 16434–16442.

Pimanda, J.E., Ottersbach, K., Knezevic, K., Kinston, S., Chan, W.Y., Wilson, N.K., Landry, J.R., Wood, A.D., Kolb-Kokocinski, A., Green, A.R., Tannahill, D., Lacaud, G., Kouskoff, V., Gottgens, B., 2007. Gata2, Fli1, and Scl form a recursively wired gene-regulatory circuit during early hematopoietic development. Proc. Natl. Acad. Sci. U.S.A 104, 17692–17697.

Potter, C.S., Pruett, N.D., Kern, M.J., Baybo, M.A., Godwin, A.R., Potter, K.A., Peterson, R.L., Sundberg, J.P., Awgulewitsch, A., 2011. The nude mutant gene Foxn1 is a HOXC13 regulatory target during hair follicle and nail differentiation. J. Invest. Dermatol. 131, 828–837.

Prud'homme, B., Gompel, N., Carroll, S.B., 2007. Emerging principles of regulatory evolution. Proc. Natl. Acad. Sci. U.S.A. 104 (Suppl. 1)), 8605–8612.

Ranade, S.S., Yang-Zhou, D., Kong, S.W., McDonald, E.C., Cook, T.A., Pignoni, F., 2008. Analysis of the Otd-dependent transcriptome supports the evolutionary conservation of CRX/OTX/OTD functions in flies and vertebrates. Dev. Biol. 315, 521–534.

Rast, J.P., Smith, L.C., Loza-Coll, M., Hibino, T., Litman, G.W., 2006. Genomic insights into the immune system of the sea urchin. Science 314, 952–956.

Rebeiz, M., Jikomes, N., Kassner, V.A., Carroll, S.B., 2011. Evolutionary origin of a novel gene expression pattern through co-option of the latent activities of existing regulatory sequences. Proc. Natl. Acad. Sci. U.S.A. 108, 10036–10043.

Rebeiz, M., Pool, J.E., Kassner, V.A., Aquadro, C.F., Carroll, S.B., 2009. Stepwise modification of a modular enhancer underlies adaptation in a *Drosophila* population. Science 326, 1663–1667.

Reed, R.D., Papa, R., Martin, A., Hines, H.M., Counterman, B.A., Pardo-Diaz, C., Jiggins, C.D., Chamberlain, N.L., Kronforst, M.R., Chen, R., Halder, G., Nijhout, H.F., McMillan, W.O., 2011. Optix drives the repeated convergent evolution of butterfly wing pattern mimicry. Science 333, 1137–1141.

Rothenberg, E.V., 2011. T cell lineage commitment: identity and renunciation. J. Immunol. 186, 6649–6655.

Ruvkun, G., Wightman, B., Burglin, T., Arasu, P., 1991. Dominant gain-of-function mutations that lead to misregulation of the *C. elegans* heterochronic gene lin-14, and the evolutionary implications of dominant mutations in pattern-formation genes. Dev. Suppl. 1, 47–54.

Schneider, I., Aneas, I., Gehrke, A.R., Dahn, R.D., Nobrega, M.A., Shubin, N.H., 2011. Appendage expression driven by the Hoxd Global Control Region is an ancient gnathostome feature. Proc. Natl. Acad. Sci. U.S.A. 108, 12782–12786.

Shapiro, M.D., Bell, M.A., Kingsley, D.M., 2006. Parallel genetic origins of pelvic reduction in vertebrates. Proc. Natl. Acad. Sci. U.S.A. 103, 13753–13758.

Shibata, Y., Sheffield, N.C., Fedrigo, O., Babbitt, C.C., Wortham, M., Tewari, A.K., London, D., Song, L., Lee, B.K., Iyer, V.R., Parker, S.C., Margulies, E.H., Wray, G.A., Furey, T.S., Crawford, G.E., 2012. Extensive evolutionary changes in regulatory element activity during human origins are associated with altered gene expression and positive selection. PLoS Genet. 8, e1002789.

Shubin, N., Tabin, C., Carroll, S., 1997. Fossils, genes and the evolution of animal limbs. Nature 388, 639–648.

Shubin, N.H., Daeschler, E.B., Jenkins, F.A., 2006. The pectoral fin of *Tiktaalik roseae* and the origin of the tetrapod limb. Nature 440, 764–771.

Slattery, M., Ma, L., Négre, N., White, K.P., Mann, R.S., 2011. Genome-wide tissue-specific occupancy of the Hox protein Ultrabithorax and Hox cofactor Homothorax in *Drosophila*. PLoS One 6, e14686.

Smith, L.C., Ghosh, J., Buckley, K.M., Clow, L.A., Dheilly, N.M., Haug, T., Henson, J.H., Li, C., Lun, C.M., Majeske, A.J., Matranga, V., Nair, S.V., Rast, J.P., Raftos, D.A., Roth, M., Sacchi, S., Schrankel, C.S., Stensvag, K., 2010. Echinoderm immunity. Adv. Exp. Med. Biol. 708, 260–301.

Steinmetz, P.R., Kraus, J.E., Larroux, C., Hammel, J.U., Amon-Hassenzahl, A., Houliston, E., Wörheide, G., Nickel, M., Degnan, B.M., Technau, U., 2012. Independent evolution of striated muscles in cnidarians and bilaterians. Nature 487, 231–234.

Steinmetz, P.R., Urbach, R., Posnien, N., Eriksson, J., Kostyuchenko, R.P., Brena, C., Guy, K., Akam, M., Bucher, G., Arendt, D., 2010. Six3 demarcates the anterior-most developing brain region in bilaterian animals. Evodevo. 1, 14.

Stultz, B.G., Park, S.Y., Mortin, M.A., Kennison, J.A., Hursh, D.A., 2012. Hox proteins coordinate peripodial decapentaplegic expression to direct adult head morphogenesis in *Drosophila*. Dev. Biol. 369, 362–376.

Suga, H., Tschopp, P., Graziussi, D.F., Stierwald, M., Schmid, V., Gehring, W.J., 2010. Flexibly deployed Pax genes in eye development at the early evolution of animals demonstrated by studies on a hydrozoan jellyfish. Proc. Natl. Acad. Sci. U.S.A. 107, 14263–14268.

Swanson, C.I., Evans, N.C., Barolo, S., 2010. Structural rules and complex regulatory circuitry constrain expression of a Notch- and EGFR-regulated eye enhancer. Dev. Cell. 18, 359–370.

Swanson, C.I., Schwimmer, D.B., Barolo, S., 2011. Rapid evolutionary rewiring of a structurally constrained eye enhancer. Curr. Biol. 21, 1186–1196.

Swaroop, A., Kim, D., Forrest, D., 2010. Transcriptional regulation of photoreceptor development and homeostasis in the mammalian retina. Nat. Rev. Neurosci. 11, 563–576.

Swiers, G., Chen, Y.H., Johnson, A.D., Loose, M., 2010. A conserved mechanism for vertebrate mesoderm specification in urodele amphibians and mammals. Dev. Biol. 343, 138–152.

Tessmar-Raible, K., Raible, F., Christodoulou, F., Guy, K., Rembold, M., Hausen, H., Arendt, D., 2007. Conserved sensory-neurosecretory cell types in annelid and fish forebrain: insights into hypothalamus evolution. Cell 129, 1389–1400.

Ting, S.B., Caddy, J., Hislop, N., Wilanowski, T., Auden, A., Zhao, L.L., Ellis, S., Kaur, P., Uchida, Y., Holleran, W.M., Elias, P.M., Cunningham, J.M., Jane, S.M., 2005. A homolog of *Drosophila* grainy head is essential for epidermal integrity in mice. Science 308, 411–413.

Tomoyasu, Y., Arakane, Y., Kramer, K.J., Denell, R.E., 2009. Repeated co-options of exoskeleton formation during wing-to-elytron evolution in beetles. Curr. Biol. 19, 2057–2065.

Tschopp, P., Duboule, D., 2011. A genetic approach to the transcriptional regulation of Hox gene clusters. Annu Rev. Genet. 45, 145–166.

Tschopp, P., Fraudeau, N., Béna, F., Duboule, D., 2011. Reshuffling genomic landscapes to study the regulatory evolution of Hox gene clusters. Proc. Natl. Acad. Sci. U.S.A. 108, 10632–10637.

Tseng, W.F., Jang, T.H., Huang, C.B., Yuh, C.H., 2011. An evolutionarily conserved kernel of gata5, gata6, otx2 and prdm1a operates in the formation of endoderm in zebrafish. Dev. Biol. 357, 541–557.

Ullrich-Lüter, E.M., Dupont, S., Arboleda, E., Hausen, H., Arnone, M.I., 2011. Unique system of photoreceptors in sea urchin tube feet. Proc. Natl. Acad. Sci. U.S.A. 108, 8367–8372.

van den Akker, W.M., Durston, A.J., Spaink, H.P., 2010. Identification of hoxb1b downstream genes: hoxb1b as a regulatory factor controlling transcriptional networks and cell movement during zebrafish gastrulation. Int. J. Dev. Biol. 54, 55–62.

Vinagre, T., Moncaut, N., Carapuco, M., Novoa, A., Bom, J., Mallo, M., 2010. Evidence for a myotomal Hox/Myf cascade governing nonautonomous control of rib specification within global vertebral domains. Dev. Cell. 18, 655–661.

Weatherbee, S.D., Halder, G., Kim, J., Hudson, A., Carroll, S., 1998. Ultrabithorax regulates genes at several levels of the wing-patterning hierarchy to shape the development of the *Drosophila* haltere. Genes. Dev. 12, 1474–1482.

Wellik, D.M., 2009. Hox genes and vertebrate axial pattern. Curr. Top. Dev. Biol. 88, 257–278.

Wellik, D.M., Capecchi, M.R., 2003. Hox10 and Hox11 genes are required to globally pattern the mammalian skeleton. Science 301, 363–367.

Werth, M., Walentin, K., Aue, A., Schönheit, J., Wuebken, A., Pode-Shakked, N., Vilianovitch, L., Erdmann, B., Dekel, B., Bader, M., Barasch, J., Rosenbauer, F., Luft, F.C., Schmidt-Ott, K.M., 2010. The transcription factor grainyhead-like 2 regulates the molecular composition of the epithelial apical junctional complex. Development 137, 3835–3845.

Wolf, L.V., Yang, Y., Wang, J., Xie, Q., Braunger, B., Tamm, E.R., Zavadil, J., Cvekl, A., 2009. Identification of pax6-dependent gene regulatory networks in the mouse lens. PLoS One 4, e4159.

Yang, X., Weber, M., Zarinkamar, N., Posnien, N., Friedrich, F., Wigand, B., Beutel, R., Damen, W.G., Bucher, G., Klingler, M., Friedrich, M., 2009a. Probing the *Drosophila* retinal determination gene network in Tribolium (II): the Pax6 genes eyeless and twin of eyeless. Dev. Biol. 333, 215–227.

Yang, X., Zarinkamar, N., Bao, R., Friedrich, M., 2009b. Probing the *Drosophila* retinal determination gene network in Tribolium (I): the early retinal genes dachshund, eyes absent and sine oculis. Dev. Biol. 333, 202–214.

Yu, Z., Lin, K.K., Bhandari, A., Spencer, J.A., Xu, X., Wang, N., Lu, Z., Gill, G.N., Roop, D.R., Wertz, P., Andersen, B., 2006. The Grainyhead-like epithelial transactivator Get-1/Grhl3 regulates epidermal terminal differentiation and interacts functionally with LMO4. Dev. Biol. 299, 122–136.

Zakany, J., Duboule, D., 2007. The role of Hox genes during vertebrate limb development. Curr. Opin. Genet. Dev. 17, 359–366.

Gene Index

Subject Index

Note: Page number "f" indicate figures.